VLSI for Wireless Communication

Second Edition

Bosco Leung

VLSI for Wireless Communication

Second Edition

 Springer

Bosco Leung
Department of Electrical
& Computer Engineering
University of Waterloo
Waterloo, ON, Canada
bleung@vlsi.uwaterloo.ca

Please note that additional material for this book can be downloaded from
http://extras.springer.com

This book was previously published by: Pearson Education, Inc.

ISBN 978-1-4899-7377-1   ISBN 978-1-4614-0986-1 (eBook)
DOI 10.1007/978-1-4614-0986-1
Springer New York Dordrecht Heidelberg London

Printed on acid-free paper

Springer is part of Springer Science+Business Media (www.springer.com)

Contents

List of Figures

Chapter 1
Communication Concepts: Circuit Designer Perspective

1.1 Introduction

What does GSM stand for? Why is the dynamic range in a digital enhanced cordless telecommunication receiver specified to be around 93 dB? What are multipath fading and Doppler shifts? How would the noise figure requirement of a low-noise amplifier change if I change my radio from one standard (DECT) to another (GSM)? Would Doppler shift be a problem in a standard such as PHS, and what circuit techniques do I have to address this problem?

These are all the types of questions integrated circuit (IC) designers may ask when dealing with circuit design for mobile wireless communication. Unfortunately, IC designers must be familiar with diverse sources of knowledge, and each of these sources is written for a specific circle of engineers, so finding and integrating relevant knowledge across these disciplines proves difficult. For example, if you need to understand modulation techniques (such as quadrature phase shift keying (QPSK)) and you consult a standard communication text, you would find that the design trade-off is typically presented using channel capacity and the demodulator's signal-to-noise ratio (SNR) as key parameters. This is of little use to a circuit designer, who is more interested in the SNR of the whole receiver than that of demodulator itself. Nor is channel capacity a relevant parameter, since circuit designers have no control over it. On the other hand, if you go to a classical radio frequency (RF) circuit text, you will be exposed to figures of merit such as input third-order intercept points (IIP3) of individual circuit subcomponents. However, you will not find any interpretation as to how they affect overall receiver performance, such as the bit error rate (BER). Thus this information may not be very useful to a modern-day integrated circuit designers who has the whole receiver to worry about and therefore would also need to interpret and relate the figures of merit of individual circuit subblocks (given in circuit terminology) to the performance measure of the whole receiver (given in communication terminology). Often, this performance measure is the only information available to an integrated circuit designer. Hence the following questions are likely: What would a poor receiver IIP3 do to the BER for

B. Leung, *VLSI for Wireless Communication*, DOI 10.1007/978-1-4614-0986-1_1,
© Springer Science+Business Media, LLC 2011

a standard using Gaussian minimum shift keying? Is modifying the modulation scheme to reduce the required IIP3 of a receiver a reasonable option? Yet a classical RF circuit text is unlikely to answer these questions. Therefore, it is the goal of this and the next chapter to address these questions and to present the answers in a language that most integrated circuit designers can understand.

1.2 Overview of Wireless Systems

In this subsection we review a basic communication system, as shown in Figure 1.1. Such a communication system consists of a transmitter, a channel (air for wireless communication, cable for wireline communication), and a receiver. In a typical digital communication system the modulator in the transmitter takes the digital bit stream from the source and modulates it on a high-frequency carrier. The back end then conditions the modulated signal to a form suitable for transmission. The channel takes this transmitted signal and adds distortion, noise, interference, and other impairments to it. The receiver consists of a front end that takes this highly impaired signal and conditions it to a form that the demodulator can demodulate.

Next let us review specifically what wireless communication (channel is air) is all about. Mobile wireless systems provide users with the opportunity to travel freely within the service area. These systems are made possible by the unique property of using radio waves as a transmission/receiving medium. Radio communication is based on using an antenna for radiating and receiving electromagnetic waves. When the user's radio transceiver is stationary over a prolonged period of time, the term *fixed radio* is used. A radio transceiver capable of being carried or moved around, but stationary during transmission, is called a *portable radio*. A radio transceiver capable of being carried and used, by vehicle or by a person on the move, is called a *mobile radio*.

A radio transceiver consists of both the receiver and the transmitter. In this book we start with the receiver side, where a detail treatment is given. We then carry the discussion over to the transmitter, concentrating on components not already covered in the receiver, such as the power amplifier.

To understand portable receiver design, we must understand the first relevant property of a radio communication channel: while travelling, a user of mobile

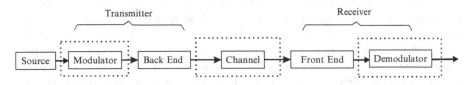

Fig. 1.1 Block diagram of a typical communication system

system must be aware of sudden changes in signal quality caused by one or more of the following:

1. movement relative to the corresponding base station
2. the surroundings
3. multipath propagation

The user must consider all these issues because unlike wireline channels such as fibre optics and cables, the radio communication channel is unprotected against natural disturbance such as lightning, temperature variation or humidity. Mathematically, the airwaves environment belongs to a class of nonstationary random fields for which the data transmission behavior is difficult to predict and model. Therefore, information degradation due to the movement of a user relative to reflection points and scatterers (buildings, trees, etc), which causes Doppler frequency shift or multipath fading, is troublesome to predict. Moreover, the channel is not protected against radio transmission in other bands. Finally, as with wireline communication systems, the signals amplitude must be large enough to overcome additive white guassian noise (AWGN) of the channel. The receiver must detect a signal whose amplitude has a large fluctuation of amplitude, buried under noise and in the presence of interference with a large amplitude. The overall signal amplitude required to achieve an acceptable BER to combat all of these hurdles is a function of the modulation scheme. This situation, as will be shown, will lead to tight sensitivity requirements.

The second relevant property is that all radio systems share the same natural resource: the airwaves (frequency bands and space). Filtering protects the channel from interference from other radio frequency bands. Furthermore, since the transmission medium is shared, we must ensure separation between the time when the portable is transmitting versus the time when it is receiving. This separation can be done in both the time and frequency domains and is called time division duplexing (TDD) or frequency division duplexing (FDD), respectively. In the FDD case, proper filtering must be done. In addition, since we have a shared transmission medium, we must provide access for individual users, including methods of multiple access, such as time division multiplexing access (TDMA), frequency division multiplexing access (FDMA), code division multiplexing access (CDMA), and orthogonal frequency division multiplexing acess, (OFDMA). In the case of FDMA, to provide channel selection, again proper filtering must be done. Finally, we must filter out jamming interference (i.e. human-made noise, adjacent channel interference, and co-channel interference inherent to cellular systems). All of these issues affect the receiver's selectivity requirements.

The hostile environment and interference also cause intersymbol interference (ISI), and the fast moving mobile nature of the system causes phase impairment. Mobile radio systems employ sophisticated coding and equalization techniques to combat ISI and diversity techniques to combat Doppler effect. Carrier and timing recovery circuits that introduce low-phase noise are also needed in the receiver.

On the other hand to understand portable transmitter design, we must be sensitive to any impacts such design will have on the receiver. For example, since the received channel is not protected against radio transmission in other bands, as mentioned above, the transmitter should be so design that the transmitted power does not spill over into other received channels. At the same time the transmitter power should be strong enough, and transmitted noise small enough, that under worst case, the received power (attenuated by channel, under fading) is large enough to overcome received noise (due to transmitted noise, channel noise).

Our goal for the rest of this chapter is to examine all the aforementioned issues and determine with the boundary conditions that such issues impose on the front end of a receiver and the back end of a transmitter. Specific radio standards are used as examples. For the receiver, starting from an ideal environment with only AWGN in the channel, we explain how to achieve an acceptable BER as a function of the modulation scheme. Next, the channel is restricted to a finite bandwidth, and we show how this causes ISI. We discuss pulse shaping techniques to combat ISI. Next, we discuss the impact of the environment on the channel, such as path loss, multipath fading, and Doppler shift, and examine how the environment can lead to a reduced received envelope amplitude and increased ISI and how the BER will be degraded. Finally, we introduce techniques to combat these impairments. Later on, in chapter 9, for the transmitter, using an example standard (with a given modulation scheme), the specs on the tolerated transmission error is translated to the required transmitted SNR.

1.3 Standards

The 3G/4G standards includes LTE (Long term evolution), where in the beginning the bandwidth will be 5MHz and the spectrum band at 2GHz. Another 3G standard in this group is the super3G experimental systems, which is aligned with the 3GPP (third generation partnership project) LTE specification. Here the bandwidth includes option of 5MHz, 10MHz and 20MHz. The downlink (DL, from base station to terminal) uses OFDMA while the uplink (UL, from terminal to base station) uses SC-FDMA (single carrier frequency division multiplexing). In spite of its name SC-FDMA can be understood by starting with OFDMA. It has an additional discrete Fourier transform (DFT) block in front of the OFDMA systems, which will be treated later. Duplex is FDD. Modulation scheme for DL is QPSK, 16QAM, 64 QAM, depending on the bandwidth. UL is QPSK, 16 QAM, again depending on the bandwidth. It also uses diversity by multi-antenna (MIMO) for DL, with options of 1×2, 2×2 and 4×4. Peak data rate for DL is 250-300Mbps and for UL is 40-50Mbps. IMT-Advanced (International Mobile Telecommunications Advanced) is a 4G standard. Here the target peak data rate is 100Mbps. Bandwidth is up to 100MHz. The bandwidth is scalable. Bandwidth for UL and DL is asymmetric. It uses layered OFDMA. Higher order MIMO is also adopted. Even

Table 1.1 Summary of DECT requirements

Cell range	50 m – 400 m
Frequency range (RF)	1880 – 1900 MHz
Carrier spacing	1.728 MHz Peak
Channels/carrier	2×12
Duplex method	TDD using two slots on the same RF carrier
Channelization	TDMA/FDMA
Speech coding	32 kbit/s ADPCM
Modulation	GMSK ($BT_b = 0.3$)
Gross data rate (R_b)	1.152 Mbit/s
BER	10^{-3}
Maximum transmitted power	250 mW

though the 3G/4G standards appear quite complicated, the design of the VLSI circuits implementing the transceiver will be illustrated in this book by using simpler standards as examples. Specifically the DECT (digital enhanced cordless telecommunications) standard is used for illustrating the receiver design, while the GSM (global system for mobile communications) standard is used for illustrating the transmitter design.

In this sub-section we cover the DECT standard, which can provide wireless access for both indoor and outdoor environment, with cell radius ranging from 50 to several hundred meters. Hence, it is suitable for the development of residential, business, and public applications. Moreover, DECT is not limited to telephony (i.e., speech) but also handles text and data as well.

The DECT system operates at a channel bit rate of 1.152 Mbps. FDMA and TDMA techniques are used to separate the different users. A Gaussian minimum shift keying (GMSK) modulation scheme converts the incoming data onto a carrier wave. There are 10 carriers with a channel spacing of 1.728 MHz occupying a band that spans from 1.88 to 1.9 GHz. Twelve channels are time division multiplexed onto each carrier, and TDD is used for each channel, resulting in 24 time slots.

Table 1.1 summarizes the DECT requirements that are important for our discussion.

1.4 Access Methods

We mentioned above different access methods. We will now go over some of the features of these accesses [5, 6].

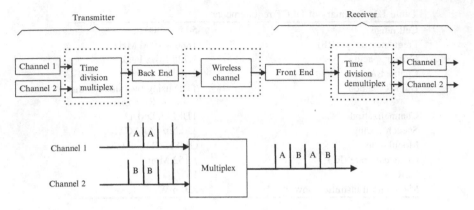

Fig. 1.2 TDMA: block diagram and timing diagram of multiplxer

1.4.1 TDMA

TDMA is provided by interleaving bit streams from different channels into one bit stream, as shown in the time division multiplex box of the transmitter part in Figure 1.2, in this case just for two users. Each stream is continuously divided into groups of bits, know as time slots and then these time slots are interleaved, as shown in the timing diagram of Figure 1.2. Each output slot occupies half the time of the input slot since the bit race is twice as large. At the time division demultiplex box of the receiver part in Figure 1.2, these bit streams are separated back into individual bit streams.

The collection of bits corresponding to precisely one time slot is known as a frame. In order to relax demultiplexing, the boundaries of time slots must be known i.e. the beginning of the frame. Additional bits for this, known as framing bits, are added. The framing bits, sometime referred to as reference burst, forces a measure of synchronization. Guard time is also included to account for different phase in a different time slot. Finally a preamble at the beginning of each traffic burst is added to allow the receiver to acquire timing and carrier phase, since each traffic burst is transmitted with an uncertain phase relative to the reference burst.

1.4.2 FDMA

FDMA has been used in analog modulation, e.g. in AM transmission. In wireless communication, transmission is typically digital, which will be covered in the next section. Here frequency division can be used where independent data streams are transmitted in non-overlapping frequency bands. FDMA is very simple, as illustrated in Figure 1.3, in this case again for just two users. The two transmitters have output power spectra in two non-overlapping bands, where they usually use passband PAM (pulse amplitude modulation, covered in modulation section). To ensure that,

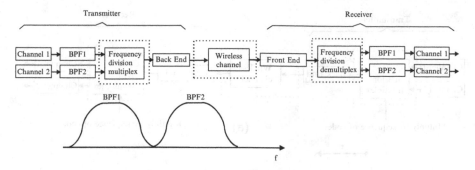

Fig. 1.3 FDMA: block diagram and frequency response of bandpass filters

it is common to put bandpass filters at output of each channel. At the receiver, similar bandpass filters eliminate all but the desired data signals. The path to the desired received signal from the undesired transmitted signal contains two bandpass filters with non-overlapping passbands, and therefore we can make the loss of the crosstalk path as large as we design for. A consideration is the nonlinearity of the amplifiers in the transmitter. These nonlinearities will create out of band energy that can interfere with other FDMA channels. It is possible to use bandpass filtering after the amplifier to reduce these spectral components. In comparison, for FDMA, filters can be expensive to implement and linearity is definitely an issue. On the other hand, for TDMA, propagation delay and crosstalk are problems.

1.4.3 CDMA

Code division multiple access (CDMA) allow multiple users to transmit signals in the same frequency band at the same time. Compared to TDMA and FDMA, the explanation of CDMA uses the concept of signal space, which can be more abstract and mathematical. Signal space borrows the idea of space from the mathematics of linear algebra, where a simple mathematical space example is a vector space (in a M-dimension vector space each element is represented by a row with M entries). In our day to day experience of a 3-dimension physical world, the corresponding vector space is a 3-dimension vector space, with each element being a 3-dimension arrow. The explanation of CDMA also exploits the idea of subspace, which is kind of a subset obtained from the original space, but with dimension reduced. Using the example a 3-dimension physical world, with a 3-dimensional vector space, an example of subspace may be a 2-dimensional vector space, with each element being a 2-dimension arrow, lying in a plane. Now let us start our explanation. In signal space suppose the available bandwidth is B Hz. We further assume this is to be distributed over N users. For a time duration of T seconds, all these waveforms form the subspace (each has bandwidth B), and will each have subspace dimension (=N) approximately given as 2BT. Therefore there is enough dimension that the

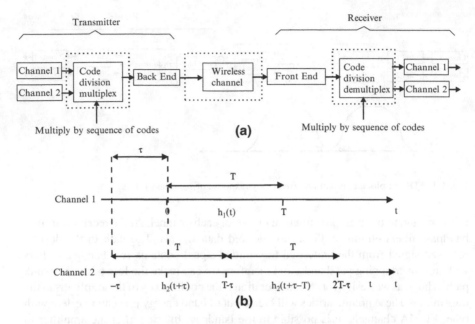

Fig. 1.4 (a) CDMA (b) partial correlation in CDMA: channel responses h_1 and h_2 over waveform duration T, which has difference in delay, τ, causing partial correlation

N different users' waveforms can stay orthogonal (roughly speaking this means separation is "perfect" i.e. not "correlated"). Specifically this means that if the received waveforms are decoded using matched filters (discussed below), each filter selects only the desired signal, while rejecting the remaining signals (because they are orthogonal). This orthogonality can be achieved as:

- time period is divided into 2BT time slots; one time slot is assigned to each user. This is essentially the TDMA concept discussed above.
- frequency band is divided into 2BT frequency bands; one band is assigned to each user. This is the FDMA concept discussed above.
- 2BT orthogonal waveforms, $h_1(t)$, $h_2(t)$...$h_{BT}(t)$ are selected; each waveform occupies bandwidth B and time T. Then each individual waveforms is assigned to each separate user for transmission. We further use one period of a shift register sequence as the transmission pulses. The sequence is so designed that they are orthogonal, and are further viewed as the "code" of each user. At the receiver, if the same sequence is used to access the incoming signal, orthogonality ensures only the user with the proper sequence/code will recover the transmitted signal; the others will be rejected because the cross-correlation product is zero. This is CDMA and is shown in Figure 1.4a

In comparison, for CDMA, there is near-far problem, which is not as severe in TDMA and FDMA. Near–far problem occurs when the interfering signals are much stronger than the desired signal. Thus it occurs when, for example, neighboring transmitters (which interfere) are closer and hence have a much higher power level than the intended transmitter, which happens to be far away. Whereas in TDMA only one transmitter is transmitting at a time (thus this is not a problem) and in FDMA, bandpass filter stopband attenuation can be big (problem greatly reduced), in CDMA orthogonality is only approximate (correlation operation is not perfect, resulting in the partial correlation problem). Recognizing that decoding by match filtering (discussed below) happens over the time duration T, inaccurate duration timing results in partial correlation. Inaccurate duration timing are due to, among other factors, inaccurate synchronization between the transmitting channels, as well as propagation delays. The phase relationship between duration timing of different transmitters can be arbitrary. Let us go to Figure 1.4b, which are for 2 channels, channel 1 and 2, with channel response h_1 and h_2, having difference in their delay given as τ. Since they can interfere each other correlation between them (partial correlations) should be zero to maintain orthogonality i.e. $\int_0^{T-\tau} h_1(t)h_2(t+\tau)dt = 0$ and $\int_{T-\tau}^{T} h_1(t)h_2(t+\tau-T)dt = 0$. In reality, since τ varies, these two partial correlations are not identically zero. The goal should thus be to reduce them under any value of τ. Since the variability in τ increases with the number of users, achieving this goal means restricting the number of users. This number is typically fewer than that in TDMA and FDMA case.

1.4.4 OFDMA

Another approach to allow multiple access is OFDM. Similar to CDMA, OFDM achieves this by exploiting the orthogonality property. Unlike CDMA, where the orthogonality is between sequences ("code") assigned to each transmission waveform, here orthogonality is between the carriers. Hence OFDM consists of multicarrier transmission. Specifically a single high rate carrier data stream is transmitted over numerous lower rate subcarriers (SC). In that sense, in addition to being an access approach, it can also be seen as a modulation technique (modulation technique is discussed in next section).

In a straightforward implementation of multicarrier transmission, the signal frequency band is divided into N disjoint frequency subchannels, each modulated with a separate symbol. The subchannels are then multiplexed in the frequency domain. Because the subchannels are nonoverlapping, interchannel interference is eliminated. The disadvantage of this is, obviously, an inefficient use of the spectrum. To overcome this, overlapping subchannels can be used instead. Such an overlapping multicarrier technique is shown in the bottom of Figure 1.5. Here there are eight subcarriers. For comparison, the conventional multicarrier approach is shown on top, where there is no overlap. It can be seen that the savings in bandwidth is close to 50%. To realize this technique, however, we need to reduce crosstalk between SC, which means orthogonality is required and hence the technique is called OFDM. This dictates that the separate frequencies follow a mathematical relationship. In contrast, in a typical FDMA system, no such relationship is required. Also shown in the bottom of Figure 1.5 is (a) an example channel

Fig. 1.5 OFDM combat
fading

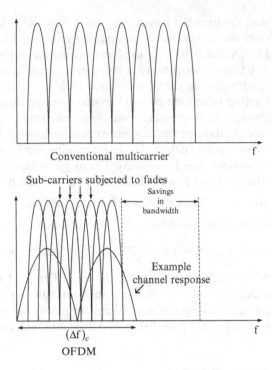

Fig. 1.6 OFDM multi user
access

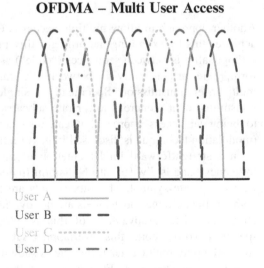

response, with coherence bandwidth $(\Delta f)_c$, (b) sub-carriers subjected to fades. As seen
only a part of the subcarriers (the middle 4) are subjected to fading under this example
channel and so OFDM is more robust towards fading. This will be explained in more
detail later, in section 1.10, after the concept of fading and coherence bandwidth is
introduced. Next we turn to Figure 1.6, which shows how OFDM can be used for

access. Let us assume there are 4 users: A, B, C, D. Users A, B are each allocated 2 bands, while users C, D are each allocated 2 bands. Notice though the multi bands for each user are not adjacent (unlike in FDMA case). This means if, again, due to fading (and coherence bandwidth is small), one of the multi bands will be affected (e.g. deep fade happen at the first band). However, 2nd and/or 3rd bands are not affected. In this sense, redundancy is built in. In addition because right now a variable number of channels can be assigned to different users, that means the bandwidth allocated to a particular user is flexible, and can be changed dynamically.

1.5   Modulation Schemes

There are many modulation schemes for transmitting digital data from the transmitter to the receiver in wireless communication. We focus on the modulation schemes that will lead us to MSK, as it is commonly used in most wireless standards (including DECT). More comprehensive treatment on modulation schemes can be found in [1].

The choice of modulation schemes influences performance, such as BER, SNR, and message bandwidth. This can affect the receiver design in both a high and a low level fashion. At a high level, modulation schemes can be chosen that will lead to a reduced message bandwidth and hence higher channel capacity. However, such design activities are usually of interest only to communication system engineers and will not be our main concern. On the other hand, modulation schemes can affect the receiver design at a low level as well. For example, modulation schemes with good BER performance increase the receiver's resistance to noise in channel, such as AWGN and resistance over destructive interference coming from multipath fading. Consequently, these better modulation schemes make the receiver design easier by lowering the required SNR (reducing the sensitivity requirement). We call these schemes low-level receiver design activities. These design activities are of significant interest to integrated circuit designers and will be our main focus.

For any modulator, demodulation can be done either in a coherent or incoherent fashion. For coherent demodulation we assume that the receiver has exact knowledge of the carrier wave's phase reference, in which case we say that the receiver is phase locked to the transmitter. In noncoherent demodulation, knowledge of the carrier wave's phase is not required. The complexity of the receiver is thereby reduced, although at the expense of inferior error performance. In the following we concentrate mainly on coherent demodulation.

1.5.1   Binary Frequency Shift Keying

With binary frequency shift keying (BFSK), the modulation happens on the frequency of the carrier. As the binary input signal changes from a logic 0 to a logic 1, and vice

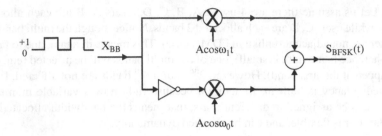

Fig. 1.7 Block diagram of a binary FSK modulator

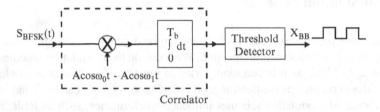

Fig. 1.8 Block diagram of a binary FSK coherent demodulator

versa, the FSK output signal shifts between two frequencies, a logic 1 frequency (f_1) and a logic 0 frequency (f_0). A possible implementation of a FSK modulator is shown in Figure 1.7.

Figure 1.8 shows a binary FSK coherent demodulator. A BFSK demodulator works by correlating the input with the function $\phi(t) = A\cos\omega_0 t - A\cos\omega_1 t$. Mathematically, this correlation is achieved by having the received signal multiplied by the correlation function $\phi(t)$ and then integrated over one bit duration T_b, as described in (1.1), and (1.2). Let us assume that the channel is ideal for the time being and so the modulator output $S_{BFSK}(t)$ is the same as the demodulator input $S_{BFSK}(t)$. Hence the correlator in Figure 1.8 can have two possible outputs depending whether a logic 0 or a logic 1 was transmitted.

$$logic\ 0: \quad s_0(t) = \int_0^{T_b} (A\cos\omega_0 t)(A\cos\omega_0 t - A\cos\omega_1 t)dt = \frac{A^2 T_b}{2} \tag{1.1}$$

$$logic\ 1: \quad s_1(t) = \int_0^{T_b} (A\cos\omega_1 t)(A\cos\omega_0 t - A\cos\omega_1 t)dt = -\frac{A^2 T_b}{2} \tag{1.2}$$

Note that the expressions (1.1) and (1.2) are exact only if $\cos\omega_0 t$ and $\cos\omega_1 t$ are orthogonal, and therefore their frequencies have to satisfy the relation $f_0 = n_0/T_b$ and $f_1 = n_1/T_b$, where n_1 and n_2 are some integers.

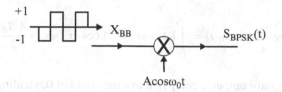

Fig. 1.9 Block diagram of a binary PSK modulator

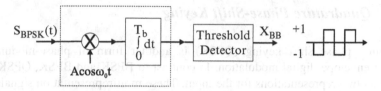

Fig. 1.10 Block diagram of a binary PSK coherent demodulator

The correlator output is then compared with a threshold, set to be 0. Thus, a simple threshold detector can determine which message was actually transmitted, yielding the baseband binary data.

1.5.2 Binary Phase Shift Keying

With binary phase shift keying (BPSK) the modulation happens on the phase of the carrier. Two output phases are possible for a single carrier frequency. One output phase represents a logic 0 and the other a logic 1. As the input digital signal changes, the phase of the output carrier shifts between two angles that are 180° out of phase. Therefore, the two signals, $s_0(t)$ and $s_1(t)$, that are used to present binary symbols 0 and 1 are defined by $s_0(t) = A\cos\omega_0 t$ and $s_1(t) = A\cos(\omega_0 t + \pi) = -A\cos\omega_0 t$ respectively. The implementation of a binary PSK modulator is shown in Figure 1.9.

A binary PSK coherent demodulator is shown in Figure 1.10. Again we assume that the channel is ideal for the time being and so the modulator output $S_{BPSK}(t)$ is the same as the demodulator input $S_{BPSK}(t)$. Since $s_0(t) = -s_1(t)$ (referred to as antipodal signals), the correlating signal in the detector is simply $\phi(t) = A\cos\omega_0 t$. Furthermore we fix the carrier frequency f_0 to be equal to n_0/T_b for some integer n_0 in order to provide a proper correlator output. Obviously, we have two possible correlator outputs depending on whether a logic 0 or a logic 1 was actually transmitted.

$$logic\ 1: \quad s_1(t) = \int_0^{T_b} (-A\cos\omega_0 t)(A\cos\omega_0 t)dt = -\frac{A^2 T_b}{2} \quad (1.3)$$

$$logic\ 0: \quad s_0(t) = \int_0^{T_b} (A\cos\omega_0 t)(A\cos\omega_0 t)dt = \frac{A^2 T_b}{2} \qquad (1.4)$$

Then the integrator output is compared to a threshold of 0, yielding the baseband binary data.

1.5.3   Quadrature Phase-Shift Keying

Quadrature phase shift keying (QPSK) is another form of phase-modulated, constant-envelope digital modulation. In contrast to BPSK and BFSK, QPSK has more than two representations for the input. These multirepresentation signals are called M-ary signals, where M stands for the possible number of representations of the signal. QPSK is an M-ary encoding technique where $M = 4$, since a QPSK output signal has four possible output phases. Note that in general we talk about M-ary signaling when the following relation is satisfied: $N = \log_2 M$, where N is the number of bits at the input of the modulator and M is the number of output conditions possible with N bits. Accordingly, with QPSK modulation four output phases are possible for a single carrier frequency. Obviously, the four different output phases must be characterized by four different input conditions. Since the digital input of a QPSK modulator is a binary signal, it takes more than a single input bit to produce four different input conditions. With 2 bits ($N = 2$), there are four possible conditions: 00,01,10, and 11. Therefore, with QPSK, the binary input data are combined into groups of 2 bits (also called dibits). Each dibit code generates one of four possible output phases.

A possible implementation of a QPSK modulator is shown in Figure 1.11. We need to add a demultiplexer, which generates a dibit sequence from a binary input data stream. The task of the demultiplexer is to separate the binary bit stream into an "upper arm" binary sequence and into a "lower arm" binary sequence. The demultiplexer separates the input sequence in such a way that 1 bit goes into the upper arm (I-channel) and the following bit goes into the lower arm (Q-channel). Obviously, the bit rate in each channel is now equal to half of the actual bit rate of the input data into the bit splitter. The binary sequence in each channel now modulates a carrier wave in the same manner as a BPSK modulator. An additional device has to make sure that the carrier signal from the I-channel ($\sin\omega_c t$) lies 90° out of phase with respect to that of the Q-channel ($\cos\omega_c t$). Finally, two corresponding BPSK signals are added together by a linear summer device to form the QPSK output signal. This output signal consists of an in-phase component $s_I(t)$ and a quadrature component $s_Q(t)$ and can be expressed as $s_{QPSK}(t) = \pm A\cos\omega_c t \pm A\sin\omega_c t$. Note that the output signal only changes after two consecutive bits are clocked in. Such a signal produces four possible message points in a two-dimensional signal space ($M = 4$).

A QPSK demodulator is shown in Figure 1.12. The QPSK demodulator consists of a pair of correlators with a common input and a corresponding pair of coherent

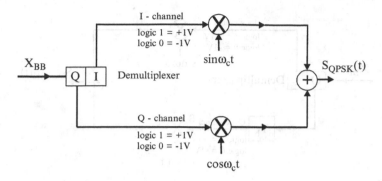

Fig. 1.11 Block diagram of a QPSK coherent modulator

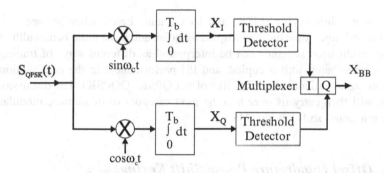

Fig. 1.12 Block diagram of a QPSK demodulator

reference signals $\phi_1(t) = \cos\omega_c t$ and $\phi_2(t) = \sin\omega_c t$ and so the modulator output $S_{QPSK}(t)$ is the same as the demodulator input $S_{QPSK}(t)$. Notice that the in-phase component $s_I(t)$ and the quadrature component $s_Q(t)$ of the received signal can be detected independently in an upper arm and in a lower arm since $s_I(t)\cos\omega_c t$ and $s_Q(t)\sin\omega_c t$ are orthogonal. Therefore, each arm produces a correlator output in the same way as the BPSK receiver. The integrator outputs, x_I and x_Q, are each compared in a threshold detector with a threshold of 0. For the I-channel if $x_I > 0$, a decision is made in favor of logic 1, and if $x_I < 0$ a decision is made in favor of logic 0. The Q-channel (with integrator output x_Q) works in the same way and the result is independent of x_I (in-phase signal). Finally, the two output binary sequences from the I-channel and the Q-channel are combined in a multiplexer to reproduce the actually transmitted message.

A few comments are now in order for QPSK. First, we can look at QPSK as an extension of BPSK and BFSK. Second, we can also look at it as the first example of a broad class of modulation, called quadrature modulation, which is characterized by subdividing the binary bit streams into pairs of 2 bits (dibits) and where each dibit is mapped onto one of four levels before modulation. Quadrature modulation is broadly divided into two classes: quadrature phase shift keying and its variants,

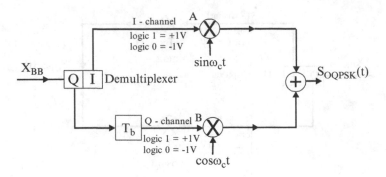

Fig. 1.13 Block diagram of a OQPSK modulator

and minimum shift keying (MSK) and its variants. Later, when we see that the channel is no longer ideal but is corrupted by noise and has a finite bandwidth, these different modulation schemes can be interpreted as different ways of trading off BER, message bandwidth occupied, and ISI performance. In the next section we will cover some variants of QPSK, like offset QPSK (OQPSK). The discussion on OQPSK will then carry us over into the next category of quadrature modulation, namely minimum shift keying (MSK).

1.5.4   Offset Quadrature Phase Shift Keying

Let us assume the channel is no longer ideal but has a finite bandwidth. Since QPSK has large phase changes at the end of each symbol, the bandwidth of the symbol is rather large. If the symbol bandwidth becomes comparable to the channel bandwidth, the transmitted symbol will be distorted by the channel. To mitigate this effect, a variant called offset QPSK (OQPSK) is introduced, whose transmitter is shown in Figure 1.13. Notice that, when compared with Figure 1.11, a time delay T_b is introduced in the Q-path so that the I, Q paths are offset in time by half the symbol period. This avoids simultaneous transitions in waveforms at nodes A and B. Instead of having a phase step of 180°, the phase step is now only 90°.

1.5.5   Minimum Shift Keying

Another method to avoid large phase changes at the end of each symbol is to adopt a modulation scheme that has continuous phase shift. One such modulation scheme is MSK. This new modulation scheme can be derived from OQPSK by applying half-sinusoids, instead of rectangular pulses, to represent the levels that are multiplied by the carriers. The resulting modulator is shown in Figure 1.14. Now ω_1 is selected such that $\omega_1=\pi/(2T_b)$. By doing so $S_{\mathrm{MSK}}(t)$ can be shown to exhibit no abrupt

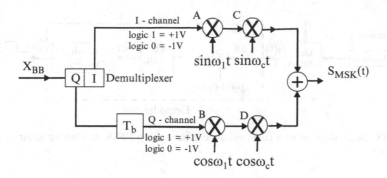

Fig. 1.14 Block diagram of an MSK modulator

change in phase. Thus, there is also no abrupt change in the slope of the $S_{MSK}(t)$. From a frequency domain point of view, this means MSK exhibits a sharper decay in its spectrum than QPSK, or a lower side-lobe signal power. Hence for a channel with a finite bandwidth, less distortion is introduced.

1.5.6 MSK: Another Viewpoint

We can also view MSK as a form of frequency shift keying, except this time it is a continuous-phase frequency shift keying (CPFSK). Essentially, MSK is binary FSK except that the logic 0 and the logic 1 frequencies are synchronized with the binary input bit rate. Synchronous simply means that there is a precise timing relationship between the two; it does not mean that they are equal. With MSK, the logic 0 and logic 1 frequencies are selected such that they are separated from the fundamental frequency by an exact odd multiple of one-half of the bit rate. In other words, f_0 and f_1 equal $nR_b/2$, where R_b is the input bit rate and n is any odd integer. This ensures that there is a smooth phase transition in the modulated signal when it changes from f_0 to f_1, and vice versa. Each transition occurs at a zero crossing and there exists no phase discontinuities. However, the disadvantage of MSK is that it requires synchronizing circuits, and the complexity of the implementation increases considerably.

1.6 Classical Channel

1.6.1 Additive White Gaussian Noise

Let us refer again to Figure 1.1. In section 1.5 we described the various modulation schemes as if the channel is ideal, except for the few cases towards the end, when we introduced the concept of a channel having finite bandwidth. In real life, the channel also has noise in it. What happens? Obviously, for a given signal amplitude A, if the

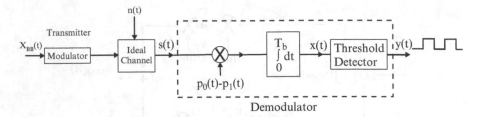

Fig. 1.15 Block diagram of a communication system using BFSK and having arbitrary pulse shape $p(t)$

noise becomes large enough, we are going to make errors. Such error cannot exceed the error as dictated for a given standard, usually specified as BER. Here we would derive the BER as a function of the SNR for a given modulation/demodulation scheme, assuming the channel is corrupted with AWGN only. We will make the derivation first for the simple modulation scheme: BFSK. We will then extend and state the result for MSK.

Let us redraw Figure 1.1 in Figure 1.15 with the following modification: the back end and front end have all been removed. Hence the transmitter is equivalent to the modulator and the receiver is equivalent to the demodulator. The channel is ideal except that there is AWGN. This is represented by having an ideal channel with noise $n(t)$ injected into it. The modulator/demodulator uses one of the modulation schemes described previously. Hence we use a general representation for the correlation signal, $p(t)$. For example, for BFSK we substitute $p_0(t) = A\cos\omega_0 t$ and $p_1(t) = A\cos\omega_1 t$. The demodulator part in Figure 1.15 would then agree with Figure 1.8, the demodulator for BFSK.

First, let us consider the signal $x(t)$ at $t = T_b$: it can have one of two values A or $-A$, with the decision threshold set at 0. Its probability distribution function (pdf) is shown in Figure 1.16(a). If there is no noise in the channel the received signal $x(T_b)$ will also have the same pdf. Now let us consider the AWGN $n(t)$. Since it is Gaussian noise, $n(T_b)$'s pdf is as shown in Figure 1.16 (b), with standard deviation σ_n. With noise in the channel $x(T_b)$ becomes $A + n(T_b)$ or $-A + n(T_b)$. Since $n(t)$ is another random variable that is independent from the symbol $x(T_b)$, when they add, their probability density functions become convolved. The pdf of $x(T_b)$ now changes to that shown in Figure 1.16(c). Notice that if the noise variance is large enough, as is the case here, then the two pdfs' that correspond to logic 0 and logic 1 overlap significantly. This means an error occurs when $A - n(T_b)$ becomes less than zero, or when $-A + n(T_b)$ is larger than zero. Now the probability of having logic 0 $(-A)$ is the same as having logic 1 (A) and equal to ½. Hence the probability of making an error when sending logic 0 $(-A)$ is just the product of ½ and the probability of making an error conditional upon this event. The probability of making an error conditional upon this event is, of course, given by the area underneath the pdf curve of $x(T_b)$ curve corresponding to logic 0 $(-A)$, but extending from 0 to ∞. This is indicated by the shaded part in Figure 1.16(c). Therefore, the

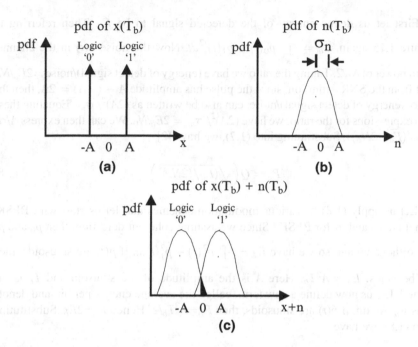

Fig. 1.16 Illustration of PDF with additive noise

total probability of making an error when transmitting logic 0, denoted by P_{e0}, is given by

$$P_{e0} = \frac{1}{2} \int_0^{\infty} \frac{1}{\sqrt{2\pi\sigma_n^2}} \exp \frac{-(x-(-A))^2}{2\sigma_n^2} dx \qquad (1.5)$$

Here σ_n is the standard deviation of the pdf for $n(t)$. Now from symmetry $P_{e1}=P_{e0}$. Hence the total error, P_e, is $2P_{e0}$.

Substituting this in (1.5) and with the proper change of variable, we have

$$P_e = \int_{\frac{A}{\sigma_n}}^{\infty} \frac{1}{\sqrt{2\pi}} \exp \frac{-y^2}{2} dy \qquad (1.6)$$

Notice that the integral is the familiar complementary error function, usually denoted as Q (not to be confused with quality factor Q). In shorthand form we have

$$P_e = Q(A/\sigma_n) \qquad (1.7)$$

The advantage of this expression is that we can see that P_e is dependent on A/σ_n, which, as will be shown, is related to the SNR. How do we derive the expression of P_e in term of SNR?

First let us define energy of the detected signal to be E_d. Then referring to Figure 1.15 again, $E_d = \int_{-\infty}^{\infty} |p_0(t) - p_1(t)|^2 dt$. Now the noise $n(t)$ in the channel has a power of $N_0/2$. Taking the ratio we have (energy of detect signal)/noise $= 2E_d/N_0$. But from the SNR definition, since the pulse has amplitude $A - (-A) = 2A$, then the ratio, (energy of detect signal)/noise, can also be written as $(2A)^2/\sigma_n^2$. Equating these two expressions for the ratio, we have $(2A)^2/\sigma_n^2 = 2E_d/N_0$. We can then express A/σ_n as $\sqrt{(E_d/(2N_o))}$. Substituting into (1.7), we have [2]

$$P_e = Q\left(\sqrt{(E_d/(2N_o))}\right) \tag{1.8}$$

Let us apply (1.8) to various modulation schemes and let us start with BFSK. What are p_0 and p_1 for BFSK? Since we assume coherent detection, then p_0 and p_1 are orthogonal and so we have $E_d = \int_{-\infty}^{\infty} p_0^2(t) + p_1^2(t) dt$. If $p(t)$ are sinusoids, then E_d becomes: $E_d = A^2 T_b$. Here A is the amplitude of the sinusoid and T_b is the period. Let us now define a new term, called the average energy per bit, and denote it as E_b. Again, if $p(t)$ are sinusoids, then $E_b = A^2 T_b/2$. Hence $E_d = 2E_b$. Substituting into (1.8), we have

$$P_e = Q\left(\sqrt{(E_b/N_o)}\right) \tag{1.9}$$

Next we turn to BPSK, where we have

$$E_d = \int_{-\infty}^{\infty} |p_0(t) - p_1(t)|^2 dt = \int_0^{T_b} (2A \cos \omega_o t)^2 dt = 2A^2 T_b$$

Again $E_b = A^2 T_b/2$, and therefore this time we have $E_d = 4E_b$. Substituting into (1.8), we have

$$P_e = Q\left(\sqrt{(2E_b/N_o)}\right) \tag{1.10}$$

We can go through similar derivations for QPSK and MSK.

Now that we have found expression of P_e in terms of E_b/N_0, we want to relate E_b/N_0 to SNR and hence express P_e in terms of SNR [3]. The SNR can be related to E_b/N_0 by the following relationship:

$$E_b/N_0 = SNR \times (f_N/R_b) \tag{1.11}$$

where f_N is the effective noise bandwidth and R_b is the effective symbol rate (or gross data rate).

Fig. 1.17 BER as a function of SNR

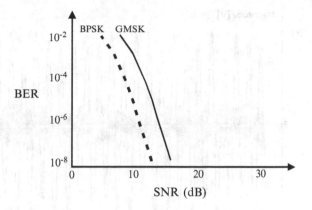

We can apply (1.11) to (1.9), (1.10), and obtain P_e expressions for BFSK and BPSK. For example, for the BPSK case, (1.10) becomes

$$P_e = Q\left(\sqrt{(2 \times SNR \times (f_N/R_b))}\right) \tag{1.12}$$

We can make one more comment: P_e is the probability of making an error when we transmit a symbol, which gives us the BER. Hence (1.12) can be interpreted as follows: for a given BER, (1.12) tells us the required SNR needed at the input of the demodulator to achieve this BER. Plots of BER versus SNR are normally derived for various modulation schemes and represent key design information. As an illustration we have given such plots for BPSK and GMSK in Figure 1.17.

1.6.1.1 Numerical Example 1.1

From Table 1.1 it can be seen that for the DECT standard a BER of 10^{-3} is needed and that it uses the modulation scheme GMSK. Apply this to Figure 1.17 we conclude that the SNR needed at the input of the demodulator is 9 dB, or

$$SNR = 9dB \tag{1.13}$$

This number will be used later in the receiver design.

The above derivation assumes that the channel impairment only comes from AWGN and that the received power is constant. For the case when the pulse is a sine wave, A, the amplitude of the receiving sine wave, is constant. In reality, this is not the case. Wireless environments suffer from path loss and multipath fading, which makes A substantially smaller during a fraction of the signaling interval. This will require, in general a much larger SNR to achieve the same BER. In subsection 1.9.2 we will rederive this SNR.

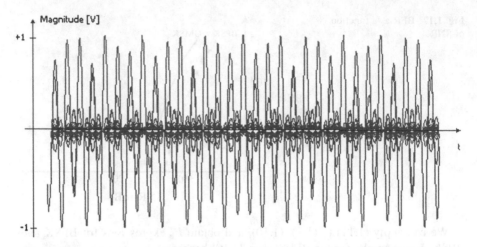

Fig. 1.18 Baseband PAM signal was modulated on a high-frequency carrier signal with unity amplitude

1.6.2 Finite Channel Bandwidth

In Table1.1, we see that DECT uses GMSK. What is GMSK? GMSK is just MSK with a particular pulse shaping. In this sub-section we discuss how finite bandwidth in channel introduces ISI and how pulse-shaping technique combats ISI. Referring back to Figure 1.1, let us imagine there are only the source/modulator/channel/demodulator blocks. The baseband signal from the source is modulated on a carrier frequency for transmission. This is called a baseband pulse amplitude modulator (PAM) signal. Thus, the baseband signal forms some sort of envelope over the high-frequency carrier signal, as shown in Figure 1.18.

If the pulse that the modulator sends is a rectangular pulse, then the envelope is rectangular. If the channel is ideal, there is no problem. However, if the channel is bandlimited, the pulses will spread in time. The pulse of each symbol will smear into the time intervals of succeeding symbols, causing ISI and thus degrading the BER. In addition, such smearing causes out-of-band radiation in the adjacent channel. Even though both of these can be improved by increasing the channel bandwidth, it is costly to do so. Alternatively, we can manipulate the RF spectrum of the transmitted symbol. Usually, this is difficult to do, and spectrum manipulation is done at the baseband instead. Such manipulation is called pulse shaping and is done inside the modulator block by the transmit filter, whereby the rectangular pulse is shaped in a particular way before being transmitted. Similarly, on the receiver side, inside the demodulator block, there is a filter, called the receive filter, that is specifically designed to receive this particular pulse shape. Henceforth, to simplify explanation, we will eliminate the carrier in our discussion and talk as if the channel works directly on the baseband pulses.

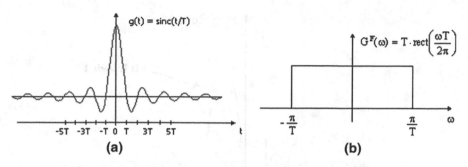

Fig. 1.19 A sinc pulse in the time domain (**a**) and its frequency response (**b**). The interval T is the same as T_b and will be taken as $T = 1$ ms in the text discussion for illustration purposes

We observe that, in general, the combined effects of the transmit and receive filters, and the propagation media, determine the pulse $p(t)$. Let us consider a pulse that has a maximum value of unity at some time t_0 and has zero value at all time instants $t_0 + kT$, where k is an integer. If we apply this pulse to the modulator in Figure 1.1, then $x_R(kT_b)$, the output of the demodulator at the k th symbol instant in the receiver, can be written as (for a noiseless channel)

$$x_R(kT_b) = \sum_{m=-\infty}^{m=+\infty} A_m p(kT_b - mT_b) \qquad (1.14)$$

$$= A_0 + \sum_{m \neq 0}^{m=+\infty} A_m p(kT_b - mT_b) \qquad (1.15)$$

Here A_m is a coded sequence of symbols (for example, a binary sequence with two levels 0 and 1), and $p(t)$ represents the pulse of this sequence. Notice that in (1.14) A_0 is the k th transmitted symbol, which is the desired received output. The second term contains all the undersirable interference and represents what we call ISI.

It can be shown that if $p(t)$ in (1.14) is a sinc pulse, as shown in Figure 1.19, neighbouring symbols will not interfere with one another, provided that we make the decision at $t = kT_b$. The sinc pulse satisfies the Nyquist criterion, which requires a pulse to have regular zero crossings spaced at multiples of the signaling interval so that at the receiver the transmitted data symbols can be detected without any mutual interference.

To explain more clearly how adopting a sinc pulse can eliminate ISI, let us go through the following example. We assume that we have a bipolar format where for the k th symbol a logic 1 is represented by $+1$ V at time instances $t = kT_b$ and a logic 0 is represented by -1 V at time instances $t = kT_b$. For ease of illustration we look at a transmission sequence that consists of only 2 bits: b_0 at -1 ms and b_1 at 1 ms, where at -1 ms a logic 0 and at $+1$ ms a logic 1 are transmitted, as shown in Figure 1.20. Let us first take a look at bit b_1, whose decision is made at $t = 1$ ms. What does the bit b_0 does to b_1? Notice after -1 ms bit b_0 still has some residual energy, although it is decaying. However, because the pulse is a sinc pulse, its crosses zero

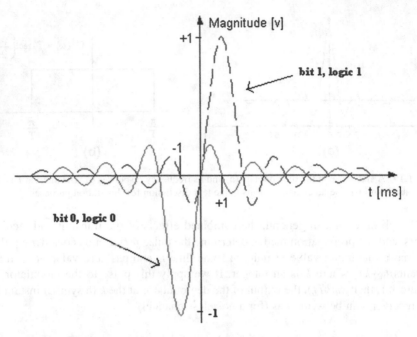

Fig. 1.20 A baseband PAM signal represented by a sinc pulse, where there is a logic 0 at −1ms and a logic 1 at 1ms

at 1 ms, which is exactly the time when we determine the value of b_1. Hence b_0's energy does not interfere with b_1's energy at b_1's decision time and there is no ISI.

Having shown that ideally a sinc pulse representation does not introduce ISI, we should be aware that there are two important practical issues to keep in mind:

1. A sinc pulse introduces additional error due to timing jitter of the sampling clock. The eye diagram illustrates the degradation (Figure 1.21). The eye diagram is easily generated in practice using an oscilloscope, where the symbol timing T_b serves as the trigger. An eye diagram consists of many overlaid traces of small sections of the received signal. Signal thresholds located at the vertical midpoints of the eyes dictate the data value decided for each sample. With ISI, the eye opening will close vertically. When there is incomplete vertical closure, the ISI will reduce the immunity against other nonideal effects, such as AWGN, and the receiver will fail to detect the actually transmitted sequence. Hence the wider the vertical eye opening, the greater the noise immunity. With AWGN present, closing of the eye will degrade the achievable BER for a given SNR. Hence the impact of ISI can be appreciated by its effect on the BER.
2. With a sinc pulse representation we cannot realize a filter in practice with such a sharp transition.

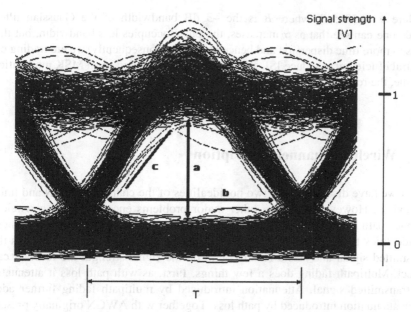

Fig. 1.21 One possible illustration of an eye diagram, where a unipolar format (the logic 0 is represented by a zero pulse) has been used to represent the baseband sequence. The vertical eye opening (**a**) indicates the immunity to all possible intersymbol interference phenomena. The horizontal eye opening (**b**) indicates the immunity to timing phase, and the slope (**c**) indicates the sensitivity to the jitter in the timing phase

Because of these two limitations, there are other pulse shapes that people adopt. One is called a raised-cosine pulse shape, which also satisfies the Nyquist criterion. It is given as

$$p(t) = \frac{\sin c(t/T_b)\cos(\pi\alpha t/T_b)}{1 - (2\alpha t/T_b)^2} \tag{1.16}$$

where $0 \leq \alpha \leq 1$ and α is the rolloff factor. This pulse has a maximum at $t=0$ and is zero at all $t=kT_b$, as desired. We observe that the raised cosine pulse is strictly band limited, but it exceeds the ideal minimal bandwidth (Nyquist bandwidth $= 1/T_b$) by a certain amount, called the excess bandwidth.

Another pulse that has been adopted is the one generated by a Gaussian pulse-shaping filter. This pulse does not satisfy the Nyquist criterion. Instead of generating zero crossings at adjacent symbol peaks, the Gaussian filter possesses a smooth transfer function but generates no zero crossing. The pulse is given by

$$p(t) = \frac{\sqrt{\pi}}{\alpha} \exp\left(-\frac{\pi^2}{\alpha^2}t^2\right) \tag{1.17}$$

Here $\alpha=0.5885/B$, where B is the -3 dB bandwidth of the Gaussian filter. Hence one can see that as α increases, the pulse occupies less bandwidth, but this leads to more time dispersion and hence more ISI. Consequently, we are trading off spectral efficiency to reduce ISI. When this pulse is applied to an MSK modulation scheme, the resulting scheme is called GMSK.

1.7   Wireless Channel Description

So far we have discussed only two nonidealities of the channel: AWGN and finite bandwidth. However, there are other major problems encountered in a wireless channel: path loss and multipath fading. In this section, we discuss how to add these nonidealities to the channel. The path loss of the channel severely attenuates the transmitted signal and sets a lower bound on the signal strength the receiver can expect. Multipath fading does a few things. First, as with path loss it attenuates the transmitted signal. Attenuation introduced by multipath fading further adds to the attenuation introduced by path loss. Together with AWGN originally present in the channel, this attenuation in the received signal strength sets a limit on the SNR required of the demodulator to achieve a certain BER. The second impairment brought about by multipath fading, distortion, introduces ISI, which limits the achievable BER. Finally, Doppler shift introduces phase impairment to the modulated signal received under multipath condition and is another error source that can limit the achievable BER. To investigate these problems properly, we need to develop more complex channel models, namely, channels having randomly time-varying impulse responses.

We now discuss the channel model for wireless communication and its impact on receiver front end design. We start by considering first the path loss and multipath fading happening in a channel, two phenomena that are closely linked. As shown in Figure 1.22, received power's variation in distance from the transmitter can be understood by observing its average value at a given distance from the transmitter as well as its local variation in close spatial proximity to that given location[4]. The first is characterized by path loss and the second by multipath fading. As such, path loss describes a large-scale propagation phenomenon and multipath fading describes a small-scale propagation phenomenon.

What leads to such received power variation? Physically, between the transmitter and the receiver there are many propagation paths, and signals traveling through these different paths interfere with one another. To illustrate this, we draw a simple picture that incorporates four of these paths (Figure 1.23).

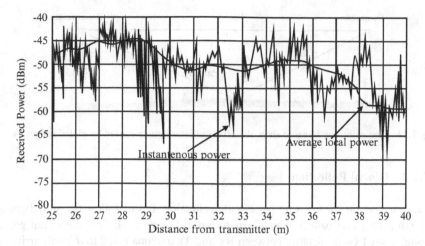

Fig. 1.22   Received power under path loss and multipath fading

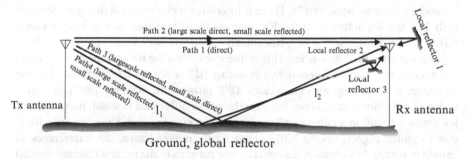

Fig. 1.23   Four paths are shown to differentiate the effect of interference caused by global and local reflectors

1.7.1   Path Environment

Referring to Figure 1.23, we would like to classify the paths according to the reflection they may undergo. The first classification is according to whether or not they suffer from global (large-scale) reflection. Paths 1 and 2 do not go through global reflections whereas paths 3 and 4 do go through such reflections. The second classification is according to whether or not they suffer from local (small-scale) reflection. Paths 1 and 3 do not go through local reflections whereas paths 2 and 4 do go through such reflections. Global and local reflections result in different interference patterns.

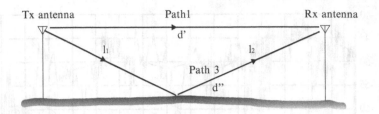

Fig. 1.24 Two paths are drawn to show the effect of interference caused by global reflectors

1.7.1.1 Global Reflection

In this subsection we focus on large-scale reflection. Let us follow the signal going through path 1 and path 3, which is redrawn in Figure 1.24. First the signal going through path 1 (with distance between Rx and Tx antenna equal to d') will arrive at the Rx antenna directly. A replica of this signal follows path 3, where it bounds off a global reflector (e.g. a hill) before it arrives at the Rx antenna (with distance between Rx and Tx antenna equal to d''). There it interferes with the signal that goes through path 1 at the Rx antenna. At the Rx antenna these two signals will have a phase difference proportional to d'-d''. Specifically, d'-d'' is very small compared with λ [4], the wavelength of the carrier. This is true even when the Rx antenna moves. Thus, the interference is always destructive in nature (d'' is always larger than d', but the difference is never large enough to cause 180° phase shift; the other 180° phase shift is due to reflection). In addition, because the phase difference is small, movement of Rx antenna results in a rather small variation of received power. Since the reflector is due to global objects (like a hill), which rarely move/change, the interference is simple in nature. In summary, the interference patterns in the present case are formed as a result of reflection due to global objects and are deterministic in nature.

1.7.1.2 Local Reflection

In this subsection we focus on small-scale reflection. Let us follow the signal going through path 1 and path 2 in Figure 1.23, which is redrawn in Figure 1.25. The signal goes along path 1 from Tx antenna to Rx antenna. A replica of this signal goes along path 2. Near the Rx antenna it bounds off a local reflector (e.g. a wall right next to the phone) before it arrives at the Rx antenna. There it interferes with the signal that travels along path 1. Depending on where Rx antenna is situated with respect to the Tx antenna, this interference can be destructive or constructive, resulting in variation of received power. In addition, since the phase difference can be large, this results in large variation of received power. Since the reflector is due to local objects (like a door, which can change, or another person, which can move), which often move/change, the interference is complicated in nature. Also, this variation can be large. In summary, the interference patterns in the present case

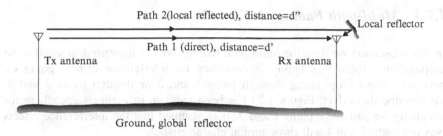

Fig. 1.25 Two paths are drawn to show the effect of interference caused by a local reflector

are formed as a result of reflection due to local objects and are random (can be constructive or destructive) in nature.

Because of differences in interference patterns in these two cases, different paths will be used in describing their different channel behaviour.

1.7.2 Path Loss: A First Glance

We start by describing path loss. We are interested in large-scale propagation, usually on the order of 5λ to 50λ. Hence this path loss is described by interference effects going on between signals propagating through paths 1, 2 and paths 3, 4. To describe this effect Figure 1.23 has been redrawn in Figure 1.24. As discussed before, only path 1 (also denoted as line-of-sight [LOS] path) and path 3 (also denoted as non-line-of-sight [NLOS] path) from Figure 1.23 are shown. Interference between paths 1 and 4, path 2 and 3, path 2 and 4 is supposed to show similar characteristics. Characteristics of the path loss phenomenon include the following:

1. Because it involves large-scale propagation, we are interested not in the instantaneous power, but in the local averaged power. However, since we are interested in an average of such power over a region of 5λ to 50λ, we can assume a constant power over this region, with its value set equal to the average.
2. Because path loss is attributed to interference between paths 1,2 and paths 3,4, we can conclude that the received power goes down as transmitter/receiver separation increases. This is as indicated in Figure 1.22. Also, the effect of movement (both due to the mobile and the reflectors) is averaged out. The effect is a simple loss in signal strength received by the receiver. Moreover, the signal loss to first order is a simple function of distance, and not of angle of incidence and other factors. Of course more sophisticated model include path loss due not only to free space propagation and reflection, but also diffraction and scattering.
3. Because it involves large-scale propagation, the number of relevant paths is small, usually boiled down to one or two. Hence the example presented in Figure 1.24 turns out to be quite representative of the real-life situation.

Path loss induces signal loss in the received signal, which lowers the SNR and hence BER.

1.7.3 Multi-path Fading: A First Glance

In this subsection we describe multi path fading. We are interested in small-scale propagation. Hence this fading is described by interference effects going on between signals propagating through paths 1 and 2 or through paths 3 and 4. To describe this effect, Figure 1.23 has been redrawn in Figure 1.25, where for simplicity we just show paths 1 and 2. It is assumed that the interference effects between paths 3 and 4 will show similar characteristics.

Characteristics of the multi path phenomenon include the following:

1. Since we are interested in the instantaneous power available to the receiver, we are interested in the instantaneous power due to the sum of path 1 and path 2.
2. Because signal loss is attributed to interference between path 1 and path 2, we can conclude that the instantaneous received power goes up and down with large variation as transmitter/receiver separation increases. This is as indicated in Figure 1.22.
3. Because of constructive and destructive interference of the carrier from multi paths, we can have the following:

   a. Fluctuation of the resultant signal's carrier's envelope's strength. This change in envelope's strength can occur rapidly as a function of separation and is called envelope fading.
   b. Distortion of the shape of the resultant signal's carrier's envelope, which leads to ISI of the passband signal. The distortion of this envelope depends on the relative delay of the various reflected signals. Analyzing such various delays works like analyzing a filter, and hence the radio channel can be characterized as having a frequency response. Depending on whether the frequency response of the radio channel is constant or not across the narrow band the signal occupies, we have flat/frequency select fading. If we have flat fading, there is no ISI, and if we have frequency selective fading, there is ISI.
   c. Movement (of the receiver itself or local reflector). Such a change would cause phase change in the signal. This causes another effect, called Doppler shift. If the channels change fast enough, we call the channel fast fading. Hence the situation can also be classified accordingly. Depending on whether the resultant signal's carrier's envelope changes slowly/rapidly with symbol time, we call it slow fade/fast fade.

4. There is a large number of reflected radio waves (much larger than 2) whose amplitudes are random and time varying in nature, both due to the random nature of the bit sequences and the movement of the reflectors/scatterers. Hence to describe and examine the phenomeon, a full-blown time-varying statistical model for the channel must be invoked.

Multi path fading lowers SNR and induces ISI, which leads to a reduced BER.

1.8   Path Loss: Detailed Discussion

1.8.1   Friis equation

As discussed in subsection 1.7.2 most wireless channels have both a LOS and a
NLOS propagation path. To get a feeling for the path loss in such an environment,
we will first calculate the path loss along the LOS path. We discuss the free space
propagation loss in such an environment, where no obstacles occur between the
transmit and the receive antenna.

To get a useful estimation for the path loss, we will make use of fundamental
antenna theory. Let us consider first an isotropic antenna; namely, one that radiates
power equally in all directions. Let us denote as P_T the power transmitted by the
transmit antenna and A_R the capture area of the receive antenna spaced r meters
apart. Since the transmit antenna is assumed to be isotropic, the received power
captured by the area A_R is:

$$P_R = P_T \frac{A_R}{4\pi r^2} \tag{1.18}$$

Obviously, in practice we would design the transmit antenna to focus its radiated
energy in the direction of the receiving antenna. Furthermore, we have to consider
that the receive antenna does not actually capture all the electromagnetic radiation
incident on it. Thus, we have to introduce some correction factors in
equation (1.18):

$$P_R = P_T \frac{A_R}{4\pi r^2} G_T \eta_R \tag{1.19}$$

where G_T is the transmit antenna gain (to account for the focusing) and η_R is the
antenna efficiency (to account for incomplete capturing).

At microwave frequencies, aperture antennas (i.e. parabolic) are typically used
for receiving, and for these antennas the achievable antenna gain is defined as

$$G = \frac{4\pi A}{\lambda^2} \eta \tag{1.20}$$

where A is the antenna capture area, λ is the wavelength of the transmission, and η
is the antenna efficiency. We observe that for this antenna type the antenna gain is a
function of the antenna dimension versus the carrier frequency. Accordingly, the
higher the carrier frequency the lower will be the requirement on the antenna
dimensions for a given antenna gain. For receiver antenna let us go to (1.20) and

set $G = G_R$, $A = A_R$. We can then solve A_R in terms of G_R. Next we substitute this expression for A_R in (1.19) and we get a relation known as the Friis equation:

$$\frac{P_R}{P_T} = G_R G_T \left(\frac{\lambda}{4\pi r}\right)^2 = G_R G_T \left(\frac{c}{4\pi r f}\right)^2 \tag{1.21}$$

Here P_R is the power received by the antenna with an antenna gain G_R, P_T is the transmitted power with an antenna gain G_T, f_c is the carrier frequency, λ is the corresponding wavelength, c is the light velocity, and r is the spacing between the transmit and receive antenna. The propagation loss L_B, expressed in decibels, is now given by expressing the Friis equation in decibels:

$$L_B[dB] = 10 \log_{10} \frac{P_R}{P_T}$$

$$= 10 \log_{10} G_R + 10 \log_{10} G_T - 20 \log_{10} f_c - 20 \log_{10} r + 147.56dB \tag{1.22}$$

In wireless communications we usually deal with carrier frequencies in the range of MHz to GHz. Equation (1.22) can be expressed adopting units of MHz for the frequency term. If we further assume we have omnidirectional transmit and receive antennas with unity gain, (1.22) becomes

$$L_B[dB] = 27.56dB - 20 \log_{10} f_c[MHz] - 20 \log_{10} \frac{r[m]}{d_o[m]} \tag{1.23}$$

From this basic LOS transmission loss equation, we note that the received power decreases by 6 dB for every doubling of distance and also for every doubling of the radio frequency. Consequently, the more we increase the carrier frequency, the more power we need to transmit the signal reliably. In addition, more repeater stations may be necessary. Note that (1.23) is only accurate for a far-field situation. The behaviour of the electromagnetic field near the transmit antenna is difficult to predict theoretically, and its models are mainly based on experimental data.

Remember that so far we have only considered ideal free space propagation loss for a LOS path. Based on empirical data, a fairly general model has been developed for NLOS propagation paths. This model is given as

$$L_A(d) \propto L_B \cdot \left(\frac{d}{d_0}\right)^{-n} \tag{1.24}$$

Here n is the path loss exponent, which indicates how fast path loss increases with distance; d_0 is the reference distance for free space propagation (unobstructed transmission distance); L_B is the corresponding propagation loss of the LOS path [refer to (1.22)]; and d is the distance between the transmit and the receive antenna. The NLOS path loss can therefore be approximated by (1.24) as

$$L_A[dB] = L_B - 10 \cdot n \cdot \log_{10} \left(\frac{d}{d_0}\right) \tag{1.25}$$

Experimental results indicate that typical NLOS outdoor cellular mobile systems have a path loss exponent n between 3.5 and 5 and that indoor systems have a path loss exponent between 2 and 4.

As discussed in subsection 1.7.2, because of reflection, the LOS path tends to cancel out (destructively interfere with) the NLOS path, making the loss larger than in the free space, and hence the exponent n larger than 2. Again, as mentioned in subsection 1.7.2, because we are talking about large-scale propagation, the path difference d'-d" is small compared with the path distance itself; hence the cancellation is rather constant, or independent of carrier wavelength. To reiterate, this is different from the small-scale propagation case. Moreover, because of this rather constant cancellation, the interference manifests itself as amplitude loss, but not ISI.

1.8.2 *Amplitude Loss and Minimum Received Signal Strength*

In general, (1.23), (1.24), and (1.25) can be used to predict amplitude loss and minimum received signal strength for a given wireless channel and a given standard. Let us present an example.

1.8.2.1 Numerical Example 1.2

For a DECT receiver, let us first assume that the transmit and the receive antennas are separated by $d = 50$ m (from Table 1.1 this is the minimum cell radius). Let us further assume that the channel has a free space loss with $d_0 = 3$ m and that the path loss exponent n is 3. This would be a real-life situation for an indoor channel. Finally, we assume that we use omnidirectional antennas with unity gain.

We start off with (1.23) to find the LOS loss:

$$L_B(d_0) = 27.56\,dB - 20\log_{10}1900MHz - 20\log_{10}\frac{3m}{3m} = -38\,dB \qquad (1.26)$$

Then for the present environment, we substitute $n = 3$, $d_0=3$m, and L_B obtained from (1.26) into (1.25) and the amplitude loss is

$$L_A(dB) = -38\,dB - 10\cdot 3\cdot\log_{10}\left(\frac{50m}{3m}\right) = -74\,dB \qquad (1.27)$$

However, we are really interested in the maximum amplitude loss. To calculate that, we assume the transmit and the receive antennas are separated by $d = 400$ m (from Table1.1 this is the maximum cell radius). In this case (1.27) becomes

$$L_A(dB) = -38\,dB - 10\cdot 3\cdot\log_{10}\left(\frac{400m}{3m}\right) = -101dB \qquad (1.28)$$

From Table 1.1 the maximum transmit power P_T was set to 250 mW (=24 dBm). Thus the received signal strength for the present example, which corresponds to the minimum received signal strength, is approximately

$$P_{Rmin} = 24dBm - 101dB = -77dBm \tag{1.29}$$

1.8.3   Minimum Separation

The Friis equation derived previously [(1.21)] also allows us to determine the minimum separation between users for a given interference level. Let us assume this time that we have two users, users 1 and 2. User 1 receives power from the base station. In addition, user 1 also receives power transmitted by user 2, which acts as interference. We would like to estimate this interference's power level. The interference propagates from user 2 to user 1, following the same path loss mechanism. From the Friis equation we know that as the separation between the two users decreases, the received interference's power level increases. Hence to calculate the maximum interference's power level received by user 1, we first assume a LOS path and apply (1.21) again

$$\frac{P_{R\max}}{P_T} = G_R G_T \left(\frac{c}{4\pi d_{\min} f_c}\right)^2 \tag{1.30}$$

Here $P_{R\max}$ is the maximum allowable interference power level received by user 1, P_T is the corresponding power level transmitted by user 2, $d_{\min}$ is the minimum separation between user 1 and user 2, and G_R and G_T are the receive and transmit gain of users 1 and 2, respectively.

Thus, solving for $d_{\min}$ in (1.30) gives

$$d_{\min} = \sqrt{\frac{P_T G_R G_T}{P_{R\max}}} \cdot \frac{c}{f_c} \cdot \frac{1}{4\pi} \tag{1.31}$$

The $d_{\min}$ calculated for an NLOS radio path is related to the $d_{\min}$ calculated for a LOS path in the same way as $L_A(\text{NLOS})$ is related to $L_B(\text{LOS})$. Hence we apply (1.24) to (1.31) and we get

$$d_{\min,NLOS} = \left[\frac{P_T G_R G_T (\lambda/4\pi d_0)^2}{P_{R\max}}\right]^{1/n} \cdot d_0 \tag{1.32}$$

where d_0 is the free space distance, λ is the wavelength of the carrier signal and n is the path loss exponent. Accordingly, (1.32) allows us to determine, for a maximum interference power level $P_{R\max}$, the minimum separation that user 2 can be from user 1, with a NLOS path.

1.9   Multipath Fading: Channel Model and Envelope Fading

In this and the following section we will present a detail discussion of multipath fading.

1.9.1   Time-varying Channel Model

In subsection 1.7.3, we stated that to describe multipath fading we need a full-blown time-varying statistical model for the channel. To describe a time-varying channel let us start by assuming that the input to the channel is a passband PAM signal. This passband PAM signal is obtained by modulating a baseband PAM signal on a high-frequency carrier. The baseband PAM signal is a sequence of pulses, which is in turn obtained by amplitude modulating the symbols on a given pulse shape. Alternately, we can view the passband PAM signal as consisting of two sinusoidal carriers at the same frequency (90° out of phase) that are modulated by the real and imaginary parts of a complex valued baseband signal. As a side note, the passband PAM scheme can be specialized to become the PSK, FSK, and QPSK modulation schemes described previously. There are different representations of a passband PAM signal. One such possible representation is the following:

$$x_T(t) = \sqrt{2}Re\left[s_l(t) \cdot e^{j2\pi f_c t}\right] \tag{1.33}$$

$$= \sqrt{2}Re\left[\sum_{m=-\infty}^{\infty} A_m g(t - mT_b) \cdot e^{j2\pi f_c t}\right] \tag{1.34}$$

Here $x_T(t)$ is the transmitted PAM signal, $s_l(t)$ is the baseband PAM signal, f_c is the carrier frequency, A_m is a coded sequence of symbols, $g(t)$ represents the pulse shape of this sequence, and T_b is the symbol period.

Next let us apply this passband PAM signal to the channel. As stated previously, since the atmosphere is inhomogeneous to electromagnetic radiation due to spatial variations in temperature, pressure, humidity, and turbulence or simply due to the fact that the mobile unit can be in motion in a wireless environment during transmission, the channel has to be modeled to be time-variant. Furthermore, we have already noted that as the most common application of wireless communication is in urban areas, there are many local obstacles to radio transmission and many opportunities for multiple reflections. As a result of changes occurring in these local reflectors, the resulting multiple propagation paths have propagation delays and attenuation factors that are time variant. Thus, the received signal, $x_R(t)$, may be expressed in the form

$$x_R(t) = \sum_i \alpha_i(t) x_T(t - \tau_i(t)) \qquad (1.35)$$

where $\alpha_i(t)$ is the attenuation factor for the signal on the i-th path with a delay $\tau_i(t)$. Substituting $x_T(t)$ from (1.33) into (1.35) yields

$$x_R(t) = \sqrt{2} Re \left[\left\{ \sum_i \alpha_i(t) s_l(t - \tau_i(t)) \cdot e^{-j2\pi f_c \tau_i(t)} \right\} \cdot e^{j2\pi f_c t} \right] \qquad (1.36)$$

Alternately, we can substitute the expanded version of $x_T(t)$ from (1.34) into (1.35) and we get

$$x_R(t) = \sqrt{2} Re \left[\left\{ \sum_i \alpha_i(t) \sum_m A_m g(t - mT_b - \tau_i(t)) \cdot e^{-j2\pi f_c \tau_i(t)} \right\} \cdot e^{j2\pi f_c t} \right] \qquad (1.37)$$

To gain more insight, let us assume that we only focus on the first bit. Accordingly, (1.37) can be simplified by setting $m = 0$. The resulting signal $x_{R,m=0}$ is:

$$x_{R,m=0}(t) = \sqrt{2} Re \left[\left\{ \sum_i \alpha_i(t) \cdot e^{-j2\pi f_c \tau_i(t)} A_0 g(t - \tau_i(t)) \right\} \cdot e^{j2\pi f_c t} \right] \qquad (1.38)$$

Let us start from (1.38). We observe that there are three potential problems in the multipath fading channel:

1. Referrring to (1.38), $x_{R,m=0}$ is seen to consist of the sum of a number of time-variant vectors (phasors) due to the multipath, each having phases $\phi_i(t) = 2\pi f_c \tau_i(t)$. Hence a significant change in the amplitude of $x_{R,m=0}$ will occur if $\phi_i(t)$ changes by 360°, because the carrier of the original signal and the carrier of the replica signal will now interfere destructively. Thus $\phi_i(t)$ will change by 360° when $\tau_i(t)$ changes by $1/f_c$. Since $1/f_c$ is a small number (in DECT it is on the order of nanoseconds), $\phi_i(t)$ can change by 360° with relatively small change in the separation distance between the Rx and Tx antennas. Hence the amplitude of $x_{R,m}=0$ fluctuates rapidly as separation changes, as stated qualitatively in subsection 1.7.3. This is known as envelope fading and will be explained in more detail in subsection 1.9.2.
2. Surprisingly if we compare (1.38) with (1.15), we note the similarity, even though (1.15) is derived for a classical channel. In (1.15) the received signal

consists of the desired transmitted symbol, A_0, plus interference caused by symbols transmitted before and after symbol period 0 (second term). In (1.38) the received signal contains the desired transmitted symbol, A_0, modulated by $e^{j2\pi f_c t}$, (the carrier), plus interference caused by delayed version of this same transmitted symbol, A_0, again modulated by $e^{j2\pi f_c t}$, (the carrier). Now following (1.15) we have explained how this interference causes ISI in the received baseband signal. We can apply the same argument and conclude that the received passband PAM signal in (1.38) also suffers from ISI. We call this frequency selective fading, which will be explained in more detail in subsection 1.10.1.

3. If the carrier frequency in (1.38) changes with time due to movement of the mobile, this will cause $x_R(t)$'s phase to change. We call this fast fading, and it will be explained in more detail in subsection 1.10.2.

1.9.2   Envelope Fading

In the discussion following (1.38), we stated that the received passband PAM signal suffers from envelope fading. We also stated in subsection 1.7.3 that there is a large number of paths involved in multipath fading. Hence it is best to describe the amplitude of the reflected signals along different paths with a probability distribution function. In fact, such amplitude variation also shows up in the amplitude variation of the resulting envelope. Hence we would focus on the pdf that describes the envelope's amplitude. It will be shown that the pdf of the amplitude of this envelope can be modeled accurately by a Rayleigh distribution. Therefore, this type of fading is also known as Rayleigh fading. Using such a model, fades of 20 dB below the root mean square (rms) value of the signal envelope occur approximately 1% of the time.

1.9.2.1   Rayleigh Distribution

To show the amplitude of the envelope follows a Rayleigh distribution, we start from (1.36), which is first repeated here and then expressed in terms of $A_i(t)$:

$$
\begin{aligned}
x_R(t) &= \sqrt{2}Re\left[\left\{\sum_i \alpha_i(t)s_l(t - \tau_i(t)) \cdot e^{-j2\pi f_c \tau_i(t)}\right\} \cdot e^{j2\pi f_c t}\right] \\
&= \sqrt{2}Re\left[\left\{\sum_i A_i(t) \cdot e^{j\varphi_i(t)}\right\} \cdot e^{j2\pi f_c t}\right]
\end{aligned}
\tag{1.39}
$$

Here $A_i(t)$ represents the attenuated, delayed baseband signal and $\phi_i(t)$ is the phase of the i th path.

We assume that the mobile unit is stationary. Then the phase shift introduced by the channel for the i th path can be expressed as

$$\phi_i = 2\pi f_c \tau_i = 2\pi \frac{c}{\lambda} \cdot \frac{s_i}{c} = 2\pi \frac{s_i}{\lambda} \qquad (1.40)$$

where λ is the wavelength of the carrier signal and s_i is the additional distance traveled by the reflected wave of the i path with respect to the direct path.

Since s_i are random, then ϕ_i are random phases. We further assume they are uniformly distributed from 0 to 2π. Writing the complex exponents in (1.39) in terms of their real and imaginary components, we get an expression of $x_R(t)$ in terms of the in-phase and the quadrature components:

$$x_R(t) = s_I(t) \cos(2\pi f_c t) - s_Q(t) \sin(2\pi f_c t) \qquad (1.41)$$

where

$$s_I(t) = \sum_i A_i(t) \cos(\varphi_i) \qquad (1.42)$$

$$s_Q(t) = \sum_i A_i(t) \sin(\varphi_i) \qquad (1.43)$$

Notice that by the central limit theorem the baseband random process $s_I(t)$ and $s_Q(t)$ are approximately Gaussian, since the different paths are supposed to be identical and independently distributed in terms of their summation. Hence the envelope of the received signal $x_R(t)$ is given by

$$x_R(t)_{envelope} = \sqrt{s_I(t)^2 + s_Q(t)^2} \qquad (1.44)$$

and its amplitude's pdf is a Rayleigh distribution.

1.9.2.2   Increased SNR Requirement

In this subsection we rederive the relationship between SNR and BER that was first derived in subsection 1.6.1 for a classical channel, except this time with multipath fading the received signal's amplitude is no longer constant but is given by a distribution. We show how this amplitude fluctutation, due to envelope fading, increases the SNR required at the demodulator input. For envelope fading, from (1.41) through (1.44), we are again interested in baseband's amplitude variation. Hence we can start from (1.36) and eliminate the $e^{j2\pi f_c t}$ term and obtain the baseband term. To simplify matters, we just develop envelope fading on a flat fading channel [4]. Envelope fading on frequency selective channels is difficult to calculate and requires computer simulation. Since ISI no longer exists, (1.36) can be simplified without the $e^{j2\pi f_c t}$ term and becomes

$$x_{R,baseband}(t) = \alpha(t)s_1(t)\exp(-j\theta(t)) \tag{1.45}$$

Here $\alpha(t)$ is the gain of the channel and $\theta(t)$ is the phase shift of the channel. Observing (1.45), we can see that multiple path fading manifests itself as a multiplicative (gain) variation in the transmitted signal $s_1(t)$.

If we incorporate AWGN in the channel, we have

$$x_{R,baseband}(t) = \alpha(t)s1(t)exp(-j\theta(t)) + n(t) \tag{1.46}$$

where $n(t)$ is AWGN of the channel.

To evaluate the probability of error, for a modulation scheme, we can follow the same procedure as in subsection 1.6.1, except that we have to recognize that the A term in (1.5) is no longer a constant. To overcome this we can use the concept of conditional probability. First the probability of an event of making an error, given the condition that the amplitude assumes a specific value, $A_{specific}$, is the joint probability of two subevents: (1) the probability of an event that the amplitude assumes that specific value, $A_{specific}$, and (2) the probability of the event that either $A_{specific}$ - $n(T_b)$ becomes less than zero, or $-A_{specific} + n(T_b)$ is larger than zero. The probability of this subevent (2) obviously depends on the value of $A_{specific}$ and the standard deviation of $n(T_b)$. Since the two subevents are assumed to be independent, the joint probability is just the product of the probability of the two subevents. The total probability, P_e, is now the integral of all these joint probabilties. Mathematically it is written as

$$P_e = \int_0^\infty p(X)P_e(X)dX \tag{1.47}$$

Here $p(X)$ is the pdf of X due to fading, which is the probability of subevent (1); $P_e(X)$ is the probability of error at a specific SNR$=X$ where X is given by $X=\alpha^2 E_b/N_0$; α is the gain, as given in (1.45), E_b and N_0 have been defined in subsection 1.6.1. Hence $P_e(X)$ can be interpreted as the probability of subevent (2).

Notice that the variable α captures the amplitude fluctuation due to fading, normalized with respect to E_b/N_0. For most flat fading, the probability distribution of the amplitude follows a Rayleigh distribution, as described in subsection 1.9.2.1. Therefore, α is given by a Rayleigh distribution. Hence α^2 and therefore X follows a chi-square distribution given by

$$p(X) = \frac{1}{\Gamma} \exp\left(-\frac{X}{\Gamma}\right) \tag{1.48}$$

Here

$$\Gamma = \frac{E_b}{N_0} \overline{\alpha^2} \tag{1.49}$$

and can physically be interpreted as the average SNR.

Of course, $P_e(X)$ is a function of the modulation scheme. As an example, let us calculate the total probability corresponding to GMSK, $P_{e,\text{GMSK}}$. We apply (1.48), and $P_e(X)$ corresponding to GMSK to (1.47) and we get

$$P_{e,GMSK} = \frac{1}{2} \left(1 - \sqrt{\frac{\delta\Gamma}{\delta\Gamma + 1}} \right) \cong \frac{1}{4\delta\Gamma} = \frac{1}{4\delta \times SNR} \tag{1.50}$$

where δ is a variable that depends on BT_b. Here B is the 99% bandwidth of the Gaussian pulse and is defined to be the bandwidth where 99% of the pulse energy is contained. T_b is the symbol period. For example $\delta=0.68$ for $BT_b =0.25$ and 0.84 for $BT_b =\infty$.

Numerical Example 1.3
From Table 1.1, DECT uses a GMSK with $BT_b =0.3$. We can take that to be close enough to 0.25 and set $\delta=0.68$. Again from Table 1.1, BER($=P_e$) for DECT$=10^{-3}$.
Substituting these into (1.50), and we have $10^{-3} = \frac{1}{4\times0.68\times SNR}$.
Solving SNR $= 367$, or

$$SNR = 25dB \tag{1.51}$$

Comparing (1.51) with (1.26), notice that the present SNR is 16 dB higher than that required for AWGN alone.

For other modulation schemes, similar equations have been derived [4]. Some of these plots are shown in Figure 1.26.

1.9.2.3 Frequency and Space Diversity

The SNR derived previously to combat amplitude fluctuation due to envelope fading is rather large. Instead of using such a large SNR, we can achieve the same BER by using diversity technique. There are basically two diversity techniques that improve the quality of the received signal (degraded due to the envelope's amplitude fluctuation) and hence help reduce the required SNR:

1. The frequency diversity approach. Frequency diversity simply involves modulating two different RF carrier waves with the same baseband signal and then transmitting both the RF signals to the same receiver. If two carrier waves use two different frequencies and if they are separated by more than $(\Delta f)_c$, then both information-bearing signals are affected differently by the channel.

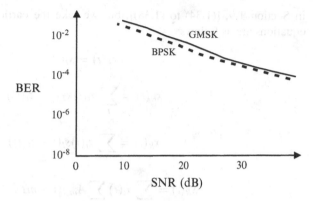

Fig. 1.26 BER as a function of SNR, envelope fading

The coherence bandwidth, $(\Delta f)_c$, will be defined in subsection 1.10.1.4. At the receiver, both RF signals are demodulated, and the one that yields a better-quality baseband signal is selected.

2. The space diversity approach. In this approach, at the receiver there is more than one antenna providing the input signal to the receiver. It is important that the antennas are separated by multiple wavelengths of the carrier signal. This is to ensure that two or more signals of the same frequency, which are attenuated in the same way by the channel, are in phase and additive. If received out of phase, they will cancel and consequently result in less received signal power than if simply one antenna system was used. Note that both diversity techniques can be combined.

3. OFDM is an example applying diversity technique.

1.10   Multipath Fading: Frequency Selective and Fast Fading

Now that we have presented how multipath fading reduces amplitude, we will show how it also introduces ISI. In addition, we discuss how it can reduce amplitude in a time-varying fashion.

1.10.1   Frequency Selective Fading

We stated in the discussion following (1.38) that under multipath fading the received passband PAM signal suffers from ISI. To see how ISI is created, we need to look at the baseband component of this passband PAM signal. We could have gone to (1.37) and started to extract the baseband component from the passband PAM signal directly. This is mathematically complex. Traditionally this complex mathematical problem is handled using tools such as correlation function, which may be less familiar for circuit designers (refer to [1] for such a treatment). Alternately (which is the approach adopted in this chapter), we can start from the channel model described in Section 1.9, [(1.34) to (1.38)], but we take the carrier term out. The resulting equations are as follows:

$$x_T(t) = s_l(t) \tag{1.52}$$

$$x_R(t) = \sum_i \alpha_i(t) x_T(t - \tau_i(t)) \tag{1.53}$$

$$x_R(t) = \sum_i \alpha_i(t) s_l(t - \tau_i(t)) \tag{1.54}$$

$$x_R(t) = \sum_i \alpha_i(t) \sum_m A_m g(t - mT_b - \tau_i(t)) \tag{1.55}$$

$$x_{R,m=0}(t) = \sum_i \alpha_i(t) \cdot e^{j2\pi f_c \tau_i(t)} A_0 g(t - \tau_i(t)) \tag{1.56}$$

Hence, for example, comparing (1.55) to (1.37), the delay due to the channel now is applied to the baseband signal (symbol) only, rather than to the carrier and the baseband signal (symbol). Of course this is only an approximation. However, the advantage of this model is that now in order to observe the effect of the channel, we need to apply attenuation and delay associated with different paths to the baseband signal (symbol) only. Hence we can concentrate our discussion only on the baseband signal. With this in mind, let us examine an example to illustrate how ISI is caused by multipath fading.

1.10.1.1 An Illustrative Example

In this example we make the following assumptions:

1. $g(t)$, the baseband pulse used in the channel [refer to (1.55), and (1.56)] is implemented using a sinc pulse.
2. We decide to send 2 bits: a logic 0 followed by a logic 1. Hence in (1.55), $A_0=0$ and $A_1=1$.
3. To simplify matters, we assume $\alpha_1 = 1$ and $\tau_1=0$; that is, path 1, the direct path, suffers no attenuation and no time delay.
4. T_b, symbol period, is set to 2 ms.

First, let us investigate a channel [represented by α_i, τ_i in (1.53)] that consists of three paths (i.e. $i = 1, 2, 3$), shown in Figure 1.27. It consists of one direct path and two reflected paths, which have specific α_i, τ_i.

To see how ISI is created in this channel and also how severe this ISI is, let us look at the following two cases:

1. Focus on path 1 and path 3: Set $\tau_3 = 0.01$ ms, $\alpha_3 = 0.8$. Therefore, attenuation and delay of path 3 are rather small. The resulting pulses are shown in Figure 1.28a.

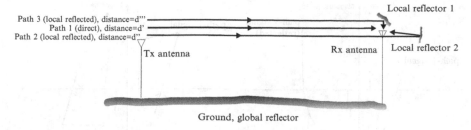

Fig. 1.27   Illustration of a channel where there exists a path (path 2) with both α and τ large
(**a**) Baseband PAM signal after multipath reception, where ISI does not occur, since reflected
signal is not delayed. (**b**) Baseband PAM signal after multipath reception, where ISI occurs, since
the reflected signal is delayed and is strong

These pulses are obtained by going through the following reasoning: the A_1 pulse
starts at symbol time of 1 ms with a peak value of 1, goes through path 1, suffers
no attenuation and no delay, and arrives at the Rx antenna at 1 ms. Meanwhile,
this same A_1 pulse, which starts at 1 ms with a peak value of 1, also goes through
path 3, gets attenuated, and results in a pulse having a peak value of 0.8.
Furthermore, it gets delayed by 0.01 ms and arrives at the Rx antenna at 1.01
ms. Hence at 1 ms, the Rx antenna receives two pulses, which add to one
another. ISI is not severe.

2. Focus on path 1 and path 2: We set $\tau_2 = 2$ ms $(=T_b)$, $\alpha_2 = 0.8$. We would want
   to see when a signal goes through such a path, which suffers little attenuation
   even with a long delay, what happens to the pulses. The resulting pulses are
   shown Figure 1.28 (b). These pulses are obtained by going through the following reasoning: the A_0 pulse starts at -1 ms with a peak value of -1, goes
   through path 2, gets attenuated to a peak value of -0.8, and gets delayed by
   2 ms and arrives at the Rx antenna at 1 ms. Meanwhile, the A_1 pulse, which
   starts at the next symbol time of 1 ms and has a peak value of 1, goes through
   path 1, suffers no attenuation and no delay, and arrives at the Rx antenna at 1
   ms. Hence at 1 ms, the Rx antenna receives two pulses, which cancel with one
   another. ISI is severe. (Note: the A_0 pulse, which starts at -1 ms with a peak
   value of -1, also goes through path 1, suffers no attenuation and no delay, and
   arrives at the Rx antenna at -1 ms. As shown, this pulse still has some residual
   energy around 1 ms. However, because this is a sinc pulse, at exactly 1 ms this
   pulse crosses zero and therefore does not interfere with the A_1 pulse that has
   gone through path 1.)

The preceding conclusions on how severe ISI is can also be seen by plotting the
eye diagram for the two different cases, shown in Figure 1.28c.

Basically, the preceding discussion highlights the fact that for this channel where
there exist paths (path 2 in Figure 1.27) that exhibit large path delays τ_i and large
attenuation factor α_i, strong ISI will occur.

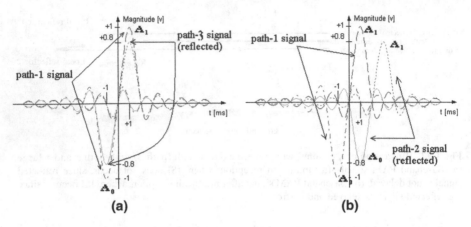

Fig. 1.28a,b (a) Baseband PAM signal after multipath reception, where ISI does not occur, since reflected signal is not delayed. (b) Baseband PAM signal after multipath reception, where ISI occurs, since the reflected signal is delayed and is strong

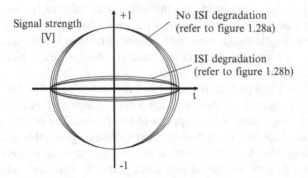

Fig. 1.28c Simplified eye diagram for comparisons on paths 1, 3 and paths 1, 2 for the channels described in Figure 1.27. I) No ISI when the delay of the reflected path is small compared to the symbol interval T_b or when attenuation is large. II) Severe ISI when the delay of the reflected path is approximately half the symbol interval T_b while attenuation is small

Now let us investigate those channels whose paths exhibit large path delays τ_i but small attenuation factor α_i. We will show that ISI is not severe for these channels. As an example of such a channel, let us modify the channel described in Figure 1.27 by changing the α_2 associated with path 2. τ_2, $(\alpha_1 \tau_1)$, $(\alpha_3 \tau_3)$ remain the same. The resulting channel is shown in Figure 1.29. To see how severe ISI is with this channel, let us do the following:

1. Focus on path 1 and path 3. Since everything is the same as in Figure 1.27, ISI is not severe.
2. Focus on path 1 and path 2: we set $\tau_2 = 2$ ms $(=T_b)$, $\alpha_2 = 0.08$. Notice that the attenuation for path 2 is severe. The resulting pulses are shown in Figure 1.30. These pulses are obtained by going through the following reasoning: the A_0 pulse

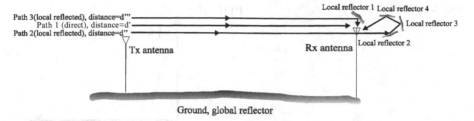

Fig. 1.29 Channel is modified from that in Figure 1.27. Specifically, path 2 is modified so that α is small

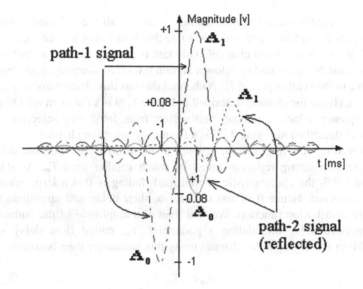

Fig. 1.30 Baseband PAM signal after multipath reception for channel in Figure 1.29. ISI does not occur since reflected signal is weak

starts at -1 ms with a peak value of -1, goes through path 2, gets attenuated to a peak value of -0.08, gets delayed by 2 ms and arrives at the Rx antenna at 1 ms. Meanwhile, the A_1 pulse starts at the next symbol time of 1 ms with a peak value of 1, goes through path 1, suffers no attenuation and no delay, and arrives at the Rx antenna at 1 ms. Hence at 1 ms, the Rx antenna receives two pulses, which hardly cancel with one another. ISI is not severe.

Hence the channel described in Figure 1.27 exhibits severe ISI while the channel described in Figure 1.29 does not. We would now want to generalize this to a typical channel that contains many more paths. We would conclude that for a channel that has a substantial portion of its paths exhibiting large path delays τ_i accompanied by large attenuation factors α_i, strong ISI will occur.

Fig. 1.31 PDF of τ_i

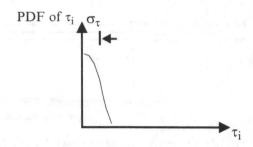

1.10.1.2 Criteria of a Flat/Frequency Selective Channel

We can elaborate the preceding conclusions further. If all the reflected signals that arrive have their amplitude attenuated at a rate that is fast enough that they does not cause ISI, then this is a good channel. This rate is crudely defined as follows: the amplitude must be attenuated to below a certain threshold when the delay becomes comparable to the symbol period T_b. A channel that has this characteristic is called a flat channel. Hence the channel described in Figure 1.29 is a flat channel. Otherwise it is a frequency selective channel that suffers from frequency selective fading. The channel described in Figure 1.27 is a frequency selective channel.

The preceding condition can be restated as the following condition: delay for paths possessing strong reflection must be much smaller than T_b. As stated in subsection 1.7.3, the characteristic of multipath fading is that a large number of paths are involved. Hence it is best to describe their delay and amplitude with a probability distribution function. We will then plot amplitude of the paths versus their respective delay and define a parameter, τ_m, called time delay spread. The condition for having a flat channel using this parameter then becomes

$$\tau_m << T_b \tag{1.57}$$

1.10.1.3 Time Delay Spread

We now discuss τ_m, the time delay spread. First let us plot the pdf of τ_i for a typical channel (Figure 1.31).

For channel described in Figure 1.27 we have $\tau_0 = 0$ (path 1), $\tau_1 = 0.01$ ms (path 3), and $\tau_2 = 2$ ms (path 2). Next we want to plot the amplitude factor α_i versus the delay factor τ_i. We can give the following first-order, qualitative observations that relate the two factors.

1. We assume that signal attenuation comes mainly from absorption of signal power that occurs during reflections. Hence we can observe that for a given path, α_i decreases as the number of reflections increases.

Fig. 1.32 Attenuation factor of paths are plotted against delays of paths; τ_m is the parameter that specifies the delay of the paths beyond which the attenuation factor is below a certain threshold

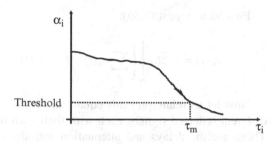

2. What is the relationship between the number of reflections and the delay experienced by the signal travelling along a particular path? We can assume, in general, that signal travelling along a longer path is more likely to encounter more reflections. Meanwhile, this longer path usually means its associated delay, τ_i, is larger. If we put the two together, we observe that the number of reflections increases when τ_i increases.

If we apply observation (2) to observation (1), then α_i decreases as τ_i increases. This result is plotted in Figure 1.32, where we also define a threshold for α_i. This threshold is defined such that for a path with α_i below its value, a signal travelling through this path does not cause any ISI. Corresponding to this threshold, we define the parameter τ_m. This τ_m can then be interpreted as follows: a signal that travels through any path with delay larger than τ_m does not cause any ISI.

The rate at which α_i decreases as τ_i increases depends on the channel. For example, the channel described in Figure 1.27 has a relatively flat plot, resulting in a large τ_m. On the contrary, the channel described in Figure 1.29 would be described with a plot having a large negative slope, resulting in a small τ_m.

How do we use Figure 1.32? As illustrated in the last paragraph, for every channel there is a plot that is similar to Figure 1.32, but with a different τ_m. From these plots the τ_m of these channels can be determined. We then go to condition (1.57) and decide if the channel is flat.

1.10.1.4   Coherence Bandwidth

Remember that in subsection 1.6.2 we discussed how a classical channel with finite bandwidth introduces ISI. In the discussion of Figure 1.27 we showed that a multipath fading channel can also introduce ISI. We would therefore like to associate the concept of finite bandwidth to a multipath fading channel as well. To do this we revisit the concept of time-varying channel model (introduced in section 1.9.1) that is used to describe a multi-path fading channel. We will show that the channel model allows us to interpret the fading channel as a filter, and therefore the concept of bandwidth can be applied to the fading channel. We further show that we can relate the bandwidth of the fading channel to τ_m.

First let us repeat (1.36):

$$x_R(t) = \sqrt{2} Re \left[\left\{ \sum_i \alpha_i(t) s_l(t - \tau_i(t)) \cdot e^{-j2\pi f_c \tau_i(t)} \right\} \cdot e^{j2\pi f_c t} \right] \tag{1.58}$$

Now let us reinterpret this equation. From (1.58) $x_R(t)$ is seen to be a sum of different reflected signals, each with their own delay τ_i and attenuation factor α_i. These are the delays and attenuation introduced by the paths of the particular channel. For example, for the channel described in Figure 1.27, we have $\tau_1 = 0$, $\tau_2 = 2$ ms, $\tau_3 = 0.01$ ms, $\alpha_1 = 1$, $\alpha_2 = 0.8$ $\alpha_3 = 0.8$; and for the channel described in Figure 1.29, we have $\tau_1 = 0$, $\tau_2 = 2$ ms, $\tau_3 = 0.01$ ms, $\alpha_1 = 1$, $\alpha_2 = 0.08$, $\alpha_3 = 0.08$. We first assume that these parameters are time invariant. Then (1.58) describes the input-output relationship similar to that of a time-invariant finite impulse response (FIR) filter. To see this more clearly, let us remove the carrier term from (1.58) and make α_i and τ_i time-invariant. Hence we have

$$x_R(t) = \sum_i \alpha_i s_l(t - \tau_i) \tag{1.59}$$

We can now identify in (1.59) the input as $s_1(t-\tau_i)$, the output as $x_R(t)$. If we further make τ_i the same for each i and identify it as the sample period, then (1.59) describes the input-output relationship of an FIR filter. The tap weight of the FIR filters are the α_i, and the length of the filter is the length of the sum. As with any filter, this filter can also be described by a transfer function and has a frequency response. If we reinsert the carrier term, we are essentially shifting the frequency response of this filter so that its center frequency is around f_c. For simplicity, we ignore the effect of the carrier in the following discussion and deal directly with the baseband signal.

Next let us assume that this filter has a certain bandwidth. To interpret this bandwidth, like the bandwidth of an ordinary filter, we will go through the following argument. Let us send a baseband signal whose symbol frequency, f_b ($=1/T_b$), is much larger than $1/\tau_m$. Then T_b is much smaller than τ_m, and condition (1.57) is violated, resulting in severe ISI. In other words, the baseband signal cannot be reproduced. If we treat the fading channel as a filter, we can roughly interpret this as saying that the baseband signal does not go through the filter. Next, we send another baseband signal whose symbol frequency f_b ($=1/T_b$), is much smaller than $1/\tau_m$. Then T_b is much larger than τ_m, and condition (1.57) is satisfied, resulting in little ISI. Again, if we treat the fading channel as a filter, we can roughly interpret this as saying that the baseband signal passes. Hence the channel behaves like a low-pass filter with a filter bandwidth of $1/\tau_m$ (band-pass filter if we reintroduce the carrier). In this context we can define the bandwidth of the channel, denoted as coherence bandwidth $(\Delta f)_c$, as

$$(\Delta f)_c \cong \frac{1}{\tau_m} \tag{1.60}$$

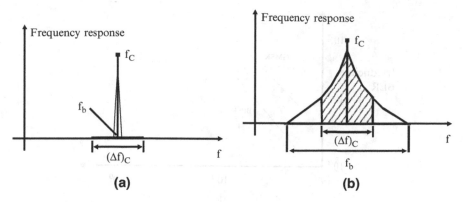

Fig. 1.33 (a) Flat fading, (b) frequency selective fading

In essence, the coherence bandwidth indicates the bandwidth over which the signal undergoes practically the same shifts and attenuation during transmission over a multipath channel.

Condition (1.57) can now be restated in terms of the coherence bandwidth as

$$f_b << (\Delta f)_c \qquad (1.61)$$

For example, we can take f_b to be 1.72MHz in DECT.

Using this new condition, a fading channel that satisfies (1.61) is called a flat fading channel and is shown in Figure 1.33 (a). A fading channel that violates this is referred to as a frequency selective channel and is shown in Figure 1.33 (b). In Figure 1.33 we have reintroduced the carrier term and so the channel is a band-pass filter.

Finally, let us revisit the assumption of making α_i and τ_i time invariant. If this is not true, then the channel becomes a time-varying channel represented by a time-varying filter, but the concept of coherence bandwidth remains valid.

1.10.1.5 ISI Degradation

The ISI caused by frequency selective fading will degrade the BER. Normally, this can be compensated by increasing the SNR. However, ISI can become severe that we reach a situation whereby the BER remains flat no matter how much we increase the SNR. This BER is called the irreducible BER. This irreducible BER is a function of how frequency selective the channel is, which in turn is quantified by the ratio $d = \tau_m / T_b$, called the normalized delay spread. A plot of this irreducible BER versus the normalized delay spread is shown in Figure 1.34, for two modulation schemes. Let us now examine an example and see how to use Figure 1.34 to find out, for a given wireless channel, whether a given standard is satisfied.

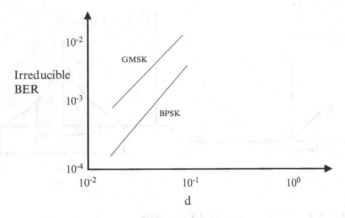

Fig. 1.34 Irreducible BER as a function of normalized delay spread for different modulation schemes, frequency selective fading

Numerical Example 1.4

For DECT from Table 1.1 we have a bandwidth of 1.7 MHz, which gives a symbol period of 0.6 us. Measurement on a given indoor channel shows a delay $\tau_m = 100$ ns, hence d=100 ns/0.6us or 0.16. From Figure 1.34 this gives a BER of about 10^{-2}. Hence this does not satisfy the DECT specification in Table 1.1.

1.10.1.6 Equalization

Numerical example 1.4 shows that the ISI is so poor that the BER requirement cannot be met. To combat ISI, an equalization technique can be adopted. Crudely speaking, equalization can be thought of as compensating for the amplitude and phase distortion of the channel with a filter whose transfer function is the inverse of the channel's transfer function. This filter is called an equalizer. If the channel is time varying, the equalizer has to be adaptive as well. This can be achieved by first sending a known training sequence to the receiver, so that the equalizer averages to a proper setting. Then when the actual signal is sent, the equalizer adopts a recursive algorithm to evaluate the channel's time-varying transfer function and adapts itself accordingly. Equalizers can be broadly classified as linear (FIR, lattice, etc.) or nonlinear (decision feedback equalizer, maximum likelihood sequence estimation equalizer, etc.). A complete treatment can be found in [4]. We assume that one such equalizer is applied to the receiver used in Numerical example 1.4 to help restore the BER to the acceptable level.

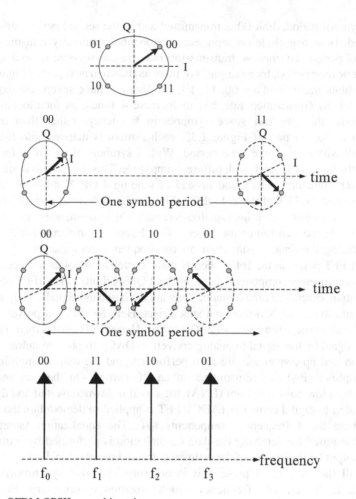

Fig. 1.35 OFDM QPSK on multi carriers

1.10.1.7   Diversity and OFDM

As data rate increases in 3G and 4G, rate of equalization has to be increased accordingly, and may become impossible to implement under a given VLSI technology. To overcome this limitation, diversity, as mentioned in section 1.9.2.3, can be adopted. Let us illustrate this with QPSK modulation, as shown in Figure 1.35. As explained in section 1.5.3, for a QPSK signal, a dibit (00, 01, 10, 11) is encoded as one dot in the 4 quadrants of a circle, as shown in the top part of Figure 1.35. Here the circle is represented in a 2-dimensional plot. The next part shows how transmission is represented, using a 3-dimensional plot. The 2-dimensional circle is tilted in the representation, with the 3rd dimension, drawn horizontally, representing time. Nominally one symbol is transmitted per symbol period, so that within each symbol period, one circle is used to represent the transmitted symbol. Thus in the first symbol period, dibit 00 is transmitted and in the second period, dibit 11 is transmitted. Here one circle on represents one symbol nominally transmitted one per symbol period. To improve transmission rate, we can squeeze several symbols into the same time period, for example, 4 of them as shown in next part of Figure 1.35. Here the dibits transmitted are 00, 11, 10, 01. This can have severe consequences though, such as equalization rate has to increase 4 times, as mentioned above. To overcome this, we can space symbols in frequency, rather than in time. As shown in bottom part of Figure 1.35, each symbol is mapped into frequency domain, all within this same time period. With 4 symbols, there are 4 frequency components: f_0, f_1, f_2, f_3, each with different amplitude. This mapping is done using a Fast Fourier Transform (FFT), and instead of sending 4 QPSK dibits in one time period, we send the FFT of these 4 QPSK dibits in the same time period. Thus date rate increases without increasing equalization rate. The implementation is shown in Figure 1.36. In the transmitter the Inverse Fast Fourier Transform (IFFT) is used for modulating the data constellation on the frequency components. If N is the number of FFT points in the IFFT, then N data constellation points are operated on (here N=4, frequency components f_0-f_3 for the 4 QPSK dibits). The IFFT generates these N output samples in time domain. They are the baseband signals. They are then mapped onto the set of N orthogonal subcarriers. Here $N=4=2^2$, a power of 2; in general N is also typically a power of 2. The mapped baseband signal is then converted to analog signal by the digital to analog converter (DAC). In-phase/quadrature (I/Q) modulation and up-conversion are then performed, and the output amplified by a power amplifier (PA) and transmitted on an RF carrier. On the receiver, upon amplified by a low noise amplifier (LNA), the signal is downconverted and digitized by the analog to digital converter (ADC). FFT is applied to demodulate the OFDM signals (here the 4 frequency components f_0-f_3).The equalization (symbol de-mapping) required for detecting the data constellation is performed by multiplying the FFT output by the inverse of the estimated channel transfer function.

With all the subcarriers present, it is not surprising that synchronization is important. Both time and frequency synchronization is necessary. Frequency synchronization is needed to align the modulator and demodulator's local oscillator frequencies, on all the subcarriers. Time synchronization is needed to identify the start of the OFDM symbol. Given that the subcarriers needed to be orthogonal, synchronization is important since loss of synchronization results in loss of ortho-gonality. Then overlapping of subcarriers is an issue, resulting in intersymbol interference and interchannel interference.

1.10.2   Fast fading

What happens if the channel is changing very fast? This time variation can come from atmospheric variation, movement of mobile receivers, or movement of reflectors. Let us look at (1.37) again. It turns out that one of the most prominent effects comes from the changing of f_c in (1.37) due to time variation. This would

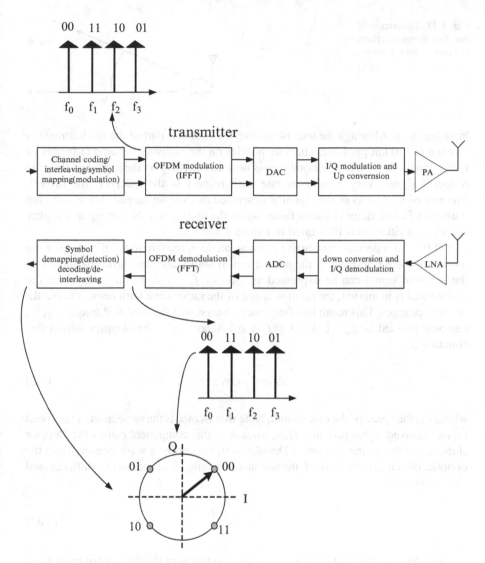

Fig. 1.36 OFDM block diagram implementation

have a pronounced effect on $x_R(t)$. To illustrate the effect due to a change in f_c, we use an example that most readers will be familiar with: the Doppler shift associated with the Doppler effect.

The Doppler effect refers to the change in frequency that occurs at a point of observation as a moving sound source passes that point. A familiar example may be the pitch change heard in a train's horn as it passes a crossing. Similar phenomena affect the received signals in wireless communication when the mobile unit is in motion. Assume that a mobile phone user is sitting in a car in a parking lot, near a

Fig. 1.37 Illustration of
possible signal reception for
the mobile unit in motion

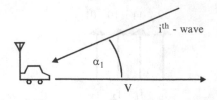

busy highway. Although the user is relatively stationary, part of the environment is moving at 120 km per hour. The automobiles on the highway become reflectors of radio signals. If the mobile phone user is now also driving on the highway, then the reflected signals vary at a faster rate with respect to the receiver (the driver). The rate of variations of the signal is described as Doppler spread. Intuitively, the multipath fading channel varies faster when the mobile unit is moving at a higher speed. This situation is illustrated in Figure 1.37.

Let the i th reflected wave arrive from an angle α_i relative to the direction of the motion of the Rx antenna. If the mobile unit is stationary, the carrier frequency of the received signal can be expressed as follows: $f_c = c/\lambda$. In the case where the mobile unit is in motion, the relative speed of the radio wave with respect to the Rx antenna changes. This result in a frequency change, and the modified frequency, f_m, can be expressed as: $f_m = (c\pm v)/\lambda = f_c \pm \Delta f$. Accordingly, the Doppler shift of this situation is

$$\Delta f_i = \frac{v}{\lambda} \cos \alpha_i \tag{1.62}$$

where v is the speed of the mobile unit. Note that $v \cdot \cos\alpha_i$ is the projection of the speed on the received signal path direction, since only this component causes the Doppler shift. Accordingly, the maximum Doppler shift occurs for a wave coming from the opposite direction toward which the antenna is moving. This maximum shift, denoted as f_d, is given by:

$$f_d = \frac{v}{\lambda} = \frac{v \times f_c}{c} \tag{1.63}$$

This Doppler spread, f_d, has a similar interpretation as the time-delay spread, τ_m. Referring to Figure 1.38 if a carrier wave is transmitted at frequency f_c, after transmission over a fast fading channel we receive a power spectrum that is spread over a certain frequency range. Note that the shape of the distribution can be calculated theoretically. It turns out that it has a specific shape under the assumption that the phase introduced by the channel is uniformly distributed over the interval $[0, 2\pi]$. The frequency range where the power spectrum is nonzero defines the Doppler spread.

So far we have shown that there is a carrier frequency change when the mobile unit is in motion. How is a change in carrier frequency related to its phase? From a phasor diagram interpretation, we observe that in the case where we have two sinusoidal signals with different frequencies they rotate with a different speed,

Fig. 1.38 Power profile of a Doppler shifted channel in the frequency domain

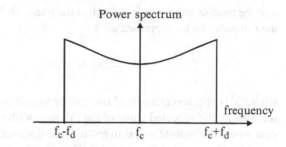

generating a time-varying relative phase. The beat frequency (frequency difference) determines the rate of change of this relative phase. In the case where the Doppler effect causes a frequency change the beat frequency would simply be the amount of Doppler frequency shift. It can be observed that this motion of the Rx antenna leads to time-varying phase shifts of individual reflected waves. Many waves arrive at the Rx antenna, all with different phase shifts, and since their relative phases change with time their effect on the amplitude and phase of the resulting composite signal is also time varying. Therefore, the Doppler effect determines the rate of change at which the amplitude and phase of the resulting composite signal change. We next examine the impact of this effect.

1.10.2.1   Impact of the Doppler Effect

Baseband Signal Amplitude Degradation

As mentioned in the previous subsection, in order to detect the baseband data sequence successfully at the receiver we have to preserve the envelope of the transmitted signal, where in the ideal case the envelope would remain constant during the transmission. To model the reception in a time-varying multipath channel we start from (1.36) again, but noting that the τ_i is time varying in the present case:

$$
\begin{aligned}
x_R(t) &= \sqrt{2} Re\left[\left\{\sum_i \alpha_i(t)s_l(t-\tau_i(t)) \cdot e^{-j2\pi f_c\tau_i(t)}\right\} \cdot e^{j2\pi f_c t}\right] \\
&= \sqrt{2} Re\left[\left\{\sum_i A_i(t) \cdot e^{j\phi_i(t)}\right\} \cdot e^{j2\pi f_c t}\right]
\end{aligned}
\tag{1.64}
$$

Here $A_i(t)$ summarizes the attenuated, delayed baseband signal and $\theta_i(t)$ is the time-varying phase of the i th path.

The phases are varying rapidly with time because very minor receiver movements are large with respect to the wavelength of the propagation. As a start, we assume that the mobile unit is stationary; then the phase shift introduced by the channel for the i th path can be expressed as

$$\phi_i = 2\pi f_c \tau_i = 2\pi \frac{c}{\lambda} \cdot \frac{s_i}{c} = 2\pi \frac{s_i}{\lambda} \tag{1.65}$$

where λ is the wavelength of the carrier signal and s_i is the additional distance traveled by the reflected wave of the i th path with respect to the direct path. In the case where the mobile unit is in motion, the distance s_i varies with time (linear with respect to the speed of the receiver). Then the time-varying phase (for the i th path) can be decomposed into a time-invariant part and a time-varying part as

$$\phi_i(t) = \phi_i + 2\pi \frac{c}{\lambda} \cdot \frac{\Delta s_i(t)}{c} \tag{1.66}$$

Here $\Delta s_i(t)$ is the time varying part of $s_i(t)$.

To simplify, we assume that $\Delta s_i(t)$ is given by its maximum value, which is $v \times t$. Then $\phi_i(t)$ is given by

$$\phi_i(t) = \phi_i + 2\pi \frac{v \cdot t}{\lambda} \tag{1.67}$$

Since v/λ represents the maximum possible Doppler frequency shift as derived in (1.63), we can model the time-varying phase as follows:

$$\phi_i(t) = (2\pi \cdot f_d)t + \phi_i \tag{1.68}$$

where ϕ_i are random phases uniformly distributed from 0 to 2π and f_d is the maximum Doppler frequency shift due to the motion of the mobile unit. Substituting (1.68) in (1.64) and writing the complex exponentials in terms of their real and imaginary components, we get an expression of $x_R(t)$ in terms of the in-phase and quadrature components:

$$x_R(t) = s_I(t) \cos(2\pi f_c t) - s_Q(t) \sin(2\pi f_c t) \tag{1.69}$$

where

$$s_I(t) = \sum_i A_i(t) \cos(2\pi f_d t + \phi_i) \tag{1.70}$$

$$s_Q(t) = \sum_i A_i(t) \sin(2\pi f_d t + \phi_i) \tag{1.71}$$

The envelope of the received signal $x_R(t)$ is given by

$$x_R(t)_{envelope} = \sqrt{s_I(t)^2 + s_Q(t)^2} \tag{1.72}$$

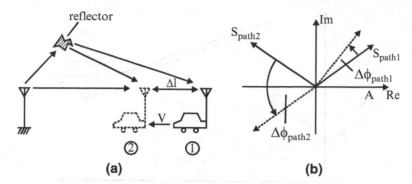

Fig. 1.39 (a) The mobile unit moves at a speed of v with respect to the Tx antenna, causing time-varying phase changes. (b) The continuous lines represent the phasors of the two signals received at time t_0 (position 1) and the dashed lines represent the phasors of two signals received at time $t_0 + \Delta t$ (position 2)

Notice that (1.70) through (1.72) are similar to (1.42) through (1.44), except that the s_I and s_Q expressions are time varying. Hence the envelope would also have a Rayleigh distribution, except the distribution becomes time varying in the present case. From (1.70) through (1.72) we observe that f_d determines the rate at which deep fading of the envelope occurs.

One possible way to quantify time-varying envelope fading is by the use of a time-varying phasor diagram. An example is illustrated in Figure 1.39b. The channel itself is shown in Figure 1.39(a), where we have assumed that there are two paths in the channel.

Referring to Figure 1.39, you may think that after each Δt we take a picture of the time-varying phasor diagram. Due to the time-varying relative phase shifts, the two paths may interfere in a destructive or in a constructive fashion. Obviously, the time interval of interest is the signaling interval T_b, since during that time we want to preserve the amplitude of the envelope. We observe from (1.70) through (1.72) that the envelope's amplitude changes (goes through a deep fade) periodically at a frequency f_d. Thus, to quantify the time-varying nature of the channel, we define the coherence time $(\Delta t)_c$, which is related to the maximum Doppler frequency shift as

$$(\Delta t)_c \approx \frac{1}{f_d} \tag{1.73}$$

Substituting (1.63) into (1.73) also allows us to establish a relation between the coherence time $(\Delta t)_c$ and the speed v of the mobile unit. What does the coherence time of a multipath channel tell us now? In the case where the mobile unit is stationary (v is zero), the coherence time $(\Delta t)_c$ is infinitely long, because there is no time-varying phase change. In the case when the mobile unit is moving during transmission, there are two scenarios here. In the first scenario the signaling interval T_b is much smaller than the coherence time $(\Delta t)_c$ of the channel. Here the channel attenuation and phase shift are essentially fixed for the duration of at least

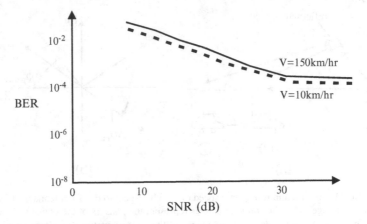

Fig. 1.40 BER as a function of SNR, with Doppler effect, for GMSK modulation

one signaling interval. We refer to this as slow fading. The important relation to ensure slow fading is

$$T_b \ll (\Delta t)_c \qquad (1.74)$$

In the second scenario we have $T_b \gg (\Delta t)_c$. Here the received signal is severely distorted, since the rate of change in the envelope's amplitude is rather large. The channel is said to be fast fading.

Random Phase and Frequency Shift

Because of the Doppler shift, the frequency is shifted and the phase is also shifted. In addition, since the Doppler shift happens in a random manner, this phase shift is also random in nature. In the case when we encode the information in phase (like MSK), such a change in phase leads to a reduction in BER. Up to a certain point, this reduction in BER can be restored by increasing SNR. However, if the fading becomes too fast, increasing SNR does not help and the BER remains fixed at an irreducible level [7], [8], and [9], as shown in Figure 1.40. This is to be expected because the pertubation caused by fast fading is multiplicative in nature and hence insensitive to signal amplitude.

Numerical Example 1.5
A car drives with $v = 120$ km/h on the highway. The driver uses a DECT mobile phone. Then the maximum Doppler frequency shift is given by (1.63):

$$f_d = \frac{120km/h \cdot 1900MHz}{3 \cdot 10^8 m/s} = 211Hz \qquad (1.75)$$

Since the Doppler frequency shift can go both ways depending on the direction of the reflected energy, the bandwidth of the received signal is approximately 106 Hz, and from (1.73) the coherence time $(\Delta t)_c \approx 9.4\text{ms}$. For DECT $T_b \approx 0.58$ us. From (1.74) we conclude that the channel is slowly fading. Hence the envelope's amplitude goes through a deep fade rather infrequently.

We will also check the impact on BER due to phase/frequency shift for GMSK modulation, by using Figure 1.40. For velocity up to 150 km/h, if SNR is larger than 25 dB, then BER is less than 10^{-3} and hence the BER requirement of DECT is satisfied.

1.10.2.2 Time Diversity

In Figure 1.40 we see that as speed increases, BER increases. To compensate for that we have to increase the SNR, which can be costly. A possible technique to improve the received signal BER without increasing SNR is called time diversity. In time diversity the same information-bearing signal is transmitted in n different time slots, where the separation between different time slots exceeds the coherence time $(\Delta t)_c$ of the channel. Consequently, the signals are affected differently by the fading channel.

1.10.3 Comparisons of Flat/Frequency Selective and Slow/Fast Fading

Figure 1.41 [4] shows a comparison of flat/frequency selective and slow/fast fading, as well as the criteria for comparison. The most benign environment of course is the slow, flat fading channel. From Figure 1.41, this is achieved by picking a signalling interval T_b such that $T_b > \tau_m$ (to achieve flat fading) and $T_b < (\Delta t)_c$ (to achieve slow fading). Other channel environments can be achieved by selecting T_b accordingly.

1.11 Summary of Standard Translation

We have applied communication concepts, such as modulation/pulse shaping and fading, to the modulator/demodulator block and channel block, respectively. This allows us to translate standards' requirements to the boundary conditions imposed on the receiver front end. These boundary conditions are functions of the wireless channel environment, modulation/demodulation scheme, and bit error rate of the standard. When we apply this translation to a specific standard, the DECT standard, we have the following:

1. Due to path loss, the receiver can expect to have a received signal, from the channel, whose minimum carrier's amplitude, $P_{\min}$, is −77 dBm. Hence the front end's $P_{\min}$ is −77 dBm.

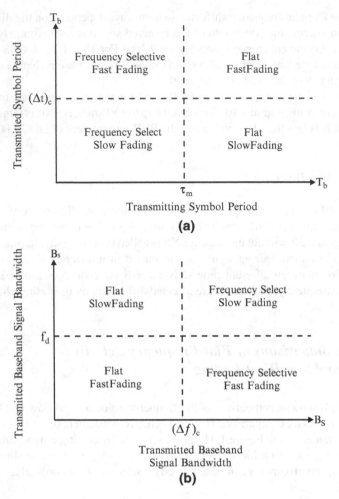

Fig. 1.41 Matrix illustrating types of fading experienced by a signal as a function of (**a**) baseband signal (symbol) period and (**b**) baseband signal bandwidth

2. Due to envelope's fading and AWGN in the channel, the SNR at the input of the demodulator, denoted as SNR_{demod_in}, is required to be 25 dB. We assume that no diversity technique is applied and hence the SNR at the output of the front end, denoted as $SNR_{rec\_front\_out}$, is required to be 25 dB.
3. Due to frequency selective fading, BER is larger than the required 10^{-3} level. We assume that equalization is applied in the demodulator block to restore the BER rate to the 10^{-3} level. Hence no extra requirement is put on the front end. The channel is slow fading: hence the envelope's amplitude goes through deep fade rather infrequently. BER for mobile velocity up to 150 km/h remains below 10^{-3} and no time diversity is needed. Hence no extra requirement is put on the front end.

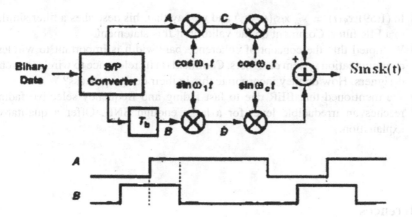

Fig. P.1

1.12 Problems

1.1 To illustrate where some of the numbers given in a standard come from, consider the DECT standard again. Derive, using proper approximation and other information given in the DECT standard, that the channel bandwidth is 1.728 MHz.

1.2 For the MSK scheme as shown in Figure P.1, assume that the signal at nodes A and B is as given. Show the corresponding waveform at nodes C, D and $S_{MSK(t)}$. Assume that the time between the dotted lines is $\frac{1}{4}T_1$, where T_1 is the time when the signal at node A is a 1.

1.3 Would the P_e versus E_b/N_0 curve be identical for BPSK and QPSK? Explain.

1.4 In this chapter we showed plots of the P_e of different modulation schemes using SNR as an independent variable (P_e vs SNR plot). On the other hand, communication texts usually show plots of P_e of different modulation schemes using E_b/N_0 as an independent variable (P_e vs E_b/N_0 plot). Comment on why, as an independent variable, SNR may be a better choice for circuit designers whereas E_b/N_0 may be a better choice for communication system engineers.

1.5 Explain, with the help of time domain graphical representations, using a sinc pulse superimposed on a carrier, how multipath fading causes fluctuation in an envelope's amplitude as well as introduces distortion in the envelope.

1.6 For the NLOS path loss calculation, show that, using a very crude approximation and involving only two paths (one direct, one reflected), the exponent n in (1.24) is 4.

1.7 In Figure 1.32, we stated that as delay increases, attenuation increases. Can you justify why this is true?

1.8 In (1.59) $x_R(t) = \sum_i \alpha_i s_l(t - \tau_i)$ and we say that this describes a filter similar to a FIR filter. Comment on the validity of this statement.

1.9 We noted that the concept of coherence bandwidth is important to wireless communication system designers. Comment on its relevance to wireless circuit designers. How do they incorporate this in their design process?

1.10 We mentioned that BER due to fast fading and frequency selective fading reaches an irreducible level for a large enough SNR. Offer a qualitative explanation.

References

1. John G. Proakis, *Digital Communications*, 3rd ed., McGraw-Hill, 1995.
2. Behzad Razavi, *RF Microelectronics*, Prentice Hall, 1998.
3. Arya Reza Behzard, Master's Thesis, U.C. Berkeley, 1995.
4. Theodore S. Rappaport, *Wireless Communications: Principles and Practice*, Prentice Hall, 1996.
5. J. R. Barry, Edward A. Lee and David G. Messerschmitt, *Digital Communication*, Kluwer Academic Publishers, 2004.
6. Ramjee Prasad, *OFDM for Wireless Communications Systems*, Artech House, 2004.
7. P. A. Bello and B. D. Nelin, "The influence of Fading Spectrum on the Binary Error Probabilities of Incoherent and Differentially Coherent Matched Filter Receivers", *IRE Transactions on Communication Systems*, Vol. CS 10, page 160–168, June 1962.
8. Liu Feher, "Performance of Non-Coherent π/4 QPSK in a Frequency Selective Fast Rayleigh Fading Channel", *IEEE International Conference on Communications*, Vol. 4, page 335.7.1–335.7.5, 1990.
9. Fung, Rappaport etc., *IEEE Transactions on Selected Areas in Communication*, page 393–405, April 1993.

Chapter 2
Receiver Architectures

2.1 Introduction

In Chapter 1 we focused on the blocks enclosed by dotted lines in Figure 1.1. They are the modulator, demodulator, and channel part of the communication system. We now redraw Figure 1.1 as Figure 2.1, with the receiver front end enclosed by dotted lines. In this chapter and the rest of the book, we emphasize this front end part of the communication system. Specifically, in this chapter we discuss the architecture of the front end as well as the filters inside, and in the rest of the book we concentrate on the design of active components inside the front end.

We first review some general philosophy on deciding the front end architecture. We then adopt specific receiver architecture and translate the boundary conditions imposed by the channel and the demodulator on front end (specified in Chapter 1 in terms of communication concepts [SNR]) to circuit concepts (gain, noise figure, distortion). Next we translate these boundary conditions into boundary conditions of the front end sub blocks. Examples of these sub blocks include low-noise amplifiers (LNA), mixers, and intermediate frequency (IF) amplifiers. In subsequent chapters, these boundary conditions will be translated into the design of various sub component specifications.

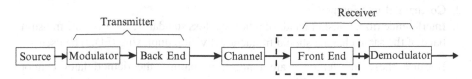

Fig. 2.1 Front end of the receiver enclosed in dotted line

B. Leung, *VLSI for Wireless Communication*, DOI 10.1007/978-1-4614-0986-1_2,
© Springer Science+Business Media, LLC 2011

2.2   Receiver Front End: General Discussion

2.2.1   Motivations

The first question we must ask is, why do we need a front end? To answer it, we repeat the boundary conditions imposed by the channel and modulator/demodulator for the front end, derived in Chapter 1 for the DECT example:

1. Due to path loss, the receiver can expect to have a received signal, from the channel, whose minimum carrier's amplitude, P_{min}, is -77 dBm. Hence the front end's P_{min} is -77 dBm.
2. Due to envelope's fading and AWGN in the channel, the SNR at the input of the demodulator, denoted as SNR_{demod_in}, is required to be 25 dB. We assume that no diversity technique is applied and hence the SNR at the output of the front end, denoted as $SNR_{rec\_front\_out}$, is required to be 25 dB.

Even though these boundary conditions have only been determined for DECT, similar situations exist for other standards as well. The difference lies usually in the numbers. For example, in GSM, the P_{min} number will be much worse due to a larger cell size. To justify the need for a front end, let us go back to the DECT standard. We would want to see, under the worst input condition as described in boundary condition (1), if boundary condition (2) would still be satisfied. In other words, when the antenna has a received power of -77 dBm and there is only AWGN in a fading channel, would the resulting SNR_{demod_in} meet the required SNR_{demod_in}? Let us assume that the AWGN comes just from the thermal noise of the antenna termination resistance of 50 Ω. As shown in Problem 2.1, the SNR_{demod_in} is a mere 5 dB. Therefore, we definitely do not meet the required SNR_{demod_in} of 25 dB. This conclusion is true in general for other wireless standards, and hence in general a front end is required.

To make the no-front-end option look even worse, let us note that in boundary condition (2) the noise source in the SNR calculation has been assumed to be from AWGN in the channel only. We now state another possible noise source in the channel: interference. Such interference in the channel comes from the following:

1. Adjacent channel interference
2. Co-channel interference
3. Interference from transmit bands of other wireless standards, and from the transmit band of the same channel should frequency division duplexing (FDD) be used

Because these interferences are random in nature, the demodulator cannot differentiate them from AWGN and will process them just like noise during the demodulation operation. With this additional noise source, SNR_{demod_in} is reduced even more, further justifying the need for a front end.

2.2.2   *General Design Philosophy*

The technique to combat a low SNR_{demod_in} is by adding a front end block, which processes (conditions) the received signal/AWGN/interference before admitting it to the demodulator. This processing can be done in several ways:

1. Reduce interference/AWGN. This can be done by putting a filter in the front end block, which filters out the interference and AWGN. Filtering for the purpose of reducing interference to an extent that the resulting SNR_{demod_in} meets the required SNR_{demod_in} is denoted as channel filtering. At what frequency do we want to do channel filtering? Remember, if we want to do channel filtering at RF this involves performing a rather narrowband filtering at RF. Doing narrowband filtering at RF, though conceptually simple, is impractical since high-frequency narrowband filtering is expensive.

2. Amplify the desired signal at RF. This can be done by putting an amplifier inside the front end block. Under this arrangement the antenna will be feeding its output into an amplifier, whose output in turn drives the demodulator. Unfortunately, the desired signal, in the worst case, is so small that a large amplification factor is necessary to boost the SNR_{demod_in} to the required level. But remember, when we amplify the desired signal, we are also amplifying the interference and AWGN. In particular, the interference's power, being rather large, when applied to an amplifier with a large gain will have the following undesirable effects:

   a) It will saturate the amplifier.
   b) The amplifier, being nonlinear, will generate intermodulation products from these interferencers. Some of these products can have the same frequency as the desired signal and corrupt it.

3. Reapply (2), but this time we want to filter out the interference first. The filtered interference has such a small amplitude that when fed to an amplifier, it will not cause undesirable effects as outlined in (2a) and (2b). This type of filtering is denoted as interference suppression filtering, to be distinguished from channel filtering. Interference suppression filtering requirement is not as stringent as channel filtering because all we need to do is to filter out enough interference so that there are no undesirable effects. Now suppose we do this and repeat (2). Can we amplify the desired signal at RF all the way up so as to achieve the required SNR_{demod_in}? It turns out that the required amplification for the desired signal at RF is so large that the accompanying interference suppression filtering needs to be very high. Hence it is not practical to do this interference suppression filtering at RF.

4. Having gone through a few alternatives, it seems that the following compromised solution holds the best promise:

   a) Do partial amplification at RF so that it is practical to implement the required interference suppression filtering at RF.

b) Frequency translates the partially amplified signal down to a lower
frequency, called the intermediate frequency (IF). We can now perform
some more amplification since it is more practical to implement the required
interference suppression filtering at a lower frequency such as IF. At this
lower frequency we can also check and see if the frequency is low enough
when channel filtering becomes practical. As a result of the combination
of this partial amplification and channel filtering the SNR_{demod_in} is
improved. This improved SNR_{demod_in} is then checked to see if it meets
the required SNR_{demod_in}. If it does not we will repeat the process until the
required SNR_{demod_in} is met. Then we feed the final amplified and filtered
signal to our demodulator. Of course in the process of frequency translation
(unless the frequency translation is down to baseband or DC) interference
that happens to be present at the image frequency (to be defined later) would
also be translated to the desired frequency and has to be filtered out before the
mixing. This is called anti-image filtering.

In a sense the process outlined in (4) reverses the attenuation the channel
imposes on the desired signal and suppresses the strong interference and noise
admitted via the channel, in a couple of steps. There are numerous variations of this
approach but they all follow a similar philosophy. To illustrate this we consider one
such architecture, called the heterodyne architecture.

2.2.3   Heterodyne and Other Architectures

A receiver front end, as we have seen, needs to achieve different objectives:
amplification, mixing, filtering, and demodulation. Demodulation may include
diversity and equalization. For DECT, given that the speed of the mobiles is
typically low, as we assumed in Chapter 1 (Section 1.8), then the resulting fading
is slow and does not induce unacceptable BER. Thus diversity (to combat fast
fading) is not needed and this simplifies the demodulator.

The resulting heterodyne architecture is shown in Figure 2.2, where we denote
all the circuitry enclosed by the dotted line's as the front end. Notice that we have,
rather arbitrarily, lumped the antenna as part of the channel. We can observe that
the front end includes three bandpass filters (BPF). One design can be such that the
bandpass filter right after the antenna (BPF1) is used primarily to select the band of
interest of the received signal and is referred to as the band selection filter.
This usually provides enough suppression on out-of-band interference that undesirable effects described in (2a) and (2b) of sub-section 2.2.2 are avoided. Note that the
band includes the entire spectrum in which the users of a particular standard are
allowed to communicate. For the DECT standard, the receive band spans from 1880
to 1900 MHz. Hence BPF1 has a bandwidth of around 20 MHz. As discussed

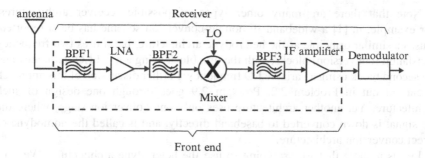

Fig. 2.2 RF receiver using a heterodyne architecture

previously the desired signal, immersed in noise (from both AWGN and interference), can be really small at the Rx antenna (on the order of microvolts) and should be amplified. This amplification should be done with minimal additional noise injected by the amplifier itself, and so a low-noise amplifier (LNA) is used. As mentioned previously, the amplification is followed by mixing, and the mixer should be preceded by an anti-image filter. This is BPF2. This heterodyne architecture adopts one single frequency translation (conversion) and hence is implemented by using a mixer with a variable local oscillator (LO) frequency and a fixed IF. Using this approach, the desired channel can be selected by choosing the proper LO frequency.

Another way to perform channel selection in this architecture consists of using a mixer that mixes down with a fixed LO frequency. The mixer is then followed by a tunable bandpass filter.

In the present approach (variable LO frequency) the mixer is followed by a fixed bandpass filter, BPF3. Since BPF3 operates at IF, we assume the quality factor (Q, not to be confused with Q for complementary error function) required of this filter is now relaxed. Hence we assume that channel filtering becomes possible and therefore BPF3 is also referred to as the channel filter. To perform channel filtering the bandwidth of this filter depends on the channel spacing accorded to each user, which is unique for a particular standard. It is one of the most important constraints on the front end design since this bandwidth is usually very narrow (e.g, 200 kHz in GSM or 1.728 MHz in DECT). The channel filter is followed by an IF amplifier to improve further the SNR_{demod_in} of the signal. The demodulator then takes this output and performs demodulation. For a DECT standard, demodulation can be carried out in a manner similar to Figure 1.7 of Chapter 1, which is a form of coherent demodulation. This demodulation can be done in analog or digital domain. In the analog approach, all the correlators, and multipliers will be performed in the analog domain. In the digital approach, analog to digital conversion is performed and all the signal processing (correlations, multiplications) will be performed using a digital signal processor. Also not shown in the diagram is the frequency synthesizer, which generates the LO signal.

Note that there are many other types of possible receiver architectures. For example, in [1] a wideband IF double-conversion scheme has been reported. It uses a similar principle as the heterodyne architecture, but it uses two frequency translation (conversion) blocks, with the first block using a fixed LO frequency and the second block using a variable LO frequency. Further discussion of this approach is carried out in Problem 2.2. Problem 2.9 goes through one design of such architecture. Yet another architecture uses only one conversion stage, where the RF signal is down converted to baseband directly, and is called the homodyne or direct conversion architecture.

Let us assume that we are going to use the heterodyne architecture. We now derive the relevant specifications of the overall receiver front end and then the specifications of the subcomponents. Basically, the interference requirement allows us to determine the bandwidth requirements of the passive subcomponents: filters. The sensitivity requirement will then allow us to determine specifications on the conversion gain, noise, and distortion, first on the front end and then on individual active subcomponents.

In the following section we discuss filter design in some details. The rest of this chapter and subsequent chapters concern the design of the active subcomponents.

2.3   Filter Design

One major technique to combat interference is to filter it out with bandpass filters. For most bandpass filters the relevant design parameters consist of the center frequency, the bandwidth (which together with center frequency defines the quality factor Q), and the out-of-band suppression. We now discuss how to derive the specifications of these parameters. Because these filters are normally not in integrated circuit form, their implementations are mentioned only briefly.

2.3.1   Band Selection Filter (BPF1)

The bandwidth of the band selection filter is typically around the band of interest (for DECT, 17.28 MHz), and the center frequency is the center of the band (for DECT, 1.89 GHz). The Q required is typically high (for DECT, 1.89 GHz/17.28 MHz $\cong$ 100) and the center frequency is high as well (as mentioned, for DECT it is 1.89 GHz). On the other hand, the suppression is typically not prohibitive. It only needs to be large enough to ensure that interference is suppressed to a point that it does not cause the undesirable effects as discussed in sub-section 2.2.2. To satisfy these specifications, BPF1 can be implemented using a passive LC filter. This LC filter can be combined with the input matching network of the LNA, which is described in Chapter 3.

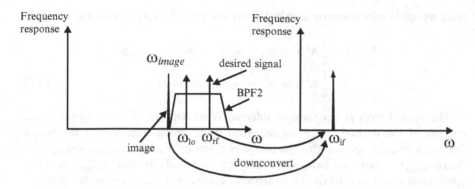

Fig. 2.3 Image rejection problem in the case where FDMA techniques are used to separate different users

2.3.2 Image Rejection Filter (BPF2)

The problem of the image due to mixing was mentioned in sub-section 2.2.2 and is now explained in more detail. As shown in Figure 2.3, during the downconversion process, the desired signal at ω_{rf}, the radio frequency, and its image at ω_{image}, the image frequency, are downconverted to the same frequency of ω_{if}, the intermediate frequency. Hence the desired signal is corrupted. For the downconverted image and downconverted desired signal to overlap, ω_{image} is spaced 2 ω_{if} apart from ω_{rf}:

$$\omega_{image} = \omega_{rf} - 2\omega_{if} \qquad (2.1)$$

where ω_{if} is from definition, given by

$$\omega_{if} = \omega_{rf} - \omega_{lo} \qquad (2.2)$$

Here ω_{lo} is the local oscillator frequency.

To show the image problem mathematically, we model the downconversion (mixing) process to be mathematically equivalent to multiplication (a more complete mathematical treatment of mixing is carried out in Chapter 4). Hence we can write the IF signal as

$$V_{if}(t) = A_{rf} \cos \omega_{rf} t \times A_{lo} \cos \omega_{lo} t \qquad (2.3)$$

where $V_{rf} = A_{rf} \cos\omega_{rf} t$ denotes the desired signal whose frequency is at ω_{rf} and $V_{lo} = A_{lo} \cos\omega_{lo} t$ is the local oscillator signal. For simplicity we assume that $A_{lo} = A_{rf} = A$ and hence we have

$$V_{if}(t) = A \cos \omega_{rf} t \times A \cos \omega_{lo} t \qquad (2.4)$$

Next we apply trigonometric manipulations and rewrite (2.4) as follows:

$$V_{if}(t) = \frac{1}{2}A^2(\cos(\omega_{rf} + \omega_{lo})t + \cos(\omega_{rf} - \omega_{lo})t)$$

$$= \frac{1}{2}A^2(\cos(\omega_{rf} + \omega_{lo})t + \cos\omega_{if}t) \qquad (2.5)$$

The second term is our term of interest. What happens if the received signal consists of the desired signal accompanied by a strong interferer at the image frequency ω_{image}? Assume that the image is given by $V_{image} = A_{image}\cos(\omega_{image}t) = A\cos(\omega_{image}t)$, where we have again assumed for simplicity that $A_{image} = A$. Then upon downconversion (mixing) the interferer generates the downconverted image:

$$A\cos\omega_{image}t \times A\cos\omega_{lo}t \qquad (2.6)$$

Next we will eliminate the term ω_{rf} in (2.1), and (2.2) and express ω_{image} as:

$$\omega_{image} = \omega_{lo} - \omega_{if}. \qquad (2.7)$$

Substituting this in (2.6) and applying trigonometric manipulations again, the downconverted image becomes

$$\frac{1}{2}A^2(\cos(\omega_{lo} - \omega_{if} + \omega_{lo})t + \cos(\omega_{lo} - \omega_{if} - \omega_{lo})t)$$

$$= \frac{1}{2}A^2(\cos(2\omega_{lo} - \omega_{if})t + \cos\omega_{if}t) = \frac{1}{2}A^2(\cos(2\omega_{lo} - \omega_{if})t) + \frac{1}{2}A^2\cos\omega_{if}t$$

$$(2.8)$$

Comparing (2.8) with (2.5) one can see that both of their second terms are at ω_{if} and so the downconverted image does indeed lie on top of the downconverted desired signal. Thus if there is a strong interference at the image frequency, then at the IF there will be a strong interference sitting on top of the desired signal, resulting in a serious degradation of SNR. Notice that no amount of filtering can help the situation after the downconversion process. Thus, we have to ensure that the image is filtered before downconversion.

When would a strong interference be present at the image frequency? Since the image frequency is only $2\omega_{if}$ from the desired signal, if ω_{if} is small (low IF), then this interference can come from adjacent channels or transmit bands of the same standard (for standards using FDD). If ω_{if} is large, this interference can come from transmit bands of other wireless standards.

To get rid of this image, we should place BPF2 in front of the mixer. The center frequency of BPF2, ω_{center_BPF2}, is typically around the LO frequency. If BPF2 has a bandwidth as shown in Figure 2.3, then the image will be filtered out. How do we determine the bandwidth of this filter?

The answer depends on whether we have a low IF or high IF front end, as the frequencies of the interferers are different. As an example we will treat the low

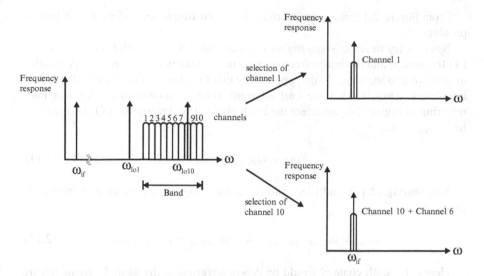

Fig. 2.4 Diagram showing problem of image and function of BPF2

IF case. In this case, the interference comes primarily from an adjacent channel. Let us redraw Figure 2.3 in Figure 2.4, but including adjacent channels. Figure 2.4 is drawn with 10 channels and illustrates the case for the DECT standard. Here different users are separated using FDMA techniques. To maintain the same IF when the desired channel frequency changes, we can change the LO frequency accordingly. We stated that ω_{center_BPF2} is approximately the same as the LO frequency. One way to implement BPF2 is to make ω_{center_BPF2} exactly the same as the LO frequency. However, since the LO frequency changes, this arrangement necessitates that ω_{center_BPF2} be variable, which is impractical because implementing bandpass filters with a variable center frequency at high frequency is very difficult. A better alternative (which is our choice here) is to let ω_{center_BPF2} remain fixed.

Next we determine the bandwidth of a BPF2 that has a fixed center frequency. Referring to Figure 2.4 we show that the 10 channels in DECT are frequency multiplexed over a band that spans from 1880 to 1900 MHz. First let us try to select channel 1. We do this by setting the LO frequency to be ω_{lo_1}. This will allow us to select the first channel. Now let us arbitrarily choose

$$\omega_{if} = 2 \times \omega_{channel} \tag{2.9}$$

where $\omega_{channel}$ is the channel frequency.

Substituting (2.9) in (2.7), we have

$$\omega_{image} = \omega_{lo\_1} - \omega_{if} = \omega_{lo\_1} - 2 \times \omega_{channel} \tag{2.10}$$

From Figure 2.4 there is no interference at that frequency and we do not have a problem.

Next we try to select some higher number channels. To do this we increase the LO frequency, and the image frequency (at $\omega_{lo} - _{if}$) starts to increase and eventually may move into the band. At that point some other channels of the band will become interferers. This is what we call adjacent channel interference. For example, referring to Figure 2.4, we select the tenth channel by setting the LO frequency to be ω_{lo_10}; hence

$$\omega_{rf} = \omega_{10-th-channel}. \tag{2.11}$$

Substituting (2.11), and (2.9) into (2.1), the image will now be at a frequency given by

$$\omega_{image} = \omega_{10-th-channel} - 4 \times \omega_{channel} = \omega_{6-th-channel}. \tag{2.12}$$

Hence the sixth channel would be downconverted to the same IF frequency as the tenth channel. To get rid of this image we should set BPF2's bandwidth to be less than $4 \times \omega_{channel}$. This means that BPF2 has a center frequency $\omega_{center\_BPF2} = \omega_{8-th-channel}$ and span from $\omega_{6-th-channel}$ upward. The lower cutoff frequency is then $\omega_{6-th-channel}$. However, this creates a problem. For example, if the user being assigned this tenth channel is later reassigned to channel 5, then $\omega_{rf} = \omega_{5-th-channel}$. This is below $\omega_{6-th-channel}$ and the user's channel will be cut off. We can avoid this problem by making sure that under the worst case, as the LO frequency varies, the resulting image frequency never falls inside the band.

What is the worst case? Referring to Figure 2.4 the worst case happens when the desired channel is at the highest number channel, or channel 10. Under this worst case we have

$$\omega_{rf} = \omega_{10-th-channel} \tag{2.13}$$

We want to place the desired signal's frequency at the upper cutoff frequency of BPF2 so the upper cutoff frequency of BPF2 is given by:

$$\omega_{upper\_cutoff} = \omega_{10-th-channel} \tag{2.14}$$

The lower cutoff frequency ω_{lower_cutoff} of BPF2, ω_{cutoff}, should be larger than the image frequency:

$$\omega_{lower\_cutoff} > \omega_{image} \tag{2.15}$$

However, in order not to cut off any channel, this lower cutoff frequency also has to satisfy

$$\omega_{lower\_cutoff} < \omega_{1-th-channel} \tag{2.16}$$

As a minimum condition we set

$$\omega_{\text{lower\_cutoff}} = \omega_{1-\text{th-channel}} \tag{2.17}$$

Meanwhile, the bandwidth of BPF2, $\omega_{\text{bw\_BPF2}}$, is given by definition as

$$\omega_{\text{bw\_BPF2}} = \omega_{\text{upper\_cutoff}} - \omega_{\text{lower\_cutoff}} \tag{2.18}$$

Then, if we substitute (2.14), and (2.17) in (2.18), we will have

$$\omega_{\text{bw\_BPF2}} = \omega_{10-\text{th-channel}} - \omega_{1-\text{th-channel}} = \omega_{\text{band}} \tag{2.19}$$

Here ω_{band} is the bandwidth of the band.

For DECT, ω_{band} is around $2\pi \times 20$ Mrad/s and ω_{rf} is around $2\pi \times 1.9$ Grad/s. We assume that ω_{lo} is close to ω_{rf} or close to $2\pi \times 1.9$ Grad/s. Hence $\omega_{\text{center\_BPF2}}$ is close to $2\pi \times 1.9$ Grad/s.

Hence we need a filter centered at close to $2\pi \times 1.9$ Grad/s with a $2\pi \times 20$ Mrad/s bandwidth. The amount of suppression must be large enough that in the presence of the strongest adjacent channel interference, the resulting $\text{SNR}_{\text{demod\_in}}$ is larger than its required value. With these specifications BPF2 is typically designed using a surface acoustic wave (SAW) filter or ceramic filter. The output matching network of the LNA can also be used to provide some of this filtering, as shown in Chapter 3.

2.3.3 Channel Filter (BPF3)

First we want to find the center frequency of BPF3, which is the same as the IF. To find IF we do the following: from (2.15) $\omega_{\text{image}} < \omega_{\text{lower\_cutoff}}$. As a minimum condition we have

$$\omega_{\text{image}} = \omega_{\text{lower\_cutoff}} \tag{2.20}$$

(the filter barely suppresses the image). Now we apply the worst case scenario as described in equations (2.13) through (2.19). Substituting (2.17) in (2.20), we have

$$\omega_{\text{image}} = \omega_{1\_\text{st\_channel}} \tag{2.21}$$

Substituting (2.13), and (2.21) into (2.1), we finally have

$$\omega_{10-\text{th-channel}} - \omega_{1-\text{st-channel}} = 2\omega_{\text{if}} \text{ or } \omega_{\text{if}} = (\omega_{10-\text{th-channel}} - \omega_{1-\text{st-channel}})/2 = \omega_{\text{band}}/2 \tag{2.22}$$

As an example, for DECT, using (2.22) the IF should be larger than $2\pi \times 10$ Mrad/s. However, this would put a stringent requirement on the rolloff of BPF2. Hence the IF is usually increased substantially beyond this bound. In the present case we set IF to be $2\pi \times 100$ Mrad/s:

$$\omega_{if} = 2\pi \times 100 \ Mrad/s \tag{2.23}$$

Therefore, the center frequency of BPF3, ω_{center_BPF3}, is

$$\omega_{center\_BPF3} = 2\pi \times 100 \ Mrad/s \tag{2.24}$$

Next we want to find the BW of BPF3. Since at this point we are at a low enough frequency to do channel filtering, without fearing an unrealizable Q, we will attempt to do so. Hence the bandwidth of BPF3 is simply the channel bandwidth, or

$$\omega_{bw\_BPF3} = \omega_{channel} = 2\pi \times 1.728 \ Mrad/s \tag{2.25}$$

For the Q required [for DECT standard and our choice of intermediate frequency (IF), we have $Q = (2\pi \times 100 \ mMrad/s)/(2\pi \times 1.728 \ Mrad/s) \cong 60$] and the center frequency specified ($2\pi \times 100$ Mrad/s), this filter may be designed with an active filter (continuous time filter) or ceramic filter. The output network of the mixer can also provide some channel filtering.

2.3.3.1 Trade-Off Between BPF2, and BPF3

We noted that the choice of IF frequency affects both BPF2 and BPF3. We discuss this in a bit more detail here. First we eliminate the ω_{band} term between (2.19) and (2.22) and show that

$$\omega_{bw\_BPF2} = 2\omega_{if} \tag{2.26}$$

Hence if ω_{if} increases then ω_{bw_BPF2} also becomes larger. Consequently, Q_{BPF2}, which is given roughly by $\omega_{lo} / \omega_{bw\_BPF2}$, becomes smaller and the design of BPF2 becomes more relaxed. Meanwhile, let us look at Q_{BPF3}, which is given roughly by $\omega_{if} / \omega_{bw\_BPF3}$. Substituting (2.25) in this equation, we have $Q_{BPF3} \cong \omega_{if} / \omega_{channel}$. Hence a larger ω_{if} means that Q_{BPF3} becomes larger, making BPF3 more difficult to design.

2.4 Rest of Receiver Front End: Nonidealities and Design Parameters

Now that we have talked about the design of filters in the receiver front, we turn our attention to the design of the rest of the components. Normally these components consist of circuits such as LNA, mixer, IF amplifier, and analog/digital (A/D) converter. Unlike filters, their relevant design parameters are different. Hence our first task is to discuss these design parameters.

We start by noting that these design parameters are intimately related to the nonidealities in these circuits. First, these circuits are nonlinear. The primary effect of this nonlinearity is that it can frequency translate interference with frequencies outside the channel frequency onto the channel frequency, thereby corrupting the desired signal. Since this interference behaves like noise, this will degrade the SNR at the output of the front end. Second, these circuits introduce noise of their own. Again, this will degrade the SNR at the front end's output. In the following subsection we translate the effects of these two nonidealities into the corresponding design parameters.

2.4.1 Nonlinearity

In this subsection we limit our analysis to nonlinearities up to the third order, because normally these are the nonlinearities of most interest in a radio environment. For simplicity we further assume that these nonlinearities are memoryless. Mathematically, memoryless third-order nonlinearities are specified by a third-order polynomial, which characterizes the input-output relation of a memoryless nonlinear system:

$$y(t) = \alpha_1 s(t) + \alpha_2 s^2(t) + \alpha_3 s^3(t) \tag{2.27}$$

Here s(t) is the input signal and y(t) is the output signal and the nonlinearity is memoryless. Because of this nonlinearity, distortion is generated.

2.4.1.1 Harmonic Distortion

Using the input-output relation (2.27) with a single tone at the input $s(t) = A \cos \omega_o t$, the output of the nonlinear systems can be viewed mathematically as

$$y(t) = \alpha_1 A \cos \omega_o t + \alpha_2 A^2 \cos^2 \omega_o t + \alpha_3 A^3 \cos^3 \omega_o t$$

$$= \frac{\alpha_2 A^2}{2} + \left(\alpha_1 A + \frac{3\alpha_3 A^3}{4}\right) \cos \omega_o t + \frac{\alpha_2 A^2}{2} \cos 2\omega_o t + \frac{\alpha_3 A^3}{4} \cos 3\omega_o t \tag{2.28}$$

Harmonic distortion is defined as the ratio of the amplitude of a particular harmonic to the amplitude of the fundamental. For example, third-order harmonic distortion (HD$_3$) is defined as the ratio of amplitude of the tone at $3\omega_0$ to the amplitude of the fundamental at ω_0. Applying this definition to (2.28) and assuming $\alpha_1 A \gg (3\alpha_3 A^3)/4$, we have

$$HD_3 = \frac{1}{4} \frac{\alpha_3}{\alpha_1} A^2 \tag{2.29}$$

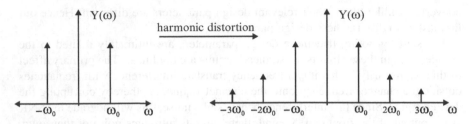

Fig. 2.5 Frequency spectrum of input, output of the nonlinear system

Next we take the Fourier transform of (2.28).

$$Y(\omega) = \alpha_2 A^2 \pi \delta(\omega) + \pi \left(\alpha_1 A + \frac{3\alpha_3 A^3}{4} \right) [\delta(\omega - \omega_o) + \delta(\omega + \omega_o)]$$

$$+ \pi \frac{\alpha_2 A^2}{2} [\delta(\omega - 2\omega_o) + \delta(\omega + 2\omega_o)]$$

$$+ \pi \frac{\alpha_3 A^3}{4} [\delta(\omega - 3\omega_o) + \delta(\omega + 3\omega_o)] \tag{2.30}$$

Equation (2.30) is plotted in Figure 2.5.

Harmonic distortion is typically not of major concern. As an example, for DECT, $\omega_o = 2\pi \times 1.9$ Grad/s. Suppose that the LNA in Figure 2.2 is nonlinear and generates a second harmonic distortion. However, this is at $2\omega_o$ or $2\pi \times 3.8$ Grad/s and will be filtered by BPF2, hence posing no harm.

2.4.1.2 Intermodulation

Intermodulation arises when more than one tone is present at the input. A common method for analyzing this distortion is the "two-tone" test. We assume that two strong interferers occur at the input of the receiver, specified by $s(t) = A_1 \cos\omega_1 t + A_2 \cos\omega_2 t$. Again, the intermodulation distortion can be expressed mathematically by applying $s(t)$ to (2.27).

$$y(t) = \alpha_1 (A_1 \cos \omega_1 t + A_2 \cos \omega_2 t) + \alpha_2 (A_1 \cos \omega_1 t + A_2 \cos \omega_2 t)^2$$

$$+ \alpha_3 (A_1 \cos \omega_1 t + A_2 \cos \omega_2 t)^3 \tag{2.31}$$

Using trigonometric manipulations, we can find expressions for the second and the third-order intermodulation products as follows:

$$\omega_1 \pm \omega_2 : \alpha_2 A_1 A_2 \cos(\omega_1 + \omega_2)t + \alpha_2 A_1 A_2 \cos(\omega_1 - \omega_2)t$$

$$2\omega_1 \pm \omega_2 : \frac{3\alpha_3 A_1^2 A_2}{4} \cos(2\omega_1 + \omega_2)t + \frac{3\alpha_3 A_1^2 A_2}{4} \cos(2\omega_1 - \omega_2)t \tag{2.32}$$

$$2\omega_2 \pm \omega_1 : \frac{3\alpha_3 A_2^2 A_1}{4} \cos(2\omega_2 + \omega_1)t + \frac{3\alpha_3 A_2^2 A_1}{4} \cos(2\omega_2 - \omega_1)t$$

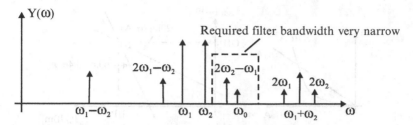

Fig. 2.6 The effect of the intermodulation distortion in the frequency domain, where the frequencies represent the following signals: ω_0: desired signal; ω_1, ω_2: strong interferers; $2\omega_1$, $2\omega_2$: harmonics of the interferers; $\omega_1 \pm \omega_2$: second order intermodulation products; $2\omega_{1,2} \pm \omega_{2,1}$: third order intermodulation products

The output spectrum in the frequency domain can be determined from (2.32) by evaluating its Fourier transform $Y(\omega)$. This is shown in Figure 2.6.

It can be seen from Figure 2.6 that the intermodulation product with frequency $2\omega_2 - \omega_1$ (denoted as the third order intermodulation product, I_{D3}) lies at ω_0 and corrupts the desired signal at ω_0. Furthermore ω_1, ω_2 are close to ω_0 and so trying to filter them out requires a filter bandwidth that is very narrow and is impractical. Hence keeping down $2\omega_2 - \omega_1$ by keeping the nonlinearity (which generates them in the first place) is the only solution.

Where do the two tones, at ω_1 and ω_2, come from? They can be any one of the interferences described in sub-section 2.2.1. Strictly speaking, the interferers are not tones but are more like narrowband noise. For simplicity, for the time being we represent the intereferer at the band from $\omega_1 - \omega_{channel}/2$ to $\omega_1 + \omega_{channel}/2$ as a single tone centered at ω_1, with the rms value of the narrowband noise set equal to the amplitude of the tone, or A_1. Similar representation is applied to the desired signal at the band from $\omega_0 - \omega_{channel}/2$ to $\omega_0 + \omega_{channel}/2$ and the interferer at the band from $\omega_2 - \omega_{channel}/2$ to $\omega_2 + \omega_{channel}/2$. To quantify this distortion we first define the third-order intermodulation distortion, IM_3, as the ratio of the amplitude of the I_{D3} to the amplitude of the fundamental output component (denoted as I_{D1}) of a linear system given by $y(t) = \alpha_1 A \cos \omega_0 t$, where α_1 is the linear small signal gain. Mathematically, this is written as

$$IM_3 = I_{D3}/I_{D1} \qquad (2.33)$$

Note that IM_3 expressed in decibels is simply the difference between the interferer's fundamental output's signal strength in decibels (at ω_0) and the interferer's intermodulated product's strength in decibels (at $2\omega_{2,1} - \omega_{1,2}$).

In order to quantify IM_3, let us simplify by assuming $A = A_1 = A_2$. Applying (2.31) and (2.32) to (2.33) we get

$$IM_3 = \frac{\frac{3}{4}\alpha_3 A^3}{\alpha_1 A} = \frac{3}{4}\frac{\alpha_3}{\alpha_1}A^2 \qquad (2.34)$$

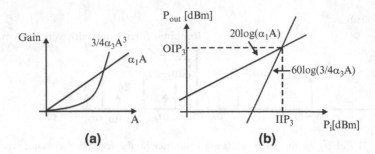

Fig. 2.7 **a)** The linear gain ($\alpha_1 A$) and the nonlinear component ($3/4\alpha_3\, A^3$). **b)** Graphical representation of the input and output third-order intercept point (IIP_3, OIP_3)

Comparing (2.29) to (2.34), it is seen that

$$IM_3 = 3HD_3. \tag{2.35}$$

Since IM_3 depends on input level and is sometimes not as easy to use, we define another performance metric, called the third order intercept point (IP_3).

2.4.1.3   Third-Order Intercept Point, IP_3

From (2.31), we note that as the input level A increases, the desired signal at the output is proportional to A (by the small signal gain α_1). On the other hand, from (2.32) we can see that the third-order product I_{D3} increases in proportion to A^3. This is plotted on a linear scale in Figure 2.7a. Figure 2.7a is replotted on a logarithmic scale in Figure 2.7b, where power level is used instead of amplitude level. As shown in Figure 2.7b the power of I_{D3} grows at three times the rate at which the desired signal I_{D1} increases. The third-order intercept point IP_3 is defined to be the intersection of the two lines.

From Figure 2.7b we can see that the amplitude (in voltage) of the input interferer at the third-order intercept point, A_{IP3}, is defined by the relation

$$20\log(\alpha_1 A_{IP3}) = 20\log\left(\frac{3}{4}\alpha_3 A_{IP3}{}^3\right) \tag{2.36}$$

From (2.36) we can solve for A_{IP3}:

$$A_{IP3} = \sqrt{\frac{4}{3}\left|\frac{\alpha_1}{\alpha_3}\right|} \tag{2.37}$$

For a 50 Ω load, we define the input third-order intercept point (IIP_3) as $IIP_3 = A_{Ip3}{}^2/50\ \Omega$. ($IIP_3$ is hence interpreted as the power level of the input interferer for a 50 Ω load at the third-order intercept point). Notice that IIP_3 can be interpreted

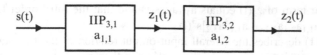

Fig. 2.8 Two cascaded nonlinear stages, where $a_{1,1}$ and $a_{1,2}$ are the linear gains of stage 1 and 2 and $IIP_{3,1}$ and $IIP_{3,2}$ denote the third-order intercept points of stages 1 and 2, $z_1(t)$ denotes the output signal of the first stage, and $z_2(t)$ denotes the output signal of the second stage

in terms of absolute value or decibels. One useful equation that relates IIP_3 to IM_3, expressed in decibels, is the following [2]:

$$IIP_3|_{dBm} = P_i|_{dBm} - \frac{IM_3|_{dB}}{2} \tag{2.38}$$

Here P_i is the power level of the input interferer and is typically defined for a 50Ω load. Both IIP_3 and P_i have been expressed in dBm, whereas IM_3 is expressed in dB.

2.4.1.4 Cascaded Nonlinear Stages

Next we investigate the impact of the nonlinearity of each stage on the overall nonlinearity performance of the front end. To get a useful design relation, we want to find the overall third-order intercept point at the input in terms of the IIP_3 and the gain of each stage. As a start, let us assume that we have a cascade of two nonlinear stages, as shown in Figure 2.8.

Using the input-output relation (2.27), we can approximate the nonlinear behavior of each subcomponent as follows:

$$z_1(t) = a_{1,1}s(t) + a_{2,1}s^2(t) + a_{3,1}s^3(t) \tag{2.39}$$

$$z_2(t) = a_{1,2}z_1(t) + a_{2,2}z_1^2(t) + a_{3,2}z_1^3(t) \tag{2.40}$$

Here $a_{i,j}$ means the i th order gain of the j th stage. For example, $a_{3,2}$ is the a_3 gain of the second stage.

Now we can write the output of the second stage $z_2(t)$ in terms of the input signal $s(t)$ by applying (2.39) to (2.40):

$$\begin{aligned} z_2(t) = &\, a_{1,2}\left(a_{1,1}s(t) + a_{2,1}s^2(t) + a_{3,1}s^3(t)\right) \\ &+ a_{2,2}\left(a_{1,1}s(t) + a_{2,1}s^2(t) + a_{3,1}s^3(t)\right)^2 \\ &+ a_{3,2}\left(a_{1,1}s(t) + a_{2,1}s^2(t) + a_{3,1}s^3(t)\right)^3 \end{aligned} \tag{2.41}$$

To determine the third-order intercept point, we need to calculate the linear and the third-order terms at the output of the second stage. From (2.41) it can be shown that the linear term of $z_2(t)$ equals $a_{1,1}a_{1,2}s(t)$ and that the third-order term equals $[a_{3,1}a_{1,2} + 2a_{1,1}a_{2,1}a_{2,2} + a_{1,1}^3 a_{3,2}]s^3(t)$.

Since (2.41) describes the overall input-output relation of the two-stage system, we can now treat it as one single stage. We can then reuse the previous formulae derived for a single stage to determine the A_{IP3} of this two-stage system. This consists of setting $\alpha_1 = a_{1,1}a_{1,2}$ and $\alpha_3 = a_{3,1}a_{1,2} + 2a_{1,1}a_{2,1}a_{2,2} + a_{1,1}^3 a_{3,2}$ in (2.37). Thus the overall A_{IP3} of the cascaded system can be expressed as

$$A_{IP3} = \sqrt{\frac{4}{3}\left|\frac{a_{1,1}a_{1,2}}{a_{3,1}a_{1,2} + 2a_{1,1}a_{2,1}a_{2,2} + a_{1,1}^3 a_{3,2}}\right|} \tag{2.42}$$

Unfortunately, the sign of the coefficient in the denominator in (2.41) is circuit dependent. Considering the worst case, we add the absolute values together. Rearranging, (2.41) becomes

$$\begin{aligned}
\frac{1}{A_{IP3}^2} &= \frac{3}{4}\frac{\left|a_{3,1}a_{1,2}\right| + \left|2a_{1,1}a_{2,1}a_{2,2}\right| + \left|a_{1,1}^3 a_{3,2}\right|}{\left|a_{1,1}a_{1,2}\right|} \\
&= \frac{3}{4}\left|\frac{a_{3,1}}{a_{1,1}}\right| + \frac{3}{2}\left|\frac{a_{2,1}a_{2,2}}{a_{1,2}}\right| + \frac{3}{4}a_{1,1}^2\left|\frac{a_{3,2}}{a_{1,2}}\right|
\end{aligned} \tag{2.43}$$

Remember, our goal is to express the overall A_{IP3} in terms of the A_{IP3} and the linear gain of each stage. Comparing the first term of (2.43) and (2.37), it can be seen that first term gives the reciproal of the square of $A_{IP3,1}$, the A_{IP3} of the first stage. Following the same reasoning, the third term is seen to consist of the product of the reciproal of the square of $A_{IP3,2}$, the A_{IP3} of the second stage and the square of $a_{1,1}$, the linear gain of the first stage. Hence (2.43) can be rewritten as

$$\frac{1}{A_{IP3}^2} = \frac{1}{A_{IP3,1}^2} + \frac{\left|3a_{2,1}a_{2,2}\right|}{\left|2a_{1,2}\right|} + \frac{a_{1,1}^2}{A_{IP3,2}^2} \tag{2.44}$$

We can make further approximations by examining the second term in (2.44). The coefficient $a_{2,1}$ in the second term only contributes to the second-order intermodulation product and harmonic distortion. As shown in Figure 2.6, since the front end operates primarily in a narrow frequency band, the second-order intermodulation products and the harmonic distortion terms lie outside the band of interest and are thus strongly attenuated by the BPFs in the front end. Hence the impact of $a_{2,1}$ in generating the overall front end's distortion component is small. Consequently, in determining the overall front end's A_{IP3}, the second term can be neglected. Accordingly, (2.44) can be further simplified as

$$\frac{1}{A_{IP3}^2} \approx \frac{1}{A_{IP3,1}^2} + \frac{a_{1,1}^2}{A_{IP3,2}^2} \tag{2.45}$$

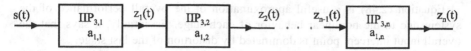

Fig. 2.9 N cascaded nonlinear stages, where $a_{1,1} \ldots a_{1,n}$ are the linear gains of stage $1\ldots n$ and $IIP_{3,1} \ldots IIP_{3,n}$ denote the third order intercept points of stage $1 \ldots n$, $z_1(t)$ denotes the ouput signal of the first stage, and $z_n(t)$ denotes the output signal of the n th stage

Equation (2.45) leads to a useful observation: since $a_{1,1}$ is large, (2.45) is essentially saying that the overall A_{IP3} is dominated by the A_{IP3} of the second stage. Physically, this is what happens: in the case where the linear gain $a_{1,1}$ of the first stage is large, the input signal to the second stage is large. Thus, the distortion of the second stage becomes more critical since it has to handle a larger input signal.

Next (2.45) can be extended to a cascade of an arbitrary number of nonlinear stages, as shown in Figure 2.9, and the resulting overall input intercept point can be derived to be

$$\frac{1}{A_{IP3}{}^2} \approx \frac{1}{A_{IP3,1}{}^2} + \frac{a_{1,1}^2}{A_{IP3,2}{}^2} + \frac{a_{1,1}^2 a_{1,2}^2}{A_{IP3,3}{}^2} + \ldots + \frac{a_{1,1}^2 a_{1,2}^2 \cdots a_{1,n-1}^2}{A_{IP3,n}{}^2} \qquad (2.46)$$

Equation (2.46) can also be restated in a form that involves IIP_3. For a 50 Ω load we stated that IIP_3 is equal $A_{IP3}{}^2/50$ and hence $IIP_{3,i}$ equals $A_{IP3,i}{}^2/50$, where i is the stage number. Substituting these relations in (2.46) we have

$$\frac{1}{IIP_3} \approx \frac{1}{IIP_{3,1}} + \frac{a_{1,1}^2}{IIP_{3,2}} + \frac{a_{1,1}^2 a_{1,2}^2}{IIP_{3,3}} + \ldots + \frac{a_{1,1}^2 a_{1,2}^2 \cdots a_{1,n-1}^2}{IIP_{3,n}} \qquad (2.47)$$

Under the assumption that the subcomponents are matched (in order to allow maximum power transfer), G_i, the power gain of each stage, can be expressed in terms of the voltage gain of each stage as

$$G_1 = Power\ gain\ of\ the\ first\ stage = a_{1,1}^2$$
$$G_2 = Power\ gain\ of\ the\ second\ stage = a_{1,2}^2$$

where i is the stage number. (As a side note, the exact relation between the power gain and the voltage gain will be clarified in Problem 2.5). Substituting this in (2.47), we get

$$\frac{1}{IIP_3} \approx \frac{1}{IIP_{3,1}} + \frac{G_1}{IIP_{3,2}} + \frac{G_1 G_2}{IIP_{3,3}} + \ldots + \frac{G_1 G_2 \cdots G_{n-1}}{IIP_{3,n}} \qquad (2.48)$$

Equation (2.48) is a useful approximation of the overall performance of the system due to the nonlinear behavior of each stage. Essentially, it says that the overall input intercept point is dominated by distortion of the last stage.

Finally, the overall output intercept point, OIP_3, can be expressed in terms of the output intercept point ($OIP_{3,i}$) and power gain (G_i) of each stage ($i = 2,3, \ldots,n$), where n = number of stages. First we start with (2.47) and divide both sides of (2.47) by $a_{1,1}{}^2 \cdot a_{1,2}{}^2 \cdot \ldots \cdot a_{1,n}{}^2$:

$$\frac{1}{a_{1,1}^2 \cdots a_{1,n}^2 IIP_3} \approx \frac{1}{a_{1,1}^2 \cdots a_{1,n}^2 IIP_{3,1}} + \frac{1}{a_{1,2}^2 \cdots a_{1,n}^2 IIP_{3,2}} + \cdots + \frac{1}{a_{1,n}^2 IIP_{3,n}} \quad (2.49)$$

Next we relate the output intercept point to the input intercept point. For the output intercept point of each stage, $OIP_{3,i}$, we have

$$OIP_{3,i} = a_{1,i}^2 IIP_{3,i}^2 \quad (2.50)$$

where $i = 1, n$. For the output intercept point of the overall front end, OIP_3, we have

$$OIP_3 = total\_gain \times IIP_3 = a_{1,1}^2 \ldots a_{1,n}^2 IIP_3 \quad (2.51)$$

Substituting (2.50) and (2.51) into (2.49), we get

$$\frac{1}{OIP_3} = \frac{1}{a_{1,2}^2 \cdots a_{1,n-1}^2 OIP_{3,1}} + \frac{1}{a_{1,3}^2 \cdots a_{1,n-1}^2 OIP_{3,2}} + \cdots + \frac{1}{OIP_{3,n}} \quad (2.52)$$

Again, under the assumption that the subcomponents are matched (in order to allow maximum power transfer), (2.52) can be expressed in terms of the power gains of each stage:

$$\frac{1}{OIP_3} \approx \frac{1}{G_2 \cdots G_n OIP_{3,1}} + \frac{1}{G_3 \cdots G_n OIP_{3,2}} + \cdots + \frac{1}{OIP_{3,n}} \quad (2.53)$$

The importance of (2.53) is that it tells us that the output intercept point of the overall front end is dominated by the output intercept point of the last stage.

2.4.1.5 Gain Compression

Another phenomeon caused by the nonlinearity of the receiver is called gain compression. When the input to an amplifier is large, the amplifier saturates, hence clipping the signal. When the strength of the input is further increased, the output signal is no longer amplified. At this point, the output is said to be compressed. We may now ask what the clipping of a signal has to do with the nonlinear behavior of a system. If we go back to (2.28), we observe that in $y(t)$ there are two terms with frequency ω_o due to the nonlinear behavior. Let us assume that the other terms in $y(t)$ have frequency outside the band of interest and hence are removed by the BPFs. Thus, $y(t)$ becomes

$$y(t) = \left(\alpha_1 A + \frac{3\alpha_3 A^3}{4}\right) \cos \omega_0 t = \left(\alpha_1 + \frac{3\alpha_3 A^2}{4}\right) A \cos \omega_0 t \qquad (2.54)$$

In the case where α_3 is negative, the second term is decreasing the gain. As the input starts to increase, the impact of the second term becomes important in the sense that it saturates the active device. To get a feeling for the input level when considerable gain compression occurs, we can use the concept of the 1-dB compression point, defined as the input level that causes the linear small-signal gain to drop by 1 dB. Thus, the $A_{1\text{-dB}}$ specifies the amplitude (in voltage) of the input signal when the linear voltage gain drops by 1 dB. From (2.54) we see that the 1-dB compression point can be expressed mathematically by

$$\left(\alpha_1 + \frac{3\alpha_3 A^2_{1-db}}{4}\right)\Bigg|_{dB} = \alpha_1|_{dB} - 1 \, dB \qquad (2.55)$$

We can rewrite (2.55) in terms of decibels:

$$20 \log\left|\alpha_1 + \frac{3\alpha_3 A^2_{1-db}}{4}\right| = 20 \log|\alpha_1| - 20 \log 1.122 \qquad (2.56)$$

From (2.56) the $A_{1\text{-dB}}$ input level is given by

$$A_{1-dB} = \sqrt{0.145 \left|\frac{\alpha_1}{\alpha_3}\right|} \qquad (2.57)$$

The idea of the 1-dB compression point is shown graphically in Figure 2.10.

2.4.1.6 Blocking

A phenomenon closely related to gain compression is blocking. So far, the compressive behavior of a nonlinear subcomponent with a single input signal has been discussed. What happens if a weak desired signal along with a strong interferer occurs at the input of a compressive subcomponent? Assume that we have the input signal $s(t) = A_0 \cos \omega_0 t + A_1 \cos \omega_1 t$ where A_1 is a strong interferer and A_0 is the

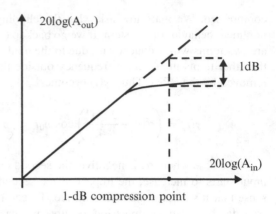

Fig. 2.10 Illustration of the 1-dB compression point

desired signal. Applying this s(t) to (2.27) we can express the output terms of interest (the term at the fundamental frequency) as

$$y(t) = \left(\alpha_1 A_0 + \frac{3\alpha_3 A_0^3}{4} + \frac{3\alpha_3 A_0 A_1^2}{2} \right) \cos \omega_o t + \dots \tag{2.58}$$

If the interferer strength is much greater than the desired signal strength (that is, $A_1 \gg A_0$), equation (2.58) can be simplified as

$$y(t) = \left(\alpha_1 + \frac{3\alpha_3 A_1^2}{2} \right) A_0 \cos \omega_o t + \dots \tag{2.59}$$

If α_3 is negative, the small signal gain is attenuated by the interferer. If the attentuation becomes so large that the overall gain drops to zero, we say that the signal is blocked. Many receivers must be able to withstand blocking signals up to 70 dB greater than the desired signal.

To summarize, the effect of gain compression and blocking is that the desired signal amplitude is reduced, and this results in a degraded SNR_{demod_in}.

2.4.2   Noise

There is circuit noise internal to the subcomponents in the front end. This noise will add on to the AWGN, cause interference, and further degrade the SNR. To explore this effect we have to treat the circuit noise aspect in a more comprehensive manner. First we talk about noise sources. Circuit noise is associated with the electrical components that build the subcomponents, such as resistors and transistors. Circuit noise is further subdivided into thermal, shot, and flicker noise.

2.4.2.1   Noise sources

In this subsection we define the effects of the circuit noise. The noise phenomena considered here are caused by the small current and voltage fluctuations that are generated within the devices themselves.

Thermal noise

Thermal noise basically arises due to the random thermally generated motion of electrons. It occurs in resistive devices and is proportional to the temperature. Fundamentally, the thermal energy of electrons causes them to move randomly, thus causing local concentrations of electrons. This net concentration of negative charge in a local spot (balanced by net concentration of positive charge in another spot, as the total charge must remain zero) will result in a local nonzero voltage. As the concentration changes randomly, the resulting voltage also changes randomly, resulting in a noiselike behavior. Note that the noise exists even though the resistor is not connected and no current is flowing through it (as opposed to shot noise, to be discussed in the next subsection). It is white and has a flat power spectral density (PSD) whose value can be given as follows (depending on whether it is modeled by its Thévenin or Norton equivalent):

$$\frac{\overline{v^2}}{\Delta f} = 4kTR \quad or \quad \frac{\overline{i^2}}{\Delta f} = 4kT\,\frac{1}{R} \tag{2.60}$$

Here $\overline{v^2}$ and $\overline{i^2}$ are the mean square noise voltage and current, respectively; k is the Boltzmann constant; T is the temperature in Kelvin; and R is the resistance value. The unit is V^2/Hz or A^2/Hz

Shot noise

Shot noise occurs in all energy barrier junctions, namely in diodes and bipolar transistors. Actually, it happens whenever a flux of carriers (possessing potential energy) passes over an energy boundary. Since the potential energy of the carrier is random, the number of carriers that possess enough energy to cross the barrier is random in nature, resulting in a flux (current if the carrier are electrons) whose density is also random in nature. This random nature gives rise to shot noise. It is thus obvious that shot noise exists only when there is a current, as opposed to thermal noise. For example, a bipolar junction transistor (BJT) is a device whose current I is composed of holes and electrons that have sufficient energy to overcome the potential barrier at the junction. Thus, the current consists of discrete charges and not of continuum of charges. The fluctuation in the current I is termed shot noise.

This current I is composed of random pulses with average value I_{DC}. Now it can be shown that shot noise is white with a flat PSD whose value is given as follows:

$$\frac{\overline{i^2}}{\Delta f} = 2qI_{DC} \qquad (2.61)$$

Here q is the charge of an electron in coulomb and I_{DC} is the value of the dc current in amperes.

Flicker noise

Flicker noise or $1/f$ noise arises from random trapping of charge at the oxide-silicon interface of MOS transistors and in some resistive devices. Obviously, the more current is flowing in a nonideal silicon interface, the higher is the rate of electrons that are trapped. On the other hand, at high frequency the electrons are varying too fast to be trapped and no noiselike current variation is caused. Consequently, the noise density is given by

$$\frac{\overline{i^2}}{\Delta f} = K\frac{I_{DC}^{\alpha}}{f} \qquad (2.62)$$

where K and α (ranging from 0.5 to 2) are constants that depend on the nature of the device. I_{DC} is the dc current in amperes and f is frequency in hertz. As we can see from (2.62), flicker noise is most significant at low frequencies. However, it can still be troublesome for frequencies up to a few megahertz.

Additive Noise Versus Phase Noise

Now we discuss the manifestation of circuit noise in the front end. The noise of a subcomponent manifests itself as either additive noise or phase noise. Additive noise is described by noise adding to the amplitude of the desired signal. Phase noise is noise adding to the phase of the desired signal. In a front end the noise inherent in a LNA, and a mixer is best described by additive noise, whereas the noise inherent in a frequency synthesizer is best described by phase noise. Both will degrade the $\text{SNR}_{\text{demod\_in}}$ and hence sensitivity of the overall receiver chain. In this chapter we will focus on the additive noise of the LNA and mixer blocks. In Chapters 7 and 8 we return to the discussion of phase noise.

2.4.2.2   Noise Figure

A parameter called noise figure (NF) is a commonly used method of specifying the additive noise inherent in a circuit or system. Use of this parameter is limited to situations where the source impedance is resistive. However, this is often the case in

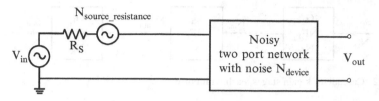

Fig. 2.11 Model used in noise figure calculations

a front end, and so this method of specifying noise is adopted here. The noise figure describes how much the internal noise of an electronic element degrades its SNR. It is often specified for a 1 Hz bandwidth at a given frequency. In this case, the noise figure is also called the spot noise figure to emphasize the very small bandwidth as opposed to the average noise figure, where the band of interest is taken into account. The interpretation should be clear from the context. Mathematically the noise figure is defined as

$$NF = \frac{SNR_{in}}{SNR_{out}} = \frac{S_{in}N_{out}}{S_{out}N_{in}} \tag{2.63}$$

where N_{in} is the input noise power and is always taken as the noise in the source resistance; and N_{out} is the output noise power including the circuit contribution and noise transmitted from the source resistance. By inserting $S_{out} = GS_{in}$ into (2.63), where G is the power gain of the corresponding stage, we get

$$NF = \frac{N_{out}}{GN_{in}} \tag{2.64}$$

Let us refer to Figure 2.11, the model for NF calculation. In Figure 2.11,

$$N_{in} = N_{source\_resistance} \tag{2.65}$$

N_{out}, the noise occurring at the output, is given by the input noise multiplied by its power gain G plus the additional device noise:

$$N_{out} = N_{Device} + G \cdot N_{Source\_resis\tan ce} \tag{2.66}$$

Substituting (2.65), and (2.66) into (2.64), we have

$$NF = \frac{N_{Device} + G \cdot N_{Source\_resistance}}{G \cdot N_{Source\_resistance}} \tag{2.67}$$

As a final remark, note that the noise figure is specified by a power ratio and given in decibels. We usually refer to the corresponding numerical ratio as the noise factor. Thus, the relation between the two is given by noise figure = 10 $\log_{10}$(noise factor). Which one we are referring to should be clear from the context.

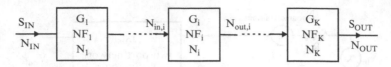

Fig. 2.12 Cascade of k gain stages, each with noise

2.4.2.3   Cascaded Noisy Stages

The noise figure of a cascade of noisy stages can be derived in terms of the noise figures of the individual blocks. Consider Figure 2.12 where a cascade of noisy stages each with available power gain G_i and noise figure NF_i is shown. N_i, $N_{in,i}$, $N_{out,i}$ specify the device noise, input noise, output noise, respectively, of the i th stage. Since we have a number of stages, we further make the assumptions that we have the same source impedance for each stage. This assures that we have the same input noise $N_{in,i}$ occurring at the input of each stage. Referring to Figure 2.12 the derivation consists of the following four steps:

1. We find the output power, which is given by the input signal power multiplied by product of the power gain of each stage.

$$S_{out} = S_{in}(G_1 G_2 \cdot \ldots \cdot G_k) \qquad (2.68)$$

2. We want to relate the noise N_i occurring in each stage to the corresponding noise figure of that stage. We start by applying (2.64) to stage 1 and we have

$$NF_i = \frac{S_{in,i}N_{out,i}}{S_{out,i}N_{in,i}} = \frac{(G_1 \cdot \ldots \cdot G_{i-1})S_{IN}N_{out,i}}{(G_1 \cdot \ldots \cdot G_i)S_{IN}N_{in,i}} = \frac{N_{out,i}}{G_iN_{in,i}} \qquad (2.69)$$

Then we apply (2.66) to the i th stage and we have

$$N_{out,i} = N_i + G_iN_{in,i} \qquad (2.70)$$

But what is $N_{in,i}$? This, being the input noise of the i th stage, comes from the input resistance of the i th stage. Since we assume that every stage has the same source resistance, then every stage has the same input noise. For ease of reference, we label all of them by the input noise of stage 1, which from Figure 2.12 is simply given by N_{IN}. Therefore, $N_{in,i}=N_{IN}$. Substitute this in (2.70), we have

$$N_{out,i} = N_i + G_iN_{IN} \qquad (2.71)$$

Now we can substitute (2.71) in (2.69) and we have

$$NF_i = \frac{N_i + G_iN_{IN}}{G_iN_{IN}} = \frac{N_i}{G_iN_{IN}} + 1 \qquad (2.72)$$

Finally, since we are interested in relating N_i to NF_i, we can rearrange (2.72) as

$$N_i = (NF_i - 1)G_iN_{\text{IN}} \qquad (2.73)$$

We want to find the output noise power of this cascade of stages. We will use (2.66) and iteratively apply this formula, starting from the first stage and then to a combination of first and second stage, until we include all k stages. The resulting output noise is the output noise from the k th stage, $N_{\text{out},k}$. From Figure 2.12 we know that $N_{\text{out},k}$ is also denoted as N_{out}. The final expression therefore is given as

$$N_{OUT} = (G_1G_2 \cdots G_K)N_{IN} + N_1(G_2 \cdots G_K) + N_2(G_3 \cdots G_K)$$
$$+ \cdots + N_{K-1}G_K + N_K \qquad (2.74)$$

We then substitute (2.73) from step 2 in (2.74) and we get

$$N_{out} = (G_1G_2 \cdots G_K)N_{in} + (NF_1 - 1)(G_1G_2 \cdots G_K)N_{in}$$
$$+ \cdots + (NF_K - 1)G_KN_{in} \qquad (2.75)$$

4.   Finally we want to derive the total noise figure of the cascaded chain. To do this we substitute (2.68) from step 1, and (2.75) from step 3 into (2.63). Canceling S_{in}, N_{in} and doing the proper simplification, we get

$$NF = NF_1 + \frac{NF_2 - 1}{G_1} + \frac{NF_3 - 1}{G_1G_2} + \cdots + \frac{NF_k - 1}{G_1G_2 \cdots G_{k-1}} \qquad (2.76)$$

So we have finally derived the equation that relates the total noise figure to the individual noise figure. This equation is called the Friis formula. Note that the NF used in the Friis formula is specified as a ratio and not in decibels. Equation (2.76) predicts that NF is dominated by the first stage NF, NF_1. When compared with (2.48) we see that both noise and distortion are dominated by one stage. The only difference is that for noise this is the first stage whereas for distortion this is the last stage. Of course, a front end with a small NF and a large IIP$_3$ is desirable.

2.5   Derivation of NF, IIP$_3$ of Receiver Front End

We have now finished discussing the NF and IIP$_3$ of the receiver front end. We have also developed formulas that relate the IIP$_3$ and NF of this receiver front end to those of the individual subcomponents. Our next step is to translate the boundary conditions on this front end, imposed by wireless standards and derived in Chapter 1, to its key design parameters. We assume that the key design parameters are IIP$_3$, NF, G. Boundary conditions such as SNR$_{\text{demod\_in}}$, and P_{min} (communication

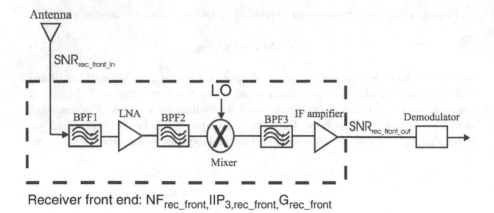

Antenna

$SNR_{rec\_front\_In}$

LO

BPF1   LNA   BPF2   BPF3   IF amplifier   Demodulator

$SNR_{rec\_front\_out}$

Mixer

Receiver front end: $NF_{rec\_front}, IIP_{3,rec\_front}, G_{rec\_front}$

Fig. 2.13 Receiver block diagram showing the G, NF, IIP$_3$ and SNR of the front end

terminology) that we derived in Chapter 1 are now translated into the required IIP$_3$ and NF (circuit terminology) of the front end.

To illustrate this process we use DECT as an example standard and we use the heterodyne architecture as the example architecture. We will have a chance to redo this using zero IF (homodyne) architecture in Problem 2.9. The heterodyne architecture is redrawn in Figure 2.13, where as in Figure 2.2 we separate it into two parts: receiver front end followed by the demodulator. The receiver front end consists of the part between the antenna and the demodulator and contains the LNA, mixer, IF amplifier, and so forth. It is characterized by having its own gain G_{rec_front}, noise figure NF_{rec_front}, and third order intercept point IIP_{3rec_front}. This front end takes an input signal from the antenna with a signal to noise ratio denoted as $SNR_{rec\_front\_in}$, processes it, and generates a signal at its output with a signal to noise ratio denoted as $SNR_{rec\_front\_out}$. As stated previously, our goal in this section is to find the required G_{rec_front}, NF_{rec_front}, and IIP_{3rec_front} of this front end. To simplify notation, unless otherwise specified all parameters used in this section are to be interpreted in terms of decibels.

2.5.1   Required NF_{rec_front}

In this subsection, we want to find the required NF_{rec_front}. Let us apply (2.63), interpreted in decibels, to Figure 2.13. We have

$$NF_{rec\_front} = SNR_{rec\_front\_in} - SNR_{rec\_front\_out} \tag{2.77}$$

Step 1: Calculate $SNR_{rec\_front\_out}$.
From Section 2.2, boundary condition (2) we know that

$$SNR_{rec\_front\_out} = 25dB. \tag{2.78}$$

Step 2: Calculate $SNR_{rec\_front\_in}$

$SNR_{rec\_front\_in}$ should be calculated under the worst-case situation. Obviously, the worst case happens when the received signal from the antenna is at its minimum and the noise is at its maximum. We have minimum signal power when the receiver is farthest away from the base station. From Section 2.2, boundary condition (1) we know that $P_{min} = -77$ dBm. Hence

$$S_{rec\_front\_in} = -77 dBm \qquad (2.79)$$

At this point we want to calculate the AWGN in the channel. This noise is hard to calculate. However, we know for sure that since the antenna has a 50 Ω resistive load, there will be thermal noise coming from this load. We further assume that this is the only AWGN in the channel, and we use (2.60) to calculate this noise. Applying (2.60), interpreted in decibels, and integrating throughout the noise bandwidth B, and we have

$$N_{rec\_front\_in} = 10 \log \overline{v^2} = 10 \log 4kTR_s B \qquad (2.80)$$

Here $R_s = 50$ Ω. How do we calculate B? Referring to Figure 2.13, we note that the receiver front end has 3 different filters. We take the one with the narrowest bandwidth to define B. From subsection 2.3.3 we note that BPF3 has the narrowest bandwidth, whose bandwidth is given by (2.25) as $2\pi \times 1.728$ Mrad/s or 1.728 MHz.

Substituting this value into (2.80) and working in decibels we have

$$N_{rec\_front\_in} = 10 log_{10}(4kTR_s) - 10 log_{10}(B/1Hz) = -174 dBm + 62 dB$$
$$= -112 \; dBm \qquad (2.81)$$

Now from the SNR definition in decibels,

$$SNR_{rec\_front\_in} = S_{rec\_front\_in} - N_{rec\_front\_in} \qquad (2.82)$$

Substituting (2.79), and (2.81) in (2.82), we have

$$SNR_{rec\_front\_in} = -77 dBm - (-112 \; dBm) = 35 dB \qquad (2.83)$$

Step 3: Calculate NF_{rec_front}

Let us substitute (2.78), and (2.83) into (2.77). We get

$$NF_{rec\_front} = 35 dB - 25 dB = 10 dB \qquad (2.84)$$

This is the required NF of the front end to satisfy the DECT standard.

Fig. 2.14 Four users are using their cellular phones at the same time, where d is the maximum range from user to the base station (dictated by standards; in our example this equals 400 m); d_0 is the free space distance; d_2 is the minimum allowable distance between user 1 and user 2, where user 2 uses the next channel; and d_3 is the minimum allowable distance between user 1 and user 3

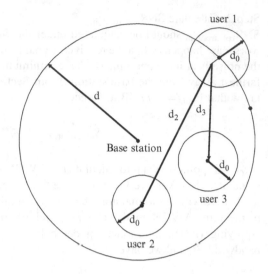

2.5.2   Required IIP$_{3,rec\_front}$

In this subsection we want to derive the required IIP$_{3,rec\_front\_end}$. The larger this IIP$_3$, the smaller is the third-order intermodulation product generated by the interferers. We denote this intermodulation product, calculated at the output of the receiver front end, as $I_{D3}{}''$. These interferers, as mentioned in subsection 2.4.1, have the same characteristics as that of narrowband noise. The maximum $I_{D3}{}''$ must be made small enough (achieved with a large enough IIP$_{3,rec\_front\_end}$) that its power is below the minimum $S_{rec\_front\_out}$ (minimum signal power at the output of the receiver front end), by a sufficiently large margin.

Step 1: State the minimum desired signal power and the maximum interferer power at the input of the front end.

To calculate the required IIP$_3$, we must look at the condition when we have the minimum $S_{rec\_front\_out}$ and the maximum I_{D3}''. This corresponds to the condition where at the input of the front end, the desired signal power is at a minimum, while the interferers are at their maximum.

From Section 2.1, boundary condition (1), the minimum desired signal power at the input of the front end is given as -77 dBm. To find the power of the interferer, we assume that the interferences are from adjacent channel interference. To understand adjacent channel interference, let us refer to Figure 2.14. Figure 2.14 shows three users, where user 1 uses the desired channel. Let us assume he is assigned the channel at 1.89 GHz. Next we assign the next two channels to users 2 and 3. Now let us consider user 3. User 3 is transmitting signal and interfering with user 1. From Figure 2.14 user 3 is closer to user 1 than user 2 and hence the interfering signal from him is larger. Hence he is placed farther away in frequency. Specifically, user 3 is assigned a channel at frequency = 1.89 GHz + 2×1.728 MHz = 1.8934 GHz. User 2 is then assigned a channel at frequency 1.8917 GHz. The maximum power

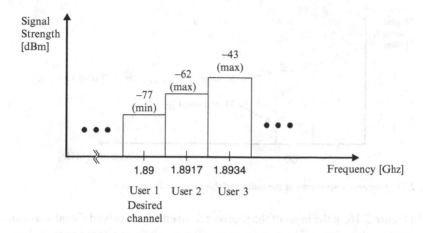

Fig. 2.15 The signal strength requirements of adjacent channels in the DECT standard

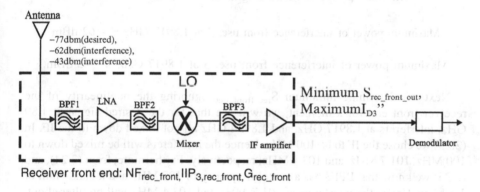

Fig. 2.16 Block diagram showing signal and interference level at input and output of receiver front end

from user 2 and user 3 is given in the DECT blocking specifications. This is shown in Figure 2.15. For example, for a channel away it is around –62 dBm [3]. Since we are interested in IIP$_3$, we need to show only the two adjacent channels. Hence Figure 2.15 summarizes the minimum desired signal power and the maximum interferer power at the input of the front end that is required to calculate IIP$_{3,\text{rec\_front}}$.

Step 2: Calculate the minimum S$_{\text{rec\_front\_out}}$
We now redraw Figure 2.13, but we specify the relevant power levels at the input and output of the front end. The resulting diagram is shown in Figure 2.16.

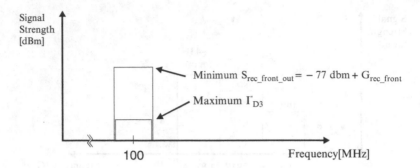

Fig. 2.17 Frequency spectrum at the output of the receiver front end

In Figure 2.16, at the input of the receiver front end the received signal is assumed to consist of the desired signal and two interferers. Their power are specified as

$$\text{Minimum power of desire dsignal at } 1.89 \text{ GHz} = -77 \text{ dBm}$$

$$\text{Maximum power of interference from user 2 at } 1.8917 \text{ GHz} = -62 \text{ dBm}$$

$$\text{Maximum power of interference from user 3 at } 1.8917 \text{ GHz} = -43 \text{ dBm.}$$

Next we determine minimum $S_{rec\_front\_out}$. Ignoring the nonlinearity of the receiver front end for a moment, we will see that for the desired signal at 1.89 GHz, interferers at 1.8917 GHz, and 1.8934 GHz all got mixed down to the IF. In (2.23) we chose the IF to be 100 MHz. Hence the interferers will be mixed down to 100 MHz, 101.7 MHz, and 103.4 MHz, respectively. Furthermore, from (2.24), and (2.25) we know that BPF3 has a center frequency of 100 MHz and a bandwidth of 1.7 MHz. Hence the interferers at 101.7 MHz and 103.4 MHz will be filtered out. The only signal at the front end output will be the one at 100 MHz. Since the receiver front end has a gain of G_{rec_front}, the signal at 100 MHz will have a power level given by -77 dBm $+ G_{rec\_front}$. Therefore we have

$$Minimum \ S_{rec\_front\_out} = -77dBm + G_{rec\_front}$$

This is shown in Figure 2.16. The frequency spectrum is shown in Figure 2.17.

Step 3: Derive the maximum $I_{D3}{}''$, in terms of the maximum power of interferers at the receiver front end input, Pi, and IIP_{3,rec_front}.

This step is subdivided into three substeps.

Step 3a: Derive equivalent block diagram for receiver front end and find the frequency spectrum for $I_{D3}{}''$.

In step 2 we have ignored the nonlinearity of the front end. Let us bring back this nonlinearity and see what happens. First let us redraw Figure 2.16 in Figure 2.18(a).

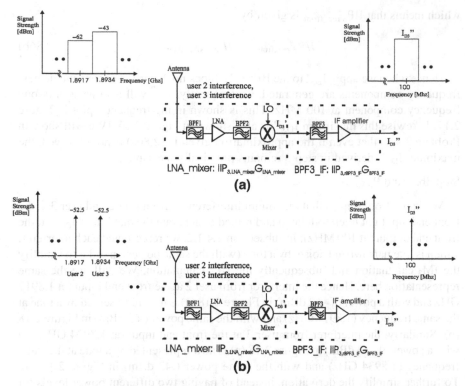

Fig. 2.18 Two equivalent representations of the front end: (**a**) Interferers represented as narrow-band noise. (**b**) Interferers represented as tones

Here we subdivide the receiver front end into two blocks: LNA_mixer block and BPF3_IF block. Each block has its own gain and IIP$_3$. Their gain relationship with $G_{\text{rec\_front}}$ is given as

$$G_{\text{rec\_front}} = G_{\text{LNA\_mixer}} + G_{\text{BPF3\_IF}} \tag{2.85}$$

We now concentrate on the LNA_mixer block, which consists of BPF1, LNA, BPF2, and the mixer. The nonlinearity in the LNA_mixer block, as characterized by IIP$_{3,\text{LNA\_mixer}}$, will take the two interferers from user 2 and user 3 (whose frequencies are at 1.8917GHz and 1.8934 GHz, respectively, and at maximum power) and generate a third-order intermodulation product, denoted as I$_{D3}$ (whose frequency is at 1.89 GHz and at maximum power). I$_{D3}$ will then be mixed down by the mixer to an IF of 100 MHz. The mixed down I$_{D3}$ is denoted as I$_{D3}$'. Now I$_{D3}$', being at 100MHz, will not be filtered by BPF3. It will pass through the filter and got amplified by the IF amplifier and generate the I$_{D3}$'' (at maximum power). To save notations, we will also use I$_{D3}$, I$_{D3}$' and I$_{D3}$'' to denote the power level of the respective intermodulation products. Which meaning they refer to should be clear from the context. Let us further assume that BPF3 and IF amplifier are very linear. Accordingly, IIP$_{3,\text{BPF3\_IF}}$ is practically given as

$$IIP_{3,\text{BPF3\_IF}} \cong \infty \tag{2.86a}$$

which means that $IIP_{3,\text{rec\_front}}$ is given by

$$IIP_{3,\text{rec\_front}} \cong IIP_{3,\text{LNA\_mixer}} \tag{2.86b}$$

Hence when we apply I_{D3}' to the BPF3_IF block no distortion occurs and no new frequency components are generated. Consequently, I_{D3}'' will also have only one frequency component at 100 MHz. This is shown in the frequency plot in Figure 2.17. Likewise this maximum I_{D3}'' is also shown in Figure 2.16. [We will show in Problem 2.4(c) that even if the approximation given in (2.86a) is not observed, the maximum I_{D3}'' derived in step 3 remains practically the same.]

Step 3b: Find P_i.

We know from step 3a that maximum interferences from user 2 and user 3 at the front end input got intermodulated and mixed to generate the maximum I_{D3}'' at the front end output (at 100 MHz). In subsection 2.4.1.2, we represented each interferer, which is like narrowband noise, by a tone (with the same power) in order to simplify the IM_3 calculation and subsequently the IIP_3 calculation. We will use the same representation here. Hence the interferer from user 2 at the front end input (at 1.8917 GHz and with a power of –62 dBm) in Figure 2.18 (a) is now represented by a tone at the same frequency (1.8917 GHz) and with the same power (-62 dBm) in Figure 2.18 (b). Similarly, the interferer from user 3 at the front end input (at 1.8934 GHz and with a power of –43 dBm) in Figure 2.18 (a) is now represented by a tone at the same frequency (1.8934 GHz) and with the same power (-43 dBm) in Figure 2.18 (b). To further simplify the derivation, instead of having two different power levels for two tones, we assign one power level, taken to be the average of the two, to both tones. We arbitrarily decide to take the geometric average of the two power levels, which equals –52.5 dBm and assign it to both tones. Hence we have

$$P_i = -52.5 dBm. \tag{2.87}$$

Step 3c: Derive maximum I_{D3}'' in terms of $IIP_{3,\text{rec\_front}}$
First we apply (2.38) to the LNA_mixer block of Figure 2.18(b) and we have

$$IIP_{3,LNA\_mixer} = P_i - \frac{IM_3}{2} \tag{2.88}$$

Remember from (2.87) $P_i = -52.5$ dBm. Rearranging (2.88) and substituting this value for P_i, we have

$$IM_{3,LNA\_mixer} = 2\left(P_i - IIP_{3,LNA\_mixer}\right) = 2\left(-52.5 dBm - IIP_{3,LNA\_mixer}\right) \tag{2.89}$$

However, from the definition as given in (2.33) (interpreted in decibels)

$$IM_{3,LNA\_mixer} = I_{D3}' - I_{D1}' \tag{2.90a}$$

To reiterate, I_{D3}' and I_{D1}' are the third-order intermodulation product and fundamental component generated by the two interferers which got mixed down and appear at the output of the mixer. In the present situation the two interferers are at the maximum power and so the intermodulation product and the fundmanetal component are also at their maximum power level. Rewriting (2.90a) under this situation, we have

$$IM_{3,\text{LNA\_mixer}} = \text{maximum } I_{D3}' - \text{maximum } I_{D1}' \qquad (2.90b)$$

Rearranging (2.90b), we have

$$\text{maximum } I_{D3}' = \text{maximum } I_{D1}' + IM_{3,\text{LNA\_mixer}} \qquad (2.91)$$

Since maximum I_{D1}' is the fundamental component generated by the maximum interferer with a power level of –52.5 dBm, it is given by

$$\text{maximum } I_{D1}' = -52.5 \ dBm + G_{\text{LNA\_mixer}} \qquad (2.92)$$

Now we can substitute (2.89), and (2.92) into (2.91), and we get

$$\begin{aligned} \text{maximum } I_{D3}' = {}& -52.5 \ dBm + G_{\text{LNA\_mixer}} \\ & +2\left(-52.5 dBm - IIP_{3,\text{LNA\_mixer}}\right) \end{aligned} \qquad (2.93)$$

In discussing (2.86a) we stated that BPF3 and IF AMPLIFIER are practically linear. When we apply maximum I_{D3}' to the input of the BPF3_IF block, the maximum I_{D3}'' generated will contain only one frequency component at 100 MHz, whose level is given by

$$\text{maximum } I_{D3}'' = \text{maximum } I_{D3}' + G_{\text{BPF3\_IF}} \qquad (2.94)$$

Substituting (2.93) into (2.94), we get

$$\text{maximum } I_{D3}'' - 52.5 \ dBm + G_{\text{LNA\_mixer}} + 2\left(-52.5 \ dBm - IIP_{3,\text{LNA\_mixer}}\right) + G_{\text{BPF3\_IF}} \qquad (2.95)$$

Applying (2.85) and (2.86b) to (2.95), we have

$$\begin{aligned} \text{maximum } I_{D3}'' &= -52.5 \ dBm + G_{\text{rec\_front}} + 2\left(-52.5 \ dBm - IIP_{3,\text{rec\_front}}\right) \\ &= -3 \times 52.5 \ dBm + G_{\text{rec\_front}} - 2 \times IIP_{3,\text{rec\_front}} \end{aligned} \qquad (2.96)$$

This gives the maximum I_{D3}'' in terms of $IIP_{3,\text{rec\_front}}$.
Step 4: Relate $SNR_{\text{rec\_front\_out}}$ to maximum I_{D3}'' and hence relate $SNR_{\text{rec\_front\_out}}$ to $IIP_{3,\text{rec\_front}}$. From the required $SNR_{\text{rec\_front\_out}}$, calculate the required $IIP_{3,\text{rec\_front}}$.

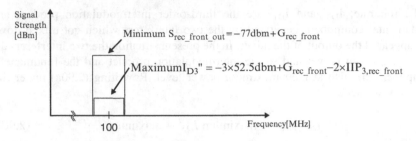

Fig. 2.19 Signal strength at the output of the receiver front end

In this step we refer again to Figure 2.18(b). As shown at the output of the receiver front end, we have I_{D3}'', a tone at 100 MHz. At this point we want to express I_{D3}'' using a narrowband noise representation again. We denote this noise as $n_{I_{D3}''}(t)$. We first assume that $n_{I_{D3}''}(t)$, like the AWGN channel noise, is also additive and has a Gaussian distribution. The power of $n_{I_{D3}''}(t)$ is, of course, the same as I_{D3}'' and is still given by (2.96). We redraw Figure 2.17 as Figure 2.19, where the value of maximum I_{D3}'' [as expressed in (2.96)] is explicitly shown.

If we assume that there is no other noise (e.g., no AWGN from the channel, antenna, or circuit noise), then at the receiver front end output we have a signal immersed in noise and the SNR is given by

$$\text{SNR}_{\text{rec\_front\_out}} = \text{minimum } S_{\text{rec\_front\_out}} - \text{maximum } I_{D3} \qquad (2.97)$$

Substituting the appropriate values for minimum $S_{\text{rec\_front\_out}}$ and maximum I_{D3}'' from Figure 2.19 into (2.97) and simplifying, we have

$$\text{SNR}_{\text{rec\_front\_out}} = 80.5 dBm + 2 \times \text{IIP}_{3,\text{rec\_front}} \qquad (2.98)$$

This signal and noise is applied to the demodulator and hence we have

$$\text{SNR}_{\text{demod\_in}} = \text{SNR}_{\text{rec\_front\_out}} = 80.5 dBm + 2 \times \text{IIP}_{3,\text{rec\_front}} \qquad (2.99)$$

Now what is the required $\text{SNR}_{\text{demod\_in}}$? To derive the required $\text{SNR}_{\text{demod\_in}}$ with this noise, $n_{I_{D3}''}(t)$, we can replace n(t) in (6.14), Chapter 1 by $n_{I_{D3}''}(t)$, so that (6.14) now becomes

$$x_R(t) = \alpha(t)s_1(t)exp(-j\theta(t)) + n_{I_{D3}''}(t) \qquad (2.100)$$

This equation is not to be interpreted in decibels. We can then start from (2.100) and go through the rest of the derivation in subsection 6.2.2, Chapter 1, and derive the required $\text{SNR}_{\text{demod\_in}}$ for DECT. Since the noise characteristics are assumed to be the same as the AWGN in the channel, the required SNR_{demod_in} for DECT should stay the same, which is given in (1.51), Chapter 1, as 25 dB. Hence if we substitute 25 dB for $SNR_{rec\_front\_out}$ in (2.99) we get

$$25 \text{ dB} = 80.5 \text{ dBm} + 2 \times IIP_{3,rec\_front} \tag{2.101}$$

Solving, we get

$$IIP_{3,rec\_front} = -27.75 \ dBm \tag{2.102}$$

This is the required IIP_3 of the front end to satisfy the DECT standard.

2.6 Partitioning of required NF_{rec_front} and IIP_{3,rec_front} into individual NF, IIP_3

Our strategy here is to start with a set of power gains, NF, IIP_3, of the individual stages, based on some existing receiver front end. This will provide us with an initial design. We then calculate the NF_{rec_front} and $IIP_{3,front\_end}$ of this existing front end and see if it satisfies the required NF_{rec_front} and $IIP_{3,front\_end}$. Iterations can be carried out if necessary. In subsequent chapters, we go through the actual design of these stages and figure out if the power gains, NF, IIP_3 of the individual stages are achievable. Further iterations can then be carried out if necessary.

Before we carry out this strategy, there is one more point to be noted. Up to now we have assumed that the power gain G_i is given as the square of the voltage gain. However, this is only true if termination resistances are the same. In general, G is given by

$$G_i = A_{v,i}^2 \cdot \frac{R_{in,i}}{R_{out,i}} \tag{2.103}$$

where G_i, $A_{v,i}$, $R_{in,i}$, $R_{out,i}$ are the power gain, voltage gain, and input and output termination resistance of the i th stage, respectively [4]. Therefore, instead of specifying G, we would specify the termination resistances and the voltage gain of the individual stages and then calculate the corresponding G.

Step 1: Specify the voltage gain and termination resistance of the subcomponents
The heterodyne architecture is redrawn in Figure 2.20 with the corresponding termination resistances. BPF1 and BPF2 both need 50 Ω termination resistance; otherwise the filters would lose their frequency responses. For this example we choose a BPF3 that has $R_{in} = 1$ kΩ and $R_{out} = 1$ kΩ. The LNA is specified to have input and output resistances of 50 Ω. The output resistance of the mixer is much lower than its input resistance. In this case it is a nice feature since it maximizes the voltage gain of the mixer. Thus the mixer is specified with an input resistance of 50 Ω

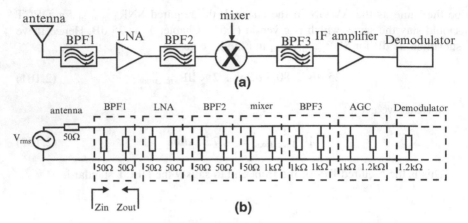

Fig. 2.20 a) The heterodyne architecture b) The heterodyne architecture with the corresponding input and output impedance of the different subcomponents

and an output resistance of 1 kΩ. Finally, the input resistance of the demodulation block is specified to be 1.2 kΩ.

The voltage gains of the various subcomponents are given in the second row of Table 2.1. The power conversion gains can now be calculated using equation (2.103) and are given in the third row of Table 2.1. Overall conversion gain G can be calculated by summing all the terms in row 3 and we have

$$G = 31.6 \ dB \tag{2.104}$$

Step 2: Specify a possible set of NF of the subcomponents that meets the required NF$_{\text{front\_end}}$.

First, we need to specify the noise figure of the individual subcomponents. An initial set of values is given in row 4 of Table 2.1. The general philosophy is for the early stages of the front end (basically the LNA) to dominate the NF (i.e., the early stages should have good NF [small values] and the latter stages can afford to have poorer NF [larger values]. The exact values (i.e. 3 dB for LNA and 12 dB for mixer) depend on the circuit and technology details and will be covered in later chapters.

Substituting the G and NF values from rows 3 and 4 of Table 2.1 into the Friis' formula (2.76) and carrying out calculation in ratio (not in decibels), we have the noise contributions calculated for individual components, which are shown in row 5 of Table 2.1.

Finally, we get the total noise figure by adding the corresponding components of the Friis formula.

$$\therefore \ \text{NF}_{\text{rec\_front}} = 11.54 \ or \ 10.6 \ dB \tag{2.105}$$

Table 2.1 Characteristics of the subcomponents used in the receiver of Figure 2.20

	BPF1	LNA	BPF2	Mixer	BPF3	IF AMPLIFIER
A_v(isolated components) [dB]	−2	12	−2	10	−2.6	30
Power Conversion Gain G (isolated components)[dB]	−2	12	−2	−3	−2.6	29.2
NF (isolated components) [dB]	2	3	2	12	6	8
Components of Friis formula	1.58	1.587	0.058	2.368	0.95	3.073
IIP_3 (isolated components) [dBm]	90	−10	90	−10	90	20
Components of overall IIP_3 formula (2.48)	ignored	$(0.15.10^{-3})^{-1}$	Ignored	$(0.01.10^{-3})^{-1}$	ignored	ignored

Comparing this value and the required NF_{rec_front} [calculated in (2.84) to be 10 dB], we see that we have selected a set of G, NF values for the subcomponents that allows the front end to meet the required NF.

Step 3: Specify a possible set of IIP_3 of the subcomponents that meets the required $IIP_{3,front\_end}$.
As in the NF case, we need to specify the IIP_3 of the individual subcomponents. The values are given in row 6 of Table 2.1. The general philosophy is for the latter stages of the receiver (basically the mixer) to dominate the distortion. The exact values (i.e. −10 dBm for LNA and −10 dBm for mixer) again depend on the circuit and technology details and will be covered in later chapters.

Substituting G from row 3 and IIP_3 from row 6 into the overall IIP_3 formula, (2.48), and carrying out calculation in ratio (not in decibels), we have the IIP_3 contributions calculated for individual components, which are shown in row 7 of Table 2.1. Finally, we get the IIP_3 by adding the corresponding components in row 7 and then taking the reciprocal.

$$\therefore IIP_{3,rec\_front} = \left(\left\{ 10log\left[1/\left((0.15.10^{-3})^{-1} + (0.01.10^{-3})^{-1} \right) \right] \right\} + 30 \right) dBm$$

$$= -20 \ dBm$$

(2.106)

Comparing this value and the required IIP_{3,rec_front} [calculated in (2.102) to be −27.75 dBm], we see that we have selected a set of G, IIP_3 values for the subcomponents that allows the front end to meet the required $IIP_{3,front\_end}$.

2.7 Problems

2.1 For DECT's standard, for worst-case reception, calculate the SNR at the input
of the demodulator (Figure 2.1). We assume no front end is used and the only
AWGN in the channel comes from a 50 Ω input resistance.

2.2 In the chapter we discussed that we want to go for a fixed f_{if} (and hence a
variable f_{lo}) scheme because this allows the use of a fixed frequency BPF2,
which is easier to implement. We showed how, with this scheme, image can
become a problem. A system designer suggests that we should instead use a
variable f_{if}, fix f_{lo} scheme, together with a fixed frequency BPF2 because he
believes this will fix the problem. Is he correct? To answer this you can follow
parts (a), (b), and (c). Assume that the receive band spans from 824 to 894 MHz.

   (a) Assume that we use a fixed f_{if} of 10 MHz and a variable f_{lo}. Draw the
   frequency spectrum, including all relevant frequencies, when f_{rf} is 894 and
   824 MHz. Repeat the case for f_{if} of 100 MHz.
   (b) Now assume that we use a variable f_{if} and a fixed f_{lo} of 760 MHz. Again,
   draw the spectrum when f_{rf} is 894 and 824 MHz. Repeat the case for f_{lo}
   of 850 MHz.
   (c) Now determine the f_{image} for all the cases in (a) and (b) and comment on
   whether scheme (b) is better than scheme (a) as far as making it easier for
   BPF2 to filter out the image.

2.3 This problem concerns the qualitative understanding of the nonlinear behaviour
of the receiver front end.

   (a) We mentioned that harmonic distortion is not an issue for the heterodyne
   architecture described in Figure 2.2 since the bandpass filters (BPF1, 2, 3)
   will filter them out. What happens if they do not filter them out completely?
   (b) We have shown mathematically what blocking is. Explain, from first
   principle and in words (no equations), the mechanism of blocking. Offer
   an explanation in words (no equations) that distinguishes how blocking and
   intermodulation of interferers affect SNR$_{rec\_front\_out}$ differently.
   (c) Is it possible to have a receiver front end such that it generates poor IM$_3$ but
   does not block?

2.4 This question clarifies subtleties encountered in Section 2.5.

   (a) At the beginning of Section 2.5, we assumed that IIP$_3$ is the only key design
   parameter as far as characterizing the receiver front end's distortion
   performance. Comment on the validity of this assumption.
   (b) In (2.81) we used 1.728 MHz (=bandwidth of BPF3) as the noise bandwidth B. Comment on the validity of this assertion.
   (c) In step 3 of subsection 2.5.2 we said that BPF3's and IF amplifier's
   nonlinearity does not matter [refer to (2.86a)]. Justify this.

2.5 In Section 2.6 we stated that power gain G is the same as the voltage gain squared a_v^2 only if $R_L = R_S$, otherwise it is given by (2.103). Derive (2.103) and show that it is only true if the input of the i th stage is matched to the output of the previous stage [the $(i\text{-}1)$ th stage].

2.6 Calculate the individual components of the Friis formula in Table 2.1.

2.7 For the heterodyne architecture whose subcomponents are described in Table 2.1, plot the change of the overall NF as a function of the isolated NF of the LNA with a fixed voltage gain of 12 dB and as a function of voltage gain of the LNA with a fixed NF of 3 dB.

2.8 Let us reexamine the architecture described in Figure 2.20 and Table 2.1 as follows: we change the IF amplifier's IIP_3 to be -20 dBm. Now the mixer, and the IF amplifier have $IIP_3 = -10$ dBm, -20 dBm respectively. Hence the IF amplifier is poorer than the mixer in terms of IIP_3. On the other hand, the OIP_3 of the IF amplifier, even with this new IIP_3 value, is 9.2 dBm. This remains better than the OIP_3 of the mixer (which is -13 dBm). Which component is better? Explain.

2.9 In this problem we deal with a different standard and a different receiver front end architecture. The standard is given as follows: the carrier frequency is 800 MHz and the channel bandwidth is 200 kHz. The input signal plus interference spans from -104 dBm to -10 dBm. Assume that a BPSK modulation/demodulation scheme is used and the required BER is 10^{-3}. Furthermore, TDMA and FDMA are used to separate the different users. To simplify matters, we will neglect the impact of fading in our calculation.

Figure P.2(a) and Figure P.2(b) show the block diagram describing a different receiver front end architecture (called zero IF or homodyne receiver architecture), together with a diagram that specifies the input and output impedance of each subcomponent (matched).

The receiver front end takes the signal, filters it with a BPF for anti-aliasing, and amplifies it with a LNA. (The LNA is assumed to have a output matching network that performs further noise rejecting filtering. This output matching network [noise rejection filter] is included in the LNA block and not explicitly shown. The noise it is rejecting comes mainly from the LNA itself.) The signal is then fed into an automatic gain circuit (AGC) which is then converted by the A/D converter. The sample and hold circuit in the A/D converter (to be discussed in Chapter 6) acts as a sampling mixer (to be discussed in Chapter 5). This sampling mixer mixes the signal to DC before getting it quantized. That is why this is called a zero IF architecture. It should be noted that the mixing to baseband (zero IF) is inherently performed as the sampling operation of the A/D converter. In this problem, subsampling (or bandpass sampling) is performed. This means that the sampling frequency in the A/D converter is much lower than the carrier frequency (800 MHz) and at least larger than the Nyquist frequency of the baseband signal (2×200 kHz). As a side note, as a result of using subsampling, noise from the channel and LNA at around multiples of the subsampled frequency will be aliased and hence there is the

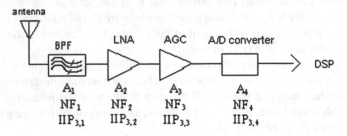

Fig. P.2(a) Receiver architecture assumed in this example, where A_i denotes the voltage gain, NF_i denotes the isolated noise figure and $IIP_{3,i}$ denotes the isolated third-order intercept point (power) at the input $(i = 1, \ldots, 4)$

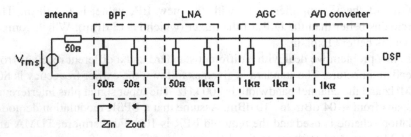

Fig. P.2(b) Input and output impedance are characterized for each subcomponent

need for BPF and noise rejection filter. Further demodulation and other signal processing is done in a digital signal processor (DSP) at DC.

(a) Find the NF required of the receiver front end.

(b) Find the NF of the A/D converter. An A/D converter is characterized by another type of additive noise: quantization noise. As shown in Chapter 6, the noise is white and the spectral density of noise is given by $\frac{\overline{v_{n,out}^2}}{\Delta f} = \frac{\Delta^2}{12 \times f_b}$ where $\Delta =$ step size of the least significant bit (LSB) and is given by $V_{FS}/2^n$. Here V_{FS} is the full-scale voltage and n is the number of bits in the A/D converter. f_b is the bandwidth of the A/D converter. Suppose $V_{FS}=3.13$V, $n=12$. What should f_b be? What is the gain of the A/D converter? What is $\frac{\overline{v_{n,out}^2}}{\Delta f}$ and what is the NF of the A/D converter?

(c) Assume that the following information for the individual stage is given:

$$NF_1 = 2\,dB \qquad A_1 = -2\,dB$$
$$A_2 = 20\,dB$$

$$NF_3 = 10\,dB \qquad A_3 = 31\,dB$$

Calculate the noise figure for the LNA.

References

1. Rudell, J.C.; Ou, J.-J.; Cho, T.B.; Chien, G.; Brianti, F.; Weldon, J.A.; Gray, P.R, "A 1.9-GHz Wide-Band IF Double Conversion CMOS Receiver for Cordless Telephone Applications," *IEEE Journal of Solid State Circuits*, Vol. 32, No. 12, P. 2071–2088, December 1997.
2. D. Johns and K. Martin, *Analog Integrated Circuit Design*, Wiley, 1997.
3. European Telcommunication Standard, ETS 300 175–1, October 1992.
4. G. Gonzalez, *Microwave Transistor Amplifiers*, 2nd ed., Prentice Hall, 1997.
5. P. Gray, P. Hurst, S. Lewis, R. Meyer, "Analysis and design of analog integrated circuits", 5th edition, Wiley, 2009.

Chapter 3
Low Noise Amplifier

3.1 Introduction

Ostensibly, one key component of any receiver chain is the low-noise amplifier (LNA) coming off the antennas. Since the signal at that point is comparatively weak, good gain and noise performance are necessary. In Chapter 2, we showed that the overall noise factor of the receiver front end is dominated by the first few stages and can be approximated according to the Friis' formula as

$$NF_{\text{rec\_front}} = \left(\frac{1}{G_{\text{LNA}}}\right)\left(NF_{\text{subsequent}} - 1\right) + NF_{\text{LNA}} \qquad (3.1)$$

where $NF_{\text{subsequent}}$ is the total input-referred noise factor of the components following the LNA, and G_{LNA}, NF_{LNA} are the gain and noise factor of the LNA itself. The noise of all subsequent stages is reduced by the gain of the LNA and the noise of the LNA is injected directly into the received signal. Thus the LNA needs both high gain and low noise. There are well-known trade-offs in amplifiers between noise and gain. Often, one is achieved at expense of the other. The inherent issues and compromises between these two are examined in this chapter.

3.1.1 General Philosophy

Low-noise amplifier design, and for that matter any high-frequency amplifier design, can be approached from one of two methods: lumped parameter or distributive parameter methods. In the lumped parameter methods, stability, gain, and noise performances are analyzed using Bode/Nyquist plots. During the design, the active devices and the input/output termination networks are treated separately as far as their impact on stability, gain, and noise performance is concerned. Here voltage and current are primary variables of interest and two port representations such as y, z,

B. Leung, *VLSI for Wireless Communication*, DOI 10.1007/978-1-4614-0986-1_3,
© Springer Science+Business Media, LLC 2011

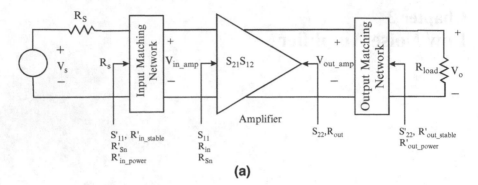

Fig. 3.1a Generalized LNA topology

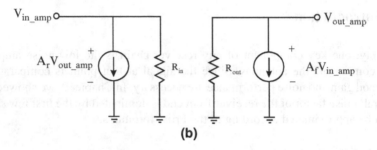

Fig. 3.1b Lumped parameter representation of amplifier

h or g parameter representations are adopted. The distributed parameter methodology, on the other hand, starts to take into consideration the distributive nature of the circuits and uses Smith charts [based on scattering (S) parameters]. Throughout the analysis both the active devices and the input/output terminations are considered together in determining the impact on circuit stability, gain, and noise performances. Here power becomes the primary variable of interest and S-parameters based analysis becomes the main tool [1]. For the same circuit the two methods will come to the same conclusion.

The general topology of any LNA can be broken down into three stages: an input-matching network, the amplifier itself, and an output-matching network [Figure 3.1a]. In Figure 3.1a the LNA and matching network are characterised by both lumped parameters (e.g R_{in}, R_{out} etc.) and S parameters (e.g. S_{11}, S_{12} etc.). Let us digress a bit here into some discussions on S parameters. There are four S-parameters of interest: S_{11}, S_{22}, S_{12}, and S_{21}. We will try to draw the analogy between them and their lumped parameters counterpart. If we redraw the amplifier in Figure 3.1a using a lumped parameter representation as shown in Figure 3.1b, it may help to think, in a very crude way, of the following analogy and relationship:

S_{21} <-> the forward gain, A_f
S_{12} <-> the reverse transmission (or leakage) factor, A_r, which is usually very small in low frequency, but can become significant at high frequency
S_{11} <-> the input impedance, R_{in}
S_{22} <-> the output impedance, R_{out}

The primary motivation for using the S-parameter is that at microwave frequencies the parameters in a lumped parameter representation are very difficult to measure. This is because the short and open circuit conditions, upon which the definitions of these representations are based, are difficult to implement over wideband, at high frequencies. In addition, high-frequency transistors are prone to oscillation under open or short circuit conditions. Consequently, this new representation, the S-parameter, whose variables are travelling waves and the power associated with them, is adopted. Having said that, our philosophy is to stay with the lumped parameter representation as much as possible, as integrated circuit designers may be more familiar with it. Only when we need to interface between the transistor and the input/output networks will we revert back to a S-parameter representation. For example, in our design example later on, we will go through the whole design process using lumped-parameters. Then we will calculate S_{11}, S_{22} based on the calculated R_{in}, R_{out} and the characteristics impedance Z_o (typically 50 Ω). In this way the S-parameter method will be made as transparent as possible.

3.1.2 Matching Networks

3.1.2.1 Objectives

Returning to the LNA, let us continue our original discussion on the input/output matching networks. The input and output networks are passive, consisting of striplines, inductors, capacitors, and resistors. These networks achieve, among other things, the following objectives:

1. They provide proper terminations to the analog filters between the antenna and the LNA as well as between the LNA and the subsequent mixer. With proper terminations, these filters' frequency responses are preserved. Sometimes the matching networks themselves are used to perform part of the filtering.
2. The input matching n4etwork ensures optimum noise performance as well as stability at the input. The output matching network ensures stability at the output.
3. The input matching network provides the proper power matching between the antenna and the LNA.

The objectives of primary interest to us are (2) and (3), which we will discuss next.

3.1.2.2 Matching for Noise and Stability

There are two separate issues in objective (2): noise and stability. Let us start by discussing the first one, the noise issue. We refer to Figure 3.1a and imagine that initially we remove the input matching network. We further assume all the noise in the amplifier is represented by the equivalent noise voltage, $\overline{v_i^2}$, and current, $\overline{i_i^2}$. From [2] it is shown that to achieve optimum noise matching, R_s's value should be made equal to the value of a fictitious resistance, called the optimum noise resistance, R_{Sn}. If $\overline{v_i^2}$ and $\overline{i_i^2}$ are uncorrelated, the value of this R_{Sn} is related to the equivalent noise sources of the amplifier by [2]

$$R_{Sn} = \sqrt{\overline{v_i^2}} \Big/ \sqrt{\overline{i_i^2}} \qquad (3.2)$$

If $\overline{v_i^2}$ and $\overline{i_i^2}$ are correlated, then [3] gives a more complete expression for the value of R_{sn}. On the other hand R_s comes from the output resistance of BPF1 (or the antenna resistance if BPF1 is not used) which will, in general, has a different value than the value of R_{Sn} (as given by (3.2) or derived in [3]). To make them equal, or match them, we have to go back to Figure 3.1a and put in the input-matching network. This input matching network should be so designed that R'_{Sn}, looking into the matching network/amplifier combination, has the same value as R_s.

Next we briefly look at the second issue of objective (2), the stability issue. Let us refer to Figure 3.1a and imagine again that initially we remove the input and output matching networks. From the definition in [1] we have

$$S_{11} = \frac{Z_o - R_{in}}{Z_o + R_{in}} \qquad (3.2a)$$

$$S_{22} = \frac{Z_o - R_{out}}{Z_o + R_{out}} \qquad (3.2b)$$

where Z_o is the characteristic impedance of the transmission line that connects R_s and R_{load} to the amplifier.

How does stability depend on these S-parameters, whose absolute value is between 0 and 1? Let us look at S_{11} first and let us assume, for example, that S_{11} is large (close to 1). From the definition of S parameters [1], we have

$$S_{11} = \frac{\text{reflected power}}{\text{incident power}}\bigg|_{input}$$

Hence having S_{11} close to 1 means that the reflected power is almost equal to the incident power. This means that whatever power we deliver to the amplifier is almost totally reflected. This reflected power, again, is going to be reflected by the voltage source V_s back to the amplifier. We can then qualitatively see that there will be large amount of power shuffling back and forth and that the amplifier is on the verge of oscillation. On the other hand, if S_{11} is small (close to 0), the reflection will be small and the amplifier becomes more stable. Therefore, in principle S_{11} should be kept as small as possible as far as stability is concerned. Ideally, S_{11} should be 0. From (3.2a) this means R_{in} should be kept close to Z_o. In cases when R_{in} is very different from Z_o we have to go back to Figure 3.1a and put in the input-matching network, which will then transform R_{in} to the required value, denoted as R'_{in_stable}. Similar observations and conclusions go for the output matching network and R_{out}, S_{22}, R'_{out_stable}. When the input and output matching networks are put back in Figure 3.1a (3.2a), and (3.2b) will be modified and will become

$$S_{11} = \frac{Z_o - R'_{in\_stable}}{Z_o + R'_{in\_stable}} \tag{3.2c}$$

$$S_{22} = \frac{Z_o - R'_{out\_stable}}{Z_o + R'_{out\_stable}} \tag{3.2d}$$

3.1.2.3 Matching for Power

We now elaborate on objective (3). Referring to Figure 3.1a, with proper power matching, V_s can deliver the maximum amount of power to the LNA. This concept is called power matching and is discussed in some detail in this sub-section.

First Figure 3.1a is redrawn in Figure 3.2a without the matching network while using Figure 3.1b to represent the amplifier. Hence $V_{out\ amp}$ is the same as V_o. In anticipation that the amplifiers to be discussed later in the chapter are MOS based, we label the input voltage of the amplifier V_{in_amp} as V_{gs}. First let us look at the case when the amplifier of Figure 3.1b is assumed to be unilateral. In this case we can set $A_r = 0$. Let us refer to Figure 3.2a and define our goal as the delivery of a maximum amount of power to R_{load}, the load resistor. Obviously, to achieve that, V_o and hence V_{gs} have to be at their maximum. This can be done by making R_{in} infinite. Now let us look at the case when the amplifier of Figure 3.1b is not unilateral. Hence the dependent current source given by $A_rV_{out\_amp}$ cannot be neglected. Under this case to deliver maximum amount of power to R_{load} we need (the detail derivation is skipped here):

$$R_{in} = R_s \tag{3.3}$$

An alternative development that leads to the same conclusion is also available if we represent the amplifier using S parameters. With S parameters, people tend to work in power (rather than voltage) as the variable. Rather than talking about power delivered to R_{load} (P_L) as being related to voltage through the expression

$$P_L = \frac{V_o^2}{R_{load}} \tag{3.4}$$

we relate this power through the expression involving G_T, the transducer power gain.

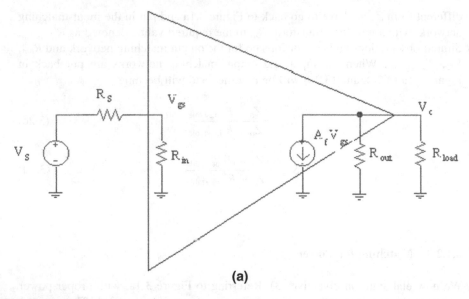

Fig. 3.2a LNA using lumped parameter representation

This is defined in [1] as

$$G_T = \frac{P_L}{P_{AVS}} = \frac{\text{power deliverd to } R_{\text{load}}}{\text{power delivered from source}} \tag{3.5}$$

Since P_{AVS} from source is fixed (e.g., for DECT, the minimum value is min $P_{AVS} = -77$ dBm), from (3.5) maximum P_L corresponds to maximum G_T. From [1] maximum G_T is obtained when both input and output are matched. To achieve this, what people do is to match R_{in} to R_s. Hence again,

$$R_{\text{in}} = R_s \tag{3.6}$$

If R_{in} does not have the same value as R_s, (3.3) or (3.6) can be satisfied by inserting a matching network and transforming R_{in} to $R'_{\text{in\_power}}$, as depicted in Figure 3.1a. This $R'_{\text{in\_power}}$ should be set equal to R_s. Similar conclusions go for R_{out}, $R'_{\text{out\_power}}$. One such example where the input matching network is done using a transformer is shown in Figure 3.2b.

Let us now refer to Figure 3.1a again and summarize what we have done so far in sub-section 3.1.2. We will concentrate on the input matching network. The conclusions essentially carry over to the output matching network. Basically, we start with a given R_{in}, build a matching network, and transform R_{in}. This transformation is used to achieve an optimum noise performance, maintain stability, or achieve matching for power. Since these are three separate goals to achieve and we only have one matching network (hence one unique transformation), a compromise has to be struck. However, as shown in [3], if we choose the objective

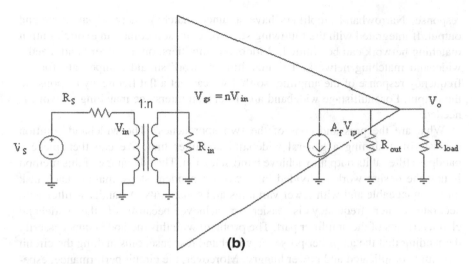

(b)

Fig. 3.2b LNA with transformer matching at the input

of maximum power transfer, we are still close to achieving the objective of achieving optimum noise performance.

3.1.2.4   Implementation

In Figure 3.2b, an ideal wideband matching element was used: a transformer. Although it is possible to utilize transformers at RF, for higher turn ratios we run into difficulty due to the interwinding capacitance limiting the frequency response of the coils. Wideband match is preferably implemented using active elements in feedback to achieve controlled impedance. Other wideband techniques include using feedback around the amplifier [3]. However, given that the signal is inherently narrowband, another form of impedance transformation is available: transformation using resonant circuits. This narrowband matching can be done using either passive *RLC* circuits [1] or feedback around active circuits. Further-more, narrowband resonant matching has another advantage: the required input and output noise-limiting filters can be folded in as part of the matching network.

3.1.3   Comparisons of Narrowband and Wideband LNA

Because of these two types of matching we can think of having two types of LNA. One uses a narrowband amplifier, and the other uses a wideband amplifier. The matching networks in these two types of LNA differ in their frequency response, one having a wideband response and the other having a narrowband response. Narrowband amplifiers have a tuned matching network at input and output. If integrated with the following stage on chip, sometimes an explicit output matching network can be omitted. Wideband amplifiers, on the other hand, need a wideband matching network. The matching network should compensate for the frequency response of the amplifier so that we can get a flat frequency response at the output. For multistage wideband amplifiers, an interstage matching network is needed.

What are the pros and cons of the two approaches? The wideband solution consists of designing a general wideband amplifier first. We can then place a bandpass filter at its output to achieve band selection. The advantage of this method is that the design work is divided into two independent steps, making each task more manageable and with fewer variables and constraints. A bandpass filter with accurate center frequency is easier to achieve because of the wideband characteristics of the amplifier part. The problem with this method is unnecessarily demanding that the amplifier possess a wideband response, thus making the circuit structure complicated and power hungry. Moreover, the circuit performance, especially the noise performance, is poor. On the other hand, for the narrowband approach, after proper narrowband impedance matching and low noise optimizing at the input, not only can we get low noise performance but we can also knock down the DC power by a substantial amount. At the output we use a *LC* tank circuit to peak the gain so that we can omit the additional gain stage and make the circuit simple. The disadvantage of the narrowband method lies in the difficulty of achieving bandpass amplification with accurate center frequency due to circuit element value variation inherent in a very large scale integration (VLSI) process. This can be taken care of by tuning the tank circuit on the chip. Also, there is a need for low-loss inductor, which has just become available in the modern VLSI process. The practical wideband and narrowband design approaches will now be discussed in the following sections.

3.2   Wideband LNA Design

Based on Table 2.1 of Chapter 2, we present a sample specification of LNA in Table 3.1.

Compared with Table 2.1, we have added a specification on S_{11} (for stability) and a power specification. Also, to simplify design we assume that $IIP_{3,LNA}$ does not dominate IIP_{3,rec_front} and hence is not considered. This LNA is supposed to obtain its input from BPF1 and drive BPF2, as shown in Figure 2.1 of Chapter 2.

First, a wideband LNA design is examined. A wideband design is simpler, since the filter and amplifier design can be decoupled. When we design a wideband amplifier, we have to consider the following issues:

- The frequency response, which is required to be flat within the specified tolerance over the entire bandwidth
- Low noise characteristics over the bandwidth

Table 3.1 Specifications
of LNA for DECT

$f_o = 1.9$ GHz

Input matching: $S_{11} < -10$ dB

Noise factor: < 3 dB

Voltage gain: 20 dB

Power at 3.3 V: 40 mW

- Input/output matching for maximum power transfer
- Stability, which must be maintained throughout the entire bandwidth

There are two ways to design a wideband amplifier. One is based entirely on the S-parameter method, which is often used by microwave engineer. Alternatively, we can use lumped parameter representation, as shown in Figure 3.1b, as a starting point for designing the core amplifier. This has the advantage of letting circuit designers perform the design task using more familiar tools. Lumped parameters tend to be simpler to apply and may lend better insight into the operation of the circuits (e.g., feedback theory can be applied rather easily). The terminal behaviour of the core amplifier is then expressed in S-parameters. The design of the matching network can then be carried out using S-parameter-based techniques as described in sub-section 3.1.2.

We now go through a wideband LNA design based in CMOS step by step. The design procedure is a synthesis and optimization process. In this design we focus briefly on the design of the matching network and concentrate on the amplifier design. For more detail on matching network design, refer to [1].

According to the specification, we first decide on a tentative amplifier circuit topology. Figure 3.3a is one such topology. We decide to select a matching network that will achieve the goal of optimum power delivery and maintaining stability. At input this matching is done by using a straightforward 50 Ω R_{match} (R_s is taken to be 50 Ω). This is done purely for simplicity. Note that noise matching is not performed. (As a side note, R_{match} itself introduces extra noise.) The matching at output is achieved by making the output resistance equals 50Ω. A pair of source followers is needed to do this matching. The first buffer stage is smaller than the second, to minimize loading on the transresistance pair; the second is designed such that $\frac{1}{g_{m2}}$ is equal to 50 Ω.

The core amplifier consists of two stages. The first stage is a simple common source (CS) transconductance structure (M_1), which is good for low-noise design. This is driving a tightly coupled transresistance amplifier with an active feedback path. This second stage can also be looked at as a CS amplifier with active shunt-shunt feedback. The feedback is introduced to modify the second-stage input and output impedance. This helps to broadband the core amplifier by reducing the Miller capacitance of M_1. This proposed amplifier topology is similar to what is

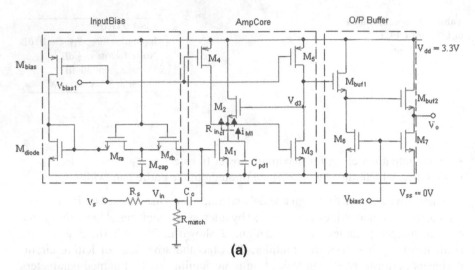

Fig. 3.3a Wideband LNA

done in conventional GaAs MMIC design [4]. As an example, we design the amplifier in a 0.8 um CMOS process. One thing to note is that even though the output resistance of the transresistance stage is reduced due to feedback, it is still difficult to make this output resistance (effectively $1/g_{M_3}$) sufficiently low such that it can achieve a 50 Ω match with BPF2. That is why M_{buf1} and M_{buf2} were used.

3.2.1 DC Bias

First, the DC bias of the entire topology is examined. One interesting bias problem involves establishing the V_{GS} of M_1 and the associated drain current. Given that the RF signal from the output of BPF1, V_s, is capacitively coupled into the amplifier at this point, it is important to prevent V_s from being injected into the current bias chains. This is accomplished by using an R-C-R filter, formed by M_{ra}-M_{cap}-M_{rb}. M_{ra} and M_{rb} are biased in the triode region and have a $W/L = 2.4um/5um$. Their resistances are calculated to be $8k\Omega$. The center capacitor M_{cap} is implemented using a $100\,um/100\,um$ FET gate. At AC, this R-C-R network forms a low-pass filter and filters V_s before it hits the gate of M_{diode}. At DC, it allows the DC bias established at the gate of M_{diode} to be copied to the gate of M_1. This bias is essentially set up by applying V_{bias1} to M_{bias}, which in turns set up $I_{D(Mdiode)}$, and hence set up $V_{G(Mdiode)}$. V_{bias1} is also applied to M_4, M_5. Due to different W/L ratios, I_{D4}, I_{D5} are different from $I_{D(Mdiode)}$ (and hence I_{D1}). In particular, I_{D4}, I_{D1} are set up such that their difference becomes I_{D2}, the bias current for M_2. V_{bias2} is used to set up the bias current in M_{buf1} and M_{buf2}.

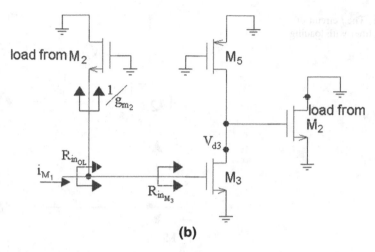

(b)

Fig. 3.3b The *a* circuit of core amplifier with loading

3.2.2   Gain and Frequency Response

Let us concentrate on the amplifier core. The coupling capacitor C_C is considered a short circuit at RF and so V_{in} is applied to the gate of M_1. V_{d3} is the output voltage.

First let us review the derivation of the low-frequency small-signal voltage gain, neglecting frequency-dependent effects from device capacitance.

$$A_v = (\text{gain of 1st stage}) \cdot (\text{gain of 2nd stage}) \qquad (3.7a)$$

From Figure 3.3a,

$$\text{gain of 1st stage} = \frac{i_{M_1}}{v_{in}} = -g_{m1} \qquad (3.7b)$$

$$\text{gain of 2nd stage} = \frac{v_{d3}}{i_{M_1}} \qquad (3.7c)$$

Now, the gain of second stage = gain of a transresistance amplifier with shunt-shunt feedback. From feedback theory,

$$\text{gain of 2nd stage} = \frac{a}{1 + af} \qquad (3.8)$$

where a = forward gain with loading, and f = feedback factor.

To calculate a let us refer to Figure 3.3b, which shows the forward amplifier with loading from the feedback network (shown as the two M_2 transistors with the drain grounded at the input and output nodes). From the figure, i_{M_1} flows into a resistance of $\frac{1}{g_{M_2}} \| R_{in_{M_3}} = \frac{1}{g_{M_2}}$ (since $R_{in_{M_3}}$, the input resistance of M_3, is infinite).

Fig. 3.3c The f circuit of
core amplifier with loading

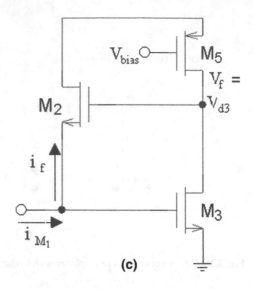

(c)

This develops a voltage of $\frac{i_{M_1}}{g_{M_2}}$. This voltage is multipled by the voltage gain of M_3
to develop v_{d3}. For this single-stage CS amplifier the voltage gain is simply $g_{M_3}r_{out}$,
where $r_{out} = r_{o3}||r_{o5}||r_{in_2} = r_{o3}||r_{o5}$ as $r_{in_2} = \infty$. Here r_{o3}, r_{o5} are the output
resistances of M_3, M_5 and r_{in_2} is the input resistance of M_2. Now we
denote $r_{o3}||r_{o5} = R_{op}$ and so

$$v_{d3} = -\left(\frac{i_{M_1}}{g_{M_2}}g_{M_3}R_{op}\right) \tag{3.9}$$

Therefore, from (3.9),

$$a = \frac{v_{d3}}{i_{M_1}} = -\left(\frac{1}{g_{M_2}}g_{M_3}R_{op}\right) \tag{3.9a}$$

This is very large, as R_{op} is large. Next let us look at Figure 3.3c and see that f is
defined as

$$\frac{i_f}{v_f} = -g_{M_2}. \tag{3.9b}$$

Substitute (3.9a), (3.9b) in (3.8) we note that since af is large,

$$\text{second stage gain} \approx \frac{1}{f} \tag{3.9c}$$

Substituting (3.9b) into (3.9c), we have

$$\text{second stage gain} \cong \frac{-1}{g_{m_2}} \tag{3.9d}$$

Substituting (3.7b), and (3.9d) into (3.7a), we have

$$A_v = g_{m_1} \frac{1}{g_{m_2}} \tag{3.10}$$

Notice that the body effect loss has been neglected.

Next let us derive the high-frequency gain. Upon detailed derivation, the high-frequency gain of the amplifier, including frequency effects due to device capacitance, can be expressed approximately as [5]

$$A_v(j\omega) = g_{M_1} \times \left(\frac{g_{M_3} - j\omega C_2}{(g_{M_3} - j\omega C_2)(g_{M_2} + j\omega C_2) + (g_{M_2} + j\omega C_3)\left(\frac{1}{R_{op}} - j\omega C_4\right)} \right) \tag{3.10a}$$

where

$$C_2 = C_{gs2} + C_{gd3} \tag{3.10b}$$

$$C_3 = C_{PD1} + C_{PD4} + C_{gs3} + C_{gd1} + C_{gs2} + C_{gd3} \tag{3.10c}$$

$$C_4 = C_{PD3} + C_{gd2} + 0.5C_{gs,buf1} + C_{PD5} + C_{gd5} + C_{gd3} \tag{3.10d}$$

$$R_{op} = r_{o3} \| r_{o5} \tag{3.10e}$$

It should be noted that capacitances C_2 through C_4 are the lumped parasitic capacitance at each node in the circuit. C_{PD} and C_{PS} are the parasitic drain/source capacitances. At low frequency (3.10a) reduces to (if $R_{op} \gg 1/g_{m3}$, which is almost always true)

$$A_v = \frac{g_{m1}}{g_{m2}} \tag{3.11}$$

which agrees with (3.10).

The –3 dB frequency can be extracted from (3.10a):

$$\omega_{-3dB} = \frac{g_{M_3} g_{M_2} R_{op}}{(g_{M_3} - g_{M_2}) R_{op} C_2 + g_{M_2} R_{op} C_4 + C_{PD1}} \tag{3.12}$$

which is equal to

$$\frac{1}{\frac{C_4}{g_{M_3}} + \frac{C_{PD1}}{g_{M_3}g_{M_2}R_{op}} + \left(\frac{1}{g_{M_2}} - \frac{1}{g_{M_3}}\right)C_2} \approx \frac{g_{M_3}g_{M_2}R_{op}}{C_{PD1}} \qquad (3.13)$$

Intuitively, this simplified expression agrees with the results from zero-valued time-constant analysis. To see why this is the case, let us refer to Figure 3.3a again and note that the frequency rolloff is controlled primarily by the parasitic capacitances from the gate of M_3 and the drain of M_1 (both included in the C_{PD1} term). Hence the dominant pole is calculated by considering the drain node of M_1. This pole can be calculated by the zero-valued time-constant method as

$$p_1 = \frac{1}{2\pi RC} \qquad (3.14)$$

where

$$C = C_{PD1} \qquad (3.15)$$

and resistance R equals the input resistance of the shunt-shunt feedback pair ($R_{in_{CL}}$) with a value given by

$$R_{in_{CL}} = \frac{1}{g_{M_2}g_{M_3}R_{op}} \qquad (3.16)$$

Substituting (3.15), and (3.16) into (3.14) gives a p_1 that agrees with ω_{-3dB} as given in (3.13). Since $R_{in_{CL}}$ is so important, let us go through the exercise and see how (3.16) is derived. Let us refer again to Figure 3.3a. To calculate $R_{in_{CL}}$, the closed-loop impedance looking into the M_2, M_3 shunt-shunt feedback loop, we first calculate $R_{in_{OL}}$, the open-loop impedance looking into the M_2, M_3 pair, including loading from the feedback path. This can be obtained by referring again to Figure 3.3b. As can be seen, $R_{in_{OL}}$ is given by

$$R_{in_{OL}} = \frac{1}{g_{M_2}} \| R_{in_{M_3}} \qquad (3.17)$$

Since $R_{in_{M_3}} = \infty$, then

$$R_{in_{OL}} = \frac{1}{g_{M_2}} \qquad (3.18)$$

From (3.9a), and (3.9b), we have

$$af = g_{M_3}R_{op}. \qquad (3.19)$$

From feedback theory $R_{\text{in}_{CL}}$ is obtained from $R_{\text{in}_{OL}}$ by

$$R_{\text{in}_{CL}} = \frac{R_{\text{in}_{OL}}}{af + 1} \approx \frac{R_{\text{in}_{OL}}}{af} \tag{3.20}$$

since $af \gg 1$. Subsituting (3.19), and (3.18) into (3.20), we have

$$R_{\text{in}_{CL}} = \frac{1}{g_{M_2} g_{M_3} R_{\text{op}}} \tag{3.21}$$

Hence (3.16) is confirmed. This reduction of impedance is performed at the point where the largest capacitance is hanging, namely, at the drain node of M_1. This drain node has a large parasitic capacitance C_{PD1} because $\left(\frac{W}{L}\right)_1$ of M_1 is large (this is, in turn, done so as to reduce noise, as will be shown in sub-section 3.2.3). This exercise highlights why shunt-shunt feedback is performed: mainly to reduce $R_{\text{in}_{CL}}$, which increases p_1 according to (3.14), and hence broaden the frequency response of the whole circuit. Finally, returning to the full expression in (3.13), the $\left(\frac{1}{g_{m_2}} - \frac{1}{g_{m3}}\right)C_2$ correction term accounts for the pole splitting introduced by C_2 shunted across the gate drain of M_3.

3.2.3 Noise Figure

In this sub-section we will derive the NF of the amplifier core. From chapter 2 noise figure is defined as

$$NF = \frac{N_{dev} + N_{\text{in}}}{N_{\text{in}}} = 1 + \frac{N_{dev}}{N_{\text{in}}} \tag{3.22}$$

where

N_{dev} = noise coming from the amplifier core
N_{in} = noise from the source resistance R_s

Noise from the amplifier core is further broken down into noise coming from the first and second stages. For the first stage the noise figure becomes that of the noise figure of a MOS transistor, which is assumed to be dominated by thermal noise of the drain current and is given by [2]

$$NF = 1 + \frac{4kT \frac{2}{3} \frac{1}{g_{M_1}}}{4kTR_s} = 1 + \frac{2}{3g_{M_1}R_s} \tag{3.23}$$

Again, for a more complete picture on NF of an MOS transistor, refer to [3].
Including the noise contribution from the second stage and the most prominent frequency-dependent effect, we have (detail derivation is covered in problem 3.2)

$$NF = 1 + \frac{2}{3g_{m1}R_s} + \frac{2g_{m2}^3 r_{o1}^2}{3g_{m1}^2 R_s} + \frac{2g_{m2}^2}{3g_{m1}^2 g_{m3} R_s}$$
$$+ \left(\frac{2}{3g_{m1}R_s} + \frac{2g_{m2}^3 r_{o1}^2}{3g_{m1}^2 R_s} + \frac{2g_{m2}^2}{3g_{m1}^2 g_{m3} R_s} \right) \omega^2 C_{gs1}^2 R_s^2 \qquad (3.24)$$

Note that noise from M_1, when input referred, is also being shaped in the frequency domain by C_{gs1} and R_s. This is apparent in the $\frac{2}{3g_M} \omega^2 C_{gs1}^2 R_s$ term in (3.24). Obviously, we could also include the parasitic capacitance of M_2 and M_3, but the equation for the noise figure would become rather messy and would lose its usefulness.

At this point we have derived all the relevant design equations. Next let us go through a design example.

Design Example 3.1.
We now illustrate how to design a wideband LNA using the topology as shown in Figure 3.3a, concentrating on the amplifier core. The specifications are given in Table 3.1 and the underlying technology is selected to be 0.8um CMOS. The equations in subsections 3.2.2, and 3.2.3 can now be used to help size the transistors and current. First we give some general philosophy on how to do the sizing.

M_2 and M_3 are necessarily small devices to minimize the impact of their parasitic capacitance. Moreover, it is desirable that M_2 be small, since the amplifier gain increases with decreasing g_{m2} and hence decreasing $\left(\frac{W}{L}\right)_2$ [see (3.10)]. However, M_1 is necessarily a very large device, to maximize gain [see (3.10)] as well as to minimize noise [see (3.23)]. Thus, the parasitic parameters and area capacitance at its drain determine the overall frequency response of the amplifier. Next we do the design steps by steps.

Step 1: Designing M_1 by determining I_{D1} and $\left(\frac{W}{L}\right)_1$ from NF and power specification
From Table 3.1 we set NF = 3 dB. Putting this in (3.23) and setting $R_s = 50\Omega$, we have

$$2 = 1 + \frac{2}{3g_{m1}50\Omega} \qquad (3.25)$$

$$\therefore \frac{1}{g_{M_1}} = 75\Omega \; or \; g_{M_1} = 0.013\Omega^{-1} \qquad (3.26)$$

Now g_{M_1} is related to bias current I_{D1} and $\left(\frac{W}{L}\right)_1$ as follows:

$$g_{M_1} = \sqrt{2I_{D1}k'\left(\frac{W}{L}\right)_1} \; or \; 0.013\,\Omega^{-1} = \sqrt{2I_{D1}k'\left(\frac{W}{L}\right)_1} \qquad (3.27)$$
$$where \; k' = \mu C_{ox}$$

We have one equation and two unknowns: I_{D1} and $\left(\frac{W}{L}\right)_1$. To determine I_{D1} and $\left(\frac{W}{L}\right)_1$, we need one more equation, which we can get from the power specs. To proceed we first observe that since noise is dominated by the first stage, minimizing noise means $g_{M_1} \gg g_{M_2}, g_{M_3}$. This in turn implies $I_{D1} \gg I_{D2}, I_{D3}$. Thus, the power of the amplifier core is dominated by current I_{D1}. Next from Figure 3.3a there are four branches (through $M_{diode}, M_1, M_{buf1}, M_{buf2}$). Let us assume that they carry the same current. Hence total current $= 4 \times I_{D1}$. Thus the total power P_{Total}, is given as $4 \times I_{D1} \times V_{dd}$. It is clear, then, with a V_{dd} of 3.3 V and a power specification of 40mW we have:

$$I_{D1} \le \frac{P_{total}}{4 \times V_{dd}} = \frac{40mW}{4 \times 3.3V} = 3mA \tag{3.27a}$$

We put in some safety margin and set

$$I_{D1} = 1.2mA \tag{3.27b}$$

Substituting this back in equation (3.27), we get

$$0.013\Omega^{-1} = \sqrt{2 \times 1.2mA \times k'\left(\frac{W}{L}\right)_1} \tag{3.28}$$

Assuming $k' = 100$ uA/V^2 in our 0.8 um CMOS process, this gives

$$\left(\frac{W}{L}\right)_1 = 737 = \frac{590um}{0.8um} \tag{3.29}$$

Step 2: Designing M_2 by determining I_{D2} and $\left(\frac{W}{L}\right)_2$ from gain specification
To determine g_{M_2}, let us go to (3.10). We know from Table 3.1 that $A_v = 20$ dB or 10, and from step 1 $g_{M_1} = 0.013\ \Omega^{-1}$. Substituting these in (3.10), the resulting equation is as follows:

$$10 = \frac{g_{M_1}}{g_{M_2}} = \frac{0.013\Omega^{-1}}{g_{M_2}}$$

Therefore, we have

$$g_{M_2} = \frac{1}{750\Omega} \tag{3.30}$$

To achieve this g_{M_2}, which is 10 times less than g_{M_1}, we have two variables: I_{D2} and $\left(\frac{W}{L}\right)_2$. Some simple choices include (1) setting I_{D2} to be $\frac{I_{D1}}{100}$ and $\left(\frac{W}{L}\right)_2 = \left(\frac{W}{L}\right)_1$; or (2) setting $\left(\frac{W}{L}\right)_2 = \frac{1}{100}\left(\frac{W}{L}\right)_1$ and $I_{D2} = I_{D1}$; or (3) setting $I_{D2} = \frac{I_{D1}}{10}$ and

$\left(\frac{W}{L}\right)_2 = \frac{1}{10}\left(\frac{W}{L}\right)_1$. We choose the third alternative. Therefore, $I_{D2} = \frac{1}{10}I_{D1} = 0.12mA$ and $\left(\frac{W}{L}\right)_2 = \frac{59um}{0.8um}$. Among the simple choices the advantage of the third alternative is that it ends up having both a small I_{D2} and $\left(\frac{W}{L}\right)_2$. We want a small I_{D2} so that our assumption in step 1, that $I_{D2} \ll I_{D1}$, still holds. Hence the power specification is still met. Also, we want $\left(\frac{W}{L}\right)_2$ small because that means a small C_{gs2}. Hence C_2 in (3.13) remains small. Therefore, the correction term in (3.13), $\left(\frac{1}{g_{m2}} - \frac{1}{g_{m3}}\right)C_2$ (due to pole splitting), is also small and ω_{-3dB} is large, which is a desirable feature. With the third simple choice one can further optimise by varying I_{D2} and $\left(\frac{W}{L}\right)_2$ as long as the combination results in a g_{M_2} that satisfies (3.29).

Step 3: Designing M_3 by determining I_{D3} and $\left(\frac{W}{L}\right)_3$ from f_0 specification
From Table 3.1, $f_0 = 1.9$ GHz. To pass the signal, $f_{-3\ dB}$ should be larger than 1.9 GHz. Here, to simplify calculations we set $f_{-3\ dB} = 1.9$ GHz.

   $f_{-3\ dB}$ is obtained from (3.13) and is given as

$$f_{-3dB} = \frac{g_{M_3} g_{M_2} R_{op}}{2\pi C_{PD1}} \tag{3.31}$$

We know g_{M_2} from step 2 [see (3.30)]. Next let us find C_{PD1}, the parasitic capacitance at the drain of M_1. Assuming C_{PD1} is dominated by the parasitic capacitance of the large transistor M_1, we have

$$C_{PD1} = C_{gd1} + C_{db1} \tag{3.32}$$

However, C_{gd1} is the drain overlap capacitance of M_1 and is given by $C_{gd1} = W \cdot L_D \cdot C_{OX}$. Assume L_D = lateral diffusion = 0.12 um and $C_{ox} = 1$ fF/um^2 in our 0.8um CMOS process. Then

$$\begin{aligned} C_{gd1} &= W \cdot L_D \cdot C_{OX} \\ &= 590\,um \times 0.12\,um \times 1\,fF/um^2 \\ &= 70.8\,fF \end{aligned} \tag{3.33}$$

C_{db1} is the drain-to-substrate capacitance and is given by

$$C_{db1} = \frac{C_{db0}}{\sqrt{1 + V_{db1}/\psi_0}} \tag{3.34}$$

$$C_{db0} = C_{jsw0} \times drain - perimeter + C_{j0} \times drain - area \tag{3.35}$$

Here, C_{jsw0} and C_{j0} are the (zero bias sidewall capacitance)/um and the (zero bias junction capacitance)/um^2 and are given in our 0.8um CMOS process to be 1 fF/um and 0.18 fF/um^2 respectively. Also, we can assume that the design rule specifies the height of the drain area to be 9 um. Subsituting this information into (3.35), we have

$$C_{db0} = 1fF/um(590\,um + 9\,um + 9\,um) + 0.18\,fF/um^2(590um \times 9um)$$
$$= 608\,fF + 956\,fF$$
$$= 1564\,fF \tag{3.36}$$

Furthermore, we note that $V_{db} = V_{gs3}$ = overdrive voltage of M_3.

$$\therefore V_{db} = \sqrt{\frac{2I_{D3}}{k'(W/L)_3}} + V_T \tag{3.36a}$$

Since we want to save power, we want to make I_{D3} as small as possible. Consequently, we make the assumption that I_{D3} is small enough that the first term in (3.36a) can be neglected. Hence,

$$V_{db} \approx V_T = 1V \tag{3.36b}$$

($V_T = 1V$ is given in our 0.8um CMOS process). Finally, ψ_0 is also given in our 0.8um CMOS process to be

$$\psi_0 = 0.65\ V. \tag{3.37}$$

Hence, substituting (3.36), (3.36b), and (3.37) into (3.34), we get

$$C_{db1} = 981fF \tag{3.38}$$

Substituting (3.33) and (3.38) into (3.32), we have

$$C_{PD1} = 981\ fF + 70.8\ fF = 1052\ fF \tag{3.39}$$

The preceding calculations highlight that for large devices, the source and the drain-to-substrate capacitances become important. We can now substitute (3.39) and (3.30) into (3.31) and solve for $g_{m_3}R_{op}$ that will achieve a $f_{-3dB} = 1.9GHz$

$$f_{-3dB} = 1.9GHz = \frac{g_{m2} \times g_{m_3}R_{op}}{2\pi C_{PD1}}$$
$$= \frac{\frac{1}{750} \times g_{M_3}R_{op}}{2\pi(1052fF)} \quad \Rightarrow \quad g_{M_3}R_{op} = 9.5 \tag{3.39a}$$

Assume that $r_{o3} \ll r_{o5}$, then from (3.10e) $R_{op} \approx r_{o3}$. Substituting this into (3.39a), we have

$$g_{m3}r_{o3} = 9.5 \tag{3.39b}$$

Next we calculate r_{o3}. To calculate r_{o3} we need I_{D3}. As stated previously, since we want to have a limited power consumption, we want a small I_{D3} [actually to solve for (3.36a) we have set $I_{D3} \approx 0$]. One simple choice of I_{D3} to achieve this is to make I_{D3} approximately equal to I_{D2} ($= 0.12$ mA, obtained from step 2). With this I_{D3} we can calculate r_{o3}:

$$r_{o3} = \frac{1}{\lambda I_{D3}} = \frac{1}{(0.2\text{V}^{-1})(0.12\text{mA})} = 41.5k\Omega \tag{3.40}$$

(the channel length modulation factor λ, is given as 0.2 V^{-1} in our 0.8um CMOS process).

Finally, we can substitute (3.40) in (3.39b), which gives us

$$g_{m3} = \frac{9.5}{r_{o3}} = \frac{9.5}{41.51 k\Omega} = \frac{1}{4.3k\Omega} \tag{3.40a}$$

Consequently, with $I_{D3} = 0.12$ mA and $g_{m3} = \frac{1}{4.3k\Omega}$, $\left(\frac{W}{L}\right)_3$ can now be determined:

$$g_{m3} = \frac{1}{4.3k\Omega} = \sqrt{2 \times 0.12\text{mA} \times k' \times \left(\frac{W}{L}\right)_3} \rightarrow \left(\frac{W}{L}\right)_3 = 2.2 = \frac{1.76um}{0.8um} \cong \frac{2um}{0.8um} \tag{3.40b}$$

Step 4: Checking that S_{11} is satisfied
Finally, by putting a $R_{\text{match}} = 50$ Ω at the input, we have transformed $R_{\text{in}}(= \infty)$ to $R'_{\text{in\_stable}}(= 50$ Ω). Hence, assuming that $Z_o = 50$ Ω, from (3.2c) $S_{11} = 0$, and the specs on S_{11} are satisfied.

At this point we have finished our design on this wideband LNA.

3.3 NarrowBand LNA: Impedance Matching

In this and the following section we look at the principle of narrowband LNA design. The design is again based in CMOS. Since the signal occupies a narrow bandwidth, we only need to provide impedance matching as well as amplification in this narrow bandwidth. This can be done rather effectively with the principle of resonance. Here the reactive part of the impedance is controlled to be nulled out at the resonance frequency, leaving only the resistive part to be matched to the resistive source resistance R_s. Since the amplification factor is a product of the device transconductance and load impedance, when the load impedance is controlled to peak up via the resonance process, the amplification factor peaks up as a result.

There are various ways of achieving resonance. One simple way is the use of a LC tank circuit. This approach has the added advantage in a practical LNA as there already exists in the circuit some parasitic capacitance. The idea, of course, is to add some extra inductance and combine it with this parasitic capacitance, hence achieving resonance. This is especially attractive at RF, since inductance in integrated form and with high enough Q becomes available. This approach basically makes good use of this undesirable parasitic capacitance inherent in the circuit by tuning it out. This is just another way of saying that one makes the resulting capacitance/inductance network possess a zero impedance at the resonance frequency (a consequence of resonance) and as a result the parasitic capacitance is rendered harmless. In this section we explain the behaviour of the impedance of this network under resonance.

As in Section 3.2, we decide to design the matching network for power and stability consideration. We further assume that this will give us close to optimum noise performance. How do we design the matching network for the present amplifier? First let us revisit the idea of power matching. In sub-section 3.1.2.3 we explained the concept of matching and how it can be achieved by making $R_{in} = R_s$. The concept was then applied to the design of a wideband LNA for DECT application whereby to achieve an impedance match across the entire 1.9 GHz band, a real physical 50 Ω resistor R_{match} is placed at the input. In the present case, we only need to achieve an impedance match in a narrowband around the input RF (or around the carrier frequency ω_c). Hence the idea is to make the input impedance look like a 50 Ω resistor around this ω_c. On the other hand, since the input impedance is not simply resistive, this complicates the matter because we have to match both the real and imaginary part (the reactance) as well. To satisfy power matching, the condition as specified in (3.6) is modified to $Z_{in} = Z_s$.

3.3.1 Matching the Imaginary Part

Let us find out what is involved in matching the reactive part of the impedance. To do matching the LNA's input impedance, Z_{in} is selected to have a conjugate match with the source resistance, whose imaginary part is zero. First let us try to do matching by placing an inductor L in front of a MOS transistor [Figure 3.4a] whose equivalent circuit is shown in [Figure 3.4b]. The C_{gs} of M_1 is relabelled C. It can be shown that the series LC circuit produces a narrowband output characteristic and eliminates the need for a separate filter. However, the LC circuit always generates an imaginary part in Z_{in} and no matching is possible for a given ω_c. An improved circuit, where the inductor is moved to degenerate the source, is shown in Figure 3.5a. Figure 3.5b is its small-signal equivalent circuit. Phasor representation is used for the voltage and current. As shown in Figure 3.5b, the input impedance Z_{in} is solved by writing KVL in its phasor form across the input loop:

$$V_{in} = I_{in}(j\omega L) + I_{in}\left(\frac{1}{j\omega C}\right) + I_a j\omega L \qquad (3.41)$$

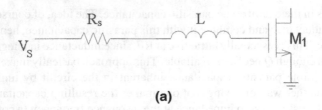

Fig. 3.4a LNA using inductor at the gate for matching

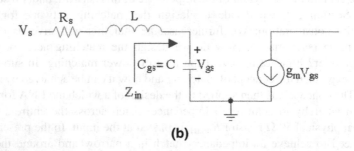

Fig. 3.4b Small-signal equivalent circuit of LNA using inductor at the gate for matching

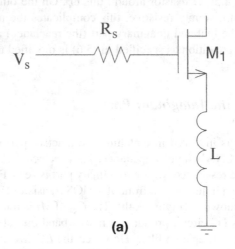

Fig. 3.5a LNA using source degrenerated inductor for matching

At the output loop

$$I_o = g_m V_{gs} = g_m I_{in} \frac{1}{j\omega C}$$ (3.42)

Substituting (3.42) into (3.41),

$$V_{in} = I_{in} \left[j\omega L + \frac{1}{j\omega C} + \frac{g_m L}{C} \right]$$ (3.43)

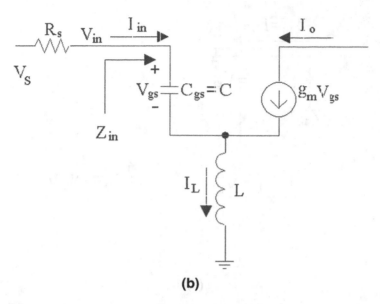

(b)

Fig. 3.5b Small-signal equivalent circuit of LNA using source degenerated inductor for matching

Therefore

$$Z_{in} = \frac{V_{in}}{I_{in}} = j\omega L + \frac{1}{j\omega C} + \frac{g_m L}{C} \tag{3.44}$$

What is surprising from the preceding equation is that we can actually get a real term, $g_m \frac{L}{C}$, out from only reactive components: L and C. The mystery of this actually lies in the voltage-controlled current source (the g_m source). To understand this better, let us first redraw Figure 3.5b in Figure 3.6a. The phasor diagram of the currents and the voltages in Figure 3.6a is then drawn in Figure 3.6b. Figure 3.5b has been redrawn in Figure 3.6a by viewing I_{in} as consisting of two components: I_L and $-I_o$. Let us start off by assuming that this input phasor I_{in} has an angle of 90°, as shown in Figure 3.6b. Referring back to Figure 3.6a this current flows through capacitor C and generates a voltage V_{gs} that is 90° lagging [as shown in Figure 3.6b]. This voltage V_{gs} generates an in-phase current I_o via the voltage control current source. Therefore I_{in} generates an I_o that is 90° out of phase. If we revisit Figure 3.6a and apply *KCL* at the source node, it can be seen that I_{in} consists of a 90° out of phase component $-I_o$, and a second component I_L. This second component of the current in I_{in} is responsible for generating a voltage across C, which we denoted as V_c. This same current flows from C into L and generates a voltage across L, which we denoted as V_L. The idea is to make these two voltages equal and cancel one another. When this happens we call the condition the resonance condition. Then the voltage across C and L comes only from the voltage developed across C due to $-I_o$. This I_o current generates a voltage that is 90° out of

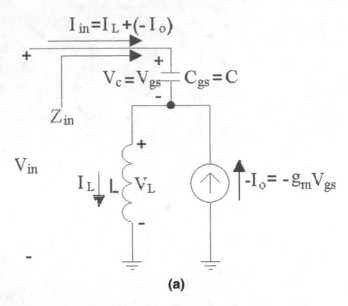

(a)

Fig. 3.6a Alternative viewpoint of small-signal equivalent circuit

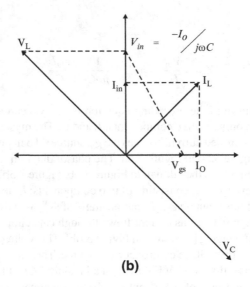

(b)

Fig. 3.6b Phasor diagram showing I_{in} and V_{in} in phase under resonance condition. This means input impedance becomes resistive

phase again (by flowing through a capacitor, which shifts phase by 90°) and now becomes in phase with the incoming I_{in}. Thus the impedance looks resistive and the imaginary part is zero.

To highlight this condition mathematically, we start off by applying *KVL* to the input loop of Figure 3.6a and write

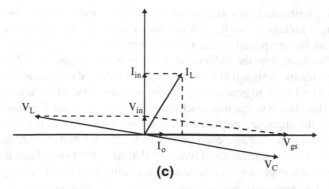

(c)

Fig. 3.6c Phasor diagram showing a situation when $V_{in} < V_{gs}$. Under this situation the resonant circuit amplifies V_{in} by a factor Q to obtain V_{gs}. Q, the quality factor, becomes the voltage gain of the matching network

$$V_{in} = \frac{I_{in}}{j\omega C} + I_L j\omega L = \frac{[I_L + (-I_o)]}{j\omega C} + I_L j\omega L = \left[\frac{I_L}{j\omega C} + I_L j\omega L\right] + \frac{-I_o}{jwC} \quad (3.45)$$

Under resonance condition the two terms in the bracket cancel out. Therefore,

$$V_{in} = \frac{-I_o}{j\omega C} \quad (3.46)$$

But from (3.42) we have $I_o = g_m V_{gs} = g_m \frac{I_{in}}{j\omega C}$; therefore, I_o and I_{in} are 90° out of phase.

Substituting this in (3.46), $V_{in} = \frac{g_m}{\omega^2 C^2} I_{in}$ and the impedance indeed looks resistive.

We can also look at the resistive interpretation from an alternative way. As stated before, Figure 3.5b was redrawn in Figure 3.6a to highlight that I_{in} contains 2 components: I_L and $-I_o$. Let us interpret these two components in an alternate manner. From Figure 3.6a,

$$I_o = g_m V_{gs} = g_m \frac{I_{in}}{j\omega C} \quad (3.47)$$

That is I_{in} generates an I_o that is 90° out of phase (I_o is 90° lagging), as shown in Figure 3.6b. If we can further make I_o generate a V_{in} that is 90° leading, then the phasor will be rotated back and V_{in}, I_{in} will be in phase again. Consequently, Z_{in}, which by definition $= \frac{V_{in}}{I_{in}}$, is in phase or is resistive.

To see how we can make I_o generate a V_{in} that is 90° leading, let us return to Figure 3.6a and observe that I_o flows into the capacitor C and generates a voltage $-\frac{I_o}{j\omega C}$. As shown in Figure 3.6b, this rotates the phasor I_o counterclockwise by 90° and hence will be in phase with I_{in}. If this is the only voltage component of V_{in}, the resistive input condition is already met (i.e., Z_{in} already looks resistive). However, from Figure 3.6a it is seen that V_{in} contains other voltage components. First, the current I_L also flows into C, generating a voltage $V_c = \frac{I_L}{j\omega C}$, whose phasor representation is shown in Figure 3.6b. This can be interpreted as the reactive component of V_{gs}. Second, the same current I_L also flows into the inductor L, generating a voltage $V_L = I_L j\omega L$, whose phasor representation is shown in Figure 3.6b. Fortunately, the capacitor rotates the I_L phasor clockwise by 90° to generate V_L and the inductor rotates this same I_L phasor in the opposite direction (i.e., counter clockwise by 90°) to generate V_c. Hence they will be pointing in the opposite direction and cancelling each other. If the cancellation is exact, then these other voltage components in V_{in} do not matter and, as discussed previously, Z_{in} does look resistive. It turns out that since the two voltage components have the same current phasor, I_L, this cancellation can be achieved by simply making $\frac{1}{j\omega C} = -j\omega L$. Of course, this occurs when $\omega_c = \frac{1}{\sqrt{LC}}$, or at resonance.

In summary, as shown in Figure 3.6a, the input current phasor I_{in} is converted to a voltage phasor V_{gs} that is rotated clockwise by 90° by a capacitor (undesired phase conversion). This capacitor voltage turns out to be a control voltage and hence is converted to a current I_o (essential operation) by the voltage-controlled current source. The current phasor is converted back to a voltage phasor that is rotated counterclockwise by 90° (correction) by making the current flow through a capacitor. This is made the sole component of V_{in} by cancelling the other components in V_{in}, which is achieved by forcing the remaining current component in the current phasor flows through an inductor. Upon cancellation, the original input current I_{in} phasor goes through two equal and opposite phase conversions, 90° clockwise followed by 90° counterclockwise, and generates a V_{in} that is in phase.

Finally, it should be noted that the input resistance is only an ac resistor (i.e., it does not generate noise). Therefore, we can terminate R_s with this input resistance without introducing thermal noise [contrasted this with the case depicted in Figure 3.3a]. In return, we suffer from the potential disadvantage that the matching occurs only at one frequency. Fortunately, the matching is nearly perfect even when we look at a narrowband around this frequency. Since the signal is narrowband enough, we can assume that matching is essentially achieved.

3.3.2   Matching the Real Part

The preceding sub-section describes only half of the matching condition: it simply means that the input impedance's reactive component is zero, which matches with the R_s's reactive component (which of course is zero). We still have to match the real part. Under resonance condition what is the real part? From (3.44) this is $g_m \frac{L}{C}$. Accordingly, for matching we set R_s equal to this. In real life the inductance L can be implemented using a spiral inductor (to be covered later). However, the inductance available from the spiral inductor structure is on the order 2 to 5 nH and may not be sufficient to satisfy the aforementioned matching criteria. Therefore, it is customary to combine the circuit in Figure 3.4a and 3.2a, which becomes the circuit as shown in Figure 3.7a. Here L_2 is the same as L before and is done using the spiral inductor.

Fig. 3.7a LNA using
inductor at the gate and
source degenerated inductor
to achieve matching

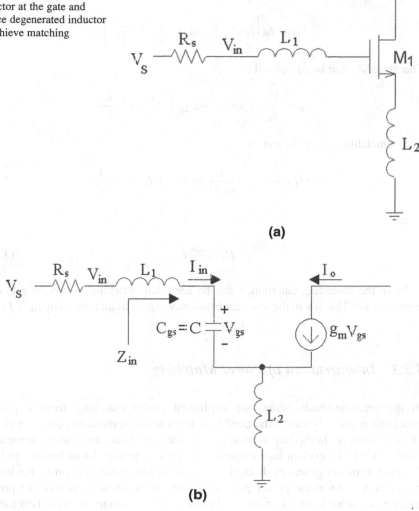

(a)

(b)

Fig. 3.7b Small-signal equivalent circuit of LNA using L_1, L_2 for matching

L_1 is done using the bond wire inductor. The reason for the choice of this inductor
will be covered later. Figure 3.7b is the small-signal equivalent circuit of Figure
3.7a. Applying *KVL* to the input loop in Figure 3.7b, we have

$$V_{in} = I_{in}(j\omega L_1 + j\omega L_2) + I_{in}\left(\frac{1}{j\omega C}\right) + I_o j\omega L_2 \qquad (3.48)$$

Independently,

$$I_o = g_m V_{gs} = g_m I_{in}\frac{1}{j\omega C} \qquad (3.49)$$

Substituting (3.49) in (3.48),

$$V_{\text{in}} = I_{\text{in}} \left[j\omega(L_1 + L_2) + \frac{1}{j\omega C} + \frac{g_m L_2}{C} \right] \qquad (3.50)$$

Therefore, Z_{in} can be calculated as

$$Z_{\text{in}} = \frac{V_{\text{in}}}{I_{\text{in}}} = j\omega(L_1 + L_2) + \frac{1}{j\omega C} + \frac{g_m L_2}{C} \qquad (3.51)$$

For matching, $Z_{\text{in}} = R_s$ and so

$$\omega_c(L_1 + L_2) = \frac{1}{\omega_c C} \ \text{ or } \ (L_1 + L_2)C = \frac{1}{\omega_c^2} \qquad (3.52)$$

and

$$R_s = \frac{g_m}{C} L_2 \qquad (3.53)$$

From the preceding equation it can be seen that matching occurs only at one frequency ω_c. This is also the resonant frequency of the circuit formed by $L_1 + L_2, C$.

3.3.3 Interpretation of Power Matching

In the preceding sub-section we explained power matching from a phasor cancellation point of view. The condition leads to the resonating condition of the *LC* tank network. In this sub-section we explain power matching from a resonance point of view. To explain how resonance helps save power, let us borrow an idea from the ac power generator discipline: the use of resonance to improve the load's power factor. As in the power generator case, we want to maximize the power transfer to our load, the capacitor *C*. However, *C* is a storage element. This means that if we apply a sinusoidal signal to this element, it will take in energy from the source in the first half-cycle and release energy back to the source in the second half-cycle. This means energy (or current) is shuffled back and forth from the source and is not contributing to anything useful, except burning power in the source resistance R_s. To remedy this situation, one approach is to provide a local energy storage element that acts in a complementary (push-pull) fashion to that of the capacitor. One convenient choice is an inductor that releases energy in the first half-cycle (when the capacitor absorbs energy) and absorbs energy in the next half-cycle (when the capacitor releases energy). If the capacitor and inductor act in unison, that is, if the release of energy from one element is synchronized with the absorption of energy from the other element then all the energy transfer is local, and no energy transfer occurs between the source and the elements. This is another interpretation of resonance where we can see that as far as the source is concerned, all energy transfer is local. Hence if we put a voltage source across the LC tank, no energy is delivered from the voltage source to the LC tank, which also means no current flows from the voltage source to the tank. Zero current delivered for a finite voltage source means impedance $Z_{in} = \frac{V_{in}}{I_{in}} = \frac{V_{in}}{0} = \infty$, which agrees with the customary interpretation of resonance (LC tank in parallel represented by $Z_{in} = \infty$ at resonance). To summarize, putting inductor there helps to null out the reactance due to the capacitance. This results in a minimum power wasted in R_s. Therefore, of the power delivered the maximum amount goes to C, the input of the device.

We can gain further insight into power matching by comparing our present case to that in sub-section 3.1.2.3. As a reminder, in sub-section 3.1.2.3, we talked about the concept of matching in the case where R'_{in_power} is real under all frequency. We want to reexplore the interpretation in the present case where R'_{in_power} is actually complex except at ω_c when it becomes real. We will find the interpretation in sub-section 3.1.2.3 to be equally applicable in this more general case. At first glance, however, it seems that as opposed to the case in sub-section 3.1.2.3, where we use a transformer to do matching, the present case does not give us the maximum power output. We show how we may reach this erroneous conclusion and then show the correct interpretation.

First, to get us to the misleading conclusion, let us compare Figure 3.5a with Figure 3.5b. In Figure 3.5a, where M_1 is not inductor degenerated, we assume that C_{gs} is an open circuit at ω_c. We further assume that C_c is a short at ω_c. Hence $V_{gs} = \frac{V_s}{2}$. In Figure 3.5b, under matching condition, we have $R_s = Re(Z_{in}) = g_m \frac{L_2}{C_{gs}}$ and so $V_{in} = \frac{1}{2} V_s$. When we apply KVL to the input loop in Figure 3.5b then $V_{in} = V_{gs} + V_L$. At first glance, we conclude from this equation that $|V_{in}| > |V_{gs}|$. Then $|V_{gs}| < \frac{1}{2} |V_s|$. That means $|V_{gs}|$ in Figure 3.5b is smaller than $|V_{gs}|$ in Figure 3.5a. Since the same transistor M_1 is used in both cases, they have the same g_m, and therefore I_o in Figure 3.5b is smaller than $|I_{M_1}|$ in Figure 3.5a, leading to a smaller output power or the erroneous conclusion. This conclusion is misleading because our assumption that $|V_{in}| > |V_{gs}|$ is wrong. At first glance this seems impossible, since $V_{in} = V_{gs} + V_L$. However, the preceding sum is a vector sum, not a scalar sum. This is the major difference between the present case and that described in sub-section 3.1.2.3. To see how $|V_{gs}| > |V_{in}|$ is possible, let us refer to Figure 3.6c, where a small V_{in} generates a large I_{in} (by setting $R = \frac{V_{in}}{I_{in}} = g_m \frac{L}{C}$ small). The large I_{in} generates a large V_{gs} (by setting $\omega_c C_{gs}$ small). Then the large V_{gs} generates a small I_o (by setting a small g_m). Since I_o is small compared with I_{in}, then I_L is almost vertical. Now V_L is leading I_L by 90°; therefore, V_L is almost horizontal and pointing in almost the opposite direction from V_{gs}. This means that their vector sum, V_{in}, is very small. To summarize, this means that even for a small V_{in}, by setting the proper component values, there can exist a large V_{gs} because V_L can be made large and pointing in the opposite direction from this V_{gs}. This will lead to a large V_{gs} and hence large I_o or large output power. This is, of course, nothing but saying that in a circuit involving LC [Figure 3.5b being an example], at resonance, the voltage (or current) across the C (or through L)

can be much larger than the applied voltage (or current) (actually Q times larger, Q being the quality factor). This is well known in a passive RLC circuit. We just manage to highlight that it is equally valid in a LC circuit involving a transistor.

3.3.4   Similarity Between Q (Quality factor) and n (Turn Ratio)

Now that we have explained matching in both wideband and narrowband LNA, it is a good opportunity to compare the two corresponding matching networks: transformers and resonance circuit. Specifically we will show the turn ratio n of a transformer plays a similar role to the quality factor Q of a resonant circuit. We can show this by first considering the transformer. Referring to Figure 3.2b, we can show that as n goes up, V_{gs} becomes larger because

$$V_{gs} = nV_{in} \qquad (3.54)$$

However, the reflected R_{in}, defined as R_{refl}, which is given by

$$R_{refl} = \frac{R_{in}}{n^2} \qquad (3.55)$$

goes down. Substitute this in the following voltage divider formula:

$$V_{in} = V_s \times \frac{R_{refl}}{R_{refl} + R_s} \qquad (3.56)$$

We can see that V_{in} becomes smaller. According to (3.54), this reduces V_{gs}. Hence an increase in n affects V_{gs} in two opposite ways. In the end V_{gs} can be shown to go to 0. The same conclusion can be shown to be true as n decreases. To see this more clearly, substitute (3.56) into (3.54) and we have

$$V_{gs} = \frac{nV_s R_{refl}}{R_{refl} + R_s} \qquad (3.57)$$

Substitute (3.55) in (3.57) and we have

$$V_{gs} = \frac{nV_s \frac{R_{in}}{n^2}}{\frac{R_{in}}{n^2} + R_s}$$

$$= \frac{\frac{V_s R_{in}}{n}}{\frac{R_{in}}{n^2} + R_s} \qquad (3.58)$$

As $n \to \infty$, from (3.58) V_{gs} is seen to go to 0. We next rearrange (3.58) so that it becomes

$$V_{gs} = nV_s \frac{1}{1 + \frac{R_s n^2}{R_{in}}} \qquad (3.59)$$

As $n \to 0$, from (3.59) V_{gs} is seen to go to 0.

The maximum V_{gs} is obtained for an n that is determined by setting $\frac{dV_{gs}}{dn} = 0$. This happens when

$$n = \sqrt{\frac{R_{in}}{R_s}} \qquad (3.60)$$

Substitute this into (3.55) and we can see that this happens when

$$R_{refl} = R_s \qquad (3.61))$$

This is just restating that maximum power transfer (which occurs when V_{gs} is at its maximum) occurs when $R_{refl} = R_s$, or when matching occurs.

Next we consider the resonant circuit. We go to Figure 3.5b and see what happens as Q changes. First we want to jump ahead of ourselves a bit and state a few formulas that will be derived later. Specifically, they are (3.80a), and (3.82), which are rewritten here in the following form:

$$Q = \frac{1}{R} \sqrt{\frac{L_1 + L_2}{C}} \qquad (3.62)$$

$$\frac{V_{gs}}{V_{in}} = \frac{Q}{j} \qquad (3.63)$$

If we compare (3.54) and (3.63), we can see that both n, Q can be interpreted as voltage gain of the matching network (with V_{in} as input and V_{gs} as output). Without loss of generality, let us set $L_1=0$ in (3.62) and we have

$$Q = \frac{1}{R} \sqrt{\frac{L_2}{C}} \qquad (3.64)$$

Substituting L_2/C from (3.79b) in (3.64), we can express R as

$$R = \frac{1}{Q^2 g_m} \qquad (3.65)$$

When comparing (3.65) to (3.55), it can be seen that the matching network in Figure 3.5b reflects the g_m and produces a reflected input resistance R. Since we are operating under resonance, then Z_{in} in Figure 3.5b becomes R. Therefore, we have the following voltage divider formula:

$$V_{\text{in}} = V_s \frac{R}{R + R_s} \tag{3.66}$$

We have now developed all the necessary formulas to see the impact of changing Q. As Q increases, from (3.63) V_{gs} increases. However, from (3.65) the reflected resistance, R, goes down. According to (3.66), we see that this makes V_{in} smaller. Referring to (3.63), a smaller V_{in} makes V_{gs} smaller.

Hence an increase in Q affects V_{gs} in two opposite ways. In the end V_{gs} can be shown to go to 0. The same conclusion can be shown to be true as Q decreases. Again, to see this more clearly substitute (3.66) in (3.63) and we have

$$V_{gs} = \frac{Q}{j} \frac{V_s R}{R + R_s} \tag{3.67}$$

Substitute (3.65) in (3.67), and we have

$$V_{gs} = \frac{Q}{j} \frac{V_s \frac{1}{Q^2 g_m}}{\frac{1}{Q^2 g_m} + R_s} \tag{3.68}$$

Rearranging, this becomes

$$V_{gs} = \frac{1}{j} \frac{\frac{V_s}{Q} \frac{1}{g_m}}{\frac{1}{Q^2 g_m} + R_s} \tag{3.69}$$

As $Q \to \infty$ from (3.69), V_{gs} is seen to go to 0.

We rearrange (3.69) so that it becomes

$$V_{gs} = \frac{QV_s}{j} \frac{1}{1 + R_s Q^2 g_m} \tag{3.70}$$

As $Q \to 0$, from (3.70) V_{gs} is seen to go to 0.

The maximum V_{gs} is obtained for a Q that is determined by setting $\frac{dV_{gs}}{dQ} = 0$. This happens when

$$Q = \sqrt{\frac{1}{g_m R_s}} \tag{3.71}$$

Substituting this into (3.65), we can see that this happens when

$$R = R_s \tag{3.72}$$

This is just restating that maximum power transfer (which happens when V_{gs} is at its maximum) occurs when $R = R_s$, or when matching occurs.

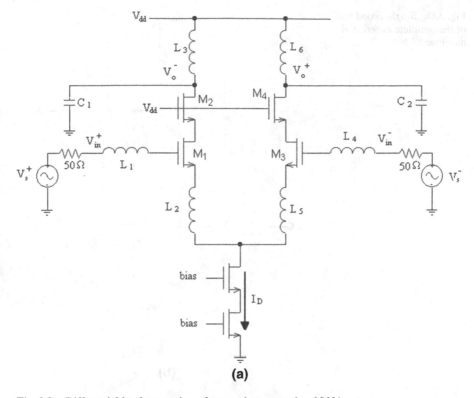

(a)

Fig. 3.8a   Differential implementation of a complete narrowband LNA

In summary when we compare the two matching networks and observe the behavior of V_{gs} as a function of n versus V_{gs} as a function of Q we note the similarity between n and Q.

As a final note, a transformer is a voltage amplification device (not a power amplification device) and a resonant circuit is also a voltage, not a power amplification device. It should also be noted the transformer is a wideband voltage amplification device whereas the resonance circuit is a narrowband voltage amplification device. Therefore, the transformer is used in the wideband LNA and the resonance circuit is used in the narrowband LNA.

3.4   Narrowband LNA: Core Amplifier

Now that we have finished discussing the matching network, we can talk about the core amplifier. In a narrowband design, because of its good high-frequency performance, cascode configuration is one of the typical LNA topology for RF application. Using differential cascode pair and narrowband LC tuning at input/output, we could further improve the circuit performance, such as noise figure and gain. Building these features into the circuit described in Figure 3.7a, we have the final circuit as shown in Figure 3.8a. The single-end schematic is redrawn in

Fig. 3.8b Single-ended half
of the complete differential
narrowand LNA

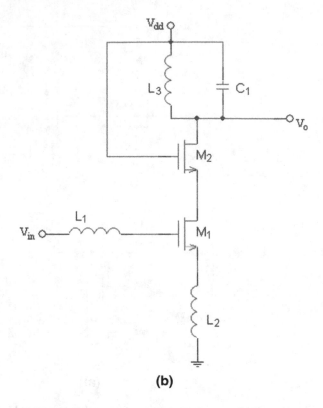

(b)

Figure 3.8b, where V_s and 50 Ω are omitted for simplicity. We would now continue our discussion by looking at design issues of this core amplifier [6].

3.4.1 Noise Figure

Let us take a look at the circuit in Figure 3.8b and derive its NF.

As shown by the Friis' formula ((4.50) of Chapter 2), the NF of any cascaded network is dominated by the first stage. Therefore, for the time being we assume that the noise is dominated by the noise from M_1. We further assume that noise from M_1 is dominated by thermal noise of the drain current. If M_1 operates in a CS configuration, it was shown in (3.23) that NF is given by

$$NF = 1 + \frac{4kT\frac{2}{3}\frac{1}{g_{M_1}}}{4kTR_s} = 1 + \frac{2}{3g_{M_1}R_s} \tag{3.73}$$

In the present case, M_1 is source degenerated with inductor L_2 and it also has an input inductor L_1. To calculate the equivalent input noise generators for the present case, we make use of the equivalent circuit shown in Figure 3.9a. Figure 3.9a shows

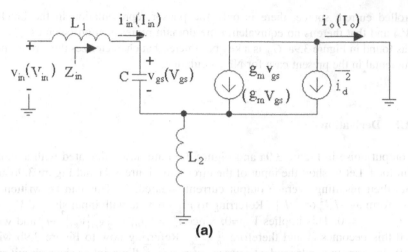

(a)

Fig. 3.9a Small-signal equivalent circuit of LNA with noise generators

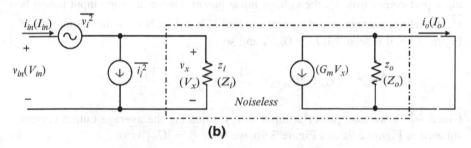

(b)

Fig. 3.9b Representation of Figure 3.9a by two input noise generators

the small-signal equivalent circuit of the amplifier shown in Figure 3.8b, with thermal noise of drain current represented by the current source $\overline{i_d^2}$. The time domain variables are shown together with their phasor representations enclosed in brackets (the noise sources, of course, do not have phasor representations and only their mean square values are shown). The time domain variables are in lowercase and the phasor representations are in uppercase. Figure 3.9b represents Figure 3.9a with equivalent noise representation $\overline{v_i^2}$, $\overline{i_i^2}$ at the input, followed by a noiseless small-signal model of the amplifier with transconductance G_m. Here we have g_m defined as the transconductance of the device and G_m as the transconductance of the whole amplifier, which takes into consideration the resonance effect caused by C, L. It turns out that G_m, which is defined as the short circuit output current divided by the input current of this new circuit, has to be a phasor because, from Figure 3.6b, current I_o and voltage V_{in} are 90° out of phase. This observation is also valid for the present circuit, although it has an extra inductor L_1. Hence G_m will be defined as the phasor representation of the transconductance of the whole amplifier. This also explains why next to the controlled current source, there is only the phasor representation in the bracket $(G_m V_x)$ and that there is no equivalent time domain representation (such as $g_m V_{gs}$), as was found in Figure 3.9a. G_m is a key parameter that characterizes the circuit. It is instrumental in the present case for NF calculation.

3.4.1.1 Derivation

The output noise in Figure 3.9a and Figure 3.9b are now calculated with a short-circuit load. Let us short the input of the circuit in Figure 3.9a and Figure 3.9b and equate their resulting average output current squared, $\overline{i_o^2}$. This can be written in phasor form as $\frac{1}{2} I_o I_o^*$ or $\frac{1}{2}|I_o|^2$. Referring to Figure 3.9a, with input shorted $V_{in}=0$, then $I_{in} = \frac{V_{in}}{Z_{in}} = 0$. This implies $V_{gs}=0$. Since $\overline{i_o^2} = \frac{1}{2}|I_o|^2 = \frac{1}{2}g_m^2 |V_{gs}|^2 + \overline{i_d^2}$ and with $V_{gs}=0$ this becomes $\overline{i_d^2}$ and therefore $\overline{i_o^2} = \overline{i_d^2}$. Referring now to Figure 3.9b with $V_{in}=0$, there are two independent sources, $\overline{v_i^2}$ and $\overline{i_i^2}$. $\overline{i_i^2}$ flows into a short circuit and develops zero power and can be neglected. Therefore, the only power source at the input port comes from $\overline{v_i^2}$, the voltage noise power. Hence, average input power is $\overline{v_i^2}$. As before, average output current squared $\overline{I_o^2}$ is $\frac{1}{2}|I_o|^2$. Now at the output port of Figure 3.9b, it is seen that $I_o = G_m V_x$ and so

$$\frac{1}{2}|I_o|^2 = \frac{1}{2} G_m V_x (G_m V_x)^* = \frac{1}{2} G_m V_x G_m^* V_x^* = \frac{V_x V_x^*}{2} G_m G_m^*$$

(since $\frac{V_x V_x^*}{2}$ = average power at input = $\overline{v_i^2}$). Equating the average output current squared in Figure 3.9a and Figure 3.9b, we have $\overline{i_d^2} = |G_m|^2 \overline{v_i^2}$ or

$$\overline{v_i^2} = \frac{\overline{i_d^2}}{|G_m|^2} \tag{3.73a}$$

Notice that this is almost identical to the expression derived for $\overline{v_i^2}$ in the wideband case, where $\overline{v_i^2} = \frac{\overline{i_d^2}}{g_m^2}$, except here g_m^2 is replaced by $|G_m|^2$. Assuming that thermal noise dominates, then

$$\overline{i_d^2} = 4kT\left(\frac{2}{3} g_m\right)\Delta f \tag{3.73b}$$

Substituting (3.73b) into (3.73a), we have

$$\overline{v_i^2} = \frac{4kT\left(\frac{2}{3} g_m\right)\Delta f}{|G_m|^2} \tag{3.73c}$$

To further simplify the preceding expression, we need to express G_m in terms of g_m. Referring to Figure 3.9a again, first a voltage V_{in} is applied to the input (with noise source $\overline{i_d^2}$ turned off), and the output is short-circuited. Hence an internal voltage V_{gs} is developed, which, upon multiplication by g_m of the device, generates a short circuit output current I_o given as

$$I_o = -g_m V_{gs} \tag{3.74}$$

Next, this same voltage V_{in} is applied to Figure 3.9b (with noise source $\overline{v_i^2}$, $\overline{i_i^2}$ turned off). With output short circuited, the short circuit output current I_o is given by

$$I_o = -G_m V_x = -G_m V_{in} \tag{3.75}$$

Since the two I_o are identical, we equate (3.74) and (3.75), resulting in

$$g_m = G_m \frac{V_{in}}{V_{gs}} \tag{3.76}$$

Let us digress a bit and find the equation that relates V_{in} to V_{gs}. To do this, we refer to Figure 3.9a again and rewrite Ohm's law in phasor representation for the input loop as

$$V_{in} = Z_{in} I_{in} \tag{3.77}$$

$$V_{gs} = \frac{1}{j\omega C} I_{in} \tag{3.78}$$

Dividing (3.78) by (3.77), we get

$$\frac{V_{gs}}{V_{in}} = \frac{1}{Z_{in} j\omega C} \tag{3.79}$$

So we get $\frac{V_{gs}}{V_{in}}$, but to go further we need to solve for Z_{in}. To find Z_{in}, let us repeat the expression of Z_{in} from (3.51):

$$Z_{in} = j\omega(L_1 + L_2) + \frac{1}{j\omega C} + g_m \frac{L_2}{C} \tag{3.79a}$$

We can redraw the representation of this Z_{in} as in Figure 3.10, which shows a passive *RLC* circuit and where we put

$$R = g_m \frac{L_2}{C} \tag{3.79b}$$

Now, from matching condition $Z_{in} = g_m \frac{L_2}{C} = R$ and so (3.79) becomes

Fig. 3.10 Representation of Z_{in} using a passive *RLC* circuit

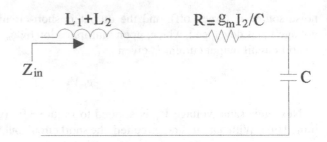

$$\frac{V_{gs}}{V_{in}} = \frac{1}{R(j\omega_c C)} \tag{3.80}$$

Let us try to express R of this resonant circuit in terms of its quality factor, Q and the resonating frequency, ω_c. For an *RLC* circuit as shown in Figure 3.10 from [3] we have

$$Q = \frac{1}{R}\sqrt{\frac{L_1 + L_2}{C}} \tag{3.80a}$$

$$\omega_c = \frac{1}{\sqrt{(L_1 + L_2)C}} \tag{3.80b}$$

From (3.80a), and (3.80b), we want to eliminate $L_1 + L_2$ and express Q in terms of ω_c. The resulting expression is

$$Q = \frac{1}{R}\sqrt{\frac{L_1 + L_2}{C}} = \frac{1}{R}\frac{1}{\omega_c\sqrt{C}}\frac{1}{\sqrt{C}} = \frac{1}{R}\frac{1}{\omega_c C} \tag{3.81}$$

Finally, we can substitute (3.81) into (3.80) and we have

$$\frac{V_{gs}}{V_{in}} = \frac{Q}{j} \tag{3.82}$$

We can then go back to (3.76) and solve for G_m in terms of g_m. This is done by substituting (3.82) into (3.76), resulting in

$$G_m = g_m\frac{Q}{j} \ or \ |G_m|^2 = g_m^2 Q^2 \tag{3.83}$$

Now, to derive $\overline{v_i^2}$, we substitute (3.83) into (3.73c) and we have

$$\overline{v_i^2} = \frac{\frac{2}{3}4kTg_m\Delta f}{g_m^2 Q^2} \tag{3.84}$$

Substituting this in (3.22) with $N_{dev} = \frac{\overline{v_i^2}}{\Delta f}$ will give a NF of

$$NF = 1 + \frac{\frac{2}{3}4kTg_m}{g_m^2 Q^2 (4kTR_s)} = 1 + \frac{2}{3}\frac{1}{g_m Q^2 R_s} \tag{3.85}$$

3.4.1.2 Insight

From (3.85) it can be seen that NF is inversely proportional to g_m and Q. Accordingly, to obtain the best NF in the present case, Q should be set to ∞. However, from (3.81) $Q = \frac{1}{R\omega_c C}$. For a given ω_c, C, setting Q to ∞ means setting $R = 0$. On the other hand, the preceding expression is defined under matching condition for maximum power transfer, which means that this R must be made equal to R_s, or R_s will become 0. In general, R_s is fixed and so the best NF condition cannot be achieved. Hence we can see that the matching conditions for optimal noise performance and optimal power performance are not identical.

To gain more insight let us re-express the NF expression in (3.85) in an alternate form, concentrating on the $g_m Q^2 R_s$ factor in the denominator. First, the matching of real parts dictates $R_s = \frac{g_m L_2}{C}$. Also, $Q = \frac{1}{R\omega_c C}$, where $R = \frac{g_m L_2}{C}$. Upon substitution, the $g_m Q^2 R_s$ factor becomes

$$g_m Q^2 R_s = g_m \left(\frac{1}{\frac{g_m L_2}{C}\omega_c C}\right)^2 \times \frac{g_m L_2}{C} = \frac{1}{L_2 \omega_c^2 C} \tag{3.86}$$

Next, we make use of the imaginary matching condition $\frac{1}{\omega_c^2 C} = L_1 + L_2$, and (3.86) becomes

$$g_m Q^2 R_s = \frac{1}{L_2}(L_1 + L_2) \tag{3.87}$$

Now we substitute (3.87) in (3.85) to get

$$NF = 1 + \frac{2}{3}\frac{L_2}{(L_1 + L_2)} = 1 + \frac{2}{3}\frac{1}{1 + \frac{L_1}{L_2}} \approx 1 + \frac{2}{3}\frac{L_2}{L_1} \tag{3.88}$$

The last approximation will be true if $L_1 \gg L_2$. Notice that we finally express NF in terms of only passive circuit components L_1, L_2. From (3.88), NF goes down as L_1 increases and L_2 decreases. This provides valuable design guidelines.

3.4.2 Power Dissipation

In this sub-section we derive the dependence of power dissipation on technology and circuit parameters under matching condition and for a given $V_{GS} - V_T$ (which is usually fixed for a design). First,

$$P = I_D V_{DD} \tag{3.89}$$

where I_D is the drain current of M_1 [shown Figure 3.8b]. Therefore

$$P \propto I_D \tag{3.90}$$

Since M_1 operates in the saturation region, we have the following I_D equation:

$$I_D = \frac{1}{2} k' \frac{W}{L} (V_{GS} - V_T)^2 \tag{3.91}$$

where $k' = \mu C_{ox}$; that is, $I_D \propto \mu C_{ox} \frac{W}{L}$ for a given $V_{GS} - V_T$ (fixed for a design, as stated at the beginning of this sub-section). Substituting (3.91) into (3.90), we have

$$P \propto \mu C_{ox} \frac{W}{L} \tag{3.92}$$

Next, let us try to express $\mu C_{ox} \frac{W}{L}$ in terms of g_m, C. We will show that $\mu C_{ox} \frac{W}{L}$ and $\frac{g_m^2}{C} \frac{L^2}{\mu}$ are proportional to one another. To show this, it is easier to start from the opposite end; that is, from the expression $\frac{g_m^2}{C} \frac{L^2}{\mu}$. We first note that $C = C_{gs}$ of M_1 and hence

$$C = \frac{2}{3} W L C_{ox} \tag{3.93}$$

Also, for M_1 its g_m can be expressed as

$$g_m = k' \frac{W}{L} (V_{GS} - V_T); \ i.e. \ g_m \propto \mu C_{ox} \frac{W}{L} \ \text{for a given} \ (V_{GS} - V_T) \tag{3.94}$$

Substituting (3.93) and (3.94) into the expression $\frac{g_m^2}{C} \frac{L^2}{\mu}$, we have

$$\frac{g_m^2}{C} \frac{L^2}{\mu} \propto \frac{\left(\mu C_{ox} \frac{W}{L}\right)^2}{W L C_{ox}} \frac{L^2}{\mu} = \frac{\mu^2 C_{ox}^2 \frac{W^2}{L^2} L^2}{W L C_{ox} \mu}$$

$$= \mu C_{ox} \frac{W}{L} \tag{3.95}$$

Reversing the proportionality operator in (3.95), we have

$$\mu C_{ox} \frac{W}{L} \propto \frac{g_m^2}{C} \frac{L^2}{\mu} \tag{3.96}$$

Substituting (3.96) into (3.92), we get

$$P \propto \frac{g_m^2}{C} \frac{L^2}{\mu} \tag{3.97}$$

Finally, we want to express g_m, C in (3.97) in terms of more basic parameters. Under matching,

$$\omega_c (L_1 + L_2) = \frac{1}{\omega_c C} \text{ which implies } C = \frac{1}{\omega_c^2} \frac{1}{(L_1 + L_2)} \tag{3.98}$$

Also,

$$R_s = \frac{g_m}{C} L_2 \text{ or } g_m = \frac{R_s C}{L_2} \tag{3.99}$$

First, substituting (3.99) into (3.97), we get

$$P \propto \left(\frac{R_s C}{L_2} \right)^2 \frac{1}{C} \frac{L^2}{\mu} = \frac{L^2}{\mu} \frac{R_s^2}{L_2^2} C \tag{3.100}$$

Then, substituting (3.98) into (3.100), we have

$$P \propto \frac{L^2}{\mu} \frac{R_s^2}{L_2^2} \frac{1}{\omega_c^2 (L_1 + L_2)} = \frac{L^2}{\mu} \frac{R_s^2}{L_2^3 \omega_c^2 \left(1 + \frac{L_1}{L_2} \right)} \tag{3.101}$$

Now we can regroup terms after the equality sign in (3.101) and rewrite (3.101) in the following form:

$$P \propto \underbrace{\frac{L^2}{\mu}}_{\text{technology}} \underbrace{\left(\frac{R_s}{\omega_c} \right)^2}_{\text{standard}} \underbrace{\frac{1}{L_2^3 \left(1 + \frac{L_1}{L_2} \right)}}_{\text{circuit parameter}} \tag{3.102}$$

As in the NF case, we finally break down clearly the dependence of P in terms of various groups of parameters: technology, standard, and circuit parameters. Examining (3.102), the first term that includes the minimum feature size L and mobility is fixed for a given technology. As L goes down (scaled technology), P goes down as well. This is satisfying because it means that as technology scales, power goes down. With a given standard, parameters in the second term, R_s and ω_c, are fixed. For example, with the DECT standard, S_{11} is given. From (3.2a) this fixes R_{in}, which through matching conditions in turn fixes R_s. In the DECT standard ω_c is also fixed and is set to $2\pi \times 1.9 GHz$. Hence, the only degrees of freedom are L_1, L_2, the circuit parameters as grouped in the third term. From (3.102) we have the following conclusion: P goes down as L_1 goes up.

To summarize, from (3.88) and (3.102) we can conclude that both NF and P go down as L_1 goes up.

3.4.3 Trade-Off Between Noise Figure and Power

In the last two sub-sections, we derived the dependence of the two most important parameters, NF and power, of a narrowband LNA on the circuit parameters. In this sub-section we compare this dependence to the wideband case as discussed previously. This will uncover some fundamental differences between the two approaches and further show why a narrowband design enjoys some inherent advantages in achieving a better NF, power trade-off than the wideband case. This may help justify its use, in view of its need to have an inductor, which is not trivial to realize in silicon.

The key observation we make in the comparison is that in the narrowband case, the average input referred voltage noise power $\overline{v_i^2} \propto g_m$, the transconductance of the device, whereas in the wideband case, $\overline{v_i^2} \propto \frac{1}{g_m}$. To show this, let us reformulate the expression for G_m in (3.83) in a different form that highlights this dependence. We start off again with (3.83):

$$G_m = g_m \frac{Q}{j} \tag{3.103}$$

From (3.81) we have, under matching conditions (i.e., $R = R_s$),

$$Q = \frac{1}{R_s \omega_c C} \tag{3.104}$$

Substituting (3.104) into (3.103), we get

$$G_m = g_m \frac{1}{R_s \omega_c C} \frac{1}{j} \tag{3.105}$$

Now we want to rewrite G_m in terms of inductors L_1, L_2. This can be done because under matching conditions R_s can be expressed as $\frac{g_m}{C} L_2$, and if this is substituted in (3.105) we can simplify the expression as

$$G_m = \frac{g_m}{\frac{g_m}{C} L_2 \omega_c C} \frac{1}{j} = \frac{1}{j \omega_c L_2} \tag{3.106}$$

Table 3.2 Comparisons of wideband and narrowband LNA

	Wideband	Narrowband
$\overline{v_i^2}\alpha$	$\frac{1}{g_m}$	g_m
$P\alpha$	g_m	g_m

Now substitute (3.106) into (3.73c):

$$\overline{v_i^2} = \frac{\frac{2}{3}(4kT)g_m\Delta f}{\left|\frac{1}{j\omega_c L_2}\right|^2} \tag{3.107}$$

Notice that once ω_c, L_2 is fixed from the resonance condition, the denominator of (3.107) is fixed and is independent of g_m. Hence, from (3.107),

$$\overline{v_i^2}\alpha\, g_m \tag{3.108}$$

Now compare this with wideband case. From (3.23) we have

$$\overline{v_i^2} = 4kT\frac{2\Delta f}{3g_m} \tag{3.109}$$

or

$$\overline{v_i^2} \propto \frac{1}{g_m} \tag{3.110}$$

Comparing (3.110), and (3.108), $\overline{v_i^2}$ (and hence NF) behaves in an opposite way in its dependence on g_m for the narrowband versus wideband cases.

We summarize in Table 3.2.

Therefore, for the wideband case, if we want less noise (i.e., smaller $\overline{v_i^2}$) we pay a price by dissipating more power P. On the other hand, in the narrowband case, power actually goes down as $\overline{v_i^2}$ goes down and the trade-off between NF and power is broken. The underlying reason is that in using the inductor we tune out C (so that it matches R_s), so the gain is dependent on g_m and therefore the input referred noise is proportional to $\frac{1}{g_m}$.

3.4.4 Noise Contribution from other Sources

We have assumed up to now that the dominant noise source is the thermal noise from M_1. It turns out that the passive devices can also contribute noise. Inductor L_1 is usually a bond wire inductor, which has a high Q factor. Accordingly, its noise

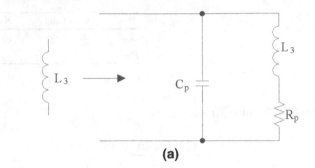

Fig. 3.11a Representation of L_3 with parasitics using an equivalent network

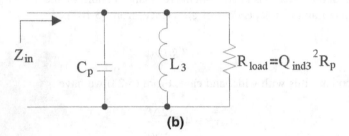

Fig. 3.11b Approximation to Figure 3.11a when $Q_{ind3} \gg 1$

contribution can be neglected. Inductors L_2 and L_3 are usually on chip spiral inductors, which are lossy and have a low Q factor. By design, the circuit needs a very small L_2, so the accompanying parasitic resistance and its noise contribution can also be ignored. Hence, the only significant noise source from the passive devices may be from L_3.

From the Friis' formula, [(4.50) in Chapter 2], we know that if gain of M_1 is large, the source resistance and M_1's thermal noise are the dominant noise sources, unless the thermal noise of M_2 and L_3 is much larger than that of M_1. We ignore noise from M_2 for the time being. Turning to L_3, we begin by referring to Figure 3.8b, where we would like to find out the noise contribution from the noise sources due to the loss in the output inductor L_3. We will show that the input referred equivalent noise resistance R_{eq} of L_3 equals $\frac{1}{|G_m|^2 Q_{ind3}^2 R_p}$, where $Q_{ind3} = \omega \frac{L_3}{R_p}$ is the quality factor of the inductor L_3. This should not be confused with Q, the quality factor of the input impedance Z_{in}. R_p is the parasitic resistance associated with L_3 due to substrate loss.

First we recognize that inductor L_3 is implemented as a spiral inductor on the silicon surface. Due to substrate loss, the ideal L_3 is replaced by its equivalent circuit, as shown in Figure 3.11a, where C_p, R_p are the parasitic resistance and capacitance associated with L_3. If $Q_{ind3} \gg 1$, Figure 3.11a becomes Figure 3.11b, where R_p is replaced by $R_{load} = Q_{ind3}^2 R_p$. This can be derived easily by equating the

Fig. 3.12a Narrowband
LNA redrawn, separating the
transconductor from the load

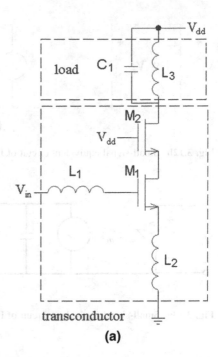

(a)

input impedance seen in Figure 3.11a and Figure 3.11b while keeping in mind that $Q_{ind3} \gg 1$ (for small R_p). The impedance in Figure 3.11a is

$$Z_{in}(s) = \frac{sL_3 + R_p}{s^2L_3C_p + sR_pC_p + 1} \cong \frac{sL_3}{s^2L_3C_p + sR_pC_p + 1} \qquad (3.110a)$$
$$(\textit{for } Q_{ind3} \gg 1)$$

The impedance in Figure 3.11b is

$$Z_{in}(s) = \frac{sL_3}{s^2L_3C_p + sL_3/R_{load} + 1} \qquad (3.110b)$$

In order to get the same impedance (in a steady state), we equate Z_{in} in (3.110a) to Z_{in} in (3.110b). We then observe that $L_3/R_{load} = R_pC_p$ at ω_c. Substituting $Q_{ind3} = \omega_c \frac{L_3}{R_p}$ and $\omega_c = \frac{1}{\sqrt{L_3C_p}}$ in this expression, it follows that the equivalent load resistance is given by

$$R_{load} = Q_{ind3}^2 R_p \qquad (3.111)$$

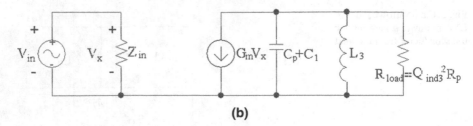

(b)

Fig. 3.12b Small-signal equivalent circuit of Figure 3.12a

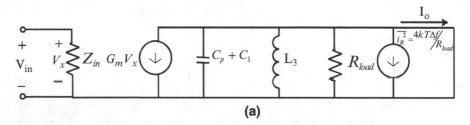

(a)

Fig. 3.13a Small-signal equivalent circuit of Figure 3.12a with noise generators

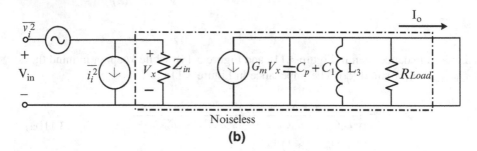

Noiseless

(b)

Fig. 3.13b Representation of Figure 3.13a by two input noise generators

Let us next redraw Figure 3.8b in Figure 3.12a. Figure 3.12a serves to highlight the noise contribution from the parasitic resistance of L_3 by separating the amplifier into the load part and the transconductor part. Then L_3 in the load part is replaced by its equivalent model, as shown in Figure 3.11b. Similarly, the transconductor part is replaced by its small-signal equivalent model. Combining, we have the resulting circuit as shown in Figure 3.12b.

Since we are now interested only in noise from resistor R_p, Figure 3.12b is next redrawn in Figure 3.13a, where the noise source $\overline{i_R^2}$ that denotes the thermal noise from R_p is explicitly shown. Following the equivalent representation of Figure 3.9a by Figure 3.9b, Figure 3.13a is now equivalently represented by Figure 3.13b, where all the noise sources have been absorbed in the two equivalent input noise source $\overline{v_i^2}, \overline{i_i^2}$. To find $\overline{v_i^2}, \overline{i_i^2}$, we short the inputs and outputs in Figure 3.13a and Figure 3.13b, find the short-circuited average output current squared $\overline{i_o^2}$, and equate them.

From Figure 3.13a,

$$\overline{i_o^2} = \overline{i_R^2} = \frac{4kT\Delta f}{R_{\text{load}}} \tag{3.112}$$

From Figure 3.13b,

$$\overline{i_o^2} = |G_m|^2 \overline{v_i^2} \tag{3.113}$$

Equating $\overline{i_o^2}$ in both (3.112) and (3.113), we have

$$|G_m|^2 \overline{v_i^2} = \frac{4kT\Delta f}{R_{\text{load}}} \ or \ \overline{v_i^2} = \frac{4kT\Delta f}{|G_m|^2 R_{\text{load}}} \tag{3.114}$$

Now we can substitute for R_{load} from (3.111) into (3.114) to get the equivalent input referred noise as

$$\overline{v_i^2} = \frac{4kT\Delta f}{|G_m|^2 Q_{\text{ind3}}^2 R_p} \tag{3.115}$$

Comparing this with noise from M_1 in (3.73c), it is seen that noise from L_3, when input referred, gets an extra attenuation due to Q_{ind3} and so normally the noise from M_1 still dominates, unless the substrate loss is substantial (meaning that $Q_{\text{ind3}}^2 R_p$ is small).

We can also find the noise-equivalent resistance R_{eq} of the lossy inductor L_3. We apply the definition of R_{eq} to (3.115) and we have

$$R_{\text{eq}} = \frac{1}{|G_m|^2 Q_{\text{ind3}}^2 R_p} \tag{3.116}$$

From (3.116) we observe that $R_{\text{eq}} \propto \frac{1}{Q_{\text{ind3}}^2 R_p}$. Substituting R_p in (3.111) in this expression means that $R_{\text{eq}} \propto \frac{1}{R_{\text{load}}}$. So in order to minimize noise, we maximize R_{load} or minimize substrate loss.

3.4.5 Gain

Let us go back to the circuit in Figure 3.12a. Assume that initially there is no L_3, at high frequency. Since we have a capacitive load C_1, it seems that C_1 is going to shunt output to ground and will reduce gain. By using L_3, we show that we can still get very large voltage gain. The gain is $A_v = G_m \left(\frac{1}{j\omega C_1} \| j\omega L_3 \right)$. Here G_m is the transconductance defined in Figure 3.12b. If we substitute G_m from (3.106), then

$$A_v = \left(\frac{1}{j\omega L_2} \right) \frac{j\omega_c L_3}{1 - \omega_c^2 L_3 C_1} = \frac{L_3}{L_2} \frac{1}{1 - \omega_c^2 L_3 C_1} \tag{3.117}$$

Fig. 3.14a Small-signal equivalent circuit of a MOS transistor including gate resistance due to high-frequency effect

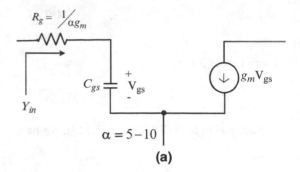

Fig. 3.14b Plot of input admittance of a MOS transistor versus frequency

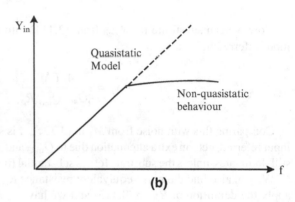

From (3.117) it can be seen that the output LC tuning circuit peaks the gain at ω close to $\frac{1}{\sqrt{L_3 C}}$. In reality, we cannot achieve this gain because we have neglected R_{load} of L_3.

3.4.6 Other Real-Life Design Considerations

We have now finished deriving essential design equations governing matching, noise, power, and gain of the circuits. In this sub-section we discuss the more important features on models for the active and passive components. They have some impact on our design procedure.

First, at high frequency the gate of an MOS transistor starts to have a resistive component, as shown in Figure 3.14a. This is due to distributive gate resistance, and the C_{gs} is replaced by a RC ladder network. This resistance R_g will change the matching criteria $(R = R_g + g_m \frac{L}{C})$. The distributive gate resistance comes about because the signal at one end of the gate has not quite traveled to the other end and the gate potential is in a non-quasistatic state. Hence, as shown in Figure 3.14b, the

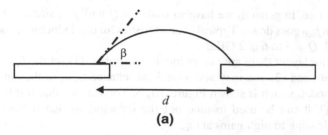

Fig. 3.15a Physical construction of a bond wire inductor

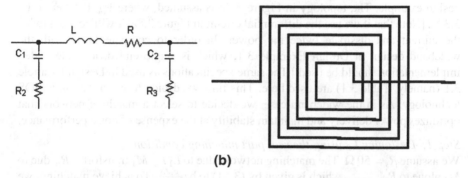

Fig. 3.15b Physical construction of a spiral inductor and its complete equivalent model

plot of input admittance Y_{in} versus frequency f is no longer a straight line (purely capacitive). Rather it bends and flattens out, indicating a resistive component. Notice that this R_g is not a real physical resistor and therefore does not contribute to noise.

Second, we briefly discuss the two types of on-chip inductors available. First there is the bond wire inductor, as shown in Figure 3.15a, where the inductance value is a function of the pad distance d, and bond angle β. Typical values in packages range from 2 nH to 5 nH, with about 1 nH/mm. The series loss is about 0.2 Ω/mm for a 1 mil diameter Al. The Q of this inductor is about 60 at 2 GHz. Even though it has a high Q, one major disadvantage is the reproducibility and predictability of this type of inductor. The second type of inductor is the spiral inductor on lossy substrate, as shown in Figure 3.15b. Also shown is its complete equivalent model [compare that with the simple equivalent model shown in Figure 3.11a]. For this type of inductor, the key design parameters are as follows:

- Inductance: $L = \sum (L_{self} + L_{mutual})$ where L_{self}, L_{mutual} are the self and mutual inductance
- Quality factor: $Q = \omega \frac{L}{R}$
- Self-resonant frequency: $f_{self} = \frac{1}{\sqrt{LC}}$

The existence of f_{self} may come as a surprise, but it comes from the fact there is capacitance associated with this inductor. Accordingly, the inductor can self-resonate, rendering it useless. This capacitance exists here because there is significant parasitic capacitance between the metal deposited on top of the silicon surface and the substrate. In general, we have to trade off Q with f_{self} since as L goes up, Q goes up but f_{self} goes down. Typical characteristics for this inductor include an $L = 1$ nH to 8 nH, $Q = 3$ to 6 at 2 GHz.

For inductor layout there are two points to remember: (1) set the f_{self} so it is far away from ω_c; and (2) you may add a load capacitance C_{load} to the output of the equivalent model, which is shown Figure 3.15b. You can see that if this capacitor has a high Q, it can be used to tune out the inductor, so that it has very high impedance (leading to high gain) at ω_c.

Design Example 3.2.

To illustrate the salient points of a narrowband LNA design, let us go through a design example. The topology in Figure 3.8b is assumed, where V_{dd} is taken to be 3.3 V. [Note that if we use the differential circuit in Figure 3.8a, it will conduct twice the current and dissipate twice the power. In order to compare fairly with the wideband design in Design Example 3.1, which is single ended, a single ended implementation should be used.] The same specifications as used in Design Example 3.1 (namely, Table 3.1) are used here. This time we decide to use a 0.5 um CMOS technology. As in the wideband case, we decide to select a matching network that optimizes power delivery and maintain stability at the expense of noise performance.

Step 1: Designing L_2 using the real part matching condition
We assume $R_s = 50\ \Omega$. The matching network due to $L_1 L_2$, M_1 transforms R_{in} due to M_1 alone to $R'_{\text{in\_power}}$, which is given by (3.51) to be $\frac{g_m L_2}{C}$. To achieve matching, we set $R'_{\text{in\_power}} = R_s$ and we have

$$50\Omega = g_m \frac{L_2}{C} = \frac{g_m}{C_{gs}} L_2 = 2\pi f_T L_2 \tag{3.118}$$

where f_T is the unity gain frequency of the MOS transistor and is given by

$$2\pi f_T = \frac{g_m}{C} \tag{3.119}$$

Therefore,

$$L_2 = \frac{50\Omega}{2\pi f_T} \tag{3.120}$$

For our 0.5 um technology we are given $f_T = 4GHz$. Hence:

$$L_2 = \frac{50\Omega}{2\pi \times 4GHz} = 2nH \tag{3.121}$$

Step 2: Designing L_1 using NF specification
Let us now apply the second specification: NF < 3 dB. To make use of this specification, let us invoke (3.88). Then we have

$$NF = 1 + \frac{2}{3}\frac{1}{1+\frac{L_1}{L_2}} < 2 \tag{3.122}$$

Solving,

$$\frac{L_1}{L_2} < \frac{1}{3} \tag{3.123}$$

Substituting (3.121) from step 1, we have

$$L_1 < 0.67\text{nH} \tag{3.123a}$$

We arbitrarily make

$$L_1 = 0.5\text{nH} \tag{3.123b}$$

Step 3: Designing $C = C_{gs}$ and hence $\left(\frac{W}{L}\right)_1$ using the imaginary part matching condition
Using the matching condition from (3.52), we have

$$\omega_c(L_1 + L_2) = \frac{1}{\omega_c C} \tag{3.124}$$

With $\omega_c = 2\pi \times 1.9Grad/s$, L_1, L_2 determined from (3.121), and (3.123b), we have

$$C = \frac{1}{\omega_c^2(L_1 + L_2)} = \frac{1}{(2\pi \times 1.9Grad/s)^2(0.5nH + 2nH)} = 2.8pF \tag{3.125}$$

Now we can determine $\left(\frac{W}{L}\right)_1$.
First we have

$$C = C_{gs} = \frac{2}{3}WLC_{ox}. \tag{3.126}$$

Secondly, we want L to be minimum. Hence we set $L = 0.5$ *um*. For this process C_{ox} is given to be 2.6 fF/*um*^{2}
Substituting (3.125), value of C_{ox} and L in (3.126), we have

$$W = \frac{C}{\frac{2}{3}C_{ox}L} = \frac{2.8pF}{1.75fF/um^2 \times 0.5um} = 3200um \tag{3.127}$$

Therefore,

$$\left(\frac{W}{L}\right)_1 = \frac{3200um}{0.5um} \qquad (3.128)$$

Step 4: Checking to see if the power specification is met
First, we find g_m from C. To do so, we re-arrange (3.119) and substitute (3.125) in it and we have

$$g_m = 2\pi f_T \times C = 2\pi \times 4GHz \times 2.8pF = 70m\Omega^{-1} \qquad (3.129)$$

Next, we find the drain current of M_1 and hence power from following equation:

$$g_m = \sqrt{2k'\left(\frac{W}{L}\right)_1 I_{D_1}} \qquad (3.130)$$

For this 0.5 *um* process we are given $k' = 100\frac{uA}{V^2}$. Re-arranging (3.130) and substituting (3.129) and (3.128) in it, we have

$$I_{D_1} = \frac{g_m^2}{2k'\left(\frac{W}{L}\right)_1} = \frac{\left(70m\Omega^{-1}\right)^2}{2 \times \frac{100\mu A}{V^2} \times \frac{3200}{0.5}} = 3.8mA \qquad (3.131)$$

Hence

$$P_{M_1} = 3.8mA \times 3.3V = 12mW \qquad (3.132)$$

This is just the power in the amplifier core.

If we assume that we have a similar bias and output buffer arrangement as in the wideband case, which is depicted in Figure 3.3a, then following similar arguments that led to (3.27a), we can assume

$$P_{total} = 4P_{M_1} = 48mW \qquad (3.133)$$

This slightly exceeds the specification.

Step 5: Using the gain equation and gain specification to find capacitance C_1
From Table 3.1, gain must be larger than 20 dB or 10. Substituting in (3.117), we have

$$A_v = \frac{L_3}{L_2(1 - \omega_c L_3 C_1)} > 10 \qquad (3.133a)$$

For simplicity, let us arbitrarily pick $L_3 = L_2$ which equals 2 nH [from (3.121)]. Hence

$$L_3 = 2nH. \qquad (3.134)$$

With this value the input referred noise from L_3's parasitic resistance R_p will be very small and, as discussed in sub-section 3.4.4, will not affect our NF.

Substituting (3.121) and (3.134) in (3.133a), we have

$$1 - (2\pi f_c)^2 (2nH)C_1 < \frac{1}{10} \qquad (3.135)$$

Rearranging,

$$C_1 > \frac{0.9}{(2\pi f_c)^2 \times 2nH} = \frac{0.9}{(2\pi \times 1.9\text{GHz})^2 \times 2nH} = 3.1pF \qquad (3.136)$$

We arbitrarily set

$$C_1 = 4\text{pF}. \qquad (3.137)$$

Finally, for simplicity, we set

$$\left(\frac{W}{L}\right)_2 = \left(\frac{W}{L}\right)_1 \qquad (3.138)$$

Step 6: Checking that S_{11} is satisfied
In step 1 we transformed R_{in} so that $R'_{in\_stable} = 50\Omega$. Hence, assuming a Z_o of 50Ω, from (3.2c) $S_{11} = 0$ and specification on S_{11} is satisfied.

At this point we have finished our design on the narrowband LNA.

3.5 Problems

3.1 The following is a differential wideband LNA.

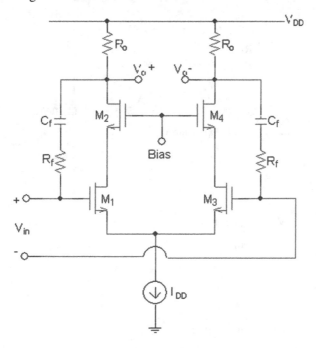

(a) Identify the type of feedback.

(b) Develop the single-ended portion only of this circuit. For this single ended circuit, under the following conditions:

    1. without feedback elements R_f, C_f

    2. with only feedback element R_f (assume C_f is a short circuit)

       (i) Calculate the equivalent input referred current and voltage noise in terms of device paramters: g_m, C_{gs}, C_{gd}, R_o, etc and circuit paramters: R_f, C_f etc. You can neglect frequency effect and assume only thermal noise.

      (ii) Calculate the −3 dB bandwidth of this circuit, using the zero valued time-constant method, in terms of device paramters: g_m, C_{gs}, C_{gd}, R_o, etc and circuit paramters: R_f, C_f etc. Simplify the expression.

(c) The C_f-R_f feedback arrangement in this LNA is particularly useful when V_{in} has source inductances (i.e., between $V_{in}{}^+$ and the gate of M_1 there is an inductor; similarly between $V_{in}{}^-$ and the gate of M_3 there is another induc-tor). Explain why, with source inductances, having this C_f - R_f feedback arrangement is better than having no feedback (but still having a resistor R_f connected from the gate of M_1 to ground and a resistor R_f connected from the gate of M_3 to ground).

3.2 For the wideband LNA shown in Figure 3.3a, rederive the NF, including noise contribution from M_2, M_3, in terms of circuit paramters: R_s, R_{op} etc., and device parameters: g_m, C_{gs}, etc., when appropriate. Include the frequency-dependent effect.

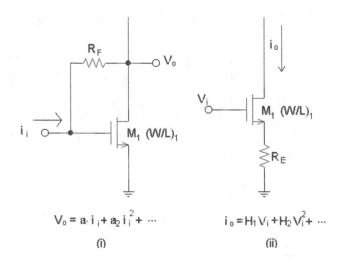

$$V_0 = a \cdot i_i + a_2 i_i^2 + \cdots$$

(i)

$$i_0 = H_1 V_i + H_2 V_i^2 + \cdots$$

(ii)

Fig. P3.1

3.3 Shown in Figure P3.1 are two wideband amplifiers using different feedback schemes.

(a) Calculate the equivalent input referred voltage and current noise sources at the input with feedback, $\overline{v_i^{*2}}$ $\overline{i_i^{*2}}$, in terms of the equivalent input referred voltage and current noise sources without feedback, $\overline{v_i^2}$ $\overline{i_i^2}$.

(b) In the above two amplifiers because M_1's I-V characteristics has only a square law dependence, there is no third order distortion. Let us now look at the differential implementations of the above two amplifiers. Due to the differential structure, there is now third order distortion. Assume these structures to be memoryless, calculate the IM$_3$ (using Taylor series) of these two differential implementations of LNA in terms of circuit paramters: R_F, R_E etc., device paramters: g_m etc., input amplitude A and the overdrive voltage, V_{GS}-V_T. (Hint: (4.32) of chapter 4 is a good start).

3.4 This question is related to some qualitative understanding of the narrowband LNA described in the chapter.

a. Implement the narrowband LNA in Figure 3.5a using another active device to replace M_1. This active device's small-signal representation is such that in Figure 3.5b, instead of having a voltage controlled current source $(g_m V_{gs})$ we have a voltage controlled voltage source (αV_{gs}). Can Z_{in} still be made resistive?

b. What is the function of C_1 in Figure 3.8b other than providing gain?

3.5 Rederive (3.106) using feedback theory.

3.6 For the narrowband LNA shown in Figure 3.8b, rederive the equivalent input referred voltage noise, this time including noise from L_3, M_2 (you should also consider the frequency effect). The answer should be expressed in terms of circuit paramters: R_s etc., device parameters: g_m, C_{gs}, etc., as well as Q (Quality factor) of the input circuit.

	Wideband	Wideband	Narrowband
	1-stage CS (common source)	CS-cascode	CS; inductor degenerated
Gain			
NF			
Matching			

3.7 In the following chart, compare the different LNA architectures according to their gain, NF, and matching performances.

3.8 Shown in Figure P3.2 is a wideband LNA with 2-stage and using feedback. Assume R_s to be very small, $R_1 \ll r_{oM_1}$, $R_2 \ll r_{oM_2}$ where r_{oM_1}, r_{oM_2} are the device output resistances of M_1, M_2. Further assume that R_1 is approximately the same as R_2, $(W/L)_1$ and $(W/L)_2$ are approximately the same.

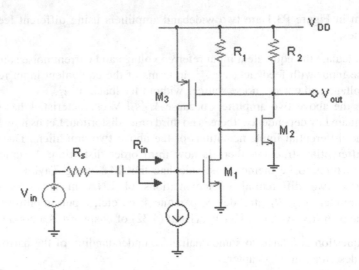

Fig. P3.2

a. Identify the type of feedback
   In terms of device parameters: g_m, r_o, C_{gs}, C_{gd}, C_{db} etc, circuit parameters: R_1, R_2 etc. calculate the following [for b), c) ignore frequency effects]:
b. Forward gain a, and feedback factor f
c. R_{in}
d. $f_{-3\ dB}$, the -3 dB frequency

References

1. G. Gonzalez, *Microelectronic Transistor Amplifiers, Analysis and Design*, 2nd ed., Prentice Hall, 1997.
2. P. Gray and R. Meyer, *Analysis and Design of Analog Integrated Circuits*, 3rd ed., John Wiley & Sons, 1993.
3. T. Lee, *The Design of CMOS Radio Frequency Integrated Circuits*, Cambridge University Press, 1998.
4. H. Fukui, *Low-noise microwave transistors and amplifiers*, IEEE Press, 1991.
5. S. Sheng, "Wideband Spread-Spectrum Digital Communications for Portable Applications," Ph.D. thesis, U.C. Berkeley, 1996.
6. J. Rudell et al., "A 1.9 GHz Wideband IF Double Conversion CMOS Receiver for Cordless Telephone Applications," *IEEE Journal of Solid State Circuits*, Vol. 32, No. 12, Pp. 2071–2088.

Chapter 4
Active Mixer

4.1 Introduction

A mixer, or frequency converter, converts a signal from one frequency (typically ω_{rf}) to another frequency (typically ω_{if}) with a certain gain. This gain is called the conversion gain (G_c) and is defined to be the output signal amplitude at ω_{if} divided by the input signal amplitude at ω_{rf}. The power gain of a mixer, G, which has already been defined in Chapter 2, is related to this G_c. In general, G is simply G_c^2, unless the mixer's switching effect is considered. Ideally, in a mixer, G_c should be large whereas distortion and noise should be low.

If we ignore distortion and noise effects, one useful first-order model of a mixer is that of a variable gain-controlled amplifier in which the local oscillator signal V_{lo} controls the gain while the RF signal V_{rf} is amplified. V_{lo} is a periodic signal oscillating at ω_{lo}. A two-port representation of such an amplifier using a voltage-controlled voltage source is shown inside the rectangular box in Figure 4.1. The voltage-controlling function can be expressed as the product of G_o, a scaling factor, and the term S, where S is a nonlinear function of V_{lo}. Hence S is written as $S[V_{lo}]$. Typically S has the shape as drawn in dotted line in Figure 4.2. To simplify matter we usually perform piecewise linearization so that $S[V_{lo}]$ is now approximated by the solid line in Figure 4.2. Using this approximation, when A_{lo}, the amplitude of V_{lo}, is large (much larger than V_+ and V_-), S is modeled by using the sgn function (called the sign function). Specifically, $S[V_{lo}] = (\text{sgn}[V_{lo}] +1)/2$. When A_{lo} is small (within V_+ and V_-), $S[V_{lo}]$ follows V_{lo} linearly. Mathematically, $S[V_{lo}] = (V_{lo} - V_-)/ (V_+ - V_-)$.

To show mathematically how mixing occurs, let us start by following the expression for V_{if} as given in Figure 4.1:

$$V_{if}(t) = V_{rf}(t)G_o S[V_{lo}(t)] \tag{4.1}$$

B. Leung, *VLSI for Wireless Communication*, DOI 10.1007/978-1-4614-0986-1_4,
© Springer Science+Business Media, LLC 2011

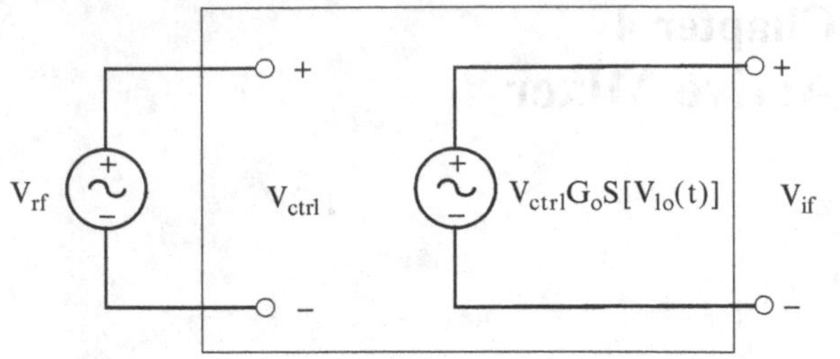

Fig. 4.1 Mixer modeled as
variable gain amplifier

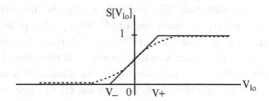

$$V_{if}(t) = V_{ctrl}G_oS[V_{lo}(t)]$$

$$= V_{rf}(t)G_oS[V_{lo}(t)]$$

$$\text{where } S[V_{lo}(t)] = 1 \text{ for } V_{lo} > V_+$$

$$S[V_{lo}(t)] = 0 \text{ for } V_{lo} < V_-$$

$$S[V_{lo}(t)] = \frac{V_{lo} - V_-}{V_+ - V_-} \text{ otherwise}$$

Fig. 4.2 Plot of the
S-function versus the control
voltage V_{lo}.

First let us assume A_{lo} is large (much larger than V_+ and V_-) and, as stated in Figure 4.1, S is modeled by the sgn function of V_{lo}. Further, we express V_{lo} as $V_{lo}(t) = A_{lo} \cos \omega_{lo}t$, then $S[V_{lo}]$ is a periodic function oscillating at ω_{lo}. Consequently, $S[V_{lo}]$ can be expanded in a Fourier series [1] with fundamental frequency ω_{lo}.

Substitute this expansion in (4.1) and V_{if} becomes

$$V_{if}(t) = V_{rf}(t)\left[\frac{G_o}{2} + \frac{2G_o}{\pi}\cos \omega_{lo}(t) - \frac{2G_o}{3\pi}\cos 3\omega_{lo}(t) + \ldots\right] \quad (4.2)$$

Let us examine the second term in (4.2). If V_{rf} is expressed as $V_{rf}(t) = A_{rf} \cos \omega_{rf}t$, this second term becomes

$$2\frac{G_o}{\pi}\cos \omega_{lo}(t)A_{rf}\cos \omega_{rf}(t)$$

Doing trigonometric expansion this term further becomes:

$$\frac{G_o \cdot A_{rf}}{\pi}\left[\cos(\omega_{rf} - \omega_{lo})t + \cos(\omega_{rf} + \omega_{lo})t\right] \quad (4.3)$$

It is now obvious that frequency translation and hence mixing occur. To obtain the mixed-down frequency component, we are interested just in the first term in (4.3). We can do so by following the mixer with a bandpass filter centered at $\omega_{if} = \omega_{rf} - \omega_{lo}$. Then we obtain the mixed down component at ω_{if}, which, for the sake of reusing old symbols, we simply denote as V_{if} again. (From now on V_{if} will be used interchangeably for the two cases: before filtering and after filtering. The proper interpretation should become clear from the context.) Hence V_{if} is given by

$$V_{if} = \frac{G_o \cdot A_{rf}}{\pi} \cos(\omega_{rf} - \omega_{lo})t = \frac{G_o \cdot A_{rf}}{\pi} \cos \omega_{if}t \qquad (4.4)$$

Applying the definition of G_c to (4.4), we get

$$G_c = \frac{G_o}{\pi} \qquad (4.5)$$

Notice from (4.4) that V_{if} is independent of A_{lo}. This is equivalent to saying that the S function in (4.1) behaves as a switching function.

Secondly let us assume A_{lo} is small (within V_+ and V_-). Under this condition, as stated in Figure 4.1, $S(V_{lo}(t)) = \frac{V_{lo}-V_-}{V_+-V_-}$. To simplify our discussion, let us for the present case set $V_+ = 1$, $V_- = 0$. Then we have $S(V_{lo}(t)) = V_{lo}(t) = A_{lo} \cos \omega_{lo}t$. Substituting in (4.1), we have

$$V_{if}(t) = G_0 V_{rf}(t) A_{lo} \cos \omega_{lo}t$$

Doing trignometric expansion V_{if} becomes:

$$\frac{G_0}{2} A_{lo} A_{rf} \left(\cos(\omega_{rf} - \omega_{lo})t + \cos(\omega_{rf} + \omega_{lo})t \right) \qquad (4.6)$$

If we examine the first term, again it is apparent that frequency translation and hence mixing occur. G_c becomes $(G_0/2)A_{lo}$. Notice from (4.6), however, that V_{if} is now dependent on A_{lo}. This is equivalent to saying that the S function in (4.1) behaves like a multiplying function. Therefore the S function's behaviour is different from the case when A_{lo} is large.

The disadvantage of having V_{if} dependent on A_{lo} stems from the fact that V_{lo} is usually generated from some frequency synthesizer (to be described in Chapter 7) and its exact amplitude is hard to control, leading to a G_c that is hard to control. As a result, the rest of the book is focused mostly on mixers that operate with a large A_{lo} (much larger than V_+ and V_-). For these mixers because the accompanying S-function behaves as a switching function, they are called switching mixers. These switching mixers can also be further classified according to whether they perform the switching in the voltage or current domain. In this chapter we concentrate on switching mixers based on current. Because the mixers we consider have G_c larger than one, we call them active mixers.

To summarize, the basis of mixing lies in having a variable gain amplifier (shown in Figure 4.1) whose gain can be controlled by an external periodic signal. Analyzing a mixer using this representation involves specifying the amplifier gain or transfer term with respect to the input signal and then expanding the gain in a Fourier series. This vewpoint highlights the periodic time-varying nature of a mixer. Taking this viewpoint one step further, it is precisely this periodic time-varying nature (rather than the nonlinear nature, as explained in some books) of a mixer that leads to mixing.

Finally, let us explore the type of electronic devices that will implement this variable gain amplifier. One of the most common electronic devices that can implement a variable gain function consists of using the transconductance of a MOS/BJT transistor biased in the saturation/active region. This transconductance varies as a function of gate/base bias. Transconductance of a source-coupled pair (SCP)/emitter coupled pair (ECP) is another choice. One other choice consists of the on resistance (R_{on}) of a MOS transistor biased in the triode region.

4.2 Balancing

In this section we will use the concept of balancing to classify active mixers.

4.2.1 Unbalanced Mixer

The simplest active mixer is an unbalanced mixer, as shown in Figure 4.3, where the IF signal (I_{if}) is taken from one branch only. This I_{if} signal flows into a resistor R_L (not shown) and develops V_{if}. V_{if} can be obtained by rewriting (4.2):

$$V_{if}(t) = A_{rf} \cos \omega_{rf} t \times G_o \left(\frac{1}{2} + \sum_{n=1}^{\infty} \frac{\sin \frac{n\pi}{2}}{\frac{n\pi}{2}} \cos n\omega_o t \right) \qquad (4.7)$$

From (4.7) it is apparent that the DC component of V_{lo} (the 1/2 term inside the bracket) is multiplied to V_{rf}, producing a scaled product at V_{if}. Hence frequency components at the RF frequency appear at the output. This phenomenon is called RF feedthrough and is undesirable.

4.2.2 Single Balanced Mixer

An improvement over the unbalanced mixer is the single balanced mixer, as shown in Figure 4.4. Here the IF signal is taken from both branches (I_{if}^{+} and I_{if}^{-}).

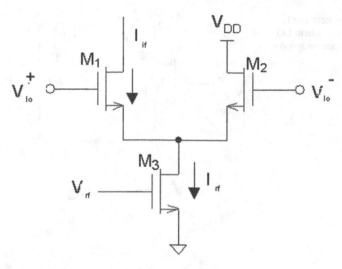

Fig. 4.3 Unbalanced mixer

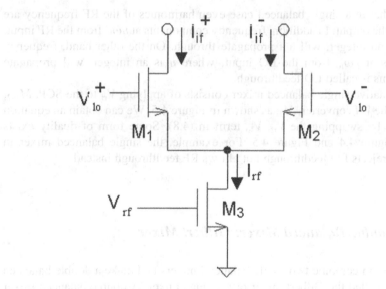

Fig. 4.4 Single balanced mixer

Each I_{if}^{+} and I_{if}^{-} flows into a separate resistor, R_L^{+} and R_L^{-} (not shown) and develops V_{if}^{+} and V_{if}^{-}. V_{if} is taken as $V_{if}^{+} - V_{if}^{-}$. V_{if} can now be written as

$$V_{if}(t) = A_{rf} \cdot \cos \omega_{rf} t \cdot 2G_o \sum_{n=1}^{\infty} \frac{\sin \frac{n\pi}{2}}{\frac{n\pi}{2}} \cos n\omega_o t. \qquad (4.8)$$

Because signals are taken from both branches, RF feedthrough from both branches cancel one another. In addition to the absence of RF feedthrough, it can

Fig. 4.5 Alternate single balanced mixer, where LO and RF ports are swapped

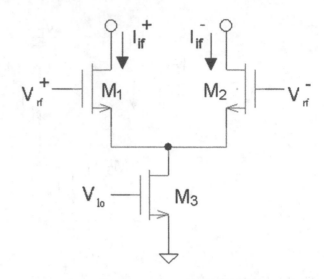

be shown that in a single balanced case even harmonics of the RF frequency are missing in the output. In addition, frequency components at $n\omega_{lo}$ from the RF input, where n is an integer, will not propagate through. On the other hand, frequency components at $n\omega_{lo}$ from the LO input, where n is an integer, will propagate through. This is called LO feedthrough.

An alternative single balanced mixer consists of applying V_{rf} to the SCP, M_{1-2}, and V_{lo} to the V-I converter, M_3, as shown in Figure 4.5. We can obtain an equation for V_{if} just by swapping the V_{rf}, V_{lo} term in (4.8). Some form of duality exists between Figure 4.4 and Figure 4.5. For example, the single balanced mixer in Figure 4.5 rejects LO feedthrough but allows RF feedthrough instead.

4.2.3   Double Balanced Mixer: Gilbert Mixer

Finally, we can combine two single balanced mixers and make a double balanced mixer, also called the Gilbert mixer or the quad mixer. A double balanced mixer rejects both RF and LO feedthroughs. The double balanced mixer is the most commonly used mixer, so we will focus on it for the rest of the chapter.

The absence of feedthrough from RF to IF ports and LO to IF ports in a double balanced mixer occurs only under ideal conditions. In practice, a finite amount of feedthrough remains. They come mainly as a result of mismatches. Finally, apart from feedthroughs between RF to IF and LO to IF ports, feedthrough between the RF and LO ports is also undesirable as it introduces reradiation. This type of feedthrough is present in a double balanced mixer.

4.3   Qualitative Description of the Gilbert Mixer

Qualitatively, a Gilbert mixer consists of a *V-I* converter, a current switching block, and an *I-V* converter. Figure 4.6 shows an implementation of the double balanced mixer using three SCP's. Two SCPs (M_{3-4} and M_{5-6}) are used to do the switching and one SCP (M_{1-2}) is used for *V-I* conversion. The two switches operate in opposite polarity since the V_{lo} signals are applied in opposite polarity. The *V-I* converter generates two current inputs to the tail nodes of the switches, which are also 180° out of phase. Notice, as discussed in Chapter 2, if used in a heterodyne architecture this is preceded by a BPF for anti-imaging. Hence a matching network is needed, which we have not shown.

To further describe this circuit, let us follow a few labeling conventions:

1. Input signals have "rf" subscripts, control signals have "lo" subscripts, and output signals have "if" subscripts.
2. For the SCP (M_1 - M_2) and the output currents, along the main line of symmetry (passing vertically through the tail node of M_1 - M_2) of the differential circuits, the branch current on the left-hand side has "+" superscript and conversely the branch current on the right-hand side has "−" superscript.

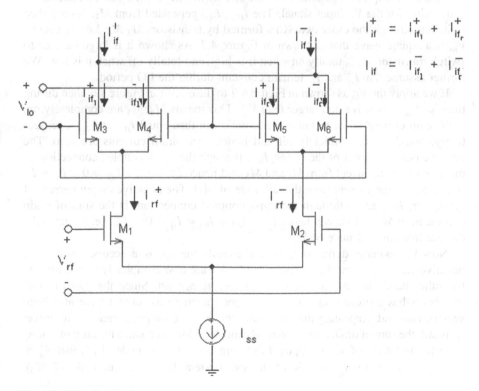

$$I_{if}^+ = I_{if_l}^+ + I_{if_r}^+$$

$$I_{if}^- = I_{if_l}^- + I_{if_r}^-$$

Fig. 4.6 Gilbert/quad mixer

Fig. 4.7 A square wave V_{lo}
for the Gilbert mixer.
The two phases are shown

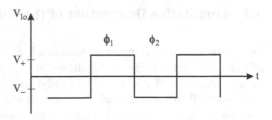

3. For the second and third SCPs (M_3-M_4 and M_5-M_6), there are two levels of
   subscript and superscript distinctions:

   3.1 Along the main line of symmetry (passing vertically between M_4 and M_5) of
       the differential circuits, the branch current on the left-hand side has subscript
       labeled as "l" and conversely the branch current on the right-hand side has
       subscript labeled as "r".
   3.2 Along the secondary line of symmetry (passing vertically through the tail
       node of M_3-M_4 and M_5-M_6) of the differential circuits, the branch current
       on the left-hand side has "+" superscript and conversely the branch current
       on the right-hand side has "−" superscript.

Next let us explain the operation of the circuit. M_1, M_2 form a SCP that does a V-I
conversion for the V_{rf} input signal. The I_{rf}^+, I_{rf}^- generated from M_1, M_2 are then
switched through the other two SCPs formed by transistors M_3-M_6. Let us assume
V_{lo} is a square wave that is shown in Figure 4.7. As shown it is large enough to
switch transistor M_{3-6} totally on when it is high and totally off when it is low. We
further assume that I_{rf}^+ and I_{rf}^- remain constant during the LO period.

If we apply the V_{lo} as shown in Figure 4.7 to the circuit in Figure 4.6, then during
time ϕ_1 V_{lo} is positive and larger than V_+. This means M_4, M_5 are completely on,
with drain current equal to tail current, which in turn equal I_{rf}^+ and I_{rf}^-, respectively. M_3, M_6 are completely off and hence their drain currents are zero. The
positive output current of the mixer, I_{if}^+, through the cross-coupled connection, is
the sum of drain current from M_3 and M_5 and hence $I_{if}^+ = I_{d3} + I_{d5} = 0 + I_{rf}^- = I_{rf}^-$
(i.e, it equals the current from the tail node of M_5). The negative output current of
the mixer, I_{if}^-, again through the cross-coupled connection, is the sum of drain
current from M_4 and M_6 and hence $I_{if}^- = I_{d4} + I_{d6} = I_{rf}^+ + 0 = I_{rf}^+$ (i.e, it equals the
current from the tail node of M_4).

Now V_{lo} switches during time ϕ_2 and exactly the opposite occurs. Since V_{lo} is
negative and is less than V_-, this means M_4, M_5 are now completely turned off. On
the other hand, M_3, M_6 are now completely turned on. Since the pair of "on"
transistors has switched, the current that got routed to the output have also been
swapped around. Repeating the procedure, the positive output current of the mixer
I_{if}^+ is still the sum of drain current from M_3 and M_5. However, since the current is now
from the tail node of M_3 as opposed to being from the tail node of M_5 then $I_{if}^+ =
I_{d3} + I_{d5} = I_{rf}^+ + 0 = I_{rf}^+$ (i.e. it equals the current from the tail node of M_3).

Fig. 4.8 Gilbert mixer output
current as a function of time

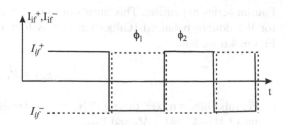

Similarly, the negative output current of the mixer I_{if}^- is still the sum of drain current
from M_4 and M_6, but again the current is now from the tail node of M_6 as opposed to
being from the tail node of M_4 and hence $I_{if}^- = I_{d4} + I_{d6} = 0 + I_{rf}^- = I_{rf}^-$.

In summary output currents have swapped polarity in synchronization with the
controlling V_{lo} and mixing has occurred. This is shown in Figure 4.8 where we have
assumed that I_{rf}^+ and I_{rf}^- remain constant during the LO period.

Assuming that I_{rf}^+ and I_{rf}^- amplitudes are equal, then the I_{if}^+, I_{if}^- waveforms as
shown in Figure 4.8 will be completely symmetrical.

We already mentioned that, when compared to the single-ended mixer, this
Gilbert mixer has the advantage that feedthrough from V_{lo}, V_{rf} will not get
propagated to the output.

In addition, this mixer's G_c is doubled compared with the single balanced case
and is four times that of the unbalanced case.

We next go through the design issues of the active mixer. From Chapter 2 we
have identified the following key parameters: power gain G, distortion (IIP3) and
noise (NF). We will be developing formulas of these key parameters for the active
mixer (except that we will determine G_c instead of G, as G_c is more general and
considers the mixer's switching effect). The idea is to translate these key
parameters into circuit component values such as W/L ratio and bias current/
voltages. Because a mixer is a periodic time-varying circuit, these parameters are
difficult to derive mathematically in one step. Hence we will first derive them by
assuming that V_{lo} is constant (not switching). We then try to derive them when V_{lo} is
switching. Furthermore in the V_{lo} switching case, V_{lo} is assumed to have zero rise/
fall time. Relaxing to a nonzero rise/fall time makes the closed-form solution
practically impossible to get, and we have to resort to simulations.

4.4   Conversion Gain

Refer to Figure 4.3 for the unbalanced mixer. Let us assume that V_{lo} is not
switching, and we have $G_c = g_{m3} \times R_L$ where R_L (not shown) is the resistor hanging
from drain of M_1 to V_{DD}. For the V_{lo} switching case let us assume that poles/zeros
frequencies from the mixer are much higher than f_{lo}. We further assume that V_{lo} is a
square wave with a 50% duty cycle. G_c can be obtained from the coefficient of the

Fourier series expansion. This turns out to be $1/\pi(g_{m3} \times R_L)$. The conversion gain for the double balanced (Gilbert) mixer is four times that. Hence, referring to Figure 4.6 we have

$$G_c = 4/\pi(g_{m1} \times R_L) \tag{4.9}$$

For this Gilbert mixer, there will be two load resistors R_L connecting between the drains of M_3-M_5, M_4-M_6 and V_{DD}.

Numerical example 4.1.
Referring to Figure 4.6 let us assume that M_1-M_6 all have a $\frac{W}{L}$ of $^{50um}/_{0.6um}$ and we make $V_{GS_1} - V_t = 0.387V$, assume $k' = 75uA/V^2$. Then $k = 6250uA/V^2$ $g_{m_1} = k(V_{GS} - V_t) = 2.4m\Omega^{-1}$. Suppose we want to design for a conversion gain of 10 dB. Substituting this in (4.9) we get

$$G_c = \frac{4}{\pi}g_{m_1}R_L = \frac{4}{\pi} \times 2.4m\Omega^{-1} \times R_L = 3.16 \tag{4.10}$$

Solving R_L=1034 Ω.

4.5   Distortion, Low Frequency Case: Analysis of Gilbert Mixer

If we no longer ignore distortion and noise effects in a mixer, the first-order model as shown in Figure 4.1 is no longer adequate. The distortion and noise effects can be included by specifying the $\mathrm{IIP_3}$ and NF of a mixer, as was done in Chapter 2. In this and the next section we are interested in taking the $\mathrm{IIP_3}$ specified for a mixer in Chapter 2 and find circuit parameters such as W/L, I_{bias} that will satisfy this specification. In section 4.7 we will repeat this process for the NF.

To relate $\mathrm{IIP_3}$ to circuit parameters we need to find the proper design equations. For the bulk of sections 4.5, 4.6 we assume that V_{lo} is not switching and so the mixer is represented by a nonlinear time invariant (NLTI) system. In this section we look at the distortion behaviour of this NLTI system at low frequency while the corresponding high frequency distortion behaviour is reserved for the next section. For any NLTI system, such as a mixer, operating at low frequency means we assume that any capacitors or inductors can be neglected. Returning to the Gilbert mixer in Figure 4.6 since we neglect capacitive effect, then individual transistors in the quad pair M_{3-6} are either completely turned on or off. Accordingly, the output current I_{if} equals either I_{rf} or $-I_{\mathrm{rf}}$. Hence the switching SCPs (M_3-M_6) in Figure 4.6 do not contribute distortion.

Therefore distortion comes primarily from the bottom SCP, which does the $V-I$ conversion. Furthermore, we assume that this distortion is dominated by the nonlinear square law I-V characteristics of the MOS transistors biased in saturation.

Referring to the SCP M_{1-2} in Figure 4.6 we can write, for transistor M_1

$$I_{rf}^+ = \frac{k}{2}(V_{gs_1} - V_t)^2 \tag{4.11}$$

Next the loop equation gives us:

$$V_{rf}^+ - V_{gs_1} = V_{rf}^- - V_{gs_2} \tag{4.12}$$

Rearranging (4.12) gives us

$$V_{gs_1} = V_{rf}^+ - V_{rf}^- + V_{gs_2} = V_{rf} + V_{gs_2} \tag{4.13}$$

Substitute (4.13) into (4.11)

$$I_{rf}^+ = \frac{k}{2}\left(v_{rf} + (v_{gs_2} - v_t)\right)^2 \tag{4.14}$$

Repeating the procedure for transistor M_2

$$I_{rf}^- = \frac{k}{2}(V_{gs_2} - V_t)^2$$

which means

$$V_{gs_2} - V_t = \sqrt{\frac{2I_{rf}^-}{k}} \tag{4.15}$$

Substituting (4.15) into (4.14),

$$I_{rf}^+ = \frac{k}{2}\left(v_{rf} + \sqrt{\frac{2\left(I_{ss} - I_{rf}^+\right)}{k'}}\right)^2 \tag{4.16}$$

Let us normalize by defining the normalized I_{rf}^+, I_{ssn} as

$$I_{rf_n}^+ = \frac{2\,I_{rf}^+}{k}$$

$$I_{ss_n} = \frac{2I_{ss}}{k}$$

Substituting the normalized variables into (4.16), we have

$$I_{f_n}^+ = \left(V_{rf} + \sqrt{I_{ss_n} - I_{rf_n}^+}\right)^2 \tag{4.17}$$

After some algebraic manipulation

$$\sqrt{I_{rf_n}^+} - V_{rf} = \sqrt{I_{SS_n} - I_{rf_n}^+}$$

$$\text{or } V_{rf} = \sqrt{I_{rf_n}^+} - \sqrt{I_{SS_n} - I_{rf_n}^+}$$

$$= \sqrt{i_{rf_n}^+ + \frac{I_{SS_n}}{2}} - \sqrt{I_{SS_n} - \left(i_{rf_n}^+ + \frac{I_{SS_n}}{2}\right)}$$

$$= \sqrt{i_{rf_n}^+ + \frac{I_{SS_n}}{2}} - \sqrt{\frac{I_{SS_n}}{2} - i_{rf_n}^+} \qquad (4.18)$$

Here, i_{rfn}^+ is the ac part of I_{rfn}^+. Notice that (4.18) represents an odd function of V_{rf} around $\frac{I_{SSn}}{2}$.

Factoring out the I_{SS_n} term we have:

$$V_{rf} = \sqrt{\frac{I_{SS_n}}{2}} \left(\sqrt{1 + \frac{2i_{rf_n}^+}{I_{SS_n}}} - \sqrt{1 - \frac{2i_{rf_n}^+}{I_{SS_n}}} \right) \qquad (4.19)$$

(4.19) gives V_{rf} in terms of i_{rfn}^+ Since V_{rf} is input and $i_{rf_n}^+$ is output, we would instead want to express $i_{rf_n}^+$ in terms of V_{rf}. Since there is no capacitive effect, each $i_{rf_n}^+$ term can be expanded as a power series (also denoted as Taylor series; the two terms will be used interchangeably) in powers of V_{rf}:

$$i_{rf_n}^+ = a_1 V_{rf} + a_2 V_{rf}^2 + a_3 V_{rf}^3 + \dots \qquad (4.20)$$

Here $a_1, a_2 \dots$ are coefficients.

Unfortunately, since it is not easy to write $i_{rf_n}^+$ explicitly in terms of V_{rf} [as evidenced in (4.19)], we would get around this difficulty by doing two expansions:

First we expand the two square root terms inside the bracket in (4.19) around $\frac{2i_{RF_n}^+}{I_{SS_n}}$:

$$V_{rf} = \sqrt{\frac{I_{SS_n}}{2}} \left[\begin{array}{l} \left(1 + \frac{1}{2}\left(\frac{2i_{rf_n}^+}{I_{SS_n}}\right) - \frac{1}{8}\left(\frac{2i_{rfn}^+}{I_{SS_n}}\right)^2 + \frac{1}{16}\left(\frac{2i_{rf_n}^+}{I_{SS_n}}\right)^3 + \dots \right) \\[2mm] - \left(1 - \frac{1}{2}\left(\frac{2i_{rf_n}}{I_{SS_n}}\right) - \frac{1}{8}\left(\frac{2i_{rf_n}^+}{I_{SS_n}}\right)^2 - \frac{1}{16}\left(\frac{2i_{rf_n}^+}{I_{SS_n}}\right)^3 + \dots \right) \end{array} \right]$$

$$= \sqrt{\frac{I_{SS_n}}{2}} \left[\frac{2i_{rf_n}^+}{I_{SS_n}} + \frac{1}{8}\left(\frac{2i_{rf_n}^+}{I_{SS_n}}\right)^3 + \dots \right] \qquad (4.21)$$

Secondly we would expand each of the $i_{rf_n}^+$ term in (4.21) using (4.20).

For simplicity we write down only the first three terms when expanding (4.20). Upon substituting in (4.21) we have:

$$V_{rf} = \sqrt{\frac{I_{SS_n}}{2}} \left[\frac{2}{I_{SS_n}} \left(a_1 V_{rf} + a_2 V_{rf}^2 + a_3 V_{rf}^3 + \ldots \right) + \frac{1}{8} \left(\frac{2}{I_{SS_n}} \right)^3 \left(a_1 V_{rf} + a_2 V_{rf}^2 + \ldots \right)^3 + \ldots \right]$$

$$(4.22)$$

Finally we can solve for the coefficients $a_1, a_2, a_3 \ldots$ by equating the coefficients of V_{rf}, V_{rf}^2, $\ldots$ on both sides of (4.22). For the V_{rf} term

$$1 = \sqrt{\frac{I_{SS_n}}{2}} \left(\frac{2}{I_{SS_n}} a_1 \right) \qquad \therefore a_1 = \sqrt{\frac{I_{SS_n}}{2}} \qquad (4.23)$$

For the V_{rf}^2 term:

$$0 = \sqrt{\frac{I_{SS_n}}{2}} \left[\frac{2}{I_{SS_n} a_2} \right] \qquad \therefore a_2 = 0 \qquad (4.24)$$

For the V_{rf}^3 term

$$0 = \sqrt{\frac{I_{SS_n}}{2}} \left(\frac{2}{I_{SS_n}} a_3 + \frac{1}{8} \left(\frac{2}{I_{SS_n}} \right)^3 a_1^3 \right) \qquad \therefore a_3 = -\frac{1}{8} \left(\frac{2}{I_{SS_n}} \right)^2 a_1^3 \quad (4.25)$$

Let us apply the definition of HD_3 as given in (2.28) to the present case

$$HD_3 = \frac{I_{rf}^+ |3rd - order \cdot term}{I_{rf}^+ |fundamental} = \frac{1}{4} \frac{a_3}{a_1} A_{rf}^2 \qquad (4.26)$$

Substituting (4.25) into (4.26), we have

$$HD_3 = \frac{1}{4} \left| -\frac{1}{8} \left(\frac{2}{I_{SS_n}} \right)^2 \right| a_1^2 \cdot A_{rf}^2 \qquad (4.27)$$

Substituting (4.23) into (4.27) we have:

$$HD_3 = \frac{1}{4} \left| -\frac{1}{8} \left(\frac{2}{I_{SS_n}} \right)^2 \right| \frac{I_{SS_n}}{2} A_{rf}^2 = \frac{1}{16} \frac{1}{I_{SS_n}} A_{rf}^2 = \frac{1}{16} \frac{k}{2I_{SS}} A_{rf}^2 = \frac{1}{32} \frac{k}{I_{SS}} A_{rf}^2 \quad (4.28)$$

From (2.35) $IM_3 = 3HD_3$ and we can substitute (4.28) into this to obtain IM_3.

In summary the distortion at the output of the mixer will be

$$HD_3 = \frac{1}{32}\frac{k}{I_{SS}}A_{rf}^2 = \frac{1}{32}\frac{\mu C_{ox}\frac{W_1}{L_1}}{I_{SS}}A_{rf}^2 \qquad (4.29a)$$

$$IM_3 = \frac{3}{32}\frac{k}{I_{SS}}A_{rf}^2 = \frac{3}{32}\frac{\mu C_{ox}\frac{W_1}{L_1}}{I_{SS}}A_{rf}^2 \qquad (4.29b)$$

Notice that for IM_3, normally we are interested in the I_{D3} generated in the desired signal frequency from the two adjacent channel interferences. These interferences are denoted as $v_{\text{interference}}$ and have amplitudes denoted as $A_{\text{interference}}$. To quantify the IM_3 in this case, we rewrite (4.29b) as follows:

$$IM_3 = \frac{3}{32}\frac{k}{I_{SS}}A_{\text{interence}}{}^2 = \frac{3}{32}\frac{\mu C_{ox}\frac{W_1}{L_1}}{I_{SS}}A_{\text{interence}}{}^2 \qquad (4.30)$$

From now on, IM_3 will be expressed in terms of $A_{\text{interference}}$. Referring to (4.29a) and (4.30), we observe that as I_{SS} goes up, distortion goes down. However, as the $\frac{W}{L}$ ratio goes up, distortion goes up. Therefore to design a low distortion mixer, one needs to burn more power (undesirable) and keeps the $\frac{W}{L}$ ratio (or size) down (there is a limit). Of course, the amplitude of the input signal, A_{rf}, or that of the interference, $A_{\text{interference}}$, should also be kept down, but their values are dictated by the overall receiver front end design.

If A_{rf} is small, then the current flowing through M_1 can be assumed to be equal to that through M_2, or half of I_{SS}. Then we can write (4.29a) as

$$HD_3 = \frac{1}{32}\frac{k}{2I_{D_1}}A_{rf}^2 = \frac{1}{32}\frac{k}{2\frac{k}{2}(V_{GS_1}-V_t)^2}A_{rf}^2 = \frac{1}{32}\frac{1}{(V_{GS_1}-V_t)^2}A_{rf}^2 \qquad (4.31)$$

If $A_{\text{interference}}$ is small, similar considerations apply to (4.30). Then (4.29a) and (4.30) can be rewritten as

$$HD_3 = \frac{A_{rf}^2}{32}\frac{1}{(V_{GS_1}-V_t)^2}$$

$$IM_3 = \frac{3A_{\text{interference}}^2}{32(V_{GS_1}-V_t)^2} \qquad (4.32)$$

We can characterize this distortion using A_{IP3}, amplitude of v_{RF} or $v_{\text{interference}}$ at the third order intercept point, as well. We start from (2.38), $IIP_3|_{dBm} = P_i|_{dBm} - \frac{IM_3|_{db}}{2}$.

As shown in Problem 4.4 this equation will lead to the equation $A_{IP_3}^2 = \frac{A_{interference}^2}{IM_3}$. Substituting IM_3 obtained in (4.32) in this equation, we have

$$A_{IP_3}^2 = \frac{A_{interference}^2}{IM_3} = \frac{32(V_{GS_1} - V_t)^2 A_{interference}^2}{3A_{interference}^2} = \frac{32(V_{GS_1} - V_t)^2}{3}$$

or

$$A_{IP_3} = 4\sqrt{\frac{2}{3}}(V_{GS_1} - V_t) \tag{4.33}$$

Finally, we look at the case when V_{lo} is switching, but with zero rise and fall time (i.e, an ideal square wave). Even though we no longer have a time-invariant system since V_{lo} is an ideal square wave, the situation can be handled by modifying the preceding results through multiplying the output by the Fourier series representation of the square wave. The effect of switching is simply shifting the frequency of the fundamental and the third-order products by ω_{lo}. The amplitudes are both reduced by the same amount, $1/\pi$, which means that HD_3, IM_3 remain unchanged.

Numerical example 4.2.
Assume that a Gilbert mixer operates under the following condition:

$$V_{GS_1} - V_t = 0.387V \quad A_{rf} = A_{interference} = 0.316V \quad \text{or } 0\,dBm$$

Assume that the LO is not switching. Find the mixer's distortion behavior: HD_3, IM_3, IIP_3.
We assume that the Gilbert mixer's distortion is dominated by the V-I converter. Substituting the foregoing values in (4.32) we have

$$HD_3 = -33.1dB$$

$$IM_3 = -23.5dB$$

From (2.38)

$$IIP_3|_{dBm} = P_i|_{dBm} - \frac{IM_3|_{db}}{2} = 0 + \frac{23.5}{2} = 12.5\ dBm$$

Numerical example 4.3.
From Table 2.1, the mixer is assigned an IIP_3 of −10 dbm. In this example we want to design this mixer with some safety margin. Let us arbitrarily set the specifications to be

$$IIP_3 = 5\,dBm$$

$$Power = 10\,mW$$

$$V_{\text{dd}} = 3.3\text{V}$$

$$\text{k' } = 100 \ u\text{A/V}^2$$

Assume that V_{lo} is not switching. Find $\left(\frac{W}{L}\right)_1$.

IIP$_3$ = 5dBm means that $5 \ dbm = 10 \ \log\frac{A_{IP_3}^2}{2\times 50}$. Solving we have

$A_{IP_3} = \sqrt{2 \times 50 \times 10^{\frac{A_{IP_3}|_{dbmV}}{10}}} = \sqrt{2 \times 50 \times 10^{\frac{5 \ dBmV}{10}}} = 0.56 \ V$. Substituting into
(4.33) we have $\left(V_{gs_1} - V_t\right) = 0.17 \ V$. Now with P $= 10\text{mW}$, $V_{\text{dd}} = 3.3\text{V}$, we
have I $= 3\text{mA}$. Substituting half of this value of current into the square law current
equation, with a given k' $= 100 \ u\text{A/V}^2$, we get $\left(\frac{W}{L}\right)_1 = 1038$.

4.6 Distortion, High-Frequency Case

In this section we look at the distortion behaviour of a mixer at high frequency.
To reiterate, for the bulk of this section we assume that V_{lo} is not switching and so
the mixer is a NLTI system. Returning to the Gilbert mixer in Figure 4.6, at high
enough frequency (e.g, 1.9 GHz), the assumption that there is no memory effect
due to capacitors (such as junction capacitors C_{sb}, C_{db}) and inductors present in
the mixer is no longer correct. These capacitors can be nonlinear capacitors,
which of course introduce their own nonlinear effect. But even if they are linear
capacitors (such as C_{gd} C_{gs}), their presence renders the calculation of distortion
due to nonlinear, memoryless components (such as that due to the square law I-V
characteristics) unamenable to simple Taylor series analysis, as described in
section 4.5. Specifically, because of memory effect we will see that (4.20),
repeated here,

$$i_{rf_n}^+ = a_1 V_{rf} + a_2 V_{rf}^2 + a_3 V_{rf}^3 + \ldots$$

is no longer correct.

To see why that is so, let us change our focus from a specific nonlinear
circuit with memory, the Gilbert mixer, to some general discussion of nonlinear
system with memory. We start off our discussion by reverting, for the time being,
to a linear memoryless system, with input x and output y. Using the same symbol a_1,
which denotes linear gain in (4.20), to denote the gain of this linear system,
we have

$$y = a_1 x \tag{4.34}$$

From (4.34) we can see that y, the output at a particular instant, depends on x, the
input at that particular instant only, and not on inputs at any other instant. From a
circuit point of view, this will only be true if the circuit contains no memory

(a resistor will be an example). As soon as the circuit has memory (e.g, a capacitor), then for this simple linear system the output at a particular instant is dependent on all the past input values. This output is then obtained by summing all the effects of these past inputs. It is easier to see this in the discrete time domain, and so y, incorporating memory effect, is first written in the discrete time domain:

$$y(n) = \sum_{\tau=-\infty}^{n.} h(\tau) \cdot x(n - \tau) \qquad (4.35)$$

where n is the time index, $h(\tau)$ is the weight that signifies how each past input sample affects the present output sample and $h(\tau)$ is, of course, simply your familiar impulse response. The sum in (4.35) is the familiar convolution sum. Next we write y, incorporating memory effect, in the continuous time domain. The sum in (4.35) becomes an integral, the familiar convolution integral, and (4.35) becomes

$$y = \int h(\tau)x(t - \tau) \, d\tau \qquad (4.36)$$

where $h(\tau)$ is the impulse response. To summarize, because of memory effect, we use (4.36), rather than (4.34), as the input-output relationship of a linear system.

The incorporation of memory effect can be extended to a nonlinear system. As an example, we start off with a system such as the one described in (4.20), which describes the input-output relationship of a nonlinear system that is memoryless. To incorporate memory in this nonlinear system, every term in (4.20) should be replaced by an integral, in much the same way that (4.34) becomes (4.36). Consequently (4.20) is no longer correct.

4.6.1 Volterra Series

We will now develop a general theory that can allow us to calculate high-frequency distortion for a NLTI system, including memory effect. This is the theory of Volterra series. When applied to circuits it shows that the high frequency effect can degrade the distortion performance easily by close to 100% more than predicted using low frequency analysis (see Numerical example 4.4). Also when applied to fully differential (balanced) circuits with no mismatch, it reveals the surprising result that there can be second harmonic distortion (HD$_2$) as high as -32db (see Numerical example 4.4), when the low frequency analysis predicts the HD$_2$ should be zero. This HD$_2$ would have come from a circuit with an equivalent mismatch as high as 2.5% and can be a major concern because it means RF feedthrough will still be present in a balanced mixer with zero mismatch. In addition to providing tools necessary for high frequency low distortion analysis/design the Volterra series also helps in verifying this design. This is because a significant component of the

Fig. 4.9 Block diagram of a
discrete time LTI system

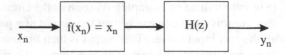

verification involves simulations of the circuit. To properly interpret results produced by the simulator the analysis tool used in the simulator should be well understood. The use of Volterra series for high frequency distortion in mixers has been adopted in many popular simulators (e.g. SpectreRF [6]). This should help motivate the interest of the readers in the following sections which, on first glance, may appear a little mathematical.

To further motivate the readers regarding the usefulness of Volterra series we can take comfort in the fact that the theory can be applied to high frequency distortion analysis of any NLTI system. In this book this includes mixers in chapter 4, 5, LNA in chapter 3, power amplifier in chapter 9 and BPF in chapter 2.

4.6.1.1   Introduction

We introduce the concept of Volterra series by going through systems of increasing complexity.

Case 1a: Linear, discrete

We start off by going back to the simplest system: a linear, discrete time system as shown in Figure 4.9.

1. First assume $h(\tau)=1$ for all τ. Then we have:

$$y_n = x_n + x_{n-1} + \dots \tag{4.37}$$

2. In general $h(\tau)$ is a function of τ, the time index and the output is a convolution sum:

$$y(n) = \sum_{\tau=-\infty}^{n} h(\tau) \cdot x(n - \tau) \tag{4.38}$$

Case 1b: Linear, continuous

We now evolve to the continuous time system, where the sum changes to an integral.

$$y(t) = \int h(\tau)x(t - \tau) \, d\tau \tag{4.39}$$

Fig. 4.10 Block diagram of a discrete time, NLTI system

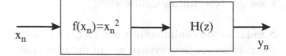

Case 2a: Nonlinear, Discrete

Next we introduce a mild nonlinearity, a bilinear nonlinearity. We again start off in the discrete time case. Hence $f(x_n) = x_n^2$. The resulting system is shown in Figure 4.10:

1. If we assume $h(\tau_1, \tau_2) = 1$ for all (τ_1, τ_2), then

$$y_n = x_n.x_n.$$
$$+x_n. \cdot x_{n-1} + x_{n-1} \cdot x_{n-1} + \ldots$$
$$+ x_n. \cdot x_{n-2} + x_{n-1} \cdot x_{n-2} + x_{n-2} \cdot x_{n-2} + \ldots$$

$$+x_n. \cdot x_0 + x_{n-1} \cdot x_0 + x_{n-2} \cdot x_0 + \ldots + x_0 \cdot x_0$$
$$= \sum_{j=0}^{n} x_n x_j + \sum_{j=0}^{n-1} x_{n-1} x_j + \ldots$$
$$= \sum \sum x_i x_j \tag{4.40}$$

Notice that this simple bilinear nonlinearity has changed the output from being a simple single sum as described in (4.37) to that of a double sum as shown in (4.40).

2. In general, $h(\tau_1, \tau_2)$ is not 1 but is a function of time indexes (τ_1, τ_2), as in Case 1a (2). Notice this has made the simple single sum shown in (4.37) become a weighted single sum as shown in (4.38). Likewise, (4.40) will become a weighted double sum, which is the convolution sum of this bilinear discrete system and is given as

$$y_n = \sum \sum h(\tau_1, \tau_2) x(n - \tau_1) x(n - \tau_2) \tag{4.41}$$

Because of the double summation, h is now a function of two time indexes: τ_1, τ_2.

Case 2b: Non-linear, Continuous

Similar to the evolution of Case 1a (2) to Case 1b, Case 2a (2) can also evolve to the continuous time system where the sum changes to an integral. Equation (4.41) becomes

$$y(t) = \int \int h(\tau_1, \tau_2) x(t - \tau_1) x(t - \tau_2) d\tau_1 d\tau_2 \qquad (4.42)$$

It just takes a little bit more imagination to figure out that in general for an nth-order nonlinearity described by $y = x^n$, (notice this n describes the order of the nonlinearity and is not to be confused with the symbol n used to describe the time index in a discrete time system. We choose to reuse this symbol so as to save using new symbol, and also because from now on we will be dealing exclusively in continuous time systems and so no confusion should arise.) equation (4.42) generalizes to

$$y(t) = \int_{-\infty}^{\infty} h_1(\tau_1) x(t - \tau_1) \, d\tau_1 + \dots$$

$$+ \int_{-\infty}^{\infty} \dots \int_{-\infty}^{\infty} h_n(\tau_1, \tau_2 \dots \tau_n) x(t - \tau_1) x(t - \tau_2) \dots x(t - \tau_n) d\tau_1 \dots d\tau_n$$

$$(4.43)$$

in which for n = 1, 2...

$$h_n(\tau_1, \dots \tau_n) = 0 \text{ for any } \tau_j < 0, j = 1, \dots n$$

Notice that the first term is similar to the convolution integral for a linear system. Equation (4.43) is the Volterra series expansion of an nth order nonlinear system.

4.6.1.2 Comparisons with Taylor Series

At this point it may be illustrative to compare the expansion form in Taylor series and in Volterra series. Let us just repeat the Taylor series expansion from (4.20), but using y and x as the output and input variables instead.

$$y = a_1 x + a_2 x^2 + \dots$$

We can interpret this form of Taylor series expansion by considering $x, x^2 \dots$ as the basis function. Similarly we can repeat the Volterra series expansion from (4.43) above

$$y(t) = \int h_1(\tau_1)x(t-\tau_1)d\tau_1 + \ldots \int \ldots \int h_n(\tau_1 \ldots \tau_n)x(t-\tau_1) \ldots x(t-\tau_1) \ldots x(t-\tau_n)d\tau_1 \ldots d\tau_n$$

$$= H_1[x(t)] + H_2[x(t)] + \ldots H_n[x(t)]$$

$$(4.44)$$

Drawing the analogy with the interpretation of Taylor series expansion, we can interpret the different order of integrals $\int, \int\int, \int\int\int \ldots$ (single integrals, double integrals, etc.) as the basis functions again. They can be viewed as operators so that $\int, \int\int, \int\int\int \ldots$ are replaced by operators $H_1, H_2, H_3 \ldots$ and so on.

4.6.1.3 Properties of a Bilinear System

We continue our investigation of the Volterra series by again considering the simplest NLTI system, the bilinear system. One property of the bilinear system that is different from the conventional LTI system is that, whereas for an LTI system for an input sinusoid consisting of one frequency the output sinusoid will consist of only one frequency, for the bilinear system, a single input frequency will generate two output frequencies.

To see this, first let us reiterate, for a bilinear system,

$$y(t, t_2) = \int_{-\infty}^{\infty} \int_{-\infty}^{\infty} h_2(t_1 - \tau_1, t_2 - \tau_2)x(\tau_1)x(\tau_3) \, d\tau_1 \, d\tau$$

Taking the Fourier transform,

$$Y(j\omega_1, j\omega_2) = H_2(j\omega_1, j\omega_2)x(j\omega_1)x(j\omega_2) \qquad (4.45)$$

where $H_2 = 2$-dim Fourier transform of the impulse response $h_2(\tau_1, \tau_2)$. The 1-dim transform then is

$$Y(j\omega) = \frac{1}{2\pi} \int_{-\infty}^{\infty} Y(j\omega_1, j\omega_2) \, d\omega_1 = \frac{1}{2\pi} \int_{-\infty}^{\infty} Y(j\omega_1, j\omega - j\omega_1) \, d\omega_1$$

Next we go through different cases corresponding to different inputs.

Case 1: One Input Exponent

For an input of $e^{j\omega_1 t}$, the output is

$$Y(j\omega) = \frac{1}{2\pi} \int_{-\infty}^{\infty} H_2(\omega_1, \omega - \omega_1)\delta(\omega_1 - \omega_0)\delta(\omega - \omega_1 - \omega_0) \, d\omega_1 \qquad (4.46)$$

Then, upon integration,

$$Y(j\omega) = \frac{1}{2\pi} H_2(\omega - \omega_0, \omega_0)\delta(\omega - 2\omega_0) = e^{j2\omega_0 t} \qquad (4.47)$$

The exponential term simply means that in the time domain the output will be a sinusoid with frequency at $2\omega_0$. Hence, as stated previously, a single input frequency generates two output frequencies, with one being two times the other. The concept of transfer function is still valid and is now $H_2(\omega_0, \omega_0)$. Furthermore, this transfer function is unique and does not depend on initial condition or signal.

Case 2: Two Input Exponents

Next we apply two exponents. Their Fourier transform becomes

$$X(j\omega) = \delta(\omega - \omega_1) + \delta(\omega - \omega_2) \qquad (4.48)$$

Then

$$Y(j\omega) = \frac{1}{2\pi} [H_2(\omega - \omega_1, \omega_1)\delta(\omega - 2\omega_1)]$$

$$+ \frac{1}{2\pi} H_2(\omega - \omega_2, \omega_2)\delta(\omega - 2\omega_2)$$

$$+ \frac{1}{2\pi} H_2(\omega - \omega_2, \omega_2)\delta(\omega - \omega_1 - \omega_2) \qquad (4.49)$$

Notice that as opposed to the one-exponent case, a cross term appears. Therefore, the principle of superposition, which is fundamental to an LTI system, does not apply. Specifically, in an LTI system if the input consists of two frequency components, ω_1, ω_2, the output consists of only the same two frequency components ω_1, ω_2. In the case of a bilinear system, the output consists not only of these two frequencies (each doubles up; i.e, $2\omega_1$, $2\omega_2$), but also the crossterm $\omega_1 + \omega_2$.

4.6.1.4   Circuit Representation 1 of an NLTI System

In order to work with circuits rather than mathematical models, we want to develop circuit representation for an NLTI system. First it can be seen that the preceding general description, where the nonlinearity and memory part of the circuit are not distinctly separated, makes the circuit representation complicated, even for a simple bilinear system. We can simplify this situation if the NLTI system allows us to lump parts that are nonlinear (but memoryless) together and separate them from the parts

Fig. 4.11 Circuit
representation 1

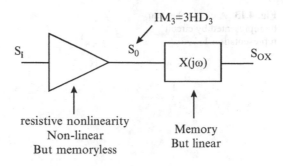

resistive nonlinearity
Non-linear
But memoryless

Memory
But linear

Fig. 4.12 A circuit that
cannot be represented by
circuit representation 1

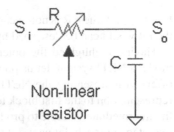

Non-linear
resistor

that have memory (but are linear). Such a representation [1] is shown in
Figure 4.11.

The circuit shown in Figure 4.11, which is the circuit representation for one such
class of NLTI system, is driven by an input signal S_i. This is applied to a nonlinearity and develops an intermediate signal S_o. This intermediate signal is then applied
to the part of the circuit with memory but is linear and develops a final output
signal S_{ox}. This second operation that involves a linear circuit with memory is
equivalent to a filtering operation. Simple and innocent as this circuit representation
may be, it is rather easy to take an existing circuit and wrongly assume that the
circuit can be represented as in Figure 4.11. The circuit shown in Figure 4.12 is one
such wrong example.

On first encounter, we may think that Figure 4.12, which consists of a nonlinear
resistor (nonlinear but memoryless) driving a linear capacitor (linear but with memory), can be described by circuit representation 1. We would probably further make
the naive assumption that we then can describe this circuit by writing the following
convolution integral that relates output to input: $S_o = \int h(t - \tau)S_i(\tau) \, d\tau$ with
$h(t - \tau) = e^{\frac{-t}{RC}}$, where R is replaced by a nonlinear R and we would then have
completely solved the problem. This is WRONG. It turns out that in the preceding
case the nonlinear resistor R in Figure 4.12 is part of the $X(j\omega)$ block in Figure 4.11
and so this simple-looking circuit cannot be described by circuit representation 1.
Instead we have to invoke the full Volterra series expression from (4.43). This also
highlights the numerous pitfalls that we can easily fall prey to when drawing
conclusions about anything that involves nonlinearity, even though the system may
look rather simple. For the sake of completeness, if we have a circuit as described in

Fig. 4.13 A circuit that can
be represented by circuit
representation 1

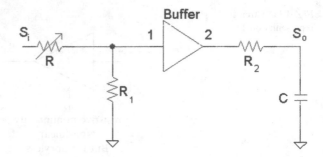

Fig. 4.13 A circuit that can be represented by circuit representation 1

Figure 4.13 (where R is nonlinear), provided that the buffer is linear and memoryless, then it would can be described by circuit representation 1.

Having highlighted the potential problem of using circuit representation 1 to model a NLTI system, let us point out its main advantage, namely, the simplicity it offers in representing the NLTI system. Here we can simply apply Taylor series representation to the first block to generate the intermediate variable. Then we filter the intermediate variable to produce the final output. To recognize the power of this two-step approach, let us use it to help illustrate one important distinction between nonlinear systems with and without memory.

Let us assume the resistive non-linearity is represented by $S_o = a_1 S_i + a_2 S_i^2 +$ ... and that the memory element $X(j\omega)$ has the magnitude and phase frequency response as shown below in Figure 4.14, [1]. Let us assume that S_i consists of two input sinusoids at ω_1, ω_2 as shown in Figure 4.14. At the output S_o has an HD_3 and an IM_3 component as shown in Figure 4.14. So far this is a memoryless system, and from (2.35) $IM_3 = 3HD_3$. Hence IM_3 is 3 times HD_3 as shown. However, upon filtering, since $X(j\omega)$ has a nonconstant frequency response, the HD_3 component is substantially attenuated (filtered out), so that at the output S_{ox}, this 3 times ratio is no longer true. If we look at the whole block in Figure 4.11 as a nonlinear block with memory, it is obvious that IM_3 is not 3 times HD_3. On the other hand, if we have analyzed the whole block by ignoring the memory, then the $IM_3 = 3HD_3$ conclusion does appear. In general, circuit representation 1 helps us gain insight into other aspects of the NLTI system as well.

4.6.1.5 Representation of a Bilinear System Using Circuit Representation 1

Case 1: Two Input Sinusoids

Now let us use circuit representation 1 to represent a bilinear system and derive output for the case with two inputs sinusoids.

Referring to Figure 4.11 let us apply

$$S_i = S_1 \cos \omega_1 t + S_2 \cos \omega_2 t$$

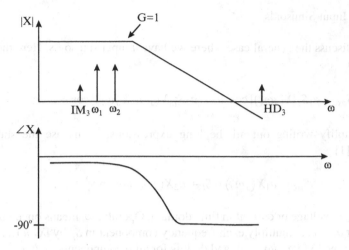

Fig. 4.14 Magnitude and phase response of $X(j\omega)$

Because it is a bilinear system we have

$$S_o = a_1 S_1 \cos \omega_1 t + a_1 S_2 \cos \omega_2 t$$
$$+ \frac{a_2}{2} \left[S_1^2 \cos((\omega_1 \pm \omega_1)t) + 2S_1 S_2 \cos((\omega_1 \pm \omega_2)t_2) + S_2^2 \cos((\omega_2 \pm \omega_2)t) \right]$$
$$+ a_3 \left[\frac{1}{4} S_1^3 \cos((\omega_1 + \omega_1 + \omega_1)t) + \dots \right]$$

$$(4.50)$$

Due to filtering:

$$S_{ox} = a_1 S_1 |X(j\omega_1)| \cos(\omega_1 t + \angle\omega_1). + a_1 S_2 |X(j\omega_2)| \cos(\omega_2 t + \angle\omega_2)$$
$$+ \frac{a_2}{2} \left[S_1^2 |X(j\omega_1 + j\omega_1)| \cos((\omega_1 + \omega_1)t + \angle X(j2\omega_1)) + S_1^2 |X(j\omega_1 - j\omega_1)| \right]$$
$$+ \frac{a_2}{2} \left[2S_1 S_2 |X(j\omega_1 \pm j\omega_2)| \cos((\omega_1 \pm \omega_2)t + \angle X(j(\omega_1 \pm \omega_2))) \right]$$
$$+ a_3 \left[\frac{1}{4} S_1^3 |X(j\omega_1 + j\omega_1 + j\omega_1)| \cos((\omega_1 + \omega_1 + \omega_1)t + \angle X(j3\omega_1)) + \dots \right]$$
$$+ \dots$$

$$(4.51)$$

Notice that this derivation generates seven frequency components that can be summarized by $\pm \omega_a \pm \omega_b$, where a, b can be equal or different and we have both the $+$ and $-$ combination.

Case 2: n Input Sinusoids

Next we discuss the general case where we have n input sinusoids. Here the output S_{ox} will be

$$S_{ox} = a_1 S_1 |X(j\omega_1)| \cos + \ldots a_2 S_1^2 |X(j\omega_1 \pm j\omega_1)| \cos + \ldots \qquad (4.52)$$

To simplify writing out all the long expressions, let us use the short-hand notation, [1]

$$S_{ox} = a_1 X(j\omega_a) \circ S_i + a_2 X(j\omega_a, j\omega_b) \circ S_i^2 + \ldots \qquad (4.53)$$

Here S_i = voltage or current in time domain. Operator $\circ$ means once you select the proper $a, b, c \ldots$ multiply each frequency component in S_i^n by $|X(j\omega_a, \ldots)|$ and shift phase by $\angle X(j\omega_a, j\omega_b \ldots)$, and do this for all n components: $a, b, c \ldots$. Notice that both $+a, -a, +b, -b, +c, -c \ldots$ appear. Also notice in doing the permutations, you should include the case when $a = b$, $b = c$, $a = b = c$, and so on. You can, of course, combine the overlapping terms.

We now go through a few examples to illustrate this short-hand notation. In the first example, we pick $X(j\omega_a, j\omega_b) \circ S_i^2$:

$X(j\omega_a, j\omega_b) \circ S_i^2 = X(j\omega_a, j\omega_b) \circ S_a S_b$ represents the following terms:

$|X(j\omega_1, j\omega_1)| S_i^2 \angle X(j\omega_1, j\omega_1)$

$|X(-j\omega_1, -j\omega_1)| S_i^2 \angle X(-j\omega_1, -j\omega_1)$

$|X(j\omega_1, -j\omega_2)| S_i^2 \angle X(j\omega_1, -j\omega_2)$

$|X(j\omega_2, j\omega_2)| S_i^2 \angle X(j\omega_2, j\omega_2)$

$|X(-j\omega_2, -j\omega_2)| S_i^2 \angle X(-j\omega_2, -j\omega_2)$

$|X(-j\omega_1, j\omega_2)| S_i^2 \angle X(-j\omega_1, j\omega_2)$

$|X(j\omega_1, -j\omega_1)| S_i^2 \angle X(j\omega_1, -j\omega_1) \qquad (4.54)$

First let us pick $a = 1$, $b = 1$, or the first term in (4.54). From $S_i = S_1 \cos \omega_1 t + S_2 \cos \omega_2 t$, we are only concerned with $S_1 \cos \omega_1 t$. But since it is multiplied by S_i^2 (that is the square term), we have to modify $S_1 \cos \omega_1 t$ to $S_1^2 \cos(\omega_1 + \omega_1)t$ and so $|X(j\omega_1, j\omega_1)| S_i^2 \angle X(j\omega_1, j\omega_1)$ becomes $|X(j\omega_1 + j\omega_1)| S_1^2 \times \cos(\omega_1 + \omega_1)t \angle X(j\omega_1 + j\omega_1)$ which equals $|X(j\omega_1 + j\omega_1)| S_1^2 \cos((\omega_1 + \omega_1)t + \angle X(j2\omega_1))$. This agrees with the first term under the first $\frac{a_2}{2}$ expression of (4.51).

Next let us pick $a = 2$, $b = 2$, then we have $|X(j\omega_2, j\omega_2)| S_i^2 \angle X(j\omega_2, j\omega_2)$, which is the fourth term in (4.53). Repeating the foregoing, this time from $S_i = S_1 \times \cos \omega_1 t + S_2 \cos \omega_2 t$ we are only concerned with $S_2 \cos \omega_2 t$. But since it is multiplied by S_i^2 (that is, the square term), that means we have to modify $S_2 \times \cos \omega_2 t$ to $S_2^2 \cos(\omega_2 + \omega_2)t$ and so finally $|X(j\omega_2, j\omega_2)| S_i^2 \angle X(j\omega_2, j\omega_2)$, becomes

$|X(j\omega_2 + j\omega_2)|S_2^2 \cos(\omega_2 + \omega_2)t \angle X(j\omega_2 + j\omega_2)$ which equals $|X(j\omega_2 + j\omega_2)|S_2^2 \cos((\omega_2 + \omega_2)t + \angle X(j2\omega_2))$.

Let us go to the third term: $|X(j\omega_1, -j\omega_2)|S_i^2 \angle X(j\omega_1, -j\omega_2)$. To get this term we would have picked $a = 1$, $b = -1$, and so we stick to this choice. According to the explanation of the short-hand notation, this means that we multiply each frequency component in S_i^2 by $|X(j\omega_1 - j\omega_2)|$ and shift phase by $\angle X(j\omega_1 - j\omega_2)$ (notice that the comma becomes a minus sign, because we have picked $(+1, -1)$. Since $S_i = S_1 \cos\omega_1 t + S_2 \cos\omega_2 t$, there are two frequency components, and we should multiply each frequency component in S_i by $|X(j\omega_1 - j\omega_2)|$ and shift by $\angle X(j\omega_1 - j\omega_2)$. Let us start from the first component, $S_1 \cos\omega_1 t$. Since it is multiplied by S_i^2 (that is the square term), we have to modify $S_1 \cos\omega_1 t$ to $S_1 S_2 \cos(\omega_1 - \omega_2)t$. Notice that we have $S_1 S_2$ as opposed to $S_1 S_1 = S_1^2$ because $a = 1$ and $b = -2$ this time. Hence the first term becomes $|X(j\omega_1 - j\omega_2)| S_1 S_2 \cos((\omega_1 - \omega_2)t + \angle X(j\omega_1 - j\omega_2))$. This agrees with the first term under the second $\frac{a_2}{2}$ expression of (4.51).

If we continue with this practice on all the seven terms in (4.54), we will generate all the terms represented by the short-hand notation. Of course, we have to eliminate any overlapping terms. The final expression will be identical to the combination of all the second-order terms in (4.51).

We next go through another example illustrating the short-hand notation. In this example, suppose we are interested in the third order term, but we still only have two frequency components as inputs (i.e, $S_i = S_1 \cos\omega_1 t + S_2 \cos\omega_2 t$). Then a possible expanded term from $a_3 X(j\omega_a, j\omega_b, j\omega_c) \circ S_i^3$ will be

$$|X(j\omega_1 - j\omega_2 + j\omega_2)|S_1 S_2 S_2$$
$$\times \cos((\omega_1 - \omega_2 + \omega_2)t + \angle X(j\omega_1 - j\omega_2 + j\omega_2)) \qquad (4.55)$$

For yet another example illustrating the short hand notation, suppose we have three frequency components as inputs (i.e, $S_i = S_1 \cos\omega_1 t + S_2 \cos\omega_2 t + S_3 \cos\omega_3 t$). Yet we are interested in the second-order term, $X(j\omega_a, j\omega_b) \circ S_i^2 = X(j\omega_a, j\omega_b) \circ S_a S_b$. Then because there are three possibilities, $a = 1$, $b = 2$ is not the only choice. We can have $a = 1, b = 2, a = 2, b = 3, a = 3, b = 1$. Starting again with $a = 1, b = 2$ we have:

$$|X(j\omega_1 + j\omega_2)|S_1 S_2 \cos((\omega_1 + \omega_2)t + \angle X(j\omega_1 + j\omega_2)) \qquad (4.56)$$

This agrees with the first term under the second $\frac{a_2}{2}$ expression of (4.51). Next we look at the case when $a = 2$, $b = 3$ and we have:

$$|X(j\omega_3 + j\omega_2)|S_3 S_2 \cos((\omega_3 + \omega_2)t + \angle X(j\omega_3 + j\omega_2))$$

Finally we look at the case $a = 3$, $b = 1$ and we have

$$|X(j\omega_3 + j\omega_1)|S_3 S_1 \cos((\omega_3 + \omega_1)t + \angle X(j\omega_3 + j\omega_1))$$

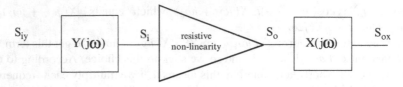

Fig. 4.15 Circuit representation 2

4.6.1.6 Circuit Representation 2 of an NLTI System [1]

To make the preceding representation more general, we can assume that the input goes into a linear element with memory first (represented as a linear filter in front). Figure 4.11 is redrawn in Figure 4.15 with this modification.

As an example of how this system works, let us start with two input sinusoids:

$$S_{iy} = S_1 \cos \omega_1 t + S_2 \cos \omega_2 t$$

First since $Y(j\omega)$ is linear then superposition applies and

$$S_i = |Y(j\omega_1)|S_1 \cos(\omega_1 t + \angle \omega_1) + |Y(j\omega_2)|S_2 \cos(\omega_2 t + \angle \omega_2) \tag{4.57}$$

To be consistent we adopt the short-hand notation that we developed in (4.53) and we have

$$S_i = Y(j\omega) \circ S_{iy} \tag{4.58}$$

Then assuming that the nonlinearity has coefficients $a_1\ a_2$ and so on, we apply S_i to this nonlinearity and we have

$$S_o = a_1 S_i + a_2 S_i^2 + a_3 S_i^3 + \ldots \tag{4.59}$$

Substituting (4.55) into (4.58) we have

$$S_o = a_1 Y(j\omega) \circ S_{iy} + a_2 \left[Y(j\omega) \circ S_{iy} \right]^2 + a_3 \left[Y(j\omega) \circ S_{iy} \right]^3 + \ldots$$

If we expand this we will find that the second-order term consists of

$$\frac{a_2}{2} \left[|Y(j\omega_1)|^2 S_1^2 \{ \cos[(\omega_1 + \omega_1)t + \angle Y(j\omega_1) + \angle Y(j\omega_1)] + \cos[(\omega_1 - \omega_1)t + \angle Y(j\omega_1) - \angle Y(j\omega_1)] \} \right]$$

$$+ \frac{a_2}{2} \left[2 S_1 S_2 |Y(j\omega_1)||Y(j\omega_2)| \{ \cos[(\omega_1 + \omega_2)t + \angle Y(j\omega_1) + \angle Y(j\omega_2)] + \cos[(\omega_1 - \omega_2)t + \angle Y(j\omega_1) - \angle Y(j\omega_2)] \} \right]$$

$$+ \frac{a_2}{2} \left[|Y(j\omega_2)|^2 S_2^2 \{ \cos[(\omega_2 + \omega_2)t + \angle Y(j\omega_2) + \angle Y(j\omega_2)] + \cos[(\omega_2 - \omega_2)t + \angle Y(j\omega_2) - \angle Y(j\omega_2)] \} \right]$$

$$\tag{4.60}$$

This is a complicated expression, but fortunately if we use the short-hand notation it can in turn be written as

$$a_2 Y(j\omega_a) \cdot Y(j\omega_b) \circ S_{iy}^2$$

with $\omega_a \pm \omega_b$ $a = 1, 2$ $b = 1, 2$

Let us do a quick check here: For the first term $a = 1$, $b = 1$, we have $a_2 |Y(j\omega_1)||Y(j\omega_1)|S_1^2(\angle Y(j\omega_1) + \angle Y(j\omega_1))$. We rewrite this term in its complete form and we have $a_2 |Y(j\omega_1)||Y(j\omega_1)|S_1^2 \cos((\omega_1 + \omega_1)t + 2\angle Y(j\omega_1))$

Looking at (4.60) one can see that under the first $\frac{a_2}{2}$ expression, the first term agrees with this term. Hence the short-hand notation is consistent.

Interesting Observation 1

Note that the output from circuit representation 2 has an important but subtle difference from the output from circuit representation 1. The phase shift contribution in the circuit representation 2 case is $2\angle Y(j\omega_1)$. On the other hand, in the circuit representation 1 case, the phase shift contribution is $\angle X(j2\omega_1)$. Physically this difference can be explained as follows: In circuit representation 1 we have applied the signal to the nonlinearity first, then a component at $2\omega_1$ is generated. This component then got filtered by $X(j\omega)$ and therefore contributes a phase shift to be calculated at $2\omega_1$, giving a phase shift contribution of $\angle X(j2\omega_1)$. On the other hand, in circuit representation 2, we do the filtering first. Therefore, the component remains at ω_1. This filtering operation contributes phase shift, which should be calculated at ω_1, giving a phase contribution of $\angle Y(j\omega_1)$. Upon going through the nonlinearity, two of these signals, each at frequency ω_1, and each contributing a phase shift of $\angle Y(j\omega_1)$, combine together to generate a component at $2\omega_1$. Therefore, the total phase shift is the sum of these two, or $2\angle Y(j\omega_1)$.

Interesting Observation 2

The linear filter and the nonlinear resistive components are not commutative. Referring to Figure 4.11 this means that S_{ox} will be different if the two blocks are swapped.

4.6.2   Analysis of Gilbert Mixer

Let us return to our main goal of Section 4.6, which is the derivation of IIP$_3$ of the Gilbert mixer at high frequency. Again we assume that at high frequency the

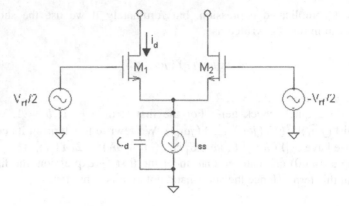

Fig. 4.16 SCP as *V-I* converter with parasitic capacitance

distortion is still dominated by the *V-I* converter. Hence let us redraw SCP, this time including parasitic capacitance.

First we want to highlight the dominant parasitics capacitors. Typically, the MOS transistor that implements the current I_{SS} has the largest size (larger than the input transistor M_{1-2} and the switching transistors M_{3-6}) and hence we assume that its drain to bulk capacitance C_{db} is dominating. In parallel to this C_{db} are the source to bulk capacitance of M_1 and M_2, and hence we lump all three of them together and denote it C_d. Again to simplify analysis we further assume that these capacitors are linear, even though in real life C_d is nonlinear. Hence the only non-linearity comes from the square law characteristics of the device. It should be noted that, similar to Figure 4.12, Figure 4.16 cannot be represented by either circuit representation 1 (Figure 4.11) or circuit representation 2 (Figure 4.15), and hence (4.53) or (4.59) cannot be used to obtain i_d. Accordingly, a complete Volterra series analysis must be applied to Figure 4.16.

4.6.2.1 Summary of Steps

First we present a summary of the steps.

All small-signals terms are in lowercase. We assume that a small-signal differential input voltage v_{rf} is applied and a small signal output current i_d is developed.

In order to determine distortion in i_d with respect to v_{rf} we do the following:

Step 1: Write *KCL* at the tail node (source node) with node voltage v_s and determine the Volterra series expansion of the intermediate variable v_s in terms of input v_{rf}, using the short-hand notation.

$$v_s = H_1 \circ v_{rf} + H_2 \circ v_{rf}^2 + H_3 \circ v_{rf}^3 + \dots \tag{4.61}$$

Step 2: Use MOS device equation in the small signal form:

$$i_d = \frac{k}{2}\left(v_{rf} - v_s\right)^2 \tag{4.62}$$

to express output current i_d in terms of input voltage v_{rf} and the intermediate variable v_s.

Step 3: Substitute v_s determined in (4.61) (from step 1) in (4.62) and derive

$$i_d = G_1 \circ v_{rf} + G_2 \circ v_{rf}^2 + G_3 \circ v_{rf}^3 + \ldots \tag{4.63}$$

G_n now becomes a function of H_n. Since H_n has been determined in step 1, G_n can now be solved. This calculation will be carried out on each G_n sequentially.

Step 4: Derive IM$_3$, HD$_3$ in terms of G_n. Since G_n has been determined in step 3, IM$_3$, HD$_3$ can now be calculated.

The step-by-step Volterra series derivation of the distortion for the SCP in Figure 4.16 is now presented.

4.6.2.2 Determine Volterra Kernel H_n

We begin step 1 by applying *KCL* at the tail node:

$$C_d \frac{dV_s}{dt} + I_{SS} - \frac{k}{2}\left[\left(V_{gs_1} - V_t\right)^2 + \left(V_{gs_2} - V_t\right)^2\right] = 0 \tag{4.64}$$

Let us define $V_{gs_1} = V_{GS} + v_{gs_1}$, $V_{gs_2} = V_{GS} + v_{gs_2}$ and substitute into (4.64). Then we have

$$C_d \frac{dV_s}{dt} + I_{SS} - \frac{k}{2}\left[\left(\left(V_{GS} - v_{gs_1}\right) - V_t\right)^2 + \left(\left(V_{GS} - v_{gs_2}\right) - V_t\right)^2\right] = 0 \tag{4.65}$$

Expanding (4.65), we have

$$C_d \frac{dv_s}{dt} + I_{SS} - \frac{k}{2}$$
$$\times \left[v_{gs_1}^2 + 2v_{gs_1}\left(V_{GS} - V_t\right) + \left(V_{GS} - V_t\right)^2 + v_{gs_2}^2 + 2v_{gs_2}\left(V_{GS} - V_t\right) + \left(V_{GS} - V_t\right)^2\right] = 0 \tag{4.66}$$

Here $\frac{dV_s}{dt}$ is replaced by $\frac{dv_s}{dt}$, because the dc term in V_s, when differentiated with respect to time, becomes 0. Expressing the input voltage in terms of differential and common mode voltage,

$$v_{gs_1} = \tfrac{v_{rf}}{2} - v_s \quad v_{gs_2} = \tfrac{-v_{rf}}{2} - v_s \qquad (4.67)$$

We can then substitute this into (4.66) and get:

$$C_d \frac{dv_s}{dt} + I_{SS} - \frac{k}{2}\left[\left(\frac{v_{rf}}{2} - v_s\right)^2 + 2\left(\frac{v_{rf}}{2} - v_s\right) \cdot (V_{GS} - V_t) + (V_{GS} - V_t)^2 \right.$$
$$\left. + \left(\frac{-v_{rf}}{2} - v_s\right)^2 + 2\left(\frac{-v_{rf}}{2} - v_s\right)(V_{GS} - V_t) + (V_{GS} - V_t)^2 \right] = 0 \qquad (4.68)$$

Expanding (4.68) we have

$$C_d \frac{dv_s}{dt} + I_{SS} - \frac{k}{2} \times \left[\frac{v_{rf}^2}{4} + v_s^2 - v_{rf} \cdot v_s + v_{rf}(V_{GS} - V_t) - 2v_s(V_{GS} - V_t) + (V_{GS} - V_t)^2 \right.$$
$$\left. + \frac{v_{rf}^2}{4} + v_s^2 + v_{rf} \cdot v_s - v_{rf}(V_{GS} - V_t) - 2v_s(V_{GS} - V_t) + (V_{GS} - V_t)^2 \right] = 0$$
$$(4.69)$$

Simplifying (4.69) we have

$$C_d \frac{dv_s}{dt} + I_{SS} - \frac{k}{2}\left[\frac{v_{rf}^2}{2} + 2v_s^2 - 4v_s(V_{GS} - V_t) + 2(V_{GS} - V_t)^2 \right] = 0 \qquad (4.70)$$

Expand v_s as a sum of its linear term, square term, and so on:

$$v_s = v_{s_1} + v_{s_2} + v_{s_3} + \ldots \qquad (4.71)$$

Substituting (4.71) into (4.70) and taking the phasor representation, we have

$$C_d j\omega(v_{s_1} + v_{s_2} + \ldots) + I_{SS} - \frac{k}{2}\left[\frac{v_{rf}^2}{2} + 2(v_{s_1} + v_{s_2} + \ldots)^2 \right.$$
$$\left. - 4(v_{s_1} + v_{s_2} + \ldots)(V_{GS} - V_t) + 2(V_{GS} - V_t)^2 \right] = 0 \qquad (4.72)$$

Next we would want to find the linear term, square term, and so on of v_s.

Linear Terms

Let us start from (4.72), keeping only the linear terms in v_s (no dc terms such as I_{SS}, $V_{GS} - V_t$, ignore higher-order terms like v_{rf}^2) and we have

$$jC_d\omega v_{s_1} + 2k v_{s_1}(V_{GS} - V_t) = 0$$
$$\therefore v_{s_1} = 0 \qquad\qquad (4.73)$$

However from (4.61) we have

$$v_{s_1} = H_1 \circ v_{rf}$$

Hence this means that

$$H_1 = 0 \qquad\qquad (4.74)$$

Second-Order Terms

Keeping only the second-order (or square) terms in (4.72), and we have

$$C_d j(\omega_1 + \omega_2)v_{s_2} - \frac{k}{2}\left[\frac{v_{rf}^2}{2} + 2v_{s_1}^2 - 4v_{s_2} \cdot (V_{GS} - V_t)\right] = 0 \qquad (4.75)$$

Notice that ω becomes $\omega_1 + \omega_2$ because we are interested in the second order terms.

Substituting (4.73) in (4.75), we have

$$j(\omega_1 + \omega_2)C_d v_{s_2} - \frac{k}{2}\left[\frac{v_{in}^2}{2} - 4v_{s_2}(V_{GS} - V_t)\right] = 0 \qquad (4.76)$$

Let us repeat (4.61) here:

$$v_s(t) = H_1(\omega) \circ v_{rf} + H_2(\omega_1, \omega_2) \circ v_{rf}^2. + ..H_3(\omega_1, \omega_2, \omega_3) \circ v_{rf}^3 + ... \qquad (4.77)$$

Since we are interested only in the second order terms, we have

$$v_{s_2} = H_2 \circ v_{rf}^2 \qquad\qquad (4.78)$$

Substituting (4.78) into (4.76), we have

$$j(\omega_1 + \omega_2)C_d H_2(\omega_1, \omega_2) \circ v_{rf}^2 - \frac{k}{2}\left[\frac{v_{rf}^2}{2} - 4H_2(\omega_1, \omega_2) \circ v_{rf}^2(V_{GS} - V_t)\right] = 0$$
$$(4.79)$$

Factoring out H_2 and v_{rf} in (4.79), we have

$$[j(\omega_1 + \omega_2)C_d + 2k(V_{GS} - v_t)]H_2(\omega_1, \omega_2) \circ v_{rf}^2 = k\frac{v_{rf}^2}{4}$$

or   (4.80)

$$H_2(\omega_1, \omega_2) = \frac{\dfrac{k}{4}}{j(\omega_1 + \omega_2)C_d + 2k(V_{GS} - V_t)}$$

Because of the operator $\circ$, (4.80) actually consists of four equations for four different cases:

$$\omega_1, \omega_2 = \pm\omega_a, \pm\omega_b$$

We can do some more simplification by noting that at intermediate frequencies when

$$j(\omega_1 + \omega_2)C_d \ll 2k(V_{GS} - V_t)$$ (4.81)

we can take the Taylor series expansion of (4.80), retain the first two terms only, and get

$$H_2(\omega_1, \omega_2) = \frac{1}{8(V_{GS} - V_t)}\left(1 - \frac{j(\omega_1 + \omega_2)C_d}{2k(V_{GS} - V_t)}\right)$$ (4.82)

For the low-frequency case we can practically set $\omega = 0$, and (4.82) becomes

$$H_2(\omega_1, \omega_2) = \frac{1}{8(V_{GS} - V_t)}$$ (4.83)

Third-Order Terms

Let us start from (4.72) and do the following: replace $j\omega$ by $j(\omega_1 + j\omega_2 + j\omega_3)$, since we are interested in the third-order terms. Equation (4.72) becomes

$$C_d j(\omega_1 + \omega_2 + \omega_3)(v_{s_1} + v_{s_2} + v_{s_3} + \ldots) + I_{SS}$$

$$-\frac{k}{2}\left[\frac{v_{rf}^2}{2} + 2(v_{s_1} + v_{s_2} + v_{s_3} + \ldots)^2.\right.$$

$$\left.-4(v_{s_1} + v_{s_2} + v_{s_3} + \ldots)(V_{GS} - V_t) + 2(V_{GS} - V_t)^2\right] = 0$$ (4.84)

First we want to expand the $2(v_{s_1} + v_{s_2} + v_{s_3} + \ldots)^2$ factor and look for third-order terms in the expansion:

$$2(v_{s_1} + v_{s_2} + v_{s_3} + \ldots)^2 = 2\left(v_{s_1}^2 + v_{s_2}^2 + v_{s_3}^2 + v_{s_4}^2 + \ldots\right.$$
$$\left. +2v_{s_1}v_{s_2} + 2v_{s_1}v_{s_3} + 2v_{s_1}v_{s_4} + \ldots 2v_{s_2}v_{s_3} + 2v_{s_2}v_{s_4} + \ldots\right) \qquad (4.85)$$

In (4.85), third-order terms come from a single v_{s_3} term or a product of a v_{s_1} term and a v_{s_2} term. Since the v_{s_3} term is inside the bracket, which goes through a square operation, then any terms it generates will be higher than third order. Hence the only third order term in this factor is

$$4v_{s_1}v_{s_2} \qquad (4.86)$$

Turning to the rest of factors in (4.84), it is obvious that there is no third-order term in the factor I_{SS}, $\frac{v_{rf}^2}{2}$, $2(V_{GS} - V_t)^2$. Hence if we keep only the third-order terms, (4.84) becomes

$$C_{dj}j(\omega_1 + \omega_2 + \omega_3)V_{s_3} - \frac{k}{2}[4v_{s_1}v_{s_2} - 4v_{s_3}(V_{GS} - V_t)] = 0 \qquad (4.87)$$

Next we take (4.61), retain the first three terms and write them down as

$$v_{s_1} = H_1 \circ v_{rf}$$
$$v_{s_2} = H_2 \circ v_{rf}^2$$
$$v_{s_3} = H_3 \circ v_{rf}^3 \qquad (4.88)$$

Substituting (4.88) into (4.87), we have

$$C_{dj}j(\omega_1 + \omega_2 + \omega_3)H_3 \circ v_{rf}^3 - \frac{k}{2}$$
$$\times \left[4(H_1 \circ v_{rf})\left(H_2 \circ v_{rf}^2\right) - 4H_3 \circ v_{rf}^3(V_{GS} - V_t)\right]$$
$$= 0 \qquad (4.89)$$

Factoring out v_{rf} and dropping it, we have

$$C_{dj}j(\omega_1 + \omega_2 + \omega_3)H_3 - \frac{k}{2}[4\overline{H_1H_2} - 4H_3(V_{GS} - V_t)] = 0 \qquad (4.90)$$

Here $\overline{H_1H_2}$ is defined as

$$\overline{H_1H_2} = \frac{H_1(\omega_1)H_2(\omega_2 + \omega_3) + H_1(\omega_2)H_2(\omega_3 + \omega_{31}) + H_1(\omega_3)H_2(\omega_1 + \omega_2)}{3}$$
$$(4.91)$$

Notice the argument of H_1 always consists of one frequency component and that of H_2 always consists of two frequency components. The bar is there to ensure that all the possible permutations are exercised.

Rearranging (4.90) we have

$$H_3[j(\omega_1 + \omega_2 + \omega_3)C_d + 2k(V_{GS} - V_t)] = 2k\overline{H_1 H_2} \tag{4.92}$$

Finally, we solve for H_3:

$$H_3(\omega_1, \omega_2, \omega_3) = \frac{2k\overline{H_1 H_2}}{j(\omega_1 + \omega_2 + \omega_3)C_d + 2k(V_{GS} - V_t)} \tag{4.93}$$

This complicated looking formula, however, has a simple answer. From (4.74), $H_1 = 0$, and hence substituting (4.74) into (4.91), we have

$$\overline{H_1 H_2} = 0 \tag{4.94}$$

Substituting (4.94) in (4.93), we have

$$H_3 = 0. \tag{4.95}$$

This brings us to the end of step 1.

Summary and Interpretation of Results from Step 1

From step 1 we have determined that

$$v_s = v_{s1} + v_{s2} + v_{s3} + \ldots = H_1 \circ v_{rf} + H_2 \circ v_{rf}^2 + H_3 \circ v_{rf}^3 + \ldots$$

and that

$$H_1 = 0, H_2(\omega_1, \omega_2) = \frac{\frac{k}{4}}{j(\omega_1 + \omega_2)C_d + 2k(V_{GS} - V_t)}, H_3 = 0 \tag{4.96}$$

Note $H_1 = 0$ is simply restating the familiar result that if we concentrate on the linear response of the circuit (or, equivalently, that the circuit is represented as a linear circuit), then the source of M_1 and M_2 is an ac ground, which means $v_s = 0$. This result, of course, is presented in any text that analyzes the SCP.

Another interesting fact is that $H_2 \neq 0$. How do we explain this? Let us neglect memory effect for the time being. Then any text on SCP will show that the output current (on either output branch) is an odd function of the input voltage, due to symmetry. It just follows from the mathematical property of an odd function that the output current cannot have even harmonics, including the second harmonics.

We may postulate that since the device characteristics have second-order nonlinearity (due to the square law), for the output current to possess no second harmonics, the gate to source voltage v_{gs} must possess second harmonics. Since the gate voltage $v_g = v_{rf}$ has no second harmonics, then the source voltage v_s must have second harmonics to "cancel" the device non-linearity. This implies that $v_{s2} \neq 0$ and therefore H_2 is nonzero.

4.6.2.3 Relating Volterra Kernel G_n to H_n

Now we turn to step 2. Let us refer to Figure 4.16 again and concentrate on transistor M_1.

To reiterate, our goal is to write i_d in the following form:

$$i_d = G_1 \circ v_{rf} + G_2 \circ v_{rf}^2 + G_3 \circ v_{rf}^3 + \ldots \tag{4.97}$$

or, alternately as

$$i_d = i_{d1} + i_{d2} + i_{d3} + \ldots \tag{4.98}$$

From the device equation, we have

$$I_{d_{M_1}} = \frac{k}{2}\left[\left(v_{gs_1} + (V_{GS_1} - V_t)\right)^2\right] = \frac{k}{2}\left[\left(v_{gs_1} + (V_{GS} - V_t)\right)^2\right]$$

(Note: Because $V_{GS1} = V_{GS2}$, we set them equal to V_{GS}.)

$$= \frac{k}{2}\left[v_{gs_1}^2 + 2v_{gs_1}(V_{GS} - V_t) + (V_{GS} - V_t)^2\right]$$

$$= \frac{k}{2}\left[\left(\frac{v_{rf}}{2} - v_s\right)^2 + 2\left(\frac{v_{rf}}{2} - v_s\right)(V_{GS} - V_t) + (V_{GS} - V_t)^2\right] \tag{4.99}$$

The small signal output current can now be written as

$$i_d = i_{d_{M_1}} = I_{d_{M_1}} - I_{D_{M_1}}$$

$$= \frac{k}{2}\left[\left(\frac{v_{rf}}{2} - v_s\right)^2 + 2\left(\frac{v_{rf}}{2} - v_s\right)(V_{GS} - V_t) + (V_{GS} - V_t)^2\right] - \frac{k}{2}\left[(V_{GS} - V_t)^2\right]$$

$$= \frac{k}{2}\left[\left(\frac{v_{rf}}{2} - v_s\right)^2 + 2\left(\frac{v_{rf}}{2} - v_s\right)(V_{GS} - V_t)\right]$$

$$= \frac{k}{2}\left[\frac{v_{rf}^2}{4} - v_{rf}(v_{s_1} + v_{s_2} + v_{s_3} + \ldots) + (v_{s_1} + v_{s_2} + v_{s_3} + \ldots)^2\right.$$

$$\left. + 2\frac{v_{rf}}{2}(V_{GS} - V_t) - 2(v_{s_1} + v_{s_2} + v_{s_3} + \ldots)(V_{GS} - V_t)\right] \tag{4.100}$$

4.6.2.4   Solving G_n

In step 3 we want to express (4.100) in the form of (4.98), where the first-, second-and third-order terms are separated. Then we can calculate G_n.

Determine G_1 from first order term i_{d1}

In this sub-section we first want to isolate the first-order terms in (4.100). The first term in (4.100) is a square term in v_{rf} and would not have contributed to a first-order term. The second term is a cross product of v_{rf} and a first- or higher-order term of v_s (v_{s1}, v_{s2}, ...) and so does not contribute either. Likewise the third term does not contribute. The fourth term, $2\frac{v_{rf}}{2}(V_{GS} - V_t)$, does contribute. Finally, the term $2v_{s_1}(V_{GS} - V_t)$, in the fifth term, also contributes. Collecting these contributions and equating them to i_{d1} we have

$$i_{d_1} = \frac{k}{2}\left[v_{rf}(V_{GS} - V_t) - 2v_{s_1}(V_{GS} - V_t)\right] \tag{4.101}$$

Next let us apply the results from step 1, sub-section 4.6.2.2, to (4.101) and simplify. Specifically, substituting (4.73) in (4.101), we have

$$i_{d_1} = \frac{k}{2}\left[v_{rf}(V_{GS} - V_t)\right] \tag{4.102}$$

Meanwhile, equating (4.97), (4.98) and we find i_{d1} is also expressed as

$$i_{d_1} = G_1 \circ v_{rf} \tag{4.103}$$

Finally, equating (4.102) and (4.103), we find that G_1 as

$$G_1 = \frac{k}{2}(V_{GS} - V_t) = \frac{1}{2}k\sqrt{\frac{I_{SS}}{k}} = \frac{1}{2}\sqrt{kI_{SS}} \tag{4.104}$$

Determine G_2 from Second-Order Term i_{d2}

As in sub-section determining G_1, we first isolate the second-order terms in (4.100). From (4.100), the first term is $\frac{v_{rf}^2}{4}$. This contributes a second-order term. For the second term the only second-order term in it is $-v_{rf}v_{s_1}$. For the third term, the second-order term in it is $v_{s_1}^2$. For the fourth term, there is no second-order term. For the fifth term, the only second-order term in it is $-2v_{s_2}(V_{GS} - V_t)$. Collecting these contributions and equating them to i_{d2} we have

$$i_{d_2}(\omega_1, \omega_2) = \frac{k}{2}\left[\frac{v_{rf}^2}{4} - v_{rf}v_{s_1} + v_{s_1}^2 - 2v_{s_2}(V_{GS} - V_t)\right] \qquad (4.105)$$

Next we can simplify (4.105) based on the results from step 1 in sub-section 4.6.2.2. First, referring to (4.73): $v_{s1} = 0$. Secondly applying (4.80) to (4.78), we obtain v_{s2}.

Substituting these values of v_{s1}, v_{s2} in (4.105), we have

$$\begin{aligned} i_{d_2}(\omega_1\omega_2) &= \frac{k}{2}\left[\frac{v_{rf}^2}{4} - 2H_2(\omega_1,\ \omega_2)(V_{GS} - V_t) \circ v_{rf}^2\right] \\ &= \frac{k}{2}\left[\frac{1}{4} - \frac{(\frac{k}{2})(V_{GS} - V_t)}{j(\omega_1 + \omega_2)C_d + 2k(V_{GS} - V_t)}\right]\circ v_{rf}^2 \qquad (4.106)\end{aligned}$$

Meanwhile equating (4.97) and (4.98) we find that i_{d2} is also expressed as

$$i_{d2} = G_2 \circ v_{rf}^2 \qquad (4.107)$$

Finally equating (4.106) and (4.107) and simplifying, we find G_2 as

$$G_2 = \frac{k}{8}\left[1 - \frac{1}{\left[1 + \frac{j(\omega_1+\omega_2)C_d}{2k(V_{GS}-v_t)}\right]}\right] \qquad (4.108)$$

As in step 1, sub-section 4.6.2.2, we can apply the intermediate frequency simplification. Hence we apply (4.81) to (4.108), take Taylor series expansion, retain the first two terms only, and we have

$$\begin{aligned} G_2(\omega_1\omega_2) &= \frac{k}{8}\left[\frac{j(\omega_1 + \omega_2)C_d}{2k(V_{GS} - V_t)}\right] \\ &= \frac{1}{16}\cdot\frac{j(\omega_1 + \omega_2)C_d}{(V_{GS} - V_t)} \qquad (4.109)\end{aligned}$$

Determine G_3 from Third Order Term i_{d3}

As in sub-sections determining G_1 and G_2, we again first isolate the third-order terms in (4.100). From (4.100) the first term is $\frac{v_{rf}^2}{4}$. This does not contribute to any third-order term. For the second term the only third order term in it is $-v_{rf}v_{s_2}$. For the third term, the only third-order term is $2v_{s_1}v_{s_2}$. For the fourth term, there is no third-order term. For the fifth term, the only third-order term in it is $-2v_3\,(V_{GS} - V_t)$. Collecting these contributions and equating them to i_{d3} we have

$$i_{d_3} = \frac{k}{2}\left[-v_{rf}v_{s_2} + 2v_{s_1}v_{s_2} - 2v_{s_3}(V_{GS} - V_t)\right] \tag{4.110}$$

Next we simplify based on the results from step 1, sub-section 4.6.2.2. Let us express all the v_s in (4.110) according to (4.88). We have

$$i_{d_3} = \frac{k}{2}\left[-v_{rf} \circ \overline{H_2(\omega_1\omega_2)} \circ v_{rf}^2 + \overline{H_1(\omega_1)H_2(\omega_1\omega_2)} \circ v_{rf}^3 - 2(V_{GS} - V_t)H_3 \circ v_{rf}^3\right] \tag{4.111}$$

One thing that is worth noting is that $H_2(\omega_1\omega_2)$ in (4.88) becomes $\overline{H_2(\omega_1\omega_2)}$ in (4.111). This is because the $H_2(\omega_1\omega_2)$ term assumes different value, depending on the combination of the pair of frequencies $(\omega_1\omega_2)$ that it is being evaluated at.

We can now refer to (4.96) and note that $H_1 = H_3 = 0$. Therefore in (4.111), the $\overline{H_1H_2}$ term in the second term is zero, which means that the second term is zero. Similarly, the H_3 term in the third term is zero and so the third term is zero as well. Hence (4.111) simplifies to

$$\begin{aligned}
i_{d_3} &= \frac{-k}{2}\overline{H_2(\omega_1\omega_2)} \circ v_{rf}^3 \\
&= \frac{-k}{2}\left[\frac{H_2(\omega_1\omega_2) + H_2(\omega_1\omega_3) + H_2(\omega_2\omega_3)}{3}\right] \circ v_{rf}^3
\end{aligned} \tag{4.112}$$

Again equating (4.97) and (4.98), we find that i_{d3} is also expressed as

$$i_{d_3} = G_3 \circ v_{rf}^3 \tag{4.113}$$

Therefore, equating (4.112), (4.113), we have

$$G_3(\omega_1\omega_2\omega_3) \equiv -\frac{k}{2}\overline{H_2(\omega_1\omega_2)} \tag{4.114}$$

Substituting H_2 from (4.96) in (4.114), we have

$$G_3(\omega_1\omega_2\omega_3) \equiv -\frac{k}{2}\overline{\frac{\frac{k}{4}}{j(\omega_1 + \omega_2)C_d + 2k(V_{GS} - V_t)}} \tag{4.115}$$

This brings us to the end of step 3.

Summary and interpretations of results from step 3

We have determined G_1, G_2, G_3:

$$G_1 = \frac{k}{2}(V_{GS} - V_t) = \frac{1}{2}k\sqrt{\frac{I_{SS}}{k}} = \frac{1}{2}\sqrt{kI_{SS}} \tag{4.116}$$

$$G_2 = \frac{k}{8}\left[1 - \frac{1}{\left[1 + \frac{j(\omega_1 + \omega_2)C_d}{2k(V_{GS} - V_t)}\right]}\right] \tag{4.117}$$

$$G_3(\omega_1\omega_2\omega_3) \equiv -\frac{k}{2j(\omega_1 + \omega_2)C_d + 2k(V_{GS} - V_t)} \tag{4.118}$$

Notice G_1 is simply the transconductance (output current/input voltage) and its value agrees with the transconductance obtained using simple linear analysis. If G_1 is normalized, it also agrees with a_1 derived using Taylor series in (4.23). This shows that the memory effect due to the capacitor C_d does not have any impact here. G_2, which corresponds to a_2 in (4.24), highlights the impact of the capacitor. Notice that a_2 is zero (due to symmetry of the SCP) but G_2 is nonzero.

Finally let us examine G_3 at low frequency: practically $\omega_1 = \omega_2 = \omega_3 = 0$.

Since $\omega_1 = \omega_2 = \omega_3 = 0$ we can see that $H_2(\omega_1, \omega_2), H_2(\omega_2, \omega_3), H_3(\omega_3, \omega_1)$ are all equal. From step 1, sub-section 4.6.2.2, we have calculated $H_2(\omega_1, \omega_2)$ under the low frequency condition in (4.83). Using that value, we have

$$H_2(\omega_1, \omega_2) = H_2(\omega_2, \omega_3) = H_3(\omega_3, \omega_1) = \frac{1}{8(V_{GS} - V_t)}$$

Substituting this in (4.114), we have

$$G_3 = \frac{-k}{16(V_{GS} - V_t)} \tag{4.119}$$

4.6.2.5 Find IM$_3$, HD$_3$

In step 4 we generalize the definition of HD$_3$ from (2.29) (derived under no memory effect) to the following definition (which is valid for cases with/without memory effect):

$$HD_3 = \frac{i_{d_3}}{i_{d_1}} = \frac{G_3 \circ v_{rf}^3}{G_1 \circ v_{rf}} = \frac{\left|G_3 \circ \left(A_{rf}\cos\omega_1 t\right)^3\right|}{\left|G_1 \circ A_{rf}\cos\omega_1 t\right|} = \frac{G_3(\omega_1, \omega_1, \omega_1)}{G_1}\frac{A_{rf}^2}{4} \tag{4.120}$$

Next let us generalize the definition of IM_3 from (2.34) (derived under no memory effect) to the following definition (which is valid for cases with/without memory effect):

$$IM_3 = \frac{\left| G_3(\omega_1, \omega_1, \omega_2) \circ A_{rf}^3 (\cos \omega_1 t + \cos \omega_2 t)^3 \right|}{\left| G_1 \circ A_{rf} \cos \omega_1 t \right|}$$

(4.121)

Notice that on the surface this expression is quite similar to the definition of HD_3 as given in (4.120). The difference stems from the fact that G_3 is now calculated at $(\omega_1, \omega_1, \omega_2)$ and because of the frequency-dependency effect, its value bears no simple relationship to its value calculated at $(\omega_1, \omega_1, \omega_1)$, which is what (4.120) uses to calculate G_3 and hence HD_3. This is the fundamental reason why with frequency-dependent effect IM_3 is no longer simply $3HD_3$, as stated in (2.35).

Next let us specialize the IM_3 definition to the case when we are interested in IM_3 caused by adjacent channel interference. It should be noted that for IM_3 we have three frequencies $(\omega_1, \omega_1, \omega_2 = -\omega_1 - \Delta\omega)$. For IM_3 caused by adjacent channel interference, ω_2 is so defined (to be at $-\omega_1 - \Delta\omega$) such that $2\omega_1 + \omega_2$ lies on the same frequency as $\omega_1 - \Delta\omega$. This guarantees that the third-order nonlinearity that takes ω_1 and ω_2 as the inputs and generates the $2\omega_1 + \omega_2$ term will end up generating a term at $\omega_1 - \Delta\omega$, which is the undesired frequency. Meanwhile A_{rf} would be replaced by $A_{\text{interference}}$. Applying these observations to (4.121), we get

$$\begin{aligned} IM_3 &= \frac{\left| G_3(\omega_1, \omega_1, \omega_2) \circ A_{\text{interference}}^3 (\cos \omega_1 t + \cos \omega_2 t)^3 \right|}{\left| G_1 \circ A_{\text{interference}} \cos \omega_1 t \right|} \\ &= \frac{\left| G_3(\omega_1, \omega_1, \omega_2) \right|}{|G_1|} A_{\text{interference}}^2 \frac{3}{4} \end{aligned}$$

(4.122)

We would now calculate HD_3 and IM_3. For low frequency (no memory effect) we can find HD_3 by substituting (4.116) and (4.119) into (4.120). We have

$$HD_3 = \frac{A_{rf}^2}{4} \frac{k}{16(V_{GS} - V_t)} \cdot \frac{1}{\frac{1}{2}\sqrt{kI_{SS}}}$$

(4.123)

Now

$$(V_{GS} - V_t)^2 = \frac{I_{SS}}{k}$$

(4.124)

Substituting this into (4.123) and simplifying, we have

$$HD_3 = \frac{A_{rf}^2 k}{32\sqrt{\frac{I_{SS}}{k}}} \cdot \frac{1}{\sqrt{kI_{SS}}} = \frac{A_{rf}^2}{32} \frac{k}{I_{SS}}$$

(4.125)

Equation (4.125) agrees with the result obtained using Taylor series expansion (4.28).

Likewise, we can do the same for IM_3 and it would agree with the Taylor series expansion case. For intermediate frequency where $j(\omega_1 + \omega_2)C_d \ll 2k(V_{GS} - V_t)$, we could have started with the expression for G_3 in (4.118) and found an intermediate frequency approximation. Alternatively, we know that in (4.114), G_3 is expressed in terms of H_2, for which, incidentally, we have found an intermediate frequency approximation [in (4.82)]. Hence we start with (4.120), substituting in (4.114) and (4.116), we have

$$
HD_3 = \frac{G_3}{G_1} \frac{A_{rf}^2}{4} = \frac{\frac{k}{2}\overline{H_2}}{\frac{1}{2}\sqrt{kI_{SS}}} \frac{A_{rf}^2}{4}
\tag{4.126}
$$

Setting $\omega_3 = \omega_2 = \omega_1$ and substituting the intermediate frequency approximation of H_2 from (4.82), we have

$$
\begin{aligned}
HD_3 &= \left| \frac{1}{\frac{1}{2}\sqrt{kI_{SS}}} \cdot \frac{k}{2} \frac{1}{8(V_{GS} - V_t)} \left[1 - \frac{j(\omega_1 + \omega_2)C_d}{2k(V_{GS} - V_t)} \right] \right| \frac{A_{rf}^2}{4} \\
&= \sqrt{\frac{k}{I_{SS}}} \frac{A_{rf}^2}{32 \cdot (V_{GS} - V_t)} \left| \left[1 - \frac{j(2\omega_1)C_d}{2k(V_{GS} - V_t)} \right] \right|
\end{aligned}
\tag{4.127}
$$

Next we can find an expression for IM_3. We start from (4.122). To obtain G_3 in (4.122) we go to (4.114) and we have

$$
G_3(\omega_1\omega_2\omega_3) = -\frac{k}{2}\overline{H_2(\omega_1\omega_2)}
$$

To avoid confusion between the subscript 1, 2, 3, we rewrite the expressions as

$$
\begin{aligned}
G_3(\omega_a\omega_b\omega_c) &\equiv -\frac{k}{2}\overline{H_2(\omega_a\omega_b)} \\
&= -\frac{k}{2} \left[\frac{H_2(\omega_a\omega_b) + H_2(\omega_b\omega_c) + H_2(\omega_c\omega_a)}{3} \right]
\end{aligned}
\tag{4.128}
$$

From (4.122) we know we are interested in $G_3(\omega_1, \omega_1, \omega_2)$, so essentially we should assign $a = 1, b = 1, c = 2$. Following the convention as described in (4.128), we assign the a, b associated with H_2's argument and we have

$$
G_3(\omega_1, \omega_1, \omega_2) \equiv -\frac{k}{2} \left[\frac{H_2(\omega_1\omega_1) + H_2(\omega_1, -\omega_2) + H_2(-\omega_2, \omega_1)}{3} \right]
\tag{4.129}
$$

At this point remember that we are dealing with the intermediate frequency case and we would use this fact for simplification. Therefore, we substitute the intermediate frequency approximation to H_2 as calculated in (4.82) into (4.129):

$$
\begin{aligned}
G_3(\omega_1, \omega_1, \omega_2) &\equiv -\frac{k}{2}\left[\frac{H_2(\omega_1\omega_1) + H_2(\omega_1, \omega_2) + H_2(\omega_2, \omega_1)}{3}\right] \\
&\equiv -\frac{k}{2}\frac{1}{8(V_{GS} - V_t)}\left(1 - \left(\frac{jC_d}{2k(V_{GS} - V_t)} \times \frac{(\omega_1 + \omega_1) + (\omega_1 + \omega_2) + (\omega_2 + \omega_1)}{3}\right)\right) \\
&\equiv -\frac{k}{2}\frac{1}{8(V_{GS} - V_t)}\left(1 - \left[\frac{jC_d}{2k(V_{GS} - V_t)} \times \frac{(\omega_1 + \omega_1) + (\omega_1 + -\omega_1 - \Delta\omega) + (-\omega_1 - \Delta\omega + \omega_1)}{3}\right]\right) \\
&\equiv -\frac{k}{2}\frac{1}{8(V_{GS} - V_t)}\left(1 - \left[\frac{jC_d}{2k(V_{GS} - V_t)} \times \frac{2(\omega_1 - \Delta\omega)}{3}\right]\right) \\
&\cong -\frac{k}{2}\frac{1}{8(V_{GS} - V_t)}\left(1 - \left[\frac{jC_d}{2k(V_{GS} - V_t)} \times \frac{2\omega_1}{3}\right]\right)
\end{aligned}
$$

$$\text{(4.130)}$$

Substituting G_1 from (4.116) and G_3 from (4.130) into (4.122), we get

$$
IM_3 = \sqrt{\frac{k}{I_{SS}} \frac{3A_{interference}^2}{32(V_{GS} - V_t)}}\left[\left|1 - \frac{jC_d}{2k(V_{GS} - V_t)} \times \frac{2\omega_1}{3}\right|\right] \tag{4.131}
$$

Furthermore, substituting (4.124) into (4.131), we finally get

$$
IM_3 = \frac{3A_{interference}^2}{32(V_{GS} - V_t)^2}\left[\left|1 - \frac{2}{3}\frac{j(\omega_1)C_d}{2k(V_{GS} - V_t)}\right|\right] \tag{4.132}
$$

It is instructive to compare (4.131) to (4.127) and note that even when we set $A_{rf} = A_{interference}$, $IM_3 \neq 3HD_3$.

Finally we look at the case when V_{lo} is switching, but with zero rise and fall time. Even though we no longer have a time-invariant system since V_{lo} is an ideal square wave, the results can still be easily obtained. In the high-frequency case the distortion still comes primarily from the bottom SCP when doing V-I conversion, which does not switch. We can again incorporate the switching effect, for a V_{lo} that is an ideal square wave, by modifying the aforementioned results through multiplying the output by the Fourier series representation of a square wave. As in the low-frequency case, the effect of switching is simply shifting the frequency of the fundamental and the third order products by f_{lo}. The amplitudes are both reduced by the same amount, $1/\pi$, and HD_3, IM_3 remain unchanged.

Numerical example 4.4.

Assume that a Gilbert mixer has a V-I converter as shown in Figure 4.16. Referring to Figure 4.16 let M_1 and M_2 both have a $\frac{W}{L}$ of $50um/0.6um$.

Let us apply a V_{rf} at 1.9 GHz and two interference signals $V_{\text{interference}}$ at 1.9017 GHz and 1.9034 GHz. Their power levels are set at 0 dBm. Again we make $V_{GS_1} - V_t = 0.387V$. Now assume $k' = 100uA/V^2$. This means $k = 6250uA/V^2$. C_d is given as $C_d = 100fF$.

Finally let us assume that V_{lo} is not switching and that the Gilbert mixer's distortion is dominated by the V-I converter.

(a) Assume intermediate frequency approximation holds, find IM_3 of this Gilbert mixer.

(b) Assume intermediate frequency approximation does not hold, find HD_2 of this Gilbert mixer.

(a) Since interference signal has a power level of 0 dbm, this means $A_{\text{interference}} = 0.316$ V. Substituting this and other relevant values into (4.132), the expression for IM_3 using intermediate frequency approximation, we have

$$IM_3 = \frac{3(0.316)^2}{32(0.387)^2}\left[\left|1 - \frac{2}{3}\frac{j(2\pi \times 1.9017G)100fF}{2 \times 6250uA/V^2(0.387)}\right|\right] = 0.0625|1 - 0.16j|$$

$$- 0.06329 = -23.97 \; db$$

(b) First we extend the definition of HD_3 given in (4.120) to HD_2. After some simplification we get

$$HD_2 = \frac{A_{rf}|G_2|}{2|G_1|}$$

Next we calculate G_2. Since the intermediate frequency approximation does not hold, we use the full expression for G_2 as given in (4.108). Substituting the proper values, we have

$$G_2 = \frac{k}{8}\left[1 - \frac{1}{\left[1 + \frac{j(\omega_1 + \omega_2)C_d}{2k(V_{GS} - V_t)}\right]}\right] = \frac{6250uA/V^2}{8}\left[1 - \frac{1}{\left[1 + \frac{j(2 \times 2\pi \times 1.9G)100fF}{2 \times 6250uA/V^2(0.387)}\right]}\right]$$

$$|G_2| = \left|\frac{6250uA/V^2}{8}\left[1 - \frac{1}{\left[1 + \frac{j(2 \times 2\pi \times 1.9G)100fF}{2 \times 6250\,uA/V^2(0.387)}\right]}\right]\right|$$

$$= \left|781\;uA/V^2\left[1 - \frac{1}{[1 + 0.493j]}\right]\right| = 781\;uA/V^2 \times 0.442$$

$$|G_2|\Big|_{\omega_1 = \omega_2 = 1.9\;GHz} = 3.45 \times 10^{-4}A/V^2$$

Next we calculate G_1 by substituting the proper values into (4.116). We have
$G_1 = \frac{k}{2}(V_{GS} - V_t) = 3125uA/V^2(0.387V) = 1209uA/V$
Finally substituting G_1, G_2 into the definition of HD_2 we have

$$HD_2 = \frac{0.316V}{2} \times \frac{3.45 \times 10^{-4}A/V^2}{1209uA/V}$$

$$= 0.045$$

$$= -26.9 \ db$$

As a comparison, let us repeat the calculation for this HD_2, except this time we ignore the memory effect. First let us extend the definition of HD_3 for a general memoryless NLTI system [given in (4.26)] to the defintion HD_2, again for a general memoryless NLTI system. We have $HD_2 = \frac{1}{2}\frac{a_2}{a_1}A_{rf}$. For a Gilbert mixer, from (4.24) $a_2 = 0$. Hence the HD_2 of a Gilbert mixer, ignoring memory effect, is 0. This is different from the HD_2 calculated incorporating memory effect. This difference may not be surprising because when we include memory effect the Gilbert mixer no longer has odd symmetry in the ac sense and therefore $HD_2 \neq 0$.

4.7 Noise

In this section we are interested in taking the NF specified for a mixer in chapter 2 and finding circuit parameters that will satisfy this specification. As opposed to section 4.5 and 4.6, in the present section the case when V_{lo} is switching is very important and we will devote considerable attention to it. For simplicity the example we choose is the unbalanced mixer. The methodology and result can be readily extended to the Gilbert mixer. Referring to the unbalanced mixer in Figure 4.3, let us assume that most of the noise comes from the input voltage to current converter M_3. In essence we have assumed the switching transistors M_{1-2} do not contribute any noise and that the load's noise contribution (not shown), when input referred, is negligible.

4.7.1 V_{lo} Not Switching Case

We assume, as a start, and to ease mathematical complication, that V_{lo} is not switching and therefore a mixer is a LTI system. Turning to the unbalanced mixer in Figure 4.3 all noise contribution from M_3 is propagated to the IF output with switching transistors M_{1-2} assumed to be on all the time. Under this assumption and with M_{1-2} assumed noiseless, the noise present at the drain of M_{1-2} is the same as the noise present at the source. Now the power spectral density (PSD) of the current noise present at the source of M_{1-2} is simply given by the PSD of M_3's input

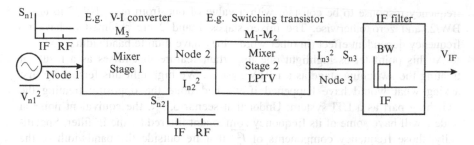

Fig. 4.17 Block diagram representation of an unbalanced mixer for noise calculation

referred voltage noise multiplied by g_{m3}^2. Assuming that there is only thermal noise present in M_3 the PSD of M_3's input referred voltage noise, denoted as S_{n1} is given by

$$S_{n1} = 4kT \times \frac{2}{3} \frac{1}{g_{m3}} \tag{4.133}$$

4.7.2 V_{lo} Switching Case

In this section we want to investigate the noise behaviour of a mixer when V_{lo} is switching periodically. For the unbalanced mixer in Figure 4.3 this means we have to include the effects due to switching of M_{1-2}.

To highlight the periodic time-varying nature of this mixer, let us represent the unbalanced mixer of Figure 4.3 in a block diagram form, as shown in Figure 4.17. Note that Figure 4.17 breaks the mixer into two stages a linear time invariant (LTI) first stage (corresponding to M_3 of Figure 4.3), followed by a linear periodic time varying (LPTV) second stage (the switching part, corresponding to M_1-M_2 of Figure 4.3). At the input to the first stage, all the noise sources from that particular stage are lumped together and input referred. This is denoted as V_{n1}^2, the equivalent input voltage noise source to stage 1. This noise gets amplified by stage 1 and generates equivalent input current noise sources I_{n2}^2 and I_{n3}^2 at nodes 2 and 3, respectively. The PSD of these noise sources are labeled as S_{n1}, S_{n2}, S_{n3}. For S_{n1}, S_{n2} they are assumed to have a flat frequency response up to the RF frequencies and beyond (i.e, we have included only the thermal noise from transistor M_3 and have neglected the 1/f noise). Our goal is to find I_{n3}^2 and S_{n3}. If the second stage is assumed to be a LTI system as in sub-section 4.7.1, S_{n3}'s frequency response is flat. However, if we include the periodic time-varying effect, this can be more complicated. Finally, the mixer output passes through an IF filter (not shown in Figure 4.3) to generate the IF signal. If we assume that this mixer is used in the architecture described in Figure 2.2 then the IF filter is the same as BPF3 described in that figure. The bandpass filter is centered at ω_{if}, with bandwidth BW. Here we assume its frequency response to be constant with a value of one from $\omega_{if} - BW/2$ to $\omega_{if} + BW/2$, and zero otherwise. The mixer stages 1 and 2 are assumed to have no frequency dependent effect (in other words, they have infinite bandwidth).

At this point let us highlight one important feature that arises as a result of treating the switching stage as a LPTV system. We highlight this feature by first asking what would have happened if we had done the opposite: treating the switching part as a LTI system. Under that scenario, $\overline{I_{n2}^2}$, the equivalent noise at node 2 will have some of its frequency component filtered by the IF filter. Specifically, those frequency components of $\overline{I_{n2}^2}$ that lie outside the bandwidth of the IF filter (i.e, frequency components of $\overline{I_{n2}^2}$ whose ω is smaller than $\omega_{if} - BW/2$ or whose ω is larger than $\omega_{if} + BW/2$) will be filtered. Next let us return to the scenario when we do treat stage 2 as an LPTV system. We assume that stage 2 is varying periodically at the LO frequency, ω_{lo}. In this case, let us again assume the PSD of $\overline{I_{n2}^2}$ also has frequency components outside the bandwidth of the IF filter. It turns out that not all of these frequency components will be filtered in the present case. Some of these frequency components will appear at the IF output. For example, frequency components of $\overline{I_{n2}^2}$ whose ω is centered around ω_{rf} (such that $\omega_{rf} = \omega_{lo} + \omega_{if}$) will appear at the IF output, even though ω_{rf} itself lies outside of the IF filter bandwidth. This is because upon switching this frequency component is translated (or aliased, if we treat switching, or mixing, as a special case of sampling) down to ω_{if} and thus lies inside the IF filter passband.

In general, noise components centered around multiples of ω_{rf}, which are normally outside the IF filter passband, can be frequency translated into the passband due to this periodic time varying property and can then pass through the IF filter. This will result in the final output noise being greater (potentially a lot greater) than that predicted by simply assuming the mixer is a LTI circuit. As a counter argument, one may argue that is not something to be worried about, as we have so far neglected the frequency response of the mixer stage 2. We may choose to argue that this stage has a frequency response with a finite bandwidth (up to now we assume that it has an infinite bandwidth) and hence would have filtered out the frequency components of these noise sources at multiples of ω_{rf} at any rate, before these frequency components even get a chance to get to the IF filter. This turns out to be overly optimistic because of the fact that by design, this mixer should be a wideband circuit so that its bandwidth is usually rather wide and such frequency filtering does not usually occur. We may further appreciate why this mixer should be a wideband circuit by recognizing the fact that the primary function of a mixer (at least for mixers used for down conversion) is to frequency translate a high-frequency RF signal to a low-frequency IF signal. This function necessitates the mixer's bandwidth to be wide enough to admit the high-frequency incoming RF signal in the first place.

4.7.2.1   Theory of Linear Periodic LPTV System

We will now develop a general theory that allows us to calculate noise for a LPTV system. Similar to Volterra series, which can be used to analyze high frequency distortion effects for any NLTI system with memory, this theory can be used to

Fig. 4.18 Block diagram of
the switching part of an
unbalanced mixer

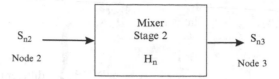

analyze noise in any LPTV system. The study of this theory also enjoys those same
benefits as the study of Volterra series, which were highlighted at the beginning of
section 4.6.1. For example, when applied to circuits it shows that the periodic time
varying effect can degrade the noise performance easily by close to 100% more
than predicted using the time invariant analysis (see section 4.7.3.3). In addition the
use of LPTV system noise analysis in mixer has been adopted in many simulators
(e.g spectreRF [7]). Meanwhile other circuits in this book where this theory can be
applied include the LC oscillator covered in chapter 7 (e.g. question 7, chapter 7
covers the basic phase noise analysis of such an LC oscillator, assuming the
oscillator is a LTI system. The theory can be used to extend the results in this
question to the case when the oscillator is treated as a LPTV system. As a matter of
fact this more sophisticated phase noise analysis has been adopted in the simulator
spectreRF [7]). The above discussions should have given the readers the proper
motivations to go through this theory, which may involve a bit of mathematics.

We introduce the concept of LPTV system by considering again the mixer stage
2 part of Figure 4.17 in Figure 4.18. In Figure 4.18, S_{n2} will be the power spectral
density function of the noise I_{n2}^2 at node 2, and S_{n3} will be the power spectral density
of noise at node 3. As discussed before due to the periodic time varying nature of
stage 2, S_{n3} will contain aliased noise. H_n is denoted to be the transfer function of
the noise from node 2 to node 3, except that since this is an LPTV system the
concept of the transfer function (as well as the input/output relationship) has to be
redefined. (We choose to use H_n to denote this transfer function because H has
customarily been used to denote the transfer function of an LTI circuit. Since this
notation is restricted to the discussion on noise, there should not be any confusion
with the use of H_n as Volterra kernel in section 4.6, which focuses exclusively on
the discussion on distortion.) In general, for such a LPTV system we have a
modified definition of input-output relationship, given as [8]

$$S_{n3}(\omega_{if}) = \sum_{n=-\infty}^{\infty} |H_n(\omega_{if})|^2 S_{n2}(\omega_{if} + n\omega_{lo}) \qquad (4.134)$$

H_n has a modified definition from the usual definition associated with the system
transfer function H of a conventional LTI system and is defined as [8]

$$H_n(\omega_{if}) = \int_{-\infty}^{\infty} \frac{1}{T} \int_0^T h(v+u,\, u)e^{jn\omega_{lo}u}du \cdot e^{-j\omega_{if}v}dv$$

$$= \frac{1}{T} \int_0^T \left[\int_{-\infty}^{\infty} h(v+u,u)e^{-j\omega_{if}v}dv \right] e^{jn\omega_{lo}u}du \qquad (4.135)$$

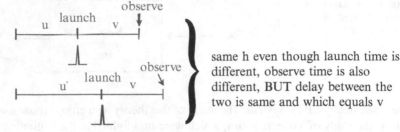

same h even though launch time is
different, observe time is also
different, BUT delay between the
two is same and which equals v

Fig. 4.19 Behavior of impulse responses of a LTI system

where $\int_{-\infty}^{\infty} h(v + u, u)e^{-j\omega_{if}v}dv$ can be interpreted as the time varying (periodic) transfer function from node 2 to node 3. Here $h(v + u, u)$ is the impulse response of the LPTV system, and T is the LO period. Now from the definition of Fourier series, it is self-evident that $\frac{1}{T}\int_0^T \left[\int_{-\infty}^{\infty} h(v+u, u)e^{-j\omega_{if}v}dv\right]e^{jn\omega_{LO}u}du$ is the nth-order coefficient of the Fourier series expansion of this impulse response. Since from (4.135) H_n is already given as $H_n = \frac{1}{T}\int_0^T \left[\int_{-\infty}^{\infty} h(v+u, u)e^{-j\omega_{if}v}dv\right]e^{jn\omega_{LO}u}du$, then H_n can now be interpreted as the nth-order coefficient of the Fourier series expansion of the impulse response.

Having introduced $h(v + u, u)$, the impulse response function of a LPTV system, first let us briefly review some of its properties. To highlight its properties, we first compare it to the impulse response of a conventional LTI system. Let us first define, in Figure 4.19, the launch time u, which is defined as the time when we launch an impulse. The observe time v, is defined as the time we wait from the moment when we launch an impulse to the moment when we observe the response of the system.

Notice that for the LTI system in Figure 4.19, the unique feature is that, no matter when you launch your impulse (in this case $u \neq u'$), as long as you wait for the same amount of time before you observe (v are the same), you see the same response, or h (=impulse response function) are identical.

On the other hand the same cannot be said of a linear time varying (LTV) system. Here if we start with different launch times, that is, $u \neq u'$, even though v is the same, h will be different. How about a LPTV system, the special case of a LTV system that is specifically applicable to that of a mixer, our present focus? As with a LTV system, in general h is not the same. However, under a special circumstance h will be the same. The special circumstance happens when the launch time differs by exactly the time T, the period with which the system is repeating itself, as shown in Figure 4.20.

Now that we have finished highlighting some qualitative differences between LTI, LTV, and LPTV systems, we will look at a more rigorous definition of the LPTV system. It should be noted that the impulse response function $h(v+u, u)$ of a LPTV system is periodic in u. Hence we can also express $h(v+u, u)$ as h(t,u). From now on we will use either expression for the impulse response, depending on which one is more convenient under the particular situation. What is the property of the Fourier transform of h? Since h is a function of two variables,

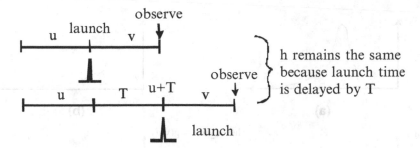

Fig. 4.20 Behavior of impulse responses of a LPTV system

its Fourier transform is also a function of two variables or it is a two-dimensional Fourier transform:

$$H(\omega_{if}, \omega_s) = \frac{1}{2\pi} \int\limits_{-\infty}^{\infty} \int\limits_{-\infty}^{\infty} h(t, u) e^{j\omega_s u} \cdot e^{-j\omega_{if} t} du dt \qquad (4.136)$$

Here ω_{if} is the IF and ω_s is any arbitrary frequency. What is the property of this two-dimensional Fourier transform? Since h is periodic in one variable, u, after taking its Fourier transform with respect to u, the resulting function is still periodic in t. Hence doing the Fourier transform of the resulting periodic function with respect to t reverts to expanding that function in a Fourier series and finding its Fourier series coefficient. Since the Fourier series coefficients are discrete (as opposed to being continuous, as in the Fourier transform case) the final function is labeled H_n, where n are integers.

In what follows we will start from (4.136) and actually derive H_n. Our goal is to show that H_n can be interpreted to be similar to H, the system transfer function of a LTI system. Before that we need to do some more clarifications on h(t,u) and also adjust its time axis though.

First let us clarify what does it mean for h(t,u) to be a function of two (rather than one) variables? Let us reassert that $h(t, u)$ is the system output that we observe at time t, where t spans from $t = 0$ to $t = \infty$, for a given launch time u. This is plotted in Figure 4.21 (a). That is, when the impulse is launched at u_1, h varies as a function of time t in one particular fashion (in our example a fast rise followed by a slow fall). When the impulse is launched at u_2, notice h also varies as a function of time, but in a different fashion, as shown in Figure 4.21 (c) (a medium rise, followed by a medium fall). This is what we mean by h(t,u) is a function of two variables.

Secondly we want to readjust the time axis of h(t,u) because the present time axis is not very convenient. Why? This is because u may be larger than 0, in which case there is a time span t from 0 to u, before the impulse is even launched, when we are observing the output. Any output being observed in this time span is meaningless. [In our example in Figure 4.21 (a), (c) we simply assume $h = 0$ at those instances, that is, $h(u_1, u_1) = 0$ and $h(u_2, u_2) = 0$.] A more convenient way is to shift the time

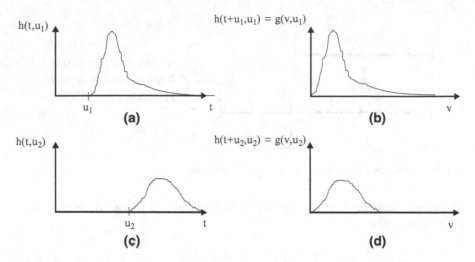

Fig. 4.21 (a), (c) Behavior of h as a function of launch time. (b), (d) Corresponding g

scale to v so that the output always start to be monitored at time zero, which is also the time when we launch the impulse. Therefore, we must have $v = t - u$. Since the time when the impulse is launched is u, which is another variable, so this shift in time axis is not constant but is a function of the launch time. Using this new time axis means that we will now be doing observation at $v > 0$, which is also the time when the output is meaningful.

Now, on this new time scale we want to define the impulse response function, denoted as g, which must equal the old impulse response function h on the old time scale. To find their relationship a convenient point is when they both must be zero. We know $h(u, u)$ must be zero (since right after we launch, the output is zero) and $g(0, u)$ must be zero (since by our choice we want this new function to always start out to be zero). Hence $g(0,u) = h(u, u)$. Since the two functions are related only by a shift of time axis and nothing else (e.g. time scale does not expand or contract), then they will be identical hereafter, each on their own time axis and so $g(0 + v, u) = h(t + u, u)$ or $g(v, u) = h(t + u, u)$. Pictorially this means that we take the original plot of $h(t, u)$, for each different u, then slide along the axis by a different amount, corresponding to the different u, and then put each of these plots individually on the v-axis. They all start out from $v = 0$ and we have created the equivalent g plot. For example the h plots in Figure 4.21 (a) and (c) are re-created as g-plots in Figure 4.21(b) and (d), respectively, following this procedure.

Having clarified h(t,u) and adjusted its time axis, let us start deriving H_n. As we stated before we start from (4.136) and substitute $v = t - u$ and $h(t, u) = g(v, u)$ and H becomes

$$H\left(\omega_{if}, \omega_s\right) = \frac{1}{2\pi} \int_{-\infty}^{\infty} \int_{-\infty}^{\infty} g(v, u) e^{j\omega_s u} \cdot e^{-j\omega_{if}(v+u)} \, du dv \qquad (4.137)$$

With a change of variable this then becomes

$$H(\omega_{if}, \omega_s) = \frac{1}{2\pi} \int\limits_{-\infty}^{\infty} \int\limits_{-\infty}^{\infty} g(v, u) e^{j(\omega_s - \omega_{if})u} \cdot e^{-j\omega_{if}v} du\, dv \qquad (4.138)$$

Since $g(v, u)$ is periodic in u with periodicity T (the period of the LO), it can be expressed as a Fourier series with angular frequency $\omega_{lo} = 2\pi/T$:

$$g(v, u) = \sum_{n=-\infty}^{\infty} g_n(v) e^{-jn\omega_{lo}u} \qquad (4.139)$$

where

$$g_n(v) = \frac{1}{T} \int\limits_0^T g(v, u) e^{jn\omega_{lo}u} du \qquad (4.140)$$

Next let us substitute (4.139) into (4.138), and upon simplification we have

$$H(\omega_{if}, \omega_s) = \int\limits_{-\infty}^{\infty} \sum_{n=-\infty}^{\infty} g_n(v)\delta(\omega_s - \omega_{if} - n\omega_{lo}) e^{-j\omega_{if}v} dv \qquad (4.141)$$

Exchanging the order of integration and summation, we can rewrite (4.141) in the following form:

$$H(\omega_{if}, \omega_s) = \sum_{n=-\infty}^{\infty} H_n(\omega_{if})\delta(\omega_s - \omega_{if} - n\omega_{lo}) \qquad (4.142)$$

where we have defined a new function H_n as

$$H_n(\omega_{if}) = \int\limits_{-\infty}^{\infty} g_n(v) e^{-j\omega_{if}v} dv \qquad (4.143)$$

Note that this function is a function of n and ω_{if}. Furthermore it is related to the Fourier coefficient g_n of the impulse response $g(v, u)$ of the system. Therefore, we call it a system function, and this is the H_n we have been seeking.

As stated above we want to show this H_n is similar to H, the system transfer function of a LTI system. To do this we have to show that H_n relates the input/output spectra of a LPTV system in a fashion similar to the way H relates the input/output spectra of a LTI system. That means we have to investigate the relationship between the output and input spectra of a LPTV system. We start off our investigation in the time domain. Similar to a LTI system, the output of an LPTV system in the time domain is related via the impulse response of the system to the input by the convolution integral:

$$y(t) = \int_{-\infty}^{\infty} h(t, u) x(u) \, du \qquad (4.144)$$

Now going back to the frequency domain the output spectrum is related to the input spectrum by taking the transform of (4.144):

$$Y(\omega) = \int_{-\infty}^{\infty} H(\omega, \omega_s) X(\omega_s) \, d\omega_s \qquad (4.145)$$

Here ω and ω_s are just two arbitrary frequencies where we want to use as the basis to observe the output frequency spectrum (analogus to two arbitrary times, t and u in the time domain). Normally though, if we are not interested in the whole spectrum, but just the response at a particular frequency, we can set the variable ω to just that frequency. In the present case the frequency response at ω_{if} is of particular interest and so we set ω equal to ω_{if} and (4.145) becomes

$$Y(\omega_{if}) = \int_{-\infty}^{\infty} H(\omega_{if}, \omega_s) X(\omega_s) d\omega_s \qquad (4.146)$$

Meanwhile $H(\omega_{if}, \omega_s)$ is related to H_n via (4.142). Therefore, we substitute (4.142) into (4.146), and after some algebraic manipulation we have

$$Y(\omega_{if}) = \sum_{n=-\infty}^{\infty} H_n(\omega_{if}) X(\omega_{if} + n\omega_{LO})$$

$$\text{and } |Y(\omega_{if})|^2 = \sum_{n=-\infty}^{\infty} |H_n(\omega_{if})|^2 |X(\omega_{if} + n\omega_{LO})|^2 \qquad (4.147)$$

This is essentially the same form as (4.134) if we map

$$|Y(\omega_{if})|^2 = S_{n3}(\omega_{if}) \text{ and } |X(\omega_{if} + n\omega_{LO})|^2 = S_{n2}(\omega_{if} + n\omega_{LO}) \qquad (4.148)$$

Examining (4.147) we finally have shown that $H_n(\omega_{if})$ so defined relates the input/output spectra of a LPTV system in a fashion similar to the way H relates the input/output spectra of a LTI system. This justifies calling H_n a system function in the sense it has similar property as the system transfer function H.

To close off the discussion, if in (4.140) we substitute $g(v, u)$ by $h(t + u, u)$ and then substitute the resulting equation into (4.143) we have H_n as:

$$
H_n(\omega_{if}) = \int_{-\infty}^{\infty} \frac{1}{T} \int_0^T h(v + u, u) e^{jn\omega_{lo}u} du \cdot e^{-j\omega_{if}v} dv
$$

$$
= \frac{1}{T} \int_0^T \left[\int_{-\infty}^{\infty} h(v + u, u) e^{-j\omega_{if}v} dv \right] e^{jn\omega_{lo}u} du \qquad (4.149)
$$

This agrees with the original form we have stipulated for H_n in (4.135).

We have now finished our discussion of a LPTV system. Next we want to apply the results derived to the specific LPTV system we are interested in, the mixer stage 2 as described in Figure 4.17 and Figure 4.18. In particular, we want to find the H_n of this specific LPTV system.

4.7.2.2 Special V_{lo} Switching Case

It is normally hard to get a close form solution for the H_n of a LPTV system. Simulators such as spectra RF [7] are typically used instead. Occasionally if the LPTV system is simple enough, close form solution is available. For example, we will attempt to find the close form solution of H_n for a simple mixer in problem 4.12(b). Alternatively when a LPTV system (such as mixer stage 2) operates under a special circumstance, its H_n can be determined rather easily.

To find out what the special circumstance is, we first derive a LTI system by applying a constant signal (its value equals the average value of V_{lo}) to the mixer stage 2, a LPTV system. This LTI system contains poles and zeroes whose lowest frequency is denoted as $f_{min\_pole\_zero}$.

If the condition

$$
f_{min\_pole\_zero} \gg f_{lo} \qquad (4.150)
$$

is satisfied, then we are operating under the special V_{lo} switching case. The LPTV system becomes a special LPTV system. H_n for this special LPTV system becomes rather easy to find.

For example, a mixer with an impulse response $h(t, u)$, described in Figure 4.22, satisfies this condition. For this mixer, once we fix the launch phase, the poles and zeroes of the resulting LTI system that corresponds to this launch phase are all at much higher frequencies than f_{lo}. (Notice that once the launch phase is fixed, the mixer is an LTI system and the concept of poles and zeroes applies). How do we know that? We find that this is true by examining the $h(t, u)$ corresponding to different launch times. Six of them are plotted in Figure 4.22, corresponding to launch times u_1 to u_6. From Figure 4.22, notice that in all six cases $h(t, u)$ rises and falls rapidly, when compared with T (the LO period). Since the rise and fall time is a

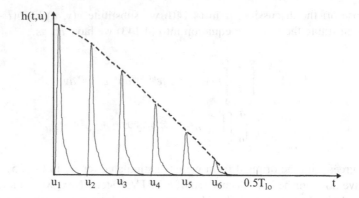

Fig. 4.22 Impulse response of a mixer operating under the special V_{lo} switching case

function of poles and zeroes, this observation indirectly tells us that the poles and zeroes are at frequencies much higher than f_{lo}. It is also precisely this property that lets the general shape of $h(t, u)$ to be rather independent of the launch time u, with only the peaks decaying. Specifically, in this example $h(t, u_1)$, $h(t, u_2)$, $h(t, u_3)$, $h(t, u_4)$, $h(t, u_5)$, $h(t, u_6)$ all have similar shapes, except that their peaks decrease as u goes from u_1 to u_6. In the circuit sense, satisfying (4.150) means that since the LO signal changes slowly, the system can be treated as a time invariant system.

Next, if mixer stage 2 of an unbalanced mixer does satisfy (4.150) and can be represented by a special LPTV system, how do we represent this special LPTV system? For illustration purposes we assume that the LO signal is a sine wave, or $V_{lo} = A_{lo} \cos(\omega_{lo} t)$, and hence the internal operating points change in a sinusoidal fashion, but change in such a way that the locations of poles/zeroes remain practically the same. Hence the shapes of the $h(t, u)$ remain the same as u changes. Of course, since the internal operating points change, the peaks of $h(t, u)$ change accordingly. Following [3], mixer stage 2, which is just a SCP, has switching characteristics as described by the S-function in Figure 4.2. It is assumed that A_{lo} is much larger than V_+ and V_-. Hence when the impulse is launched at a time when the LO signal is at its minimum, the peak of the impulse response is at its minimum (0). Conversely, when the impulse is launched at a time when the LO signal is at its maximum, the peak impulse response is at its maximum (peak_max). If we plot the impulse response of mixer stage 2, denoted as $h_{mixer\_stage2}(t, u)$, then it will be shown to exhibit a trend similar to the $h(t, u)$ of the mixer as described in Figure 4.22. For example, let us assume that an impulse is launched at the same time as u_6 in Figure 4.22. We assume the mixer described in Figure 4.22, and also the mixer stage 2 use the same V_{lo} and have the same LO period. Hence, similar to Figure 4.22, we assume that this launch time is close to 1/2 of T_{lo}. At that time the LO signal is close to $A_{lo} \cos(180°)$ or close to $-A_{lo}$. The LO signal is at its minimum value. Hence a minimum LO signal is applied to mixer stage 2 in Figure 4.17. Since mixer stage 2 corresponds to the SCP M_{1-2} of the unbalanced mixer in Figure 4.3,

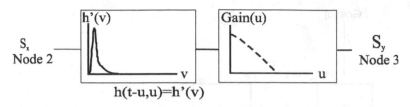

Fig. 4.23 An equivalent representation of mixer stage 2 in the special V_{lo} switching case

a minimum LO signal means that $V_{lo}{}^{+}$ is at its most negative value and $V_{lo}{}^{-}$ is at its most positive value. Consequently transistor M_1 in Figure 4.3 is off. Correspondingly the impulse injected at the input of mixer stage 2 in Figure 4.17(which corresponds to the source of M_1) does not get propagated through and the impulse response should have a small peak (close to 0) or $h_{\text{mixer\_stage2}}(t, u_6)$ has a small peak. This is similar to $h(t, u_6)$ as described in Figure 4.22.

To further demonstrate the similarity between $h_{\text{mixer\_stage2}}(t,u)$ and $h(t,u)$ we now launch an impulse to mixer stage 2 at the same time as u_1 in Figure 4.22. As in Figure 4.22, this launch time is close to the beginning of T and so the corresponding LO signal is close to $A_{lo} \cos(0°)$ or close to A_{lo}. The LO signal is at its maximum value and transistor M_1 is fully on. The impulse gets propagated through and the impulse response should have a maximum peak (close to peak_max) or $h_{\text{mixer\_stage2}}(t, u_1)$ has a maximum peak. This is similar to $h(t, u_1)$ as described in Figure 4.22. Now for impulses launched from times u_2-u_5, the SCP $M_{1\text{-}2}$ in Figure 4.3 is operating in the linear region of Figure 4.2 (between V_+ and V_-) and M_1 is partially on. Accordingly, the degree in which M_1 is on changes. Hence the peak response also decreases accordingly from u_2 to u_5, which exhibits a similar trend to the $h(t, u)$ of the mixer as shown in Figure 4.22, where the decreasing peaks are connected by a dotted line.

To summarize, the peak of the impulse response, $h_{\text{mixer\_stage2}}(t, u)$ for mixer stage 2, corresponding to an impulse launched at time u, is proportional to $S[V_{lo}(u)]$ or $S[A_{lo} \cos \omega_o u]$ in Figure 4.2. To model this behavior, the mixer stage 2 is now represented by the equivalent system as shown in Figure 4.23. This equivalent system consists of a cascade of two blocks. The first block is a LTI system whose impulse response, h_{LTI}, is obtained from mixer stage 2 by fixing the launch time at $u = u_q$, or $h_{\text{LTI}}(v) = h_{mixer\_stage2}(v, u_q)$. Here u_q stands for launch time when mixer stage 2 is in its quiescent state. We further assume that mixer stage 2 is at its quiescent state when V_{lo} is midway between its minimum and maximum value. (Physically to obtain h_{LTI}, we take mixer stage 2, apply a constant value that equals the average of V_{lo} at the LO port, apply an impulse at the input and the response obtained equals h_{LTI}.) Returning to Figure 4.23 the output from this LTI system is fed to the second block, a gain block whose gain, gain(u), is a function of launch time u. This gain block is used to model the dependency of the peak of the impulse response on V_{lo} and hence gain (u) is also proportional to $S[V_{lo}(u)]$. With this equivalence, we can set $h_{mixer\_stage2}(v + u, u)$ to $h_{\text{LTI}}(v)\text{gain}(u)$.

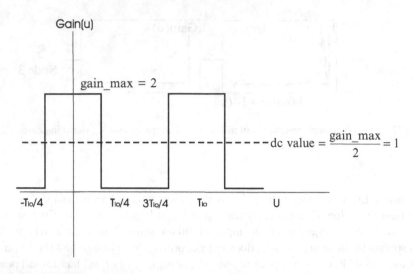

Fig. 4.24 Mixer stage 2 is represented as a special LPTV system. The gain part of this special LPTV system is plotted as a function of launch time

Using this equivalence model, we can calculate H_n of mixer stage 2. Let us start from (4.149) and set $h(v + u, u)$ to $h_{LTI}(v)gain(u)$. Hence

$$
\begin{aligned}
H_n(\omega_{if}) &= \frac{1}{T}\int_0^T \left[\int_{-\infty}^{\infty} h_{LTI}(v)gain(u)e^{-j\omega_{if}v}dv\right]e^{jn\omega_{lo}u}du \\
&= \left[\int_{-\infty}^{\infty} h_{LTI}(v)e^{-j\omega_{if}v}dv\right]\frac{1}{T}\int_0^T gain(u)e^{jn\omega_{lo}u}du \\
&= H_{LTI}(\omega_{if})\frac{1}{T}\int_0^T gain(u)e^{jn\omega_{lo}u}du
\end{aligned}
\tag{4.151}
$$

Remember that $gain(u)$ is proportional to $S[A_{lo}\cos\omega_o u]$. What is $S[A_{lo}\cos\omega_o u]$?

We have stated before that the characteristics of mixer stage 2 of the unbalanced mixer is described by Figure 4.2. If we apply $A_{lo}\cos(\omega_o u)$ to the LO port of mixer stage 2 of this unbalanced mixer, $S[A_{lo}\cos\omega_o u]$ can be obtained from Figure 4.2. Specifically, if

$$
A_{lo} \gg V_+ \text{ and } A_{lo} \gg V_-,
\tag{4.152}
$$

then from Figure 4.2 the mixer stage 2 is seen to behave in such a way that for almost 50% of the time M_1 in Figure 4.3 is turned off, and the gain is 0; and for almost the other 50% of the time, M_1 is fully on, and the gain is at its maximum value, gain_max. Therefore, $gain(u)$ varies as a square wave as shown in Figure 4.24. Now we want to verify that gain_max is 2, as depicted in Figure 4.24. This is done by applying (4.151) to the case when the mixer is not switching.

Conceptually, nonswitching means that $H_n = H_{LTI}$. This means that in (4.151) the left-hand side becomes H_{LTI}. How about the right-hand side? Note that for the nonswitching case, conceptually this also means $\omega_{lo} = 0$. Then the right-hand side becomes $H_{LTI} \cdot \frac{1}{T} \int_0^T gain(u)\, du$. Applying the function for gain(u) as shown in Figure 4.24 to this expression, it becomes $H_{LTI} \frac{gain\_max}{2}$. Hence (4.151) becomes $H_{LTI} = H_{LTI} \cdot \frac{gain\_max}{2}$. Solving,

$$gain\_max = 2 \tag{4.153}$$

Using this value of gain_max, gain(u) can be written as:

$$gain(u) = 2\Pi\left(\frac{t}{T}\right) + 1 \tag{4.154}$$

where $\Pi\left(\frac{u}{T}\right) = 1/2$ for u between $-T/4$ and $T/4$ and $-1/2$ between $T/4$ and $3/4T$.

Substituting (4.154) into (4.151) and solving [4], we have

$$H_n = H_{LTI}\operatorname{sin}c\frac{n}{2} \tag{4.155}$$

In summary we have determined H_n for mixer stage 2 under the special V_{lo} switching case.

Looking at (4.155) we can see that H_n goes down rather rapidly as n increases. As an example, $n = 0$, $H_0 = H_{LTI}$. When $n = 1$, we have $\operatorname{sinc}\frac{1}{2} = \frac{\sin \pi \frac{1}{2}}{\pi \frac{1}{2}} = \frac{2}{\pi} \approx 0.63$. Accordingly, $H_1 \approx 0.63 H_{LTI}$ or $\frac{H_1}{H_0} = \frac{2}{\pi} \approx 0.63$

Therefore,

$$H_0 = H_{LTI} \quad |H_1| \approx 0.63 H_{LTI} \tag{4.156}$$

4.7.3 Analysis of Noise in Unbalanced Mixer (V_{lo} not Switching and Special V_{lo} Switching Cases)

Assume that we are investigating the unbalanced mixer as shown in Figure 4.3.

Furthermore, assume that this unbalanced mixer operates at

$$RF = 1.9 \text{ GHz}, \quad BW = 1.7 \text{ MHz}$$

$$LO \text{ frequency} = 1.8 \text{ GHz}, \quad IF = 100 \text{ MHz}$$

and $V_{lo} = A_{lo} \cos \omega_o u$. Hence its average value is 0. A_{lo} is set to be much larger than V_+ and V_-.

The mixer's V-I converter M_3 has $V_{GS_3} - V_t = 5.1\ V$, switch M_{1-2} has an overdrive voltage, $V_{GS1}\text{-}V_t = 1V$;

$$M_{1-2} \text{ has a } \frac{W}{L} \text{ of } 50\ um/0.6um;\ M_3 \text{ has a } \frac{W}{L} \text{ of } 3.8\ um/0.6um.$$

The unbalanced mixer is built with a 0.6 um CMOS technology that has the following characteristics:

$$k' = 75\ uA/V^2$$

$$C_{ox} = 2fF/um^2$$

In this sub-section we are interested in finding this mixer's NF.

Let us represent this unbalanced mixer as shown in Figure 4.17. Again we assume that all the noise comes from stage 1 and neglect any noise contribution from stage 2. We further divide our discussion into two scenarios:

1. Mixer stage 2 is operating under the V_{lo} not switching case.
2. Mixer stage 2 is operating under the special V_{lo} switching case.

The calculation of NF is divided into three steps.

4.7.3.1 Step 1: Calculate S_{n2}

In step 1 we calculate the PSD of the noise from stage 1. This is because for both scenarios 1 and 2 the primary noise source comes from this stage. Now stage 1 consists of a V-I converter, so let us go through a simple example of noise's PSD calculation for this V-I converter.

From (4.133) the PSD of this V-I converter input referred voltage noise is given by

$$S_{n1} = \frac{8kT}{3g_{m_3}} \tag{4.157}$$

We are given that $\left(\frac{W}{L}\right)_3$ is 3.8um/0.6um. We are also given $V_{GS_3} - V_t = 5.1V$, $k' = 75\ uA/V^2$. Hence

$$g_{m_3} = k'(W/L)_3(V_{GS3} - V_t) = 2.4m\Omega^{-1} \tag{4.158}$$

Substituting (4.158) into (4.157) and taking the square root, we have

$$\sqrt{S_{n1}} = \sqrt{\frac{8kT}{3g_{m3}}}$$

$$= 2.8\sqrt{\frac{4x10^{-21}}{3}} \times \frac{1}{2.4 \times 10^{-3}} \frac{V}{\sqrt{Hz}}$$

$$\cong \frac{2.1\ nV}{\sqrt{Hz}} \tag{4.159}$$

Referring to Figure 4.17, we have

$$S_{n2} = S_{n1} \times g_{m_3}^2 = \left(\frac{2.1\ nV}{\sqrt{Hz}}\right)^2 \times \left(2.4\ m\Omega^{-1}\right)^2 = \frac{(5\ pA)^2}{Hz} \tag{4.160}$$

4.7.3.2 Step 2: Find H_{LTI}

In step 2 we are supposed to first find the average value at the LO port in order to calculate H_{LTI} for both scenarios (1) and (2). Fortunately this average value is given at the beginning of sub-section 4.7.3 to be 0 and so there is no need to calculate.

With this average value H_{LTI} for both scenarios (1) and (2) are the same and can be found by applying an average value of 0 V at the LO port. If we do that and examine M_{1-2} in Figure 4.3, we find that M_{1-2} is a SCP that is balanced. Using the half-circuit concept, each transistor operates like a common gate (CG) stage, with current input and current output. The magnitude of the low-frequency gain is ½ since the current splits evenly between two branches and i_{if} only gets ½ of i_{rf} and so

$$|H_{LTI}(0)| = \frac{1}{2} \tag{4.161}$$

Also, for a CG stage with current input and output, there is one dominant pole whose frequency is given approximately by

$$\omega_t = \frac{g_{m1}}{C_{gs1}} \tag{4.162}$$

Therefore the −3dB frequency (f_{-3db}) of H_{LTI} is the same as this pole frequency. To calculate numerically this pole frequency, we first find C_{gs1}

$$C_{gs1} = W_1 L_1 C_{ox} \tag{4.163}$$

We are given a $(W/L)_1$ of 50u/0.6u and a C_{ox} of 2fF/um^2. Substituting we have

$$C_{gs1} = 50um \times 0.6um \times 2\ fF/um^2 = 60\ fF \tag{4.164}$$

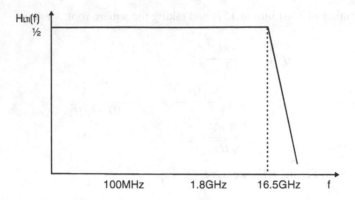

Fig. 4.25 Mixer stage 2 is represented as a special LPTV system. Frequency response of the LTI system part of this special LPTV system is shown

Next we calculate g_{m1}. We are given an overdrive voltage V_{GS1} - V_t of 1 V, $(W/L)_1$ of $50um/0.6um$ and $k' = 75$ umA/V^2. Hence

$$g_{m_1} = k'(W/L)_1(V_{GS1} - V_t) = 6.2m\Omega^{-1} \tag{4.165}$$

Substituting (4.164), (4.165) in (4.162), we finally find the numerical value of the pole frequency

$$\omega_t = \frac{g_{m1}}{C_{gs1}} \approx 100 Grad/s \, or f_t \approx 16.5 GHz \tag{4.166}$$

Using (4.161), (4.166) we can obtain the Bode plot for the magnitude response of H_{LTI}, as shown in Figure 4.25.

4.7.3.3   Step 3: Find NF

In step 3 we calculate the NF for both scenarios: V_{lo} not switching and special V_{lo} switching cases.

Scenario (1): V_{lo} not switching case

Mixer stage 2 can be treated as a LTI circuit whose frequency response is the same as that of H_{LTI}, which is shown in Figure 4.25. Hence we have

$$S_{n3} = |H_{LTI}(100MHz)|^2 S_{n2} \tag{4.167}$$

From Figure 4.25, $H_{LTI}(100\text{ MHz}) = H_{LTI}(0)$ which in turn is given in (4.161) to be ½. From (4.160) $S_{n2}(100\text{ MHz}) = \frac{(5pA)^2}{Hz}$. Substituting in (4.167) we have

$$S_{n3} = 0.25 \times \frac{(5pA)^2}{Hz} \tag{4.168}$$

Now what is the NF of the mixer for this scenario, assuming a 50 Ω R_s? N_{dev}, input referred noise of the mixer, is given by

$$N_{dev} = \frac{S_{n3}}{\left(H_{LTI}(100MHz) \times g_{m_3}\right)^2} = \frac{0.25 \times (5pA)^2/Hz}{(0.5 \times g_{m_3})^2} = \frac{0.25 \times (5pA)^2/Hz}{\left(0.5 \times 2.4m\Omega^{-1}\right)^2} \tag{4.169}$$

Using (2.67) and noting N_{dev} in (4.169) equals N_{Device}/G in (2.67) we have

$$NF = 1 + \frac{N_{dev}}{N_{Source-resis\tan ce}} = 1 + \frac{N_{dev}}{4kTR_s} = 1 + \left(\frac{0.25 \times (5pA)^2/Hz}{\left(0.5 \times 2.4m\Omega^{-1}\right)^2}\right)$$

$$\times \frac{1}{4 \times 4 \times 10^{-21} \times 50\Omega/Hz}$$

$$= 1 + 5.625 = 6.625 = 8.2dB \tag{4.170}$$

This is slightly better than the 12 dB of NF assigned to this mixer in Chapter 2 (see Table 2.1, row 4, column 5).

Scenario (2): Special V_{lo} Switching case

Let us refer to Figure 4.17 again, where mixer stage 2 is a LPTV system with S_{n2} as the input noise PSD and S_{n3} as the output noise PSD. Using (4.134) and we have

$$S_{n3}(\omega_{if}) = \sum_{n=-\infty}^{\infty} |H_n(\omega_{if})|^2 S_{n2}(\omega_{if} + n\omega_{lo})$$

$$= |H_0(\omega_{if})|^2 S_{n2}(\omega_{if}) + |H_1(\omega_{if})|^2 S_{n2}(\omega_{if} + \omega_{lo}) + |H_{-1}|^2 S_{n2}(\omega_{if} - \omega_{lo}) + \ldots \tag{4.171}$$

In scenario (1) we have derived a LTI system from this LPTV system by applying a constant signal to its LO port. Its transfer function has a frequency response as given in Figure 4.25. Since this LTI system has only one pole, its $f_{min\_pole\_zero}$ is given by that pole frequency. This was calculated in (4.166) to be 16.5 GHz.

Substitute this value and a given f_{lo} of 1.8 GHz in (4.150) and it is seen that (4.150) is justified ($f_{min\_pole\_zero}$ is about 10 times higher than f_{lo}). Hence mixer stage 2 is operating under the special V_{lo} switching case (sub-section 4.7.2.2) and therefore we can find H_n by using (4.151).

In addition, we are given at the beginning of sub-section 4.7.3 that $A_{lo} \gg V_+$ and $A_{lo} \gg V_-$ and hence (4.152) is satisfied. Then (4.155) applies, which is repeated for reference:

$$H_n = H_{LTI} \sin c \frac{n}{2} \tag{4.172}$$

It can be seen that H_n goes down rapidly as n increases, so as a simplification we include only the H_0, H_1 and H_{-1} terms and ignore higher orders terms in (4.171). H_0, H_1 and H_{-1} can be obtained from (4.156), repeated here for reference:

$$H_0 = H_{LTI} \quad |H_1| \approx 0.63 H_{LTI} \quad |H_{-1}| \approx 0.63 H_{LTI} \tag{4.173}$$

Substitute (4.173), (4.160) into (4.171), ignoring higher order terms, we can then find S_{n3} at 100 MHz as follows:

$$S_{n3} = |H_{LTI}(100 \ MHz)|^2 \frac{(5 \ pA)^2}{Hz} + |0.63 H_{LTI}(100 \ MHz)|^2 \frac{(5 \ pA)^2}{Hz} \times 2 \tag{4.174}$$

We have calculated in scenario (1) that H_{LTI} (100 MHz) = ½. Substituting in (4.174), we have

$$S_{n3} = 0.25 \frac{(5pA)^2}{Hz} + 0.2 \frac{(5pA)^2}{Hz}$$

$$= 0.45 \times \frac{(5pA)^2}{Hz} \tag{4.175}$$

Compared to (4.168), one can see that including the periodic time varying effect introduced in the special V_{lo} switching case actually increases the noise by about 80%. Of course, if we have significant noise contribution at the frequencies where H_1, H_{-1} will alias the noise in (in this case at 1.9 GHz, -1.7 GHz), then we can see from (4.171) that including the periodic time varying effect is crucial. Hence under those circumstances where there is significant wideband noise (not filtered out by any previous filters), adopting the noise calculation incorporating periodic time varying effect is particularly useful.

Finally, what is the NF of the mixer for this scenario? (We assume that the concept of equivalent input noise, already derived for an LTI system, just carries over to the LPTV system and therefore the concept of NF simply carries over.

The conversion gain is simply assumed to be the same as the conversion gain in the LTI case.) N_{dev}, input referred noise, is given by

$$N_{dev} = \frac{S_{n3}}{\left(H_{LTI}(100MHz) \times g_{m_3}\right)^2} = \frac{0.45 \times (5pA)^2/Hz}{\left(0.5 \times g_{m_3}\right)^2} = \frac{0.45 \times (5pA)^2/Hz}{\left(0.5 \times 2.4\,m\Omega^{-1}\right)^2} \quad (4.176)$$

Hence

$$NF = 1 + \frac{N_{dev}}{N_{Source-resis\tan ce}} = 1 + \frac{N_{dev}}{4kTR_s} = 1 + \left(\frac{0.45 \times (5\,pA)^2/Hz}{\left(0.5 \times 2.4\,m\Omega^{-1}\right)^2}\right)$$

$$\times \frac{1}{4 \times 4 \times 10^{-21} \times 50\,\Omega/Hz}$$

$$= 1 + 10.12 = 11.12 = 10.5\,dB$$

$$(4.177)$$

This is slightly better than the 12 dB of NF assigned to this mixer in Chapter 2. We will show in problem 4.13) that if there is a strong noise component at 1.9 GHz, NF will become substantially worse.

4.8 A complete Active Mixer

Let us take Figure 4.6 and add the proper biasing circuitry and other auxiliary circuits and we get Figure 4.26 where M_1, M_2, M_{17}, M_3-M_6 form the basic quad pair (or Gilbert mixer). Cascode devices M_7, M_8 are inserted between M_{3-6} and M_{1-2} to provide isolation between the LO and RF ports. M_9-M_{10} form the current sources. Together with resistors formed by biasing M_{T1}-M_{T2} in the triode region they are used as loads. M_{T1}-M_{T2} usually have resistance values much smaller than the resistance values of M_9-M_{10}. To bias M_{T1}-M_{T2} in the triode region, we use I_{gain}, M_{14} and V_{bias1} to set V_{G14}, which is in turn used to bias the gate of M_{T1}-M_{T2}. If I_{gain} increases V_{G14} decreases, then $V_{GS_{MT1}}$, $V_{GS_{MT2}}$ increase and the resistor's resistance decreases. The conversion gain, which equals $\frac{4}{\pi} \times g_{M_1} \times (R_{MT_1} \| R_{M_9}) \cong \frac{4}{\pi} \times g_{M_1} \times R_{MT_1}$, increases. These two resistors can sometimes be formed by P-diffusion. M_{T1}-M_{T2} are also used in the common mode feedback (CMFB) loop formed from M_{12}, M_{13}, M_{11}, M_9, M_{10}. When V_{out}^+ increases and V_{out}^- decreases, V_{center} stays constant and so the loop does not respond to differential mode output change. If due to mismatch between the top and bottom transistors (e.g. M_9 and M_1 or M_{10} and M_2) both V_{out}^+ and V_{out}^- increase, then V_{center} increases by the same amount. This steers the current I_{CM} into M_{13} and away from M_{12}, leading to a corresponding reduction of the current in M_{11}. Hence there is a reduction of $V_{GS_{11}}$ and hence V_{GS_9}, $V_{GS_{10}}$ decrease. Therefore, M_9, M_{10} pump less

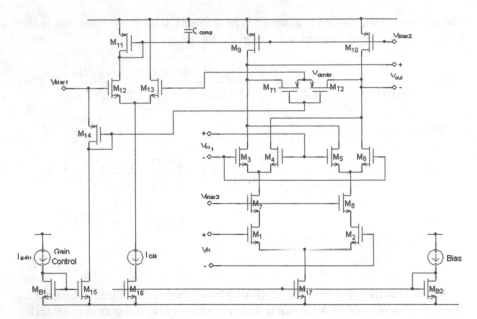

Fig. 4.26 A complete active mixer

current into M_{T1}, M_{T2} and V_{out}^+, V_{out}^- decrease, thus opposing the original increase in V_{out}^+, V_{out}^-. V_{center} in steady state is set to V_{bias1}. C_{comp} is the compensation capacitor used to stabilize the CMFB loop.

4.9 Problems

4.1. Since the nature of the signals, V_{lo} and V_{rf}, are different, mixers in Figure 4.4 and Figure 4.5 are not exactly the dual of one another. If one includes parasitic effects, such as mismatch between M_1, M_2, and parasitics capacitance, such as C_{gd}, which mixer has larger feedthrough (from LO to IF)? How about reradiation (from LO to RF)?

4.2. Show mathematically that the mixer output current shown in Figure 4.8 is balanced and does not contain any RF feedthrough.

4.3. Comment on the validity of equation (4.9). Under what circumstance would the equation not be true?

4.4. Show that from the equation $IIP_3|_{dBm} = P_i|_{dBm} - \frac{IM_3|_{dB}}{2}$ we can obtain the equation $A_{IP3}^2 = A_{interference}^2 / IM_3$.

4.5. Repeat Numerical example 4.3, this time with an IIP_3 of −10 dbm and a power of 5 mW.

4.6. Repeat cases 1 and 2 as discussed in sub-section 4.6.1.3, except this time use sinusoidal inputs. Show that there are seven terms in the case for two input sinusoids.

4.7. We have shown that the shorthand notation $a_2 Y(j\omega_a) \cdot Y(j\omega_b) \circ S_{iy}^2$ does generate terms for $a = 1$, $b = 1$ that agree with (4.60).

    a. Repeat that for $a = 1$, $b = 2$ and $a = 2$, $b = 2$. Show the terms generated.

    b. Repeat the case when there are three frequency components at the input: ω_1, ω_2, ω_3.

    c. We are interested in the third-order term in (4.59), $a_3 [Y \circ S_{iy}]^3$. Suppose there are three input frequencies: ω_1, ω_2, ω_3. Show the terms generated for all the cases corresponding to relevant assignments of a, b, c. Repeat for two input frequencies.

4.8. In sub-section 4.6.2.4 we isolated the first-order term by working on (4.100) and deriving (4.101). Try to isolate the first-order term from (4.99) instead. Show and justify which of the three terms in (4.99) do/do not contribute to the first-order term and rederive (4.101).

4.9. In Numerical example 4.4, change V_{GS1} - V_T to 0.00387 V while keeping the rest of the parameters the same. What is IM$_3$? Would IM$_3$ still be 3HD$_3$?

4.10. After (4.118) we mention that G_1 in (4.116) (derived using Volterra series), when normalized, agrees with a_1 in (4.23) (derived using Taylor series). Verify this. What is the normalizing factor?

4.11. This problem examines some subtleties of an LPTV system

    a. For an LPTV system, let us launch an impulse at u and observe the response at v. Now let us delay by T and observe the response at $v + T$, where T is the period with which the system is repeating itself. Would we get the same response?

    b. Show how (4.147) is obtained from (4.142) and (4.146).

4.12. At the end of Section 4.1 we mention that we can use R_{on} of a MOS transistor biased in the triode region as a variable gain amplifier and hence a mixer. This mixer will be treated in more detail in Chapter 5. However, because of its simplicity, we can use it to illustrate some principles of an LPTV system. The following are two different implementations of this mixer: one case loaded with a resistor (Figure P4.1) and the other case with a capacitor (Figure P4.2). Assume that the V_{lo} is a square wave. Furthermore, assume that this mixer has a $f_{min\_pole\_zero}$ that satisfies (4.150).

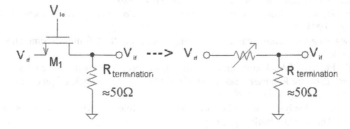

Fig. P4.1

a. The mixer loaded with a resistor $R_{termination}$, together with the equivalent network, is shown in Figure P4.1. Notice that in the equivalent network, the resistor with an arrow is used to indicate a linear but time-varying resistor. This is not to be confused as a nonlinear, time-invariant resistor. What do you think h, the impulse response of this equivalent network, is? How about H_1/H_0?

b. The mixer loaded with a capacitor C, together with the equivalent network, is shown in Figure P4.2. What do you think h, the impulse response of this equivalent network, is? How about H_1/H_0?

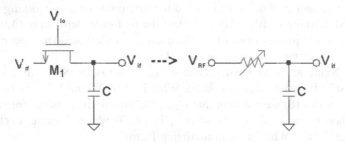

Fig. P4.2

4.13. Repeat the calculation in sub-section 4.7.3.3, scenario 2: special V_{lo} switching case, except this time assume that that there is a strong noise component at 1.7 GHz, 1.9 GHz whose PSD around 1.7 GHz, 1.9 GHz is given by $(50\,pA)^2/Hz$. Calculate the NF and show that it is significantly worse.

References

1. R. Ziemer, W. Tranter, D. Fannin, "Signals and systems: continuous and discrete", Macmillan, 1983
2. R. Meyer, "Advanced Integrated Circuits for Communications, " course notes, EECS 242, U.C. Berkeley, 1994
3. P. Gray, R. Meyer, "Analysis and design of analog integrated circuits", Wiley and Sons, 1993
4. Carlson, "Communication systems", McGraw Hill, 1986
5. Cynthia D. Keys, "Low Distortion Mixer for RF Communication", Doctoral Thesis, U.C. Berkeley, 1994
6. http://www.cadence.com/datasheets/dat_pdf/pdistoapp.pdf, Affirma RF Simulator (SpectreRF) user guide
7. Joel R. Phillips, "Analyzing time-varying noise properties with SpectreRF", Affirma RF simulator (Spectre RF) user guide appendix I, Cadence Openbook IC product documentation, Cadence Design Systems, 1998
8. C. D. Hull, "Analysis and optimization of monolithic RF RF downconversion receivers", Doctoral Thesis, U.C. Berkeley, 1992

Chapter 5
Passive Mixer

5.1 Introduction

In this chapter we focus on passive mixers. One of the most obvious trade-offs between an active and a passive mixer is that of gain versus distortion. Active mixers provide gain and dissipate quiescent power. A Gilbert mixer is such an example that achieves gain through an active predriver (the *V-I* converter). This *V-I* converter is highly nonlinear and hence the Gilbert mixer distortion performance is worse. Passive mixers, on the other hand, require only dynamic power. They have a conversion gain of less than one (conversion loss) but can achieve excellent distortion performance.

An active mixer is typically operated in the continuous time domain. The Gilbert mixer discussed in Chapter 4 is a good example. For passive mixers, some structures operate in the continuous time domain and some structures operate in the sampled data domain. A passive mixer that operates in the continuous time mode is denoted as a switching mixer. On the other hand a passive mixer that operates in sampled data domain is called a sampling mixer. Because of the sampling nature, a sampling mixer can also operate in the subsampling mode. Structurally a switching mixer is usually terminated with a resistor and may or may not have a capacitor. On the contrary, a sampling mixer is always terminated with only a capacitor.

This chapter begins by studying the switching mixer, followed by the sampling mixer. Finally to understand the sampling mixer, another viewpoint is presented in the appendix: the sampling mixer is structurally the same as an existing circuit, called the sample and hold (SAH). This SAH will be used in Chapter 6, as part of the analog-to-digital converter.

B. Leung, *VLSI for Wireless Communication*, DOI 10.1007/978-1-4614-0986-1_5,

5.2   Switching Mixer

5.2.1   Unbalanced Switching Mixer

The principle of operation of a simple switching mixer is explained by referring to Figure 5.1. When V_{lo} is positive M_1 is on, and $V_{if} = V_{rf}$. When V_{lo} is negative M_1 is off, $V_{if} = 0$. The output signal appears to be "chopped." Since the amplitude of V_{lo} exceeds V_+ and V- the output signal can be mathematically described as the input signal multiplied by the function $S(t)$, where $S(t)$ is described in Figure 1.2 of chapter 4. Given the mixer input signal

$$V_{rf}(t) = V_{RF} \cos \omega_{rf} t$$

the output voltage is

$$V_{if}(t) = V_{rf}(t)S(t) \tag{5.1}$$

$$= V_{RF} \cos \omega_{rf} t \left(\frac{1}{2} + \sum_{n=1}^{\infty} \frac{\sin \frac{n\pi}{2}}{\frac{n\pi}{2}} \cos(n\omega_{lo} t) \right) \tag{5.2}$$

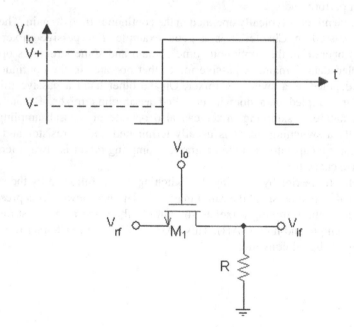

Fig. 5.1 Unbalanced switching mixer

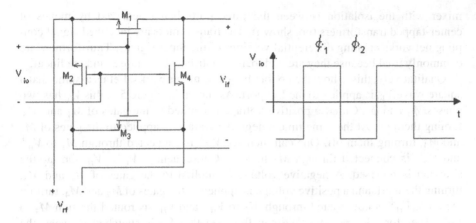

Fig. 5.2 Double balanced switching mixer

Notice that this equation is essentially the same as (1.2) of Chapter 4, with $G_0 = 1$. As in Chapter 4, we use an appropriate bandpass filter to obtain the mixed-down component at $\omega_{if} = \omega_{rf} - \omega_{lo}$, which we denote as V_{if} again:

$$V_{if}(t) = \frac{1}{\pi} V_{RF} \cos\left(\omega_{rf} - \omega_{lo}\right)t \tag{5.3}$$

Applying the definition of G_c as given in Chapter 4 to (5.3), we have

$$G_c = \frac{1}{\pi}. \tag{5.4}$$

Notice (5.1)-(5.4) bear resemblance to (1.1) through (1.5) of Chapter 4. This highlights the fact that the switching mixer can also be derived from the variable gain model of the mixer (Figure 1.1, Chapter 4) and should convince readers of the wide applicability of this model.

As with active mixers, switching mixers can be classified using the balance concept. Under this classification the simple switching mixer of Figure 5.1 is denoted as an unbalanced switching mixer.

5.2.2 Single and Double Balanced Switching Mixer

A single balanced switching mixer can be constructed by connecting a pair of unbalanced switching mixers. If, in turn, we connect a pair of single balanced switching mixers, we get a double balanced switching mixer, which is a counterpart to the Gilbert mixer. The mixer shown in Figure 5.2 is a double balanced switching mixer with the isolation between the three ports being achieved by means of center-tapped transformers (not shown). The transformers are idealized signal coupling networks creating differential versions of the input signals. Transformers are commonly used because they are wideband, contribute little noise, and are linear.

Qualitatively, this is how the double balanced mixer works. Let us assume that a square wave V_{lo} is applied to the LO port. As shown in Figure 5.2, this V_{lo} has two phases, ϕ_1 and ϕ_2. On ϕ_1 a positive voltage is applied to the gates of M_1 and M_3, turning them on. At the same time a negative voltage is applied to the gates of M_2 and M_4, turning them off. One can then see V_{rf}^+ is connected through M_1 to V_{if}^+ and V_{rf}^- is connected through M_3 to V_{if}^-. Consequently, $V_{if} = V_{rf}$. On ϕ_2 the situation is reversed. A negative voltage is applied to the gates of M_1 and M_3, turning them off, and a positive voltage is applied to the gates of M_2 and M_4, turning them on. V_{rf}^+ is now routed through M_2 to V_{if}^- and V_{rf}^- is routed through M_4 to V_{if}^+. Therefore, $V_{if} = -V_{rf}$. We can thus see that V_{if} is switched between the positive and negative value of V_{rf} and a switching operation is realized. Notice that this switching action is very similar to the switching action described in Figure 3.3 of Chapter 4. The major difference is that here voltage (V_{rf}), as opposed to current (I_{rf}), is being switched.

5.2.3 Nonidealities

The followings are some of the nonidealities of switching mixers and their impact.

5.2.3.1 Nonlinearity

Let us refer to Figure 5.1 again. Since the switch M_1 has finite resistance upon closure, any nonlinearity associated with this resistor is the primary source of distortion. Assuming that M_1 is on, at low frequency, similar to (5.10), Chapter 4, the output signal in Figure 5.1 can likewise be expanded in terms of the input signal. Using Taylor series expansion

$$V_{if}(t) = a_1 V_{rf}(t) + a_2 V_{rf}^2(t) + a_3 V_{rf}^3(t) + \cdots. \tag{5.5}$$

Now when M_1 is switching, the output is:

$$V_{if}(t) = \left[a_1 V_{rf}(t) + a_2 V_{rf}^2(t) + a_3 V_{rf}^3(t) + \cdots \right] S(t) \tag{5.6}$$

The coefficients of the series can then be used to calculate distortion in the mixer. As with other mixer design, the objective is to reduce a_3, which is the primary coefficient that determines the third-order intermodulation. At high-frequency distortion will be described by Volterra series.

5.2.3.2 Feedthrough

As with active mixers, under ideal conditions, feedthroughs from RF to IF port and from LO to IF port in a double balanced switching mixer are zero. In practice, around −40 dB of feedthrough occurs. In switching mixers, there are mainly two mechanisms that result in feedthroughs. The first one is capacitive feedthrough. This is particular acute in MOS transistors, where parasitic capacitance is significant. One major coupling path is through the gate to source/drain capacitors. The second mechanism is due to transistor mismatches. Mismatches can come from W/L ratios, V_T, and other process parameters, as well as the actual physical layout. As the mismatch between one transistor to its "balanced" counterpart increases, the even-order coefficients [a_2 in (5.5)] differ and do not completely cancel, similar to what happens in any fully differential circuits.

5.2.3.3 Finite Bandwidth

Ideally the speed with which the transistor can be turned off is just a function of how fast the channel charge can be released and sets an upper limit on the LO frequency. This turns out to be the transit time, which decreases as the square of the channel length. Hence for a given technology (channel length), the maximum frequency of operation is dictated.

If the frequency of operation gets close to this maximum frequency, the mixer suffers from the tracking error due to finite bandwidth imposed by the RC time constant of the switch and the holding capacitance. This error translates into a reduction in conversion gain.

For a given technology the use of small R to improve the bandwidth leads to large gate capacitance, resulting in increasing rise time of the LO waveform and hence distortion. The rise time is dictated by the gate capacitance of the switch, the voltage swing of the LO signal, and the driving capability of clock or LO drivers (although this driving capability is also a function of channel length, in an indirect sense). Intermodulation distortion and conversion gain both degrade rapidly with increased rise time, due to the transistor nonlinear resistance's dependence on the gate drive voltage. This, in conjunction with the transistor nonlinear resistance's dependence on the changing V_{rf}, is the largest contribution toward distortion. We should also note that for a given technology a small R leads to a large W/L, and so the associated C_{sb} and C_{db}, nonlinear junction capacitors, have a higher contribution to distortion. By using short channel length technologies one can reduce R while keeping both the gate and junction capacitance small. This will allow for higher LO frequencies while satisfying the intermodulation distortion requirement.

Next we derive design equations that address some of the aforementioned issues. Specifically, we develop the design equations that quantify the three design parameters: conversion gain (G_C), distortion (IIP_3), and noise (NF). Furthermore, all cases will be covered for the cases when V_{lo} is/is not switching.

5.3   Distortion in Unbalanced Switching Mixer

Referring to Figure 5.1 when M_1 is on it may be modeled as shown in Figure 5.3. The transistor operates in the triode region and is replaced by a nonlinear resistor R_{on}. This is the primary source of intermodulation distortion at low frequency. Note that the arrow on the resistor symbol for R_{on} now indicates nonlinearity, not time-varyingness. Other components of importance include the nonlinear source and drain to bulk capacitors, C_{sb} and C_{db}. They dominate the distortion performance at higher frequency. The resistor R can come from the 50 Ω termination resistor.

To find the intermodulation distortion of this mixer, we make the following assumptions on transistor M_1.

5.3.1   Assumptions on Model

a. **On mobility:** When on, the transistors in the mixer operate in the triode region with $V_{ds} = 0$. The tangential electric field is low and velocity saturation is assumed negligible. The assumption of long channel theory is still reasonable.
b. **On threshold voltage:** In actuality, threshold voltage V_t is modulated by the source-to-bulk voltage and will introduce additional non-linearity. For the time being the threshold voltage is assumed to be constant.
c. **On capacitors:** The capacitor coefficients are found from the expansions of the junction and sidewall capacitance equations. For C_{sb} we have

$$C_{sb} = \frac{C_{jsw} \times ps}{\left(1 + \frac{V_{sb}}{\phi}\right)^{M_{jsw}}} + \frac{C_j \times as}{\left(1 + \frac{V_{sb}}{\phi}\right)^{M_j}} \tag{5.7}$$

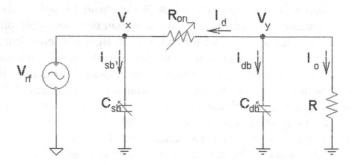

Fig. 5.3   Complete distortion model for the unbalanced switching mixer

Fig. 5.4 Low-frequency
distortion model for the
unbalanced switching mixer

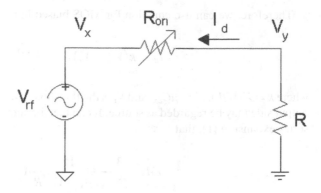

where

C_j = junction capacitance/area
C_{jsw} = sidewall capacitance/length
as = area of source
ps = perimeter of source
V_{sb} = source to bulk voltage
M_j = junction grading coefficient
M_{jsw} = sidewall grading coefficient
ϕ = bulk potential

Notice that this equation applies equally well to C_{db}, drain to bulk capacitor, provided that all the parameters are given for the drain node.

With the preceding assumptions it can be shown that at sufficiently low frequencies the distortion is mainly determined by the nonlinear R_{on} of M_1 and that distortion due to C_{sb} and C_{db} is quite small for typical values of *MOS* model parameters. Even though we have chosen to ignore body effect for simplicity in analysis, this choice does not modify our method of distortion analysis or the conclusions.

Since detailed analysis of the mixer in Figure 5.3 has been carried out in [1], only the results presented in this chapter. It would be relatively straightforward for interested readers to apply the analysis in Sections 5 and 6, Chapter 4, to the present case and reproduce this detailed analysis.

5.3.2 Low-Frequency Case

First we assume V_{lo} is not switching and so the mixer is represented by an NLTI system. Let us look at the behavior of this NLTI system at low frequency. At low frequency the model in Figure 5.3 can be simplified to that in Figure 5.4, where we neglect all reactive element contributions.

Therefore, we can use equation for MOS biased in the triode region:

$$I_d = k(V_{gs} - V_t)V_{ds} - \frac{k}{2}V_{ds}^2 \tag{5.8}$$

where $k = k'(W/L)$, $k' = \mu C_{ox}$, and V_t is the threshold voltage. Either the input or the output side may be regarded as source. It can then be shown, by applying the Taylor series expansion [1], that

$$IM_3 = \frac{3}{8}\frac{1}{k(V_{GS} - V_t)^3 R}A_{rf}^2 \tag{5.9}$$

where A_{rf} is the amplitude of the RF input to the mixer.

Note that for most receiver applications the undesirable I_{D3} is not the one generated by the RF signal. Rather it is the I_{D3} that is generated by the interferers. Hence the IM_3 that corresponds to this I_{D3} is given by

$$IM_3 = \frac{3}{8}\frac{1}{k(V_{GS} - V_t)^3 R}A_{int\,erference}^2 \tag{5.10}$$

where $A_{interference}$ is the amplitude of the interference signal.

When comparing (5.10) to (4.32) of Chapter 4, notice the difference in dependence on $V_{GS} - V_t$.

Now by definition $R_{on} = \frac{1}{\left(\frac{dI_d}{dV_{ds}}\right)}$. We can differentiate (5.8), apply the result to this definition and we have

$$R_{on} = \frac{1}{k(V_{gs} - V_t - V_{ds})} \approx \frac{1}{k(V_{GS} - V_t - V_{DS})}. \tag{5.11}$$

If we further assume V_{DS} is small and therefore neglect the V_{DS} term, then we have

$$R_{on} = \frac{1}{k(V_{GS} - V_T)} \text{ or } k = \frac{1}{R_{on}(V_{GS} - V_t)} \tag{5.12}$$

In (5.10), IM_3 is seen to depend on k. In (5.12) k is seen to depend on R_{on}. Hence IM_3 depends on R_{on}.

Finally, we look at the case when V_{lo} is switching. First let us assume that V_{lo} is an ideal square wave, which means it has zero rise time and fall time. The situation can be handled by modifying the foregoing results through multiplying the output by the Fourier series representation of the square wave. The effect is simply shifting the frequency of the fundamental and the third order products by f_{lo}. The amplitudes are both reduced by the same amount, $1/\pi$, with IM_3 remaining unchanged.

When V_{lo} has finite rise and fall time, at low frequency we may assume that the rise/fall time takes up a small percentage of the total period. Then even with finite rise/fall time the distortion formulae remain practically the same.

5.3.3   High-Frequency Case

First we assume that V_{lo} is not switching and so the mixer is represented by an NLTI system. Let us look at the behavior of this nonlinear system at high frequency. At high frequency, the reactive elements coming from the junction capacitors contribute to third-order distortion and Volterra series must be invoked. To include capacitor effects we should go back to Figure 5.3. To simplify the analysis let us arbitrarily assume for the time being that in Figure 5.3 the non-linear capacitor C_{db} at the drain is replaced by a linear capacitor C (C_{sb} at the source, of course, does not matter, since it is driven by a voltage source V_{rf}). Furthermore, let us assume that

$$R \gg R_{on} \tag{5.13}$$

and

$$\frac{1}{j\omega_{rf}C} \gg R_{on} \text{ or}$$

$$\frac{1}{2\pi R_{on}C} \ll f_{rf} \tag{5.14}$$

Applying these simplifying assumptions to the model in Figure 5.3, an expression for IM_3 due to interferers can be found by applying Volterra series analysis [1]:

$$IM_3 = \frac{A^2_{\text{interference}}\sqrt{\left(\frac{3}{R}\right)^2 + \left(\omega_{rf}C\right)^2}}{8k(V_{GS} - V_t)^3} \tag{5.15}$$

Note that this reverts to (5.10) when $\omega_{rf} = 0$, as expected. IM_3 can also be related to $IM_3|_{\text{low\_frequency}}$ via

$$IM_3 = IM_3|_{\text{low\_frequency}}\sqrt{1 + \left(\frac{\omega_{rf}CR}{3}\right)^2} \tag{5.16}$$

where $IM_3|_{\text{low\_frequency}}$ is the predicted third-order intermodulation at low frequency as given in (5.10).

Finally, we look at the case when V_{lo} is switching. Again, we first assume that V_{lo} is an ideal square wave which means it has zero rise time and fall time. As in the low-frequency case, the situation can be handled by modifying the foregoing results through multiplying the output by the Fourier series representation of the square wave. The effect of the switching is simply shifting the frequency of the fundamental and the third-order products by f_{lo}. The amplitudes are both reduced by the same amount $1/\pi$, with IM_3 remains unchanged.

Now let us look at the situation when V_{lo} has finite rise and fall time. Since at high frequency the rise/fall time no longer takes up a small percentage of the total period, we may have to resort to a technique called the "variable state approach" [1] to calculate the distortion. This is based on a two-dimensional Volterra series. For a simple circuit like a switching mixer, one general observation can be obtained [1]: It is seen that IM_3 increases as the ratio t_f/T, where t_f is the fall time and T is the LO period.

Numerical example 5.1.

Suppose an IIP_3 (third order output intercept point) of $+35$ dBm is required for a mixer and we want to design this mixer to operate at low and high frequency ($f_{rf} = 1.9$GHz). We are given $R = 50\Omega$, and $V_{GS} - V_t = 4$ V, $L = 1$ um, k' $= 100$ µA/V^2. C_{db} is assumed to be a linear capacitor with $C_{jsw} = 0$, $C_j = 2$ fF/um^2, with its area ad given by ad = W × drain_height. Drain_height is given to be 6um.

a. Low frequency case

For simplicity let us assume that V_{lo} is not switching. Since the mixer is operating at low frequency, we use (5.10) to design for the W/L. To use (5.10) we need to relate IM_3 to IIP_3. This could have been done by applying (2.38) from Chapter 2. Alternatively, we can simply apply the definition that when $P_{interference} = IIP_3$,

$$IM_3 = 1 \tag{5.17}$$

Since we are given $IIP_3 = 35$ dBm, then $P_{interference} = 35$ dBm when condition (5.17) applies. Hence when condition (5.17) applies, $A_{interference}$ is given by

$$+35 \; dBm = 10 \log \frac{A_{interference}^2}{2 \times 50\Omega \times 10^{-3}}$$

or

$$A_{interference} = 17.8V \tag{5.18}$$

Substituting (5.17), (5.18), and other given parameters into (5.10), we have

$$1 = \frac{3}{8} \frac{(17.8V)^2}{100uA/V^2 \times \frac{W}{L} \times 4^3 \cdot 50\Omega} \tag{5.19}$$

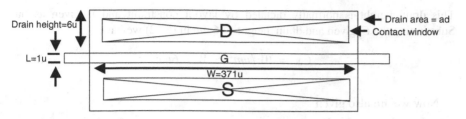

Fig. 5.5 Layout of switch M_1

Solving, we have

$$\frac{W}{L} = 371 \tag{5.20}$$

b. High frequency case
For simplicity let us assume that V_{lo} is not switching. Since the mixer is operating at high frequency we can use (5.15) to design for the W/L.

To use (5.15) let us first determine C, or C_{db} of transistor M_1. This can be found from (5.7), with source replaced by drain in all the parameters, which is rewritten as:

$$C_{db} = \frac{C_{jsw} \times pd}{\left(1 + \frac{V_{db}}{\phi}\right)^{M_{jsw}}} + \frac{C_j \times ad}{\left(1 + \frac{V_{db}}{\phi}\right)^{M_j}} \tag{5.21}$$

where

C_j = junction capacitance/area
C_{jsw} = sidewall capacitance/length
ad = area of drain
pd = perimeter of drain
V_{db} = drain to bulk voltage
M_j = junction grading coefficient
M_{jsw} = sidewall grading coefficient
ϕ = bulk potential

We are given $C_{jsw} = 0$ and hence the first term is gone. Also C_{db} is assumed to be a linear capacitor and hence the denominator of the second term becomes one. Consequently, the preceding equation becomes

$$C_{db} = C_j \times ad \tag{5.22}$$

To calculate ad, we refer to the layout of transistor M_1, which is shown in Figure 5.5. Here L is seen to be 1um, as given. Also we can see $ad = W \times$ drain_height.

This drain_height is normally dictated by layout constraint and is given as 6u. Substituting the C_j given and drain_height $= 6\ um$ in (5.22) we have

$$C_{db} = 2\text{fF}/um^2 \times W \times 6um \tag{5.23}$$

Now we are also given

$$f_{rf} = 1.9GHz \tag{5.24}$$

Substituting (5.18), (5.23), (5.24), and other relevant parameters into (5.15), we have

$$1 - \frac{(17.8V)^2 \sqrt{\left(\frac{3}{50\Omega}\right)^2 + (2\pi \times 1.9GHz \times W \times 6u \times 2fF/u^2)^2}}{8 \times 100uA/V^2 \times \frac{W}{1u}(4V)^3} \tag{5.25}$$

Solving $W = 762um$. Hence

$$\frac{W}{L} = \frac{762}{1} \tag{5.26}$$

This is about twice as big as the M_1 designed using the low frequency formula [see (5.20)].

Finally, we should check whether the assumptions leading to the high-frequency model as described in (5.14) are valid.

5.4   Conversion Gain in Unbalanced Switching Mixer

We proceed by developing the conversion gain formula when V_{lo} is not switching. Then we develop the conversion gain formula when V_{lo} is switching.

5.4.1 V_{lo} Not Switching Case

The mixer in Figure 5.1 is modeled such that the channel resistance R_{on} is replaced by a linear resistor in series with an ideal switch. The nonlinear substrate capacitors are omitted since they do not have an impact on performance except at very high frequencies. The final model is shown in Figure 5.6 [1], where C is the equivalent

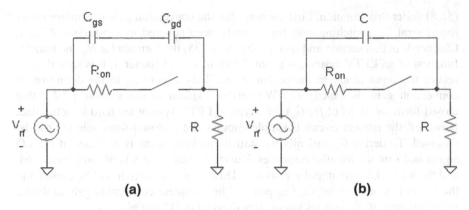

Fig. 5.6 Conversion gain model for the unbalanced switching mixer

capacitance of C_{gs} with C_{gd}. Notice that this C is different from the C in sub-section 5.3.3. We first develop the conversion gain by assuming that the switch is on and treating the mixer as a simple LTI circuit. Hence G_c becomes

$$G_c = \left| \frac{R}{R_{ON} || \frac{1}{j\omega_{rf}C} + R} \right| \tag{5.27}$$

5.4.2 Special V_{lo} Switching Case

Now with switching, if the poles/zeroes frequencies are much higher in frequency than f_{lo}, the mixer satisfies (4.150) of Chapter 4. Hence the conversion gain is modified by the Fourier series coefficient and it becomes

$$G_c = \left| \frac{1}{\pi} \frac{R}{R_{ON} || \frac{1}{j\omega_{rf}C} + R} \right| \tag{5.28}$$

Taking finite rise and fall time into consideration, expression (5.28) is modified, which involves the factor (t_r/T) [1], where t_r is the rise time and T is the LO period. The impact is usually small.

5.4.3 General V_{lo} Switching Case

When f_{lo} increases such that the poles/zeroes frequencies become comparable to f_{lo}, the effect of poles/zeros must be considered. This is because the mixer no longer satisfies (4.150) of Chapter 4. Our goal in this sub-section is to see what happens to

(5.28) under this situation. First we note that the conversion gain of a mixer under the general V_{lo} switching case has actually been covered in sub-section 4.7.2.1, Chapter 4. In that section and specifically in (4.135) the formula for H_n, the transfer function of an LPTV system, is given. From (4.135), Chapter 4, it is seen that H_1 relates the input at ω_{lo} to the output at ω_{if}. This, according to the definition of conversion gain, is exactly G_c. We further explain in sub-section 4.7.2.1 that closed-form solutions of $H_1(G_c)$ for a typical LPTV system are hard to determine. However, the present circuit is simple enough that a closed-form solution can be obtained. To derive G_c basically we start by looking at the two phases of the LO signal and sum the impulse responses. During ϕ_1, the switch is off, in Figure 5.6a, and the mixer has one impulse response. During ϕ_2, the switch will be closed and the mixer has another impulse response. The complete conversion gain is then a weighted sum of the two responses. It is derived in [1] and given as

$$G_c = \left| \frac{j}{\pi}\left(1 - \frac{\tau_2}{\tau_1}\right)\left\{\frac{1 + \imath f_{lo}[\tau_2(1+K) + \tau_1(1-K)]}{(1 + j2\pi f_{lo}\tau_1)(1 + j2\pi f_{lo}\tau_2)}\right\}\right| \tag{5.29}$$

where

$$K = \frac{e^{-\frac{1}{2f_{lo}\tau_1}} - e^{-\frac{1}{2f_{lo}\tau_2}}}{1 - e^{-\frac{1}{2f_{lo}}\left(\frac{1}{\tau_1}+\frac{1}{\tau_2}\right)}}, \text{ and}$$

$\tau_1 = CR$, $\tau_2 = C(R/\!\!/R_{ON})$. The terms τ_1, τ_2 correspond to the two time constants during the two phases, when the switch is open and closed respectively. For DECT application, G_c as calculated by (5.29) can be substantially less than that calculated by (5.28). This is particularly significant for passive mixers since G_c is small to begin with.

The magnitude of K is never greater than one. The range is $0 \le |K| \le 1$ and is positive for $\tau_1 > \tau_2$ and negative for $\tau_1 < \tau_2$.

As a special case, let us start decreasing f_{lo}. Then T starts to increase. If T has increased to such a point that the time constants are much shorter than the period of the LO signal then τ_1/T and $\tau_2/T \to 0$ or $\tau_1 f_{lo}$ and $\tau_2 f_{lo} \to 0$. Equation (5.29) reduces to

$$G_c = \frac{1}{\pi}\left(1 - \frac{\tau_2}{\tau_1}\right) \tag{5.30}$$

$$= \frac{1}{\pi}\frac{R}{R_{ON}+R} \tag{5.31}$$

Decreasing f_{lo} means that the poles/zeroes frequencies are once again becoming much higher in frequencies than f_{lo} and we are reverting to the scenario as depicted in sub-section 5.4.2. We should then get the same G_c formula. Let us revisit the G_c formula in sub-section 5.4.2, which is given by (5.28). As f_{lo} decreases,

f_{rf} decreases. In (5.28) the $1/\omega_{rf}C$ term becomes much longer than the R_{on} term and can be neglected. Equation (5.28) then becomes

$$G_c = \frac{1}{\pi} \frac{R}{R_{ON} + R}$$

and indeed agrees with (5.31).

The expression incorporating finite rise and fall time was not derived because it is too complicated and involves infinitely many phases and circuit configurations.

As a summary of Section 5.4 it should be noted that G_c derived in sub-section 5.4.1 through 5.4.3 are all less than one, which is consistent with the fact that the mixer is passive. This is also the essential difference from an active mixer, such as a Gilbert mixer, whose gain is larger than one.

5.5   Noise in Unbalanced Switching Mixer

We proceed by developing the noise expression without switching. Then we develop the noise expression with switching.

5.5.1 V_{lo} not Switching Case

To calculate the noise without switching, we normally assume that the switch is closed. Then from Figure 5.6b, the equivalent resistance is $R_{on}||R$. Now the output noise PSD can be calculated from (4.34) of Chapter 2 and is given by

$$S_n = 4kT(R_{on}||R) \tag{5.32}$$

Input referred noise PSD, N_{dev}, is then given by

$$N_{dev} = \frac{S_n}{G_c^2} \tag{5.33}$$

where G_c is given from (5.27) and S_n is given by (5.32). Finally, we want to calculate NF. First we repeat (2.67) of Chapter 2:

$$NF = \frac{N_{Device} + G_c \cdot N_{Source\_resis\,tan\,ce}}{G_c \cdot N_{Source\_resis\,tan\,ce}} \tag{5.34}$$

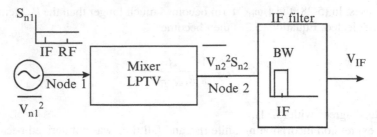

Fig. 5.7 Noise model for the unbalanced switching mixer

Substituting $N_{\text{dev}} = N_{\text{device}}/G$ into (5.34), we have

$$NF = 1 + \frac{N_{dev}}{N_{Source\_resis\tan ce}} \qquad (5.35)$$

Substitute the N_{dev} we obtained in (5.35), assuming a source resistance of 50 Ω, and the NF is determined.

5.5.2 Special V_{lo} Switching Case

First let us follow the discussion in Figure 4.17, Chapter 4 and represent the present switching mixer in a similar way. The resulting representation is shown in Figure 5.7,

Where there is an input noise PSD, denoted as S_{n1}, and an output noise PSD, denoted as S_{n2}. Applying (4.134), Chapter 4, to Figure 5.7, we have

$$S_{n2}(\omega_{if}) = \sum_{n=-\infty}^{\infty} |H_n(\omega_{if})|^2 S_{n1}(\omega_{if} + n\omega_{lo}) \qquad (5.36)$$

In sub-section 4.7.2.2, Chapter 4, we have outlined how to calculate H_n under the special V_{lo} switching case; that is, when the poles/zeroes frequencies are much higher than f_{lo}. Readers can apply the same principle to the present case and calculate H_n, and hence S_{n2}, in (5.36).

5.5.3 General V_{lo} Switching Case

What about the case when the mixer behaves as a general LPTV system; that is when the poles/zeroes frequencies become comparable to f_{lo}?

Usually a closed-form solution for H_n is rather difficult to calculate in this case. However, the present circuit is simple enough that a closed-form solution can indeed be obtained. Reference [1] has worked out these H_n expressions, and the resulting S_{n2} is calculated to be

$$S_{n2} = 4kT \left\{ \frac{R_1 + R_2}{2} - f_{lo}C(R_1 - R_2)^2 \left[\frac{\left(1 - e^{-\frac{1}{2f_{lo}\tau_1}}\right)\left(1 - e^{-\frac{1}{2f_{lo}\tau_2}}\right)}{1 - e^{-\frac{1}{2f_{lo}}\left(\frac{1}{\tau_1} + \frac{1}{\tau_2}\right)}} \right] \right\} \quad (5.37)$$

Here $R_1 = R$ (corresponding to resistance across the capacitor during phase 1, when the switch is connected) and $R_2 = R_{on}||1R$ (corresponding to resistance across the capacitor during phase 2, when the switch is open). S_{n2} can also be expressed in terms of a equivalent resistance R_{eq} as

$$S_{n2} = 4kTR_{eq} \quad (5.38)$$

As a special case notice when $f_{lo} = 0$, (5.37) reduces to

$$S_{n2} = 4kT\frac{R_1 + R_2}{2} = 4kT\frac{R + (R_{on}||R)}{2} \quad (5.39)$$

What does $f_{lo} = 0$ mean? It means that either the switch is on or off all the time. Notice, from (5.32), that the output noise PSD for the case when the switch is on is $4kT(R_{on}||1R)$.

What about the case when the switch is off? If we assume that the switch is off, then the output noise PSD is given by

$$S_{n2} = 4kTR \quad (5.40)$$

Hence the S_{n2} for these two cases are $4kT(R_{on}||1R)$ and $4kTR$ respectively, and therefore the total S_{n2} should be the average of the two, which agrees with (5.39).

Again input-referred noise, N_{dev}, is given by

$$N_{dev} = \frac{S_{n2}}{G_c^2} \quad (5.41)$$

where G_c this time is given by (5.29) and S_{n2} is given by (5.37). NF can now be calculated by substituting the N_{dev} we obtained in (5.35).

Finally, the expression incorporating finite rise and fall time was not derived as it is too complicated and involves infinitely many phases and circuit configurations.

5.6   A practical Unbalanced Switching Mixer

A practical realization of the unbalanced switching mixer is shown in Figure 5.8. This mixer is shown together with an IF amplifier (M_2) in the figure. Transistor M_1 is the basic switch. The incoming RF signal is chopped by the switching action of the LO signal, turning M_1 on and off. The IF signal is produced by M_1 and amplified

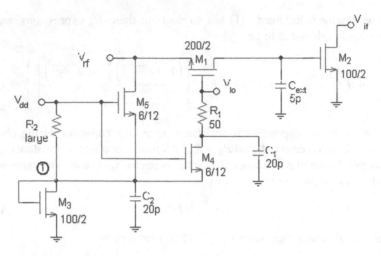

Fig. 5.8 Switching mixer and IF amplifier

by M_2. When M_1 is on, it behaves like a small resistor ($< 10\,\Omega$). Transistor M_3 sets the bias voltage. For a V_{dd} of 5 V V_{D3} is around 1.3 V. M_4 and M_5 are biased in the triode region and hence behave as large resistors (since W/L are very small).

Now let us see how the circuit operates. First let us take a look at V_{GS1}. We look at V_{G1} first. Assume first that the voltage at the LO port is zero and therefore does not provide any DC current, then $I_{D4} = I_{R_1} = I_{G1} = 0$. Hence $V_{G1} = V_{D4} = V_{D3} = 1.3$ V. Next let us look at V_{S1}. Assume that the RF port does not provide any DC current. Then $I_{D5} = 0$ and so $V_{S1} = V_{S5} = V_{D3} = 1.3$ V, again. Therefore, $V_{GS1} = 0$ V and M_1 is nominally off.

Second, let us assume that the voltage at the LO port goes up. M_1 is turned on as soon as V_{lo} goes up and becomes at least one V_t above 1.3V. Notice the full driving capability of V_{lo} is not used (only the $V_{lo} - V_t$ portion is used).

Thirdly, let us now ask why we need $M_4, M_5, C_1, C_2, C_{ext}$. They are necessary for a couple of reasons. To examine these reasons let us replace M_4, M_5, C_1, C_2 by their equivalent impedance operating at a frequency of 1.9 GHz (using DECT as an example again). The resulting circuit is shown in Figure 5.9.

Now we can give the reasons as follows:

1. The LO port cannot be connected directly to node 1. It has to be coupled through M_4. Imagine if the LO port (gate of M_1) is connected directly to node 1. Then the resistance seen by V_{lo} is practically dominated by C_1, C_2. At 1.9 GHz this is about 2 Ω, and 50 Ω matching is not achieved. To overcome this, V_{lo} is connected through R_1, M_4 to node 1. As shown in Figure 5.9 the resistance at the LO port is now practically dominated by R_1, or 50 Ω. Hence a 50 Ω match is achieved.

2. The network C_1, R_1, M_4, C_2 is there to help attenuate the LO to RF feedthrough. Going from the LO port to the RF port, we can see that the network consists of a

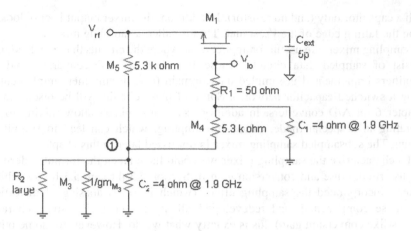

Fig. 5.9 Switching mixer and IF amplifier redrawn with M_3, M_4, M_5, C_1, C_2 replaced by their equivalent impedance

2 RC cascaded network. Neglecting resistance $\frac{1}{g_{mM_3}}$, R_2, R_{M_5} (which are assumed large compared to the impedance of C_2), the 2 RC cascaded network consists of $C_2 R_{M_4}$ driving $C_1 R_1$. To simplify the analysis, let us ignore the loading effect between these 2 RC sections. Hence the two poles of this network can be approximately given as $p_1 = \frac{1}{2\pi R_1 C_1} = 159\ MHz$ and $p_2 = \frac{1}{2\pi R_{M_4} C_2} = 1.5\ MHz$. These poles are at sufficiently low frequencies that they will provide significant attenuation for the LO signal switching at close to 1.9 GHz. Notice that since the amplitude of the LO signal is large and that of the RF signal is small, it is only necessary to provide filtering in one direction (from the LO port to the RF port).

3. C_{ext} is used to improve RF to IF isolation. Referring to Figure 5.8 notice that when M_1 is on, there is an RC filter from the RF port to the IF port. Here R is the R_{on} of switching transistor M_1 and C consists of C_{ext} in parallel with C_{db1} and C_{gs2}. This RC filter helps filter out the RF component. Obviously, the larger C_{ext} is, the more effective the filtering is. Hence what is limiting the value of this C_{ext}? It turns out that distortion leads to a ceiling. Referring to (5.15), we can get the IM_3 for the present unbalanced mixer by setting R in that equation to infinity. The resulting expression is $IM_3 = \frac{A_{interference}^2 \omega_{rf} C}{8k(V_{GS}-V_t)^3}$. Here C consists of C_{ext} in parallel with C_{db1} and C_{gs2}. $V_{GS} = V_{GS1}$. Notice IM_3 increases with increasing C_{ext}. Hence for a given IM_3 (from specifications) the maximum C_{ext} is set.

5.7 Sampling Mixer

We have previously investigated switching mixer and discussed it in an application as a mixer preceding an IF amplifier. For this mixer the output is processed in the continuous time domain. An alternative structure consists of terminating the output with a capacitor only (and no resistor). In addition, the mixer output is only looked at on the falling edge of the V_{lo} signal. This is called a sampling mixer.

Sampling mixers are useful for applications where the circuits that they feed into consist of sampled data circuits. These may include A/D converters and IF amplifiers implemented in sampled data domain (e.g, IF amplifiers implemented using a switched capacitor network). Some of these circuits will be discussed in Chapter 6 on A/D converters. In addition, sampling mixers allow mixing to be operating in a special mode, called subsampling, which can lead to simplified design. The subsampled sampling mixer is discussed later in this chapter.

To characterize the sampling mixer we could have taken the formulae derived for distortion, noise and conversion gain in Section 5.3 through 5.5 for switching mixers, incorporated the sampling effect (which involves aliasing of distortion and noise components), and redeveloped all the formulas. For some characteristics(like conversion gain) this is exactly what we do. However, for some other characteristics (like distortion and noise) this can become rather cumbersome and we decide not to do that. Instead, since the sampling mixer operates in the sampled data domain, we choose to develop an understanding of such characteristics of the circuit directly in the time domain. This seems natural, as we should remember that one of the original reasons for performing a frequency domain-based analysis on the switching mixer stems from the fact that a switching mixer operates in the continuous time domain. Another reason for the change in the domain of analysis for some characteristics stems from the fact that as we are only interested in the output value at the sampling instant, the behavior of the sampling mixer during the falling edge of the LO waveform becomes critical. The behavior along the falling edge can affect the circuit more significantly than in the case of a switching mixer. It turns out that the value of distortion and noise due to finite fall time alone can become totally dominant, and hence we cannot afford to neglect these effects. Since the effect of finite fall time is more naturally handled in time domain, this change in the domain of analysis is fully justified.

5.7.1   Qualitative Description

As a start, let us look at Figure 5.10 which shows a single-ended sampling mixer. This is similar to the unbalanced switching mixer. Analogous to the single and double balanced switching mixers, there are differential and cross-coupled differential sampling mixers [3]. Referring again to Figure 5.10, notice that the input V_{rf} consists of a baseband signal superimposed on a carrier. Initially (during the track mode when V_{lo} is high), V_{if} (dotted line) tries to follow V_{rf}. At the end of the track mode, V_{lo} drops to 0, M_1 is switched off and the voltage V_{if} on the capacitor C_{load} is held (solid horizontal line). This stays until V_{lo} goes high again, when we enter the next track mode and the operation repeats itself. Let us now look at V_{if} at the sampling instant only [V_{if} (sampled), solid dots] and record those values. Then this V_{if} (sampled) is seen to exhibit a much lower frequency than the carrier and

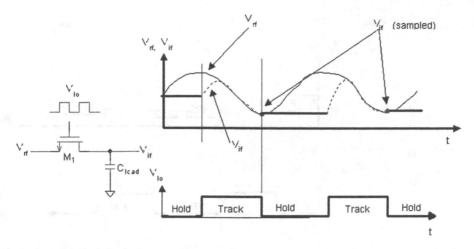

Fig. 5.10 Single-ended sampling mixer and sampling operation

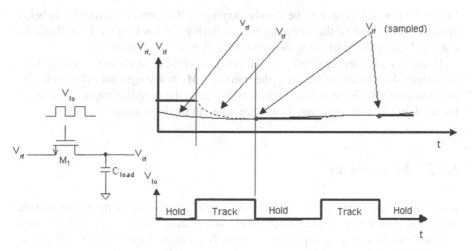

Fig. 5.11 Single-ended sampling mixer and sampling operation: only envelope of input shown

effectively a frequency translation (mixing) has occurred. On closer examination it can be shown that this V_{if}(sampled) is exactly the baseband signal of the original V_{rf} and therefore V_{if} (sampled) is essentially the IF signal we want.

Ideally the envelope (the baseband signal), rather than the carrier of V_{rf}, contains all the essential information, and hence let us for the time being neglect the carrier and focus on the concentrate on the envelope. We try to redraw Figure 5.10 in Figure 5.11 with this simplification. First, the negative peaks of V_{rf} in Figure 5.10 are extrapolated and used to construct an envelope of V_{rf}. Next to save inventing new symbols this envelope is also denoted as V_{rf} from now on. This V_{rf} is shown in

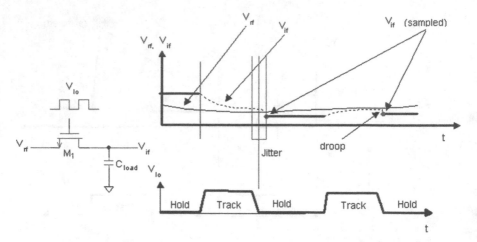

Fig. 5.12 Single-ended sampling mixer and sampling operation: nonideal effects highlighted

Figure 5.11, which is seen to be slowly varying. If this slowly varying V_{rf} is being applied to the input of the sampling mixer, during the track mode V_{if} follows V_{rf} closely. During the hold mode the solid dots show the V_{if} (sampled).

Because V_{rf} is varying slowly, within one LO period V_{rf} is almost constant. Thus to analyze this circuit we can treat the transistor M_1 as a simple on off switch with an almost constant input voltage. It passes charge to the sampling capacitor when in the track mode, and preserves the charge when in the hold mode.

5.7.2 Nonidealties

Under ideal condition V_{if}(sampled) should equal V_{rf} at the sampling instant and the mixing operation recovers the baseband signal perfectly. There are, however, a few nonidealities that make this perfect recovery impossible. Figure 5.11 is redrawn in Figure 5.12 to illustrate this.

5.7.2.1 Finite Bandwidth

As seen in Figure 5.12, during tracking V_{if} follows V_{rf} closely, but not perfectly. This is because of the non-zero drain-source resistance R_{on} of the MOS device. The effect of R_{on} is to create an R-C lowpass filter with capacitance equal to C_{load} in parellel with C_{db}. This introduces a delay. Hence at the sampling instant V_{if} (sampled) is not the same as V_{rf} at that instant and an error is introduced. This effect is similar to the finite bandwidth effect of switching mixers described in subsection 5.2.3.3, except that now we are interested in the error at the sampling instant only. Again, this error leads to a reduction in conversion gain.

5.7.2.2 Nonlinearity

During the track mode, the nonlinear R_{on} introduces distortion in much the same way as described in sub-section 5.2.3.1. As a result, the IM_3 can be predicted by developing similar equations. For example, (5.15) can be modified to predict IM_3 of a sampling mixer at high frequency by setting $R = \infty$ and $C = C_{db} + C_{load}$ in that equation.

It can be seen in Figure 5.12 that, as opposed to the ideal case, V_{lo} has a finite fall time during the sampling instant. As the switch moves from track mode to hold mode during this finite fall time, it will introduce extra distortions. As opposed to switching mixers (which also suffer from extra distortions due to finite fall time), extra distortions can dominate the total distortion. This is because these extra distortions occur around the sampling instant, a moment most critical to sampling mixers. These extra distortions one discussed in more detail later.

5.7.2.3 Feedthrough

This nonideality is similar to the feedthrough described in sub-section 5.2.3.2, except that the critical feedthrough is the one that happens during the sampling instant. Hence the feedthrough from LO at the sampling instant is most important. At that instant feedthrough from the LO is due to the existence of the capacitive divider, which consists of the gate to drain capacitance C_{gd} and C_{load}. On the falling edge of V_{lo} the change in V_{lo} (assumed going from V_{dd} to 0) results in a corresponding change of V_{if} given by

$$V_{feedthrough} = \left(C_{gd}/\left(C_{load} + C_{gd}\right)\right)V_{dd} \tag{5.42}$$

This error is one of the factors that contribute to the droop of V_{if} at the sampling instant. The droop is evident in the V_{if} waveform around the sampling instant in Figure 5.12. Now since this error is fixed, the resulting error is a DC offset. For differential and cross- coupled differential mixers, this is not an issue.

5.7.2.4 Charge Injection

This nonideality is unique to sampling mixers. The concept is borrowed from sampled data circuits where the circuit designers are concerned with the effect of the channel charge dumping upon the turn-off of the switch. This error charge is dumped onto the storage capacitor, C_{load}, and adds an offset voltage from the actual V_{rf}. The amount of charge stored in the channel is a function of V_{rf} (the magnitude varies with the frequency). When the switch moves from an on to an off state with a

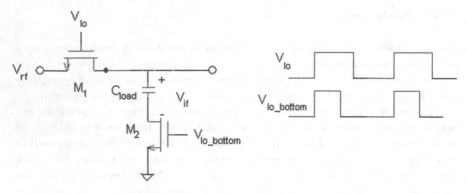

Fig. 5.13 Bottom-plate single-ended sampling mixer

slow V_{lo} edge transition, roughly half of the charge is transferred to the load. This results in an error in V_{if} [and V_{if} (sampled)] equal to

$$V_{ch\,arg\,e} = \frac{Q_{chan}}{2C_{load}} = \frac{W_1 L_1 C_{OX}(V_{dd} - V_{rf} - V_t)}{2C_{load}} \tag{18}$$

Here Q_{chan} is the channel charge. This error is the other factor that contributes to the droop of V_{if} at the sampling instant. This error is dependent on V_{rf}, so this error is translated into signal-dependent distortion. Differential and cross-coupled differential mixers still suffer from this distortion.

5.7.2.5 Jitter

Notice in Figure 5.12 that during the turn-off process, in addition to having a finite slope, the clock edge is uncertain. This uncertainty in clock edge is denoted as jitter in Figure 5.12. The uncertainty causes the exact instant of storing the V_{rf} to have an uncertainty as well. This uncertainty manifests itself as noise in the sampling mixer. This is an aspect unique to a sampling mixer.

5.7.3 Bottom-Plate Sampling Mixer

To combat charge injection error, an improved sampling mixer topology is as shown in Figure 5.13. In this circuit a technique known as bottom-plate sampling is employed to remove the signal dependence in the error term.

The input is V_{rf} and V_{if} is taken as the voltage across C_{load}. The LO signal V_{lo} is applied to switch M_1 and a signal V_{lo_bottom} is applied to switch M_2. Both switches are implemented using MOS transistors. Notice both V_{lo} and V_{lo_bottom} are the same, except that V_{lo_bottom} has its falling edge slightly advanced compared to that of V_{lo}. During tracking, both M_1 and M_2 are on, so V_{rf} is sampled onto C_{load}. When the

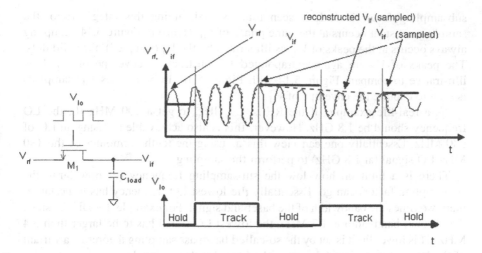

Fig. 5.14 Subsampling operation

mixer goes from the track mode to the hold mode, because of the slightly advanced falling edge of V_{lo_bottom}, M_2 is opened first, which effectively freezes the amount of charge stored on C_{load}. Then the top switch M_1 opens, which fixes the voltage level across C_{load}. Since M_2 is opened slightly before M_1, the M_2 switch determines the sampling instant. Since both the drain and source of transistor M_2 are at ground, no signal-dependent charge injection is introduced. When M_1 is subsequently turned off, the charge in the channel sees an open circuit on the side of C_{load} and therefore all the signal-dependent charge that is injected must flow to the other side of the transistor or to the input source (and again no error is introduced). Thus, the droop can be made constant. Similar to the feedthrough case, we can now employ differential techniques, and this constant offset simply becomes a harmless common-mode shift. Note this technique introduces yet another transistor M_2, whose nonlinear R_{on} has to be considered. However, as noted above if V_{rf} can be assumed to be almost constant (see Figure 5.11), then the voltage at the bottom plate of C_{load} is almost constant and distortion introduced by R_{on} of M_2 is negligible.

5.7.4 Sub-Sampled Mixer

Finally there is a special type of sampling mixer called subsampling mixer, shown in Figure 5.14. To illustrate subsampling the complete V_{rf} (baseband signal superimposed on a carrier) is shown.

Structurally this mixer is seen to be the same as the one in Figure 5.10. The only difference is that the LO frequency is an integer submultiple of the normal LO frequency. In Figure 5.14, there are six V_{rf} cycles in 1 LO period and so the subsampling factor is 6. It is seen that by maintaining this integer ratio, the sampling constant occurs at the same phase of V_{rf}. Hence in Figure 5.14, sampling always occurs at the peaks of V_{rf}, as illustrated by the V_{if} (sampled), the solid dots. The peaks of V_{rf} are again extrapolated to construct an envelope of V_{rf}. It is illustrative to compare Figure 5.14 with Figure 5.10. In both cases V_{if} (sampled) is seen to capture the envelope of V_{rf}.

As a real-life example, if the RF is 1.9 GHz, to get a 100 MHz IF, the LO frequency should be 1.8 GHz. However, this is also achievable by using an LO of 180 MHz. Essentially one can view this as using the tenth harmonics of the 180 MHz LO signal (at 1.8 GHz) to perform the sampling.

There is a limit on how low the subsampling frequency (or how large the subsampling factor) can go. Essentially the lowest LO frequency has to be larger than two times the bandwidth of the baseband signal. For example, for DECT, since the channel bandwidth is 1.7 MHz, the lowest LO signal has to be larger than 3.4 MHz. This lower limit is set by the so-called bandpass sampling theorem – a variant of the Nyquist sampling theorem applied to a bandpass signal.

The main advantage of this type of mixer is that the time it takes for the mixer output voltage V_{lo} to settle is reduced substantially. Using DECT as the example again, instead of having to settle in a period given by $1/(2 \times 1.8 \text{ GHz})$, we can settle in a period of $1/(2 \times 0.18 \text{ GHz})$. This will relax the requirement of the frequency synthesizer that drives the mixer. The disadvantage of this is that all the noise at the RF port, centered around multiples of 0.18 GHz, is now aliased into the IF port. Since you have 10 times more images, the aliased noise power increases accordingly.

A bottom-plate version of this subsampled version can also be implemented rather easily.

Next, as with switching mixers we develop design equations that address some of these issues. In the process we quantify the three relevant parameters: conversion gain (G_c), distortion (IIP_3), and noise (NF). Furthermore, all cases are covered for the scenarios when V_{lo} is/is not switching. In addition, for noise and distortion analysis, in all the cases when V_{lo} is switching, the particular case when V_{lo} is switching with a finite rise/fall time is given special treatment. In the sampling mixer, this is necessary and important since it is precisely at the falling edge of V_{lo} that the output value is determined. This makes the noise and distortion analysis quite different from the switching mixer case. Most of the analysis is focused on the single-ended sampling mixer.

5.8 Conversion Gain in Single-Ended Sampling Mixer

We proceed by developing the conversion gain with V_{lo} not switching. Then we develop the conversion gain with V_{lo} switching.

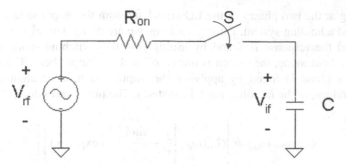

Fig. 5.15 Conversion gain model for the single-ended sampling mixer

5.8.1 V_{lo} not Switching Case

The model for the single-ended sampling mixer as described by Figure 5.10 is developed in Figure 5.15.

In this figure the channel resistance is modeled as a linear resistor R_{on} in series with an ideal switch. The nonlinear junction capacitor C_{db} is lumped together with C_{load} and is denoted as C. We first develop the conversion gain by assuming that the switch is on. Hence G_c becomes

$$G_c = \left| \frac{\frac{1}{j\omega_{rf}C}}{\frac{1}{j\omega_{rf}C} + R_{on}} \right| \tag{5.43}$$

5.8.2   Special V_{lo} Switching Case

As with the switching mixer case, if the poles/zeroes frequencies are much higher than f_{lo}, the sampling mixer satisfies (4.150), Chapter 4. The conversion gain is modified by the Fourier series coefficient and becomes

$$G_c = \left| \frac{1}{\pi} \frac{\frac{1}{j\omega_{rf}C}}{\frac{1}{j\omega_{rf}C} + R_{on}} \right| \tag{5.44}$$

Taking finite rise and fall time (t_r, t_f) into consideration, expression (5.44) is modified by (t_r/T) [1] and the impact is usually small.

5.8.3   General V_{lo} Switching Case

When the poles/zeroes frequencies become comparable to f_{lo}, conversion gain is found in a manner similar to the conversion gain case for switching mixer, except that we are interested in sampled (rather than continuous) output. Basically we start by looking at the two phases of the LO signal and sum the response to a tracking system and a holding system. During tracking, we are doing natural sampling of a signal, and the response is found by multiplying the switching function to V_{rf}. During the hold mode, the switch is turned off and the output held. The response during this phase is found by applying the impulse samples, weighted by the sampled values, to the RC filter (an LTI system). The final expression [1] is

$$G_c(\omega_{if}, \omega_{lo}) = \left| G_{c0}(\omega_{if}) \left[\frac{j}{\pi} - \frac{\sin \frac{\pi}{2} \frac{\omega_{if}}{\omega_{lo}}}{\pi \frac{\omega_{if}}{\omega_{lo}}} \left(\exp^{-\frac{\pi}{2}\frac{\omega_{if}}{\omega_{lo}}} \right) \right] \right|$$

$$G_{c0}(\omega_{IF}) = \frac{\frac{1}{\omega_{if}C}}{\frac{1}{\omega_{if}C} + R_{on}} \tag{5.45}$$

Here ω_{if} is the IF of the mixer and is related to ω_{rf}, for a downconversion mixer, by $\omega_{if} = \omega_{rf} - \omega_{lo}$. Notice that compared with (5.43) and (5.44) G_c here also depends on ω_{lo}. The expression incorporating finite rise and fall time was not derived as it is too complicated and involves an infinite number of phases.

5.9   Distortion in Single-Ended Sampling Mixer

We assume that distortion due to charge injection is effectively eliminated by bottom- plate sampling. We analyze the distortion of this bottom plate sampling mixer by first analyzing the unbalanced simple mixer. To do this we redraw Figure 5.12 in Figure 5.16 where the finite rise/fall time is highlighted. The fall

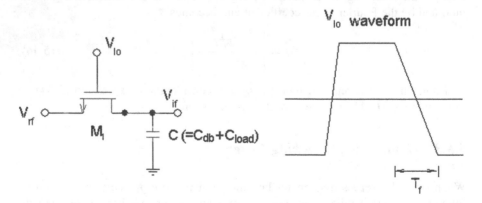

Fig. 5.16 Distortion model for single-ended sampling mixer (high frequency, finite fall time)

time is labeled T_f. Now let us concentrate on the causes of distortion. Sources include the following:

a. Traditional nonlinear resistance, R_{on}, of switches (see Section 5.3).
b. Distortion due to finite fall time, which though briefly mentioned in Section 5.3, has become significant in the present case and deserves special attention. This is because as a result of the sampling operation, the output value is determined precisely at the falling edge of the LO waveform.

We concentrate on high-frequency analysis of the sampling mixer, as this is the frequency range of interest. (Low-frequency analysis can be performed easily by repeating this high- frequency analysis.) As mentioned at the beginning of Section 5.7, it is best to perform the analysis in the time domain, rather than in frequency domain. Hence we apply the Volterra series calculation, this time in time domain.

5.9.1   High Frequency (V_{lo} not Switching) Case

To familiarize the reader in dealing with Volterra series in the time domain, we first treat the simple case where the sampling mixer is operated with the V_{lo} on (and constant) all the time. We show that distortion formulas derived under this operating condition agree with the distortion formulas derived in sub-section 5.3.3, where Volterra series in the frequency domain is used. This is not surprising because under this operation condition (V_{lo} on and constant) sampling does not occur and hence the sampling mixer operates in identically the same way as the switching mixer (with V_{lo} on and constant as well). We further note that the sampling mixer is described by a time-invariant circuit and denote distortion generated under this condition as continuous time distortion.

Let us begin the analysis by repeating the assumptions we made in sub-section 5.3.1. We again assume that the distortion is mainly determined by the nonlinear R_{on} of the MOS transistor and that distortion due to bias-dependent junction capacitance is quite small for typical values of MOS model parameters. We also choose to ignore body effect for simplicity in analysis. Therefore, in deriving the Volterra series coefficient, we assume that in Figure 5.16 the transistor operating in the triode region is represented by the most basic MOS equation:

$$I_d = k(V_{gs} - V_t)V_{ds} - \frac{k}{2}V_{ds}^2, \tag{5.46}$$

where $k = \frac{\mu C_{ox} W}{L}$, and V_t is the threshold voltage. The differential equation that describes Figure 5.16 thus becomes

$$gV_{if} + C\frac{dV_{if}}{dt} = gV_{rf} - \frac{k}{2}\left(V_{rf}^2 - V_{if}^2\right) \tag{5.47}$$

where g is the conductance of M_1 and is defined as $g = k(V_G - V_t)$.

Next we express $V_{if}(t)$ in its Volterra series in the time domain:

$$V_{if}(t) = v_1(t) + v_2(t) + v_3(t) + \dots \tag{5.48}$$

where $v_n(t) = H_n[V_{rf}(t)]$. Here H_n are the Volterra coefficients and should not be confused with the conversion gain. Now we substitute this expansion into (5.47), keeping only first three orders, and we get

$$g(v_1 + v_2 + v_3) + C\frac{v_1 + v_2 + v_3}{dt} = gV_{rf} - \frac{k}{2}\left(v_{rf}^2 - v_1^2 - 2v_1v_2\right). \tag{5.49}$$

Note that V_{rf} is first order since the input is a pure sinusoidal; the term v_1^2 is second order; and v_1v_2 is third order. Now, equating the coefficients,

$$\begin{cases} gv_1 + C\dfrac{dv_1}{dt} = gV_{rf} \\[2mm] gv_2 + C\dfrac{dv_2}{dt} = \dfrac{k}{2}\left(v_1^2 - V_{rf}^2\right) \\[2mm] gv_3 + C\dfrac{dv_3}{dt} = -kv_1v_2 \end{cases} \tag{5.50}$$

The solutions to this set of differential equations are the Volterra coefficients H_1, H_2, and H_3. Since we assume a constant gate voltage, this leads to a constant g, or the sampling mixer can be described as a time-invariant circuit. Hence we can apply time-invariant Volterra series theory to the preceding equations and derive the Volterra coefficients as

$$H_1(\omega_1) = \frac{g}{g + j\omega_1 C} \approx 1 \tag{5.51}$$

$$H_2(\omega_1, \omega_2) = \frac{\frac{k}{2}(H_1(\omega_1)H_1(\omega_2) - 1)}{g + j(\omega_1 + \omega_2)C} = \frac{j(\omega_1 + \omega_2)C}{2k(V_G - V_t)^2} \tag{5.52}$$

$$H_3(\omega_1, \omega_2, \omega_3) = \frac{-k \cdot H_1(\omega_1) \cdot H_2(\omega_1, \omega_3)}{g + j(\omega_1 + \omega_2 + \omega_3)C} = \frac{j(\omega_1 + \omega_2 + \omega_3)C}{3k(V_G - V_t)^3} \tag{5.53}$$

where ω_1, ω_2, ω_3 are the frequencies of the input signals. Here the assumption of small time constant $(g \succ\succ j\omega C)$ is used, and the expressions are symmetrized.

From (5.51) through (5.53) we get the harmonic and intermodulation distortion. If the mixer is operating at ω_{rf}, then we have

$$HD_2 = \frac{A_{rf}}{2} \cdot \frac{j\omega_{rf}C}{k(V_G - V_t)^2} \tag{5.54}$$

$$HD_3 = \frac{A_{rf}^2}{4} \cdot \frac{j\omega_{rf}C}{k(V_G - V_t)^3} \tag{5.55}$$

$$IM_2 = 0 \tag{5.56}$$

$$IM_3 = \frac{A_{interference}^2}{4} \cdot \frac{j\omega_{rf}C}{k(V_G - V_t)^3} \tag{5.57}$$

Note that if the R term in (5.15) is eliminated, then (5.57) becomes similar to (5.15).

5.9.2 High-Frequency (Special V_{lo} Switching) Case; V_{lo} has Finite Rise and Fall Time

It must be emphasized that the preceding analysis assumes a constant gate voltage. This applies to the case of a perfect square wave LO and when the mixer's RC time constant is much smaller than T. However, if the LO voltage deviates from the ideal square wave, several new distortion effects start to emerge.

First the nonlinear R_{on}, which is a function V_G, now becomes time varying during the fall time. This introduces time-varying distortion. Second, the precise instant of sampling depends not only on the gate falling slew rate but also on the input amplitude and frequency. This leads to a situation where the final sampled voltage on the sampling capacitor when M_1 turns OFF deviates from the steady-state voltage on the capacitor and leads to sampling distortion.

The preceding effects can no longer be predicted with time-invariant Volterra series system theory. This leads us to develop the theory of time-varying Volterra series in the next sub-section.

5.9.2.1 Time-Varying Distortion

As mentioned in the previous sub-section, it is necessary to generalize Volterra series to the time-varying case in order to analyze a sampling mixer with an arbitrary LO waveform. In this section, we develop a time-varying Volterra series, explore its frequency response, and show that time-varying Volterra series can be applied in the sampled data domain to solve for time-varying distortion exactly.

Impulse Response of Time-Varying Systems

The notion of impulse response can be easily generalized to time-varying systems by the addition of another time variable. Alternately, we can start from an LTV system, as discussed in sub-section 4.7.2.1, Chapter 4, and add nonliniearity to it. Just to review, the impulse response of an LTV system is a function of two variables, $h_1(t, \tau)$, which is defined as the response of the system for an input of $\delta(t - \tau)$. In this case, the system response for an arbitrary input becomes

$$y(t) = \int_{-\infty}^{\infty} h_1(t, \tau)x(\tau)d\tau \qquad (5.58)$$

Time variable t is called the observation time, and τ the launch time because $h_1(t, \tau)$ represents the output observed at time t for an impulse launched at time τ. Under general continuity conditions, a linear time-varying system can be completely characterized by its two dimensional impulse response, or its kernel.

Higher-order kernels for nonlinear LTV systems are generalized in a similar way. For example, a second-order kernel has two launch time variables plus an observation time variable. $h_2(t, \tau_1, \tau_2)$is the system response to two impulses launched at time instants τ_1 and τ_2. In general, a mildly nonlinear time-varying system has the following Volterra series expansion:

$$\begin{aligned}
y(t) = &\int_{-\infty}^{\infty} h_1(t, \tau_1)x(\tau_1)d\tau_1 \\
&+ \iint_{-\infty}^{\infty} h_2(t, \tau_1, \tau_2)x(\tau_1)x(\tau_2)d\tau_1 d\tau_2 \\
&+ \iiint_{-\infty}^{\infty} h_3(t, \tau_1, \tau_2, \tau_3)x(\tau_1)x(\tau_2)x(\tau_3)d\tau_1 d\tau_2 d\tau_3 + \ldots \qquad (5.59)
\end{aligned}$$

Recall that in the linear time-invariant case, the impulse response may be found in either the time domain or the frequency domain. In fact, the frequency domain solution is often much easier to obtain than the time domain impulse response. However, for time-varying systems, frequency response is no longer well defined, and hence it is necessary to apply impulses at the system input and find the output expression in the time domain. This amounts to solving the differential equation directly, which is not trivial in general. But for first-order differential equations, such as the sampling mixer equation, this method is feasible.

Suppose that in (5.50), g varies with time. In particular, g goes to 0 when the transistor turns off. The zero state response of the first-order equation

$$gv_1 + C\frac{dv_1}{dt} = gv_{rf}$$

is obtained using integrating factor:

$$v_1(t) = \int_0^t e^{-\int_\mu^t \frac{g(\xi)}{C}d\xi} \frac{g(\mu)}{C} v_{rf}(\mu)\, d\mu \qquad (5.60)$$

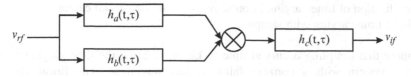

Fig. 5.17 Composition of second-order system from first-order systems

Setting $v_{rf}(\mu) = \delta(\mu - \tau)$ gives the impulse response of the first-order system:

$$h_1(t, \tau) = \begin{cases} e^{-\int_\tau^t \frac{g_s^2}{C} d\xi} \dfrac{g(\tau)}{C} & \text{if } \begin{matrix} t \geq \tau \\ t \leq \tau \end{matrix} \\ 0 & \end{cases} \tag{5.61}$$

Next, we calculate the second-order kernel. In general, a second-order system can be thought of as the composition of linear systems as shown in Figure 5.17. In particular, if the impulse responses of the linear systems are $h_a(t, \tau), h_b(t, \tau)$, and $h_c(t, \tau)$ respectively, the second order kernel can be computed as follows:

$$h_2(t, \tau_1, \tau_2) = \int_{-\infty}^{\infty} h_c(t, \tau) h_a(\tau, \tau_1) h_b(\tau, \tau_2) \, d\tau \tag{5.62}$$

Expression for the third-order system is similar. If in Figure 5.17 h_a is second order, and therefore the overall system is third order, then the overall system kernel is

$$h_3(t, \tau_1, \tau_2, \tau_3) = \int_{-\infty}^{\infty} h_c(t, \tau) h_a(\tau, \tau_1, \tau_2) h_b(\tau, \tau_3) \, d\tau \tag{5.63}$$

Therefore, as was done with the time-invariant case, h_2 and h_3, can be solved consecutively. For example, the second-order kernel is found by substituting $v_1(\tau)$ from (5.60) and setting input as an impulse in the solution to the second-order equation in (5.50):

$$v_2(t) = \int_0^t e^{-\int_\tau^t \frac{g(\xi)}{C} d\xi} \frac{g(\tau)}{C} \left(\frac{k}{2}\right) \left(v_1^2(\tau) - v_{rf}^2(\tau)\right) d\tau \tag{5.64}$$

Since the differential equations are linear and first order, there is no theoretical difficulty in obtaining the solution. Following this method, the complete time domain solution to the system can be obtained. However, the mathematic is very tedious.

Now, in a sampling mixer where the output points of interest are at the sampled points, as opposed to continuous time output, it will be shown in the next sub-section (by exploring the frequency domain interpretation and by taking advantage of the fact that output is in the sample data domain) that the problem can be simplified significantly. Closed-form solution can be obtained.

Simplification of time varying impulse response in the special case:
periodic time varying with sampling

Suppose that sampling occurs at time 0. Let $h_1(t, \tau)$ be the time varying kernel
of the system with a nonzero fall-time LO waveform. The linear term in
the sampled output voltage is

$$y(0) = \int_{-T}^{0} h_1(0, \tau)x(\tau)\, d\tau + ZIR \qquad (5.65)$$

where T is the sampling period, and *ZIR* is the zero input response due to the initial
condition. Since we have assumed that the *RC* time constant for the sampling mixer
is much smaller than the sampling period, *ZIR* is negligible. This is equivalent to
saying that any poles/zeroes of the sampling mixer have frequencies much higher
than f_{lo} or that (4.150), Chapter 4, is satisfied. Again, because of the small time
constant, we may replace the lower limit of integration in (5.65) by $-\infty$:

$$y(0) = \int_{-\infty}^{0} h_1(0, \tau)x(\tau)\, d\tau \qquad (5.66)$$

Next we make a few observations.

Observation 1:

What is the value of $y(T)$? Replacing 0 by T in (5.66), we have

$$y(T) = \int_{-\infty}^{T} h_1(T, \tau)x(\tau)\, d\tau \qquad (5.67)$$

We do not know $h_1(T, \tau)$ in general. However, we do know that the system is
not only time varying, but periodically time varying in time T. This follows from
the fact that the LO waveform is periodic in T which implies that the system is
periodic in T.

$$\therefore h_1(t + u, u) = h_1(t + u + T, u + T) \qquad (5.68)$$

which is the basic property of periodic time-varying system.
    Substituting (5.68) in (5.67), we have

$$y(T) = \int_{-\infty}^{T} h_1(0, \tau - T)x(\tau)\, d\tau \qquad (5.69)$$

Let us redefine $\tau_1 = \tau - T$:

$$y(T) = \int_{-\infty}^{0} h_1(0, \tau_1) x(\tau_1 + T) \, d\tau_1 \tag{5.70}$$

Hence, $y(T)$ can be obtained by using the same impulse response, h_1, that is used to generate $y(0)$.

Just to highlight the unique position that T assumes, we can see that $y(T/2)$, for example, cannot be obtained by this method. This is because:

$$h_1(t + u, u) \neq h_1\left(t + u + \frac{T}{2}, u + \frac{T}{2}\right)$$

The usefulness of expressing $y(T)$ in (5.70) will become even more apparent if we start to express $y(2T)$, $y(3T)$ ...in a similar fashion. Following the preceding method,

$$y(2T) = \int_{-\infty}^{0} h_1(0, \tau) x(\tau + 2T) \, d\tau$$

Now if we define a new system, where we collect $y(0)$, $y(T)$, $y(2T)$... and so on. As the output of the new system, $x(0)$, $x(T)$, $x(2T)$ as the corresponding inputs, we can see the following:

a. They are related by the same convolution integral with the same impulse function $h_1(0, \tau)$ and, most important, the impulse response is a one-dimensional impulse response.
b. The relationship is time invariant [e.g. as $x(t)$ is shifted to $x(t + T)$, it is the same $h_1(0, \tau)$ that is used to generate $y(t)$ and $y(t + T)$].

Therefore, this describes an LTI system. Extending this argument, we can likewise transform a bilinear time-varying system to a bilinear time-invariant system if we are interested only in the sampled point. Therefore, if $h_1(t, \tau_1, \tau_2)$ describes a bilinear periodic time varying system the sampled point can be described by the equivalent bilinear time invariant system $h_1(0, \tau_1, \tau_2)$.

Observation 2:

Since the output voltage of interest is at the sampling instant, the time varying characteristic of importance occurs during the time within a few time constants before the sampling instant. This means that the same Volterra kernel applies to every sampling instant except that the kernels are shifted by nT, integer multiples of the LO period. To simplify computation, we want to keep the Volterra kernel identical for each sampling period. Therefore, instead of shifting the kernel, we shift the input signal backward in time by nT and keep the sampling instant at time 0.

This effectively keeps the functional form of the kernel identical for all samples. Further, there is nothing special about sampling points nT. Hence if sampling occurs at an arbitrary time t, the sampled voltage can be represented as

$$y(t) = \int_{-\infty}^{0} h_1(0, \tau)x(\tau + t)\, d\tau \tag{5.71}$$

This $y(t)$ is a fictitious signal, whose value at t represents the sampled value of the output if sampling is to occur at t. If sampling occurs at instants nT, the output voltage samples are just $y(T), y(2T), \ldots, y(nT)$. Hence the nonideal sampling of the output signal is reduced to the ideal sampling of $y(t)$, where $y(t)$ is related to the input $x(t)$ by (5.71). This is convenient because harmonic distortion for the ideal sampled $y(nT)$ is the same as the harmonic distortion of the continuous signal $y(t)$. So the problem of calculating distortion in nonideal sampled output is reduced to the problem of calculating the distortion in the continuous signal $y(t)$.

The transformation from (5.65) to (5.71) also reduces a time varying system to a time- invariant system, as the relation between x and y in (5.71) is time invariant. This situation is analogous to the analysis of discrete control systems where although a zero-order hold is not a time-invariant operation in the continuous time domain, the input output relation is nevertheless time invariant in the sampled data domain.

Now that we have made some key observations, it remains to find the impulse response of the linear time-invariant system described in (5.71). Let us denote its impulse response in time and frequency domain by $\hat{h}_1(t)$ and $\hat{H}_1(\omega)$ respectively. To find $\hat{h}_1$, set $x(t) = \delta(t)$ in (5.71). It follows that

$$\hat{h}_1(t) = \begin{cases} h_1(0, -t) \\ 0 \end{cases} \quad \text{if } \begin{matrix} t \geq 0 \\ t \prec 0 \end{matrix} \tag{5.72}$$

The frequency response is its Fourier transform:

$$\hat{H}_1(\omega) = \int_{-\infty}^{0} h_1(0, \tau)e^{j\omega\tau}\, d\tau \tag{5.73}$$

where again, $h_1(t, \tau)$ is the time varying kernel.

The same technique applies to higher order systems. For a second order system, the sampled data domain response is:

$$y(t) = \int_{-\infty}^{0} \int_{-\infty}^{0} h_2(0, \tau_1, \tau_2)x(\tau_1 + t)x(\tau_2 + t)\, d\tau_1\, d\tau_2 \tag{5.74}$$

and the frequency response is

$$\hat{H}_2(\omega_1, \omega_2) = \int_{-\infty}^{0} \int_{-\infty}^{0} h_2(0, \tau_1, \tau_2)e^{j\omega_1\tau_1}e^{j\omega_2\tau_2}\, d\tau_1\, d\tau_2 \tag{5.75}$$

Again, the second-order kernel may be obtained from its composition from first-order systems as in (5.62).

The frequency domain expression is of most interest. It is, in fact, possible to bypass the time domain expression and obtain the frequency domain result directly. Assuming a composition form as in Figure 5.17, substituting (5.62) into (5.75), but noting t = 0, then

$$\hat{H}_2(\omega_1, \omega_2) = \int_{-\infty}^{0} \int_{-\infty}^{0} \int_{-\infty}^{0} h_c(0, \tau_3) h_a(\tau_3, \tau_1) h_b$$

$$\times (\tau_3, \tau_2) e^{j\omega_1 \tau_1} e^{j\omega_2 \tau_2} d\tau_3 \, d\tau_1 \, d\tau_2 \qquad (5.76)$$

More succinctly and for convenience only, define

$$\hat{H}_1(\omega, t) = \int_{-\infty}^{t} h_1(t, \tau) e^{j\omega\tau} \, d\tau \qquad (5.77)$$

and

$$\hat{H}_2(\omega_1, \omega_2, t) = \int_{-\infty}^{t} \int_{-\infty}^{t} h_2(t, \tau_1, \tau_2) e^{j\omega_1 \tau_{11}} e^{j\omega_2 \tau_2} d\tau_1 \, d\tau_2 \qquad (5.78)$$

Then, again substituting (5.62) into (5.78) and simplifying the resulting expression using (5.77) and we arrive at the formula

$$\hat{H}_2(\omega_1, \omega_2, t) = \int_{-\infty}^{t} h_c(t, \tau) H_a(\omega_1, \tau) H_b(\omega_2, \tau) \, d\tau \qquad (5.79)$$

In particular, setting $t = 0$ gives the sampled frequency domain kernel for the system Figure 5.17

$$\hat{H}_2(\omega_1, \omega_2) = \int_{-\infty}^{0} h_c(0, \tau) H_a(\omega_1, \tau) H_b(\omega_2, \tau) \, d\tau \qquad (5.80)$$

The third order frequency domain representation is similar. If h_a in Figure 5.17 is second order, then the resulting Volterra kernel becomes

$$\hat{H}_3(\omega_1, \omega_2, \omega_3, t) = \int_{-\infty}^{t} h_c(t, \tau) H_a(\omega_1, \omega_2, \tau) H_b(\omega_3, \tau) \, d\tau \qquad (5.81)$$

and the sampled frequency domain kernel is

$$\hat{H}_3(\omega_1, \omega_2, \omega_3) = \int_{-\infty}^{0} h_c(0, \tau) H_a(\omega_1, \omega_2, \tau) H_b(\omega_3, \tau) \, d\tau \qquad (5.82)$$

Derivation of Volterra coefficents for time-varying distortion

In this sub-section, we analyze the sampling mixer of Figure 5.16 in the sampled data domain. Assuming that the input side is source, the output drain, from the basic MOS equation, we have

$$k(V_g - V_s - V_t)(V_d - V_s) - \frac{k}{2}(V_d - V_s)^2 + C\frac{dV_d}{dt} = 0 \qquad (5.83)$$

(Assuming the output side being source will give a slightly different differential equation, but the form of the answer is the same.)

Remember the M_1 conductance $g = k(V_{gs} - V_t)$. M_1 enters cut-off region at $V_{gs} - V_t = 0$, so the value of g goes to zero at each sampling point. Now, since we have assumed that the RC time constant is small, the region of most importance is the time right before the sampling instant. Therefore, we can approximate g by a linear function prior to each sampling point. This assumption is most valid if the system time constant is smaller than the fall time of the gate voltage. In practical cases where this is not true, the analysis yields a limiting case and provides a useful bound. Now, the slope of the linearly varying g is just the slope of the cutting edge V_g minus the slope of the input signal V_s, to a first- order approximation. So if we define $\beta = \frac{(2kV_G)}{T_f}$, where V_G and T_f are as indicated in Figure 5.17, and define $g = -\beta t$, the MOS equation (5.83) becomes

$$\left(g - k\frac{dV_s}{dt} \cdot t\right)(V_d - V_s) - \frac{k}{2}(V_d - V_s)^2 + C\frac{dV_d}{dt} = 0 \qquad (5.84)$$

Let $V_s = V_{rf}$, $V_d = V_{if}$, expand V_{if} into its Volterra series as before and collect the first to third-order terms. We have

$$\begin{cases} gv_1 + C\dfrac{dv_1}{dt} = gV_{rf} \\[2mm] gv_2 + C\dfrac{dv_2}{dt} = k\dfrac{V_{rf}}{dt}t(v_1 - V_{rf}) + \dfrac{k}{2}(v_1 - V_{rf})^2 \\[2mm] gv_3 + C\dfrac{dv_3}{dt} = k\dfrac{V_{rf}}{dt}tv_2 + k(v_1 - V_{rf})v_2 \end{cases} \qquad (5.85)$$

To calculate H_1, we use (5.77), where the impulse response $h_1(t, \tau)$ has already been solved in (5.61):

$$\begin{aligned} \hat{H}_1(\omega, t) &= \int_{-\infty}^{t} h_1(t, \tau)e^{j\omega\tau}\, d\tau \\[2mm] &= \int_{-\infty}^{t} e^{-\int_{\tau}^{t}\frac{g(\xi)}{C}d\xi}\frac{g(\tau)}{C}e^{j\omega\tau}\, d\tau \\[2mm] &= \int_{-\infty}^{t} e^{-\kappa^2(\tau^2 - t^2)}(-2\kappa^2\tau)e^{j\omega\tau}\, d\tau \end{aligned} \qquad (5.86)$$

where the substitution of $g = -\beta t$ and $\kappa^2 = \frac{\beta}{2C}$ are made. This integral does not have closed-form solutions. However, we are interested in the value of the integral when t is close to 0, so the upper limit of the integral is close to zero. The integrand is a rapidly increasing function of τ as τ approaches 0, so, effectively, the only relevant portion of the integration is when τ is close to 0. Therefore, we can assume that $e^{j\omega\tau} \approx 1 + j\omega\tau$. In this case, the integral becomes

$$\hat{H}_1(\omega, t) = 1 + j\omega t - j\omega\sqrt{\frac{\pi}{4}}\cdot\frac{1}{\kappa}\cdot e^{\kappa^2 t^2}(erf(\kappa t) + 1) \tag{5.87}$$

The error function is defined as $erf(t) = \frac{2}{\sqrt{\pi}}\int_0^t e^{-x^2}dx$
So,

$$\hat{H}_1(\omega) = 1 - j\omega\sqrt{\frac{\pi}{4}}\cdot\frac{1}{\kappa} \tag{5.88}$$

The second-order term is evaluated using (5.79), where the time-varying kernel is obtained from its composition from first order terms. Mechanically, v_1 is substituted by $\hat{H}_1$, and v_{rf} is substituted by $e^{j\omega t}$.

$$\hat{H}_2(\omega_1, \omega_2, t) = \int_{-\infty}^t \frac{e^{-\int_\tau^t \frac{g(\xi)}{C}d\xi}}{C}k\cdot j\omega_1\cdot e^{j\omega_1\tau}\cdot\tau\cdot\left(\hat{H}_1(\omega_2, \tau) - e^{j\omega_2\tau}\right)d\tau$$
$$+ \int_{-\infty}^t \frac{e^{-\int_\tau^t \frac{g(\xi)}{C}d\xi}}{C}\left(\frac{k}{2}\right)\left(\hat{H}_1(\omega_1, \tau) - e^{j\omega_1\tau}\right)$$
$$\times\left(\hat{H}_1(\omega_2, \tau) - e^{j\omega_2\tau}\right)d\tau \tag{5.89}$$

Again, approximate $e^{j\omega t}$ by its Taylor expansion, substitute (5.87) to (5.89):

$$\hat{H}_2(\omega_1, \omega_2, t) = \frac{\omega_1\omega_2}{C}\cdot k\cdot\sqrt{\frac{\pi}{4}}\cdot\frac{1}{\kappa}\cdot e^{\kappa^2 t^2}\int_{-\infty}^t \tau(erf(\kappa\tau) + 1)^2 d\tau + \frac{\omega_1\omega_2}{C}$$
$$\cdot\frac{k}{2}\cdot\left(\frac{\pi}{4}\right)\frac{1}{\kappa^2}\cdot e^{\kappa^2 t^2}\int_{-\infty}^t e^{\kappa^2\tau^2}(erf(\kappa\tau) + 1)^2 d\tau \tag{5.90}$$

Fortunately, at $t = 0$, the last two integrals may be evaluated numerically, and we have:

$$\hat{H}_2(\omega_1, \omega_2) = -\omega_1\omega_2\left(\frac{k}{C}\right)\cdot\frac{b}{\kappa^3} \tag{5.91}$$

where $b = 0.234$ comes from the evaluation of definite integrals. (The relative contributions from the two integrals are actually comparable.) Finally, $\hat{H}_3$ is obtained in the same way:

$$\hat{H}_3(\omega_1, \omega_2, \omega_3, t) = \int_{-\infty}^{t} \frac{e^{-\int_\tau^t \frac{g(\xi)}{C}d\xi}}{C} k \cdot j\omega_1 \cdot e^{j\omega_1\tau} \cdot \tau \cdot \hat{H}_2(\omega_1, \omega_2, \tau) \, d\tau$$

$$+ \int_{-\infty}^{t} \frac{e^{-\int_\tau^t \frac{g(\xi)}{C}d\xi}}{C} k\left(\hat{H}_1(\omega_3, \tau) - e^{j\omega_3\tau}\right)\hat{H}_2$$

$$\times (\omega_1, \omega_2, \tau) \, d\tau \tag{5.92}$$

We need to evaluate the above expression for $t = 0$. Because $\hat{H}_2$ terms contains two integrals, there are in total four definite integrals. This is tedious but not impossible. Numerically, the answer turns out to be:

$$\hat{H}_3(\omega_1, \omega_2, \omega_3) = j\omega_1\omega_2\omega_3 \cdot \left(\frac{k}{C}\right)^2 \cdot \frac{c}{\kappa^5} \tag{5.93}$$

where $c = 0.0913$ numerically. Equations (5.88), (5.91) and (5.93) are the final solutions to the sampled time-varying Volterra series. To summarize the results in terms of circuit parameters,

$$\hat{H}_1(\omega) = 1 - j\omega \cdot \sqrt{\frac{C}{k}}\left(\frac{T_f}{V_G}\right)^{1/2} \cdot a \tag{5.94}$$

$$\hat{H}_2(\omega_1, \omega_2) = -\omega_1\omega_2 \cdot \sqrt{\frac{C}{k}}\left(\frac{T_f}{V_G}\right)^{3/2} \cdot b \tag{5.95}$$

$$\hat{H}_3(\omega_1, \omega_2, \omega_3) = j\omega_1\omega_2\omega_3 \cdot \sqrt{\frac{C}{k}}\left(\frac{T_f}{V_G}\right)^{5/2} \cdot c \tag{5.96}$$

where again $a = 0.866$, $b = 0.234$, $c = 0.0913$ numerically. Recall that this result is obtained by assuming a linearly decreasing g. In reality, g is nearly constant for a large part of the sampling period. Therefore, this solution is an asymptotic case. The closed- form solution for HD_3 and IM_3, assuming a linear falling edge, can now be derived and are shown in (5.97) and (5.98):

$$HD_3 = -j \cdot \frac{A_{rf}^2\omega_{rf}^3}{8}\left(\frac{\pi}{4}\right)^{3/2}\left(\frac{C}{k}\right)^{1/2}\left(\frac{T_f}{V_G}\right)^{5/2} \cdot \delta \tag{5.97}$$

$$IM_3 = j \cdot \frac{3A_{interference}^2\omega_{rf}^3}{8}\left(\frac{\pi}{4}\right)^{3/2}\left(\frac{C}{k}\right)^{1/2}\left(\frac{T_f}{V_G}\right)^{5/2}\delta \tag{5.98}$$

where T_f is the fall time and δ is a constant with a numerical value 0.1036.

If we compare the Volterra coefficients for the nonideal MOS sampling mixer with the ideal sampling case, we find that the nonlinearity due to the time-varying nature is much smaller than the ideal sampling case at low frequency and will exceed other nonlinearity at high frequency. Notice that the time-varying distortion varies with cube of frequency, so theoretically at high frequency the time-varying distortion may overtake the continuous time distortion. For typical sampling mixer design this will happens at an f_{rf} of around 1 GHz. The fact that time-varying distortion is small at lower frequency can be explained intuitively. There are two sources of nonlinearity from the M_1 governing equation:

$$I_d = k\left(V_{gs} - V_t\right)V_{ds} - \frac{k}{2}V_{ds}^2 \qquad (5.99)$$

An obvious source of nonlinearity is the square term V_{ds}^2, but this effect is secondary for the ideal sampling case, where the major contributing factor is the input signal dependent conductance $k(V_{gs} - V_t)$. However, in a non-ideal sampling mixer, the first order signal dependence in g is eliminated because the sampling always occurs when $V_{gs}-V_t = 0$; hence the local behavior of g prior to cutoff is approximately the same for each sampling point. Hence the only nonlinear factors left are the signal dependence in the derivative of the conductance and the V_{ds}^2 term, which are significantly smaller. However, we are not really getting something for free here. Although in the case of the nonideal mixer the $k(V_{gs} - V_t)$ term causes much less input dependence in g, the V_{gs} term manifests as an altogether different source of distortion. Because cutoff occurs when $V_{gs} - V_t = 0$, the cutoff time is now input signal dependent. In other words, the distortion in the amplitude domain is translated into the time domain by nonideal sampling. The signal-dependent cut-off time is called sampling error, which is the familiar distortion caused by finite fall time reported for an SAH circuit. Its effect is similar in the sampling mixer case. For the sake of consistency, we will re-derive that using the Volterra series in the next sub-section. The result should be similar to that reported in the literature.

5.9.2.2 Sampling Distortion

In the development of time-varying distortion analysis in the sampled data domain, the output voltage is assumed to be sampled at equally spaced instants. This assumption is not exactly true given that the gate voltage has nonzero fall time. The instance of sampling is when $V_{gs} - V_t = 0$, so the time at which sampling occurs depends not only on gate voltage but also on the input voltage. The signal-dependent sampling time introduces additional distortion term, which can be quantified using the Volterra series again (strictly speaking, this distortion can be calculated without the use of the Volterra series. The Volterra series is used here mainly for consistency).

Referring to Figure 5.18, suppose that the input signal V_{rf} is smooth, and we intend to sample the input at time 0. Let the gate voltage V_{lo} have a finite-slope falling edge with slope α, where $\alpha = \frac{2V_G}{T_f}$ as shown in Figure 5.18. The falling edge

Fig. 5.18 Diagram showing sampling error

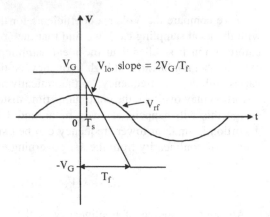

has the form $V_{lo}(t) = V_G - \alpha t$. Since the MOS transistor enters the cutoff region at $V_g - V_s = V_t$, and the input is applied at the source side, sampling occurs at the instant T_s when $V_{lo}(T_s) - V_t = V_{rf}(T_s)$. Now, using the approximation $V_{rf}(T_s) \approx V_{rf}(0) + V'_{rf}(0)t$, we can solve for the cutoff time:

$$T_s = \frac{V_g - V_t - f(0)}{\alpha + f'(0)} \approx \frac{V_g - V_t - f(0)}{\alpha}\left(1 - \frac{f'(0)}{\alpha} + \frac{f'(0)^2}{\alpha^2}\right) \qquad (5.100)$$

where we assume that α is large. Next, substituting the above expression for T_s into Taylor expansion for $V_{rf}(T_s)$ around 0, and collect the first three order terms, we have:

$$V_{rf}(t) = V_{rf}(0) - \frac{V'_{rf}(0)V_{rf}(0)}{\alpha} + \left(\frac{V''_{rf}(0)V^2_{rf}(0)}{2\alpha^2} + \frac{V'_{rf}(0)^2V_{rf}(0)}{\alpha^2}\right) \qquad (5.101)$$

where again the assumption of large α is used. Therefore, due to the finite fall time, the sampled value $V_{rf}(T_s)$ differs from the desired vale $V_{rf}(0)$ by additional signal dependent terms, which produces distortion. Since the sampling time is arbitrary, (5.95) is valid if time 0 is replaced by an arbitrary sampling time.

The signal dependence can be analyzed by the Volterra series. The linear term in (5.101) is $V_{rf}(0)$, so $H_1 = 1$. The second-order term is the product of the derivative with the function itself. Using composition from linear terms and using symmetrization, the Volterra series coefficient is found to be

$$H_2(\omega_1, \omega_2) = -\frac{j(\omega_1 + \omega_2)}{2\alpha} \qquad (5.102)$$

and for the third order,

$$H_3(\omega_1, \omega_2, \omega_3) = -\frac{(\omega_1 + \omega_2 + \omega_3)^2}{6\alpha^2} \qquad (5.103)$$

Table 1.1 Summary of distortion behavior of a single-ended sampling mixer

Sources	HD$_3$	IM$_3$
Continuous time distortion	$\frac{A_{rf}^2}{4} \frac{j\omega_{rf}C}{k(V_G-V_t)^3}$	Same as HD_3
Time varying distortion	$-j\frac{A_{rf}^2\omega_{rf}^3}{8}\left(\frac{\pi}{4}\right)^{3/2}\left(\frac{C}{k}\right)^{1/2}\left(\frac{T_f}{V_G}\right)^{5/2}\cdot 0.0913$	$3\cdot HD_3$
Sampling distortion	$\frac{3}{8}\left(\frac{A_{rf}\omega_{rf}T_f}{2V_G}\right)^2$	$\frac{1}{3}HD_3$

where ω_1, ω_2, ω_3 are the frequencies of the input signals. If the mixer is operating at ω_{rf}, using the proper definitions, harmonic distortion and intermodulation can be calculated exactly:

$$HD_3 = \frac{A_{rf}}{4}\left(\frac{j\omega_{rf}T_f}{V_G}\right) \tag{5.104}$$

$$HD_3 = \frac{3A_{rf}^2}{32}\left(\frac{\omega_{rf}T_f}{V_G}\right)^2 \tag{5.105}$$

$$IM_2 = 0 \tag{5.106}$$

$$IM_3 = \frac{A_{interference}^2}{32}\left(\frac{\omega_{rf}T_f}{V_G}\right)^2 \tag{5.107}$$

Comparing the sampling distortion as given by (5.104) through (5.107) with the continuous time distortion as given by (5.54) through (5.57), it can be shown that the sampling distortion becomes comparable to the continuous time distortion when the fall time is in the order of $\sqrt{\tau T_{rf}}$, where τ is the RC constant formed by the MOS resistor and the load capacitor, and T_{rf} is the period of the input signal V_{rf}. Therefore, sampling error is a bigger problem at high frequency.

Table 5.1 summarizes the distortion behavior of an unbalanced sampling mixer.

For any given T_f we have three 3 different sources of distortion as summarized in Table 5.1. We should calculate the IM_3 for this given T_f and do a comparison. Whichever is the largest dominates, and we call that the IM_3 of the mixer.

5.9.3 High-Frequency Case in Bottom-Plate Single-Ended Sampling Mixer

The preceding analysis is now applied to the bottom-plate sampling mixer as shown in Figure 5.13. First Figure 5.13 is redrawn in Figure 5.19. to show the parasitic capacitance. Here C_j, C_p are the parasitic capacitance associated with M_1 and M_2, respectively.

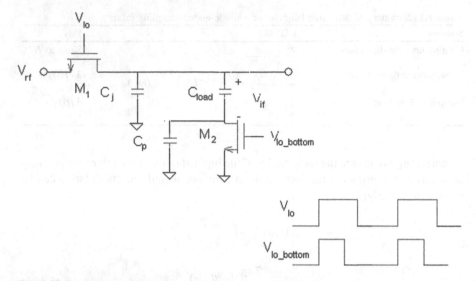

Fig. 5.19 Bottomplate single-ended sampling mixer with parasitics capacitance

The derivation of the distortion formula is given in [1]. We skip the details and just present the final closed-form solutions. For typical MOS device parameters, except at frequencies higher than 1 GHz, we neglect the time-varying distortion. It is seen that as the bottom switch M_2 opens, distortion due to sampling distortion is small. This is because the source and drain of the bottom switch are kept almost constant at the ground level, which eliminates signal-dependent sampling error. Therefore, distortion is due mainly to continuous time distortion. When the top switch opens, the capacitor is primarily due to parasitic capacitance. With the much smaller capacitance, the continuous time distortion is small. The sampling distortion, however, is the same as the case of the single-ended simple sampling mixer (Figure 5.10) and hence dominates. Therefore, the total distortion is a combination of the two (each weighed by the proper capacitor ratios due to the capacitor divider) as follows:

$$HD_3 = \left(\frac{C_{load}}{C_{load}+C_p}\right) \cdot \frac{A_{rf}^2}{4} \cdot \frac{j\omega_{rf}\left(C_{load}+C_j\right)}{k_1\left(V_G-V_T\right)^3} + \left(\frac{C_p}{C_{load}+C_p}\right) \cdot \frac{3A_{rf}^2}{32}\left(\frac{\omega_{rf}T_f}{V_G}\right)^2$$

(5.108)

$$IM_3 = \left(\frac{C_{load}}{C_{load}+C_p}\right) \cdot \frac{A_{interference}^2}{4} \cdot \frac{j\omega_{rf}\left(C_{load}+C_j\right)}{k_1\left(V_G+V_T\right)^3} + \left(\frac{C_p}{C_{load}+C_p}\right)$$
$$\cdot \frac{A_{interference}^2}{32}\left(\frac{\omega_{rf}T_f}{V_G}\right)^2$$

(5.109)

These formulas are similar to the case of a single-ended simple sampling mixer. However, the major difference is that the distortion due to sampling (when the top switch opens) is reduced by a factor of $\frac{C_p}{(C_{load}+C_p)}$, which is significant. Therefore, sampling distortion is less of a problem in a bottom-plate sampling mixer than in the simple sampling mixer. Finally, for a bottom-plate subsampled mixer, the distortion formulas remain the same.

5.10 Intrinsic Noise in Single-Ended Sampling Mixer

5.10.1 V_{lo} not Switching Case

The noise calculation is just a repetition of the switching mixer, except that instead of terminating the mixer with R_{load}, it is now terminated with C_{load}. All the formulae can be rederived with this change. We will skip the details and present the final answers.

In this sub-section we follow the notation developed in sub-section 5.5.1. We assume that the switch is on and hence the output noise PSD is given by

$$S_n = 4kTR_{on} \tag{5.110}$$

Input referred noise, N_{dev}, is then given by

$$N_{dev} = \frac{S_n}{G_c} \tag{5.111}$$

where S_n is given by (5.110) and G_c is given by (5.43). Hence NF can be calculated by substituting the N_{dev} so obtained in (5.35).

As a side note notice the noise bandwidth f_b is given as

$$f_b = \pi/2 \times (bandwidth\ of\ RC\ filter) = \pi/2 \times (1/(2\pi R_{on}C_{load})) \\ = 1/(4 \times R_{on}C_{load}) \tag{5.112}$$

If we integrate S_n as given by (5.110) throughout the entire bandwidth of interest ($=f_b$) to find the variance of the noise, σ_n^2, we note that R_{on} is canceled and that σ_n^2 is given as

$$\sigma_n^2 = \int_0^{f_b} S_n\ df = f_b \times 4kTR_{on} = \frac{1}{4R_{on}C_{load}} \times 4kTR_{on} = \frac{kT}{C_{load}} \tag{5.113}$$

5.10.2 V_{lo} Switching, Both Special and General Cases

First the output noise PSD in the special V_{lo} switching case is presented. The output noise PSD in the general V_{lo} switching case will be studied in Q7.

Let us follow the notation in sub-section 5.5.2 and Figure 5.7. We also represent the sampling mixer by Figure 5.7. With V_{lo} switching, the noises at node 1 in Figure 5.7 are just aliased on top of one another to form the noise at node 2. The PSD of noise at node 1 is given by (5.110) and repeated as

$$S_{n_1} = 4kTR_{on} \tag{5.114}$$

The noise bandwidth is given as f_b and the switching frequency is given by f_{lo}. Due to aliasing, all noises centered at frequencies that are around multiples of f_{lo} and below f_b will fall on top of one another. Hence the output noise PSD will increase by a factor of f_b/f_{lo}. In addition, due to switching there is a difference in conversion gain that has to be accounted for. Comparing (5.44) with (5.43), we note that because of switching the conversion gain is reduced by $1/\pi$. In summary, the output noise PSD at node 2 can be calculated by applying an increase of f_b/f_{lo} and a reduction of π to S_{n1} as given in (5.114). Hence S_{n2} is given by [2]

$$S_{n_2} = 4kTR_{on} \frac{1}{\pi} \times (f_b/f_{lo}) \tag{5.115}$$

We can also calculate the variance of the noise, σ_{n2}^2. First notice that f_b is still $1/(4 \times R_{on}C_{load})$. However, because of switching the entire bandwidth of interest is f_{lo} and not f_b. If we substitute (5.112) to (5.115) and integrate S_{n2} throughout the entire bandwidth of interest ($= f_{lo}$) to find the variance of the noise, σ_{n2}^2, we note that R_{on}, f_{lo} are canceled and that σ_{n2}^2 is

$$\sigma_{n_2}^2 = \int_0^{f_{lo}} S_{n2} \, df = f_{lo} \times \frac{4kTR_{on}}{\pi} \times \left(\frac{f_b}{f_{lo}}\right) = \frac{4kTR_{on}}{\pi} \times \frac{1}{4R_{on}C_{load}}$$

$$= \frac{1}{\pi} \times \frac{kT}{C_{load}} \tag{5.116}$$

Note that this is similar to the σ_n^2 given in (5.112) except that they differ by a factor of $(1/\pi)$ because of the reduction in conversion gain in a sampling mixer due to switching.

Next we can find input referred noise, N_{dev}, by substituting S_{n2} derived in (5.115) and G_c derived in (5.44) into the following formula:

$$N_{device} = \frac{S_{n2}}{G_c} \tag{5.117}$$

Substituting N_{dev} so obtained into (5.35) will allow us to get the NF.

As a final note, for a subsampling mixer the f_b/f_{lo} factor in (5.115) is much larger than that for a normal sampling mixer. Hence S_{n2} is substantially larger, which means that a subsampling mixer has a much poorer noise performance.

5.11 Extrinsic Noise in Single Ended Sampling Mixer

There are two primary noise sources in any mixer. One is the intrinsic noise from the mixer itself, which has been the primary focus of our discussion in Chapters 4 and 5 so far. In addition, any extrinsic noise from the RF/LO port will directly affect the output noise. In this section we will briefly examine this extrinsic noise as well. As it turns out, extrinsic noise from the RF port can be analyzed in much the same way as intrinsic noise of the mixer. Accordingly, we concentrate our discussion on the extrinsic noise from the LO port.

Extrinsic noise from the LO port can come, for example, from phase noise of the frequency synthesizer. In both switching/sampling mixers, this noise, which comes from the LO port, causes the MOS conductance to fluctuate, hence producing noise at the output and thus degrading the NF of the mixer. Notice that this will happen even if we assume that the MOS conductance itself does not have thermal noise; that is, there is no intrinsic noise. In addition if there is adjacent channel interference at the input (RF port), this extrinsic noise from the LO port can mix this interference down to the same IF as the desired signal, thus degrading the SNR.

In this chapter we focus our discussion on the first impairment. The issue involving mixing the interference by LO noise investigated further in Chapter 8.

5.11.1 V_{lo} not Switching Case

In general, without switching, the mixer (both active and passive) can be described as an LTI circuit. Hence the analysis is best carried out in the frequency domain. Therefore, the output noise PSD at the IF port can be calculated through multiplying the LO noise PSD by the power gain from the LO port to the IF port.

Let us assume that the LO port has an input spectral density given by $S_{n\_lo\_input}$. Let us further assume that there is no RF signal. Then, as an example, $S_{n\_lo\_output}$, the output noise PSD of the sampling mixer described in Figure 5.10 due to this noise at the LO port, is given by

$$S_{n\_lo\_output} = S_{n\_lo\_input} g^2_{m\_triode} \left(\frac{1}{\omega_{lo} C_{load}} \right)^2 \tag{5.118}$$

Here g_{m_triode} is the transconductance of M_1 operating in the triode region.

5.11.2 V_{lo} Switching, Both Special and General Cases

Now let us take a look at the case when V_{lo} is switching. For the Gilbert mixer and switching/sampling mixers, we can apply (4.134) of Chapter 4 and (5.36) of this chapter, respectively, to find the output noise PSD, $S_{n\_lo\_output}$. S_{n2} in (4.134) of Chapter 4 and S_{n1} in (5.36) are replaced by $S_{n\_lo\_input}$. Similarly, S_{n3} in (4.134) of Chapter 4 and S_{n2} in (5.36) are replaced by $S_{n\_lo\_output}$.

5.11.3 Special V_{lo} Switching Case; V_{lo} has Finite Rise and Fall Time

Even though $S_{n\_lo\_output}$ for switching/sampling and Gilbert mixers incorporating finite rise and fall time can all be derived, the $S_{n\_lo\_output}$ of the sampling mixer incorporating finite fall time is particularly important because its output value is determined exactly at the falling edge of the LO waveform. How do we derive this $S_{n\_lo\_output}$?

Let us first observe that for a sampling mixer, since the mixer works in the sampled domain, we are actually more interested in the variance of the output noise process at the sampling instant rather than $S_{n\_lo\_output}$ throughout the entire LO period. We should now denote the noise process at the LO port as N_{lo_input} and the resulting output noise process at the IF port as N_{lo_output}. We would further denote the variance (or mean square voltage) at the LO port as $\sigma_{n\_lo\_input}{}^2$ and the resulting output variance (or mean square voltage) at the IF port as $\sigma_{n\_lo\_output}{}^2$. The variances have the unit V^2. It is $\sigma_{n\_lo\_output}{}^2$ that we are interested in.

Some important properties of this variance as well as N_{lo_output} can be derived using time domain method based on the stochastic differential equations (SDE) approach. For example, N_{lo_output} is a cyclostationary process. This is not surprising as the mixer is an LPTV system. The $\sigma_{n\_lo\_output}{}^2$, the output variance, is varying at twice the input frequency at the RF port, ω_{rf}.

For this mixer we assume that (4.150), Chapter 4, is satisfied. We present a few results on the LO noise of this mixer. Let us denote the sample time as 0 and let us assume that V_{rf} is given by $V_{rf} = A_{rf} \cos(\omega_{rf}(t - t_0))$. Hence V_{rf0} is the input signal at the sampling instant and is given by $V_{rf0} = A_{rf} \cos(\omega_{rf}t_0)$. The DC term of $\sigma_{n\_lo\_output}{}^2$, $\sigma_{n\_lo\_output\_dc}{}^2$, is given as [2]

$$\sigma_{n\_lo\_output\_dc}^2 = \frac{A_{rf}^2}{4} \frac{\omega_{rf}^2 C_{load}}{k(V_G - V_t)^2} \sigma_{n\_lo\_input}^2 \tag{5.119}$$

Here V_G is the gate voltage.

The important quantity, however, is the sampled output noise's variance due to the LO noise. We denote this as $\sigma_{n\_lo\_output,s}{}^2$. The sampling occurs during the falling edge of V_{lo}. During the falling edge M_1's resistance drops, in proportion to the falling rate of the LO voltage. Now since there is noise on the LO port, the

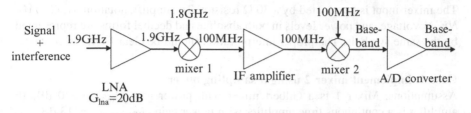

Fig. 5.20 Receiver front end architecture for Numerical example 5.2

voltage source is noisy. This noisy voltage source not only causes the value of R_{on} to decrease, but also fluctuates. The mixer output voltage tries to track the input voltage, but governed by the $R_{on}C_{load}$ time constant. Since R_{on} is now fluctuating, the output voltage, during tracking, fluctuates as well. At the sampling instant, this uncertainty is stored. This uncertainty, or noise, is the $\sigma_{n\_lo\_output,S}^2$ defined previously. Again let us denote the sample time as 0 and let us assume that V_{rf} is given by $V_{rf} = A_{rf} \cos(\omega_{rf} (t - t_0))$. $\sigma_{n\_lo\_output,S}^2$ is given as [2]:

$$\sigma_{n\_lo\_output,S}^2 = \left(A_{rf}^2 - v_{rf0}^2\right) \sqrt{\frac{k}{C_{load}}} \left(\frac{T_f}{V_G}\right)^{\frac{3}{2}} \omega_{rf}^2 \sigma_{n\_lo\_input}^2 0.443 \qquad (5.120)$$

where T_f is the fall time and V_G is the gate voltage. The preceding results for the sampling mixer [(5.119) and (5.120)] depend on the use of SDE. The derivations use a time domain (as opposed to a frequency domain) method. Due to the extensive background needed on SDE, detailed derivations of the preceding results lie outside the scope of this book (although it is interesting to note part of the idea overlap with that used to derive time-varying distortion using the time-varying Volterra series, discussed in sub-section 5.9.2.1). Interested readers are referred to the reference for the derivation [2].

Numerical example 5.2.
This numerical example applies formulas derived in Sections 5.8 through 5.11. In this example, we calculate G_c, IM_3, and NF of both a 100-MHz IF single-ended sampling mixer and a 1.9 GHz RF single-ended sampling mixer in a DECT application. The architecture under consideration is shown in Figure 5.20. This is a modified form of the heterodyne architecture discussed in Figure 2.13 of Chapter 2. The main modification consists of adding a second-stage mixer and an A/D converter. The filters and demodulator are not shown.

It is assumed that the input consists of an -83 dBm signal together with a -33 dBm interference.

The following parameters are assumed in the mixer calculation:

$C = 0.352$ pF, $W/L = 110um/0.8um$, $k' = 100$ uA/V^2, $V_G = 2.5$ V, $V_t = 1.05$ V, $T_f = 0.4$ ns.

The mixer input is terminated by a 50 Ω resistor. To simplify notation, G, G_c, IM_3, NF, N, voltage, and power levels in both absolute and decibel forms are represented by the same symbols. The proper interpretation should be clear from the context.

Case 1: implement mixer 2 using a sampling mixer
Assumptions: Mixer 1 is a Gilbert mixer with power gain $G_{mixer1} = 0$ dB; IF amplifier is a continuous time amplifier with power gain $G_{IF\_amplifier} = 13$ dB.

$$\omega_{if} = 2\pi \times 100 \times 10^6 \text{ rad/s}$$

Since mixer 2 takes an IF input and generates a baseband output, the input variables will have IF subscripts.

IM_3
From Figure 5.20. we can see that at the input to mixer 2 we have two frequency components. One is the signal and the other is the interference, which is the one of interest for IM_3 calculation. The power of the interference, denoted as $P_{interference\_if}$, is given by

$P_{interference\_if} =$ Input interference $+ G_{lna} + G_{mixer1} + G_{IF\_amplifier} = -33$ dBm $+ 20$ dB $+ 0$ dB $+ 13$ dB $= 0$ dBm

Hence $A_{interference\_if} = 0.316$ V
Following Table 5.1, we have

Continuous time distortion:

$$IM_3 = \frac{A^2_{int\ erference\_if}}{4} \frac{j\omega_{if}C}{k(V_G - V_t)^3}$$

$k = k' \times W/L = 100 \text{ } uA/V^2 \times (110um/0.8um) = 0.0138 A/V^2$, $V_{GS} - V_t = 1.45$ V
Substituting, $IM_3 = -77.64$ dB

Time-varying distortion:

$$IM_3 = \frac{3A_{int\ erference\ \_if^2}\omega_{if}^3}{8} \left(\frac{\pi}{4}\right)^{3/2} \left(\frac{C}{k}\right)^{1/2} \left(\frac{T_f}{V_G}\right)^{5/2} \cdot 0.0913$$

Substituting, $IM_3 = -120.3$ dB.

Sampling distortion:

$$IM_3 = \frac{1}{8} \left(\frac{A^2_{int\ erference\_if}\omega_{if}T_f}{2V_G}\right)^2$$

Substituting, $IM_3 = -90$ dB
Comparing, IM_3 is dominated by continuous time distortion and so $IM_3 = -77.64$ dB.

G_c and NF

We will calculate G_c, NF for the V_{lo} not switching case.

Applying (5.43) to the present case, ω_{rf} becomes ω_{if}. Hence we have

$$G_c = \left| \frac{\frac{1}{j\omega_{if}C}}{\left|\frac{1}{j\omega_{if}C}\right| + R_{on}} \right|$$

From (5.11) R_{on} is given by

$$R_{on} = \frac{1}{k(V_{GS} - V_t - V_{DS})} = \frac{1}{100uA/V^2 \frac{110u}{0.8u}(1.45V - 0V)} = 50.15\Omega$$

Notice that we have assumed V_{DS} to be practically 0 V since the sampling mixer is terminated by a capacitor. Substituting all the relevant parameters into the G_c formula, we have $G_c = 0$ dB.

NF is given from (5.34) as:

$$NF = \frac{N_{Device} + G \cdot N_{Source\_resis\tan ce}}{G \cdot N_{Source\_resis\tan ce}} = 1 + \frac{N_{device}}{G \cdot N_{source\_resis\tan ce}}$$

From (5.110), $N_{Device} = 4kTR_{on}$. $N_{source\_resistance} = 4kTR_s$. Since $G_c = 1$, $G = 1$. Substituting all these values into the NF formula, we have $NF = 1 + \frac{R_{on}}{R_s}$.

We are given that the input resistance is 50 Ω; hence $R_s = 50\ \Omega$. Substituting, we have NF $\cong 1 + 1 = 2$, or 3 dB.

Case 2: implementing mixer 1 using a sampling mixer

In the present case $\omega_{rf} = 2\pi \times 1.9 \times 10^9$ rad/s. As a side note, in this case the IF amplifier is a sampled data amplifier and mixer 2 is another sampling mixer.

IM$_3$

The power of the interference is

$$P_{interference\_rf} = \text{Input interference} + G_{lna} = -33 \text{ dBm} + 20 \text{ dB} = -13 \text{ dBm}$$

Following Table 5.1, we have

Continuous time distortion:

$$IM_3 = \frac{A^2_{interference\_rf}}{4} \frac{j\omega_{rf}C}{k(V_G - V_t)^3}$$

Substituting, IM$_3 = -78.06$ dB

Time-varying distortion:

$$IM_3 = \frac{3A^2_{\text{interference\_}rf}\omega^3_{rf}}{8}\left(\frac{\pi}{4}\right)^{3/2}\left(\frac{C}{k}\right)^{1/2}\left(\frac{T_f}{V_G}\right)^{5/2}\cdot 0.0913$$

Substituting, $IM_3 = -69.6$ dB.

Sampling distortion:

$$IM_3 = \frac{1}{8}\left(\frac{A^2_{\text{interference\_}rf}\omega_{rf}T_f}{2V_G}\right)^2$$

Substituting $IM_3 = -64.8$ dB.

At this frequency sampling distortion dominates. Accordingly, $IM_3 = -64.8$ dB. Notice that when we go from case 1 to case 2, $A_{\text{interference}}$ has gone down, but ω has gone up. Since time-varying and sampling distortion increase as ω^3 and ω^2, hence they finally become dominant.

G_c and NF

Again we calculate G_c, NF for the V_{lo} not switching case. We apply (5.43) to the present case. Hence we have

$$G_c = \left|\frac{\frac{1}{j\omega_{rf}C}}{\frac{1}{j\omega_{rf}C}+R_{on}}\right|$$

Substituting, we have $G_c = 0.82$ or -1.7 dB.

NF is given from (5.34) as

$$NF = \frac{N_{Device}+G\cdot N_{\text{Source\_resistance}}}{G\cdot N_{\text{Source\_resistance}}} = 1 + \frac{N_{Device}}{G\cdot N_{\text{source\_resistance}}}.$$

$G_c = 0.82$ means that $G = 0.67$. Substituting all the relevant values into this NF formula, we have $NF = 1 + \frac{R_{on}}{0.67R_s}$. With $R_s = 50\ \Omega$ and $R_{on} = 50.15\ \Omega$, we have NF $\cong 1 + 1.5 = 2.5$ or 4 dB.

5.12 Appendix: Comparisons of Sample and Hold and Sampling Mixer

There is an existing circuit called the sample and hold (SAH) that is similar to the sampling mixer. Although an SAH shares many of the properties of sampling mixer, it also differs in some important ones. We visit the difference and show that even though structurally the sampling mixer is similar to SAH and some design

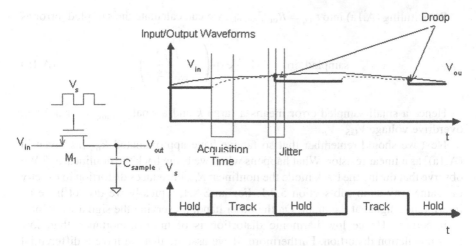

Fig. A.1 Single-ended sample and hold (SAH) and sampling operation

criteria are common, there are other design criteria that are markedly different. To simplify comparisons, all the design formulas are based on the assumption that V_{lo} (or sampling clock V_s for SAH) is not switching.

To start, let us show the operation of a single-ended SAH, as depicted in Figure A.1. In a SAH a sampling clock, V_s, with frequency f_s, is applied to the gate. Let us review some application for such an SAH, particularly for high-speed operation. In high-speed operation the input consists of a wide baseband signal (frequency f_{in}) rather than a narrow baseband signal superimposed on a high-frequency carrier, as in the sampling mixer case. To satisfy the Nyquist criteria, f_s is around 2 times the highest frequency of this wide baseband signal. Hence within one sampling clock period, $T_s (= 1/f_s)$, the V_{in} does not vary too much. This is shown in Figure A.1.

As with the sampling mixer, because of R_{on} of M_1 and C_{sample}(we are neglecting C_{db} of M$_1$), a SAH has a finite bandwidth and introduces tracking error during the track mode. Looking in the time domain, this finite bandwidth poses an RC time constant constraint on the acquisition time t_{acq}, defined as the time needed for the V_{out} to settle to V_{in} (the degree of settling is, of course, dictated by the required resolution). This is of primary importance in an SAH because here we are interested in the absolute accuracy of the sampled value. The requirement can be very high (on the order of 13 to 14 bits), which means that sampled error must be small. The $t_{acq} (= R_{on}C_{sample})$ has to be smaller than $T_s/2$. Usually T_s is rather small and hence t_{acq} must be made small, making t_{acq} a primary design constraint. Accordingly, to satisfy the sampled error requirement, both R_{on} and C_{sample} should be small.

For a MOS transistor biased in the triode region, R_{on} is given by (5.12), which is repeated here:

$$R_{on} = \frac{1}{k(V_{GS} - V_t)} = \frac{1}{\mu C_{OX}\left(\frac{W}{L}\right)(V_{GS} - V_t)} \tag{A.1a}$$

Substituting (A.1a) into $t_{acq} = R_{on}C_{sample}$, we can calculate the sampled_error as

$$\text{sampled\_error} = 1 - \exp\left(-\frac{T_s/2}{\frac{C_{sample}}{k(V_{GS}-V_t)}}\right) \tag{A.1b}$$

Hence a small sampled error means a large k and a small C_{sample} for a given overdrive voltage $V_{GS}-V_t$.

Next we should remember that so far we have approximated R_{on} [as given in (A.1a)] as a linear resistor. What happens when we bring back the nonlinearity? We observe that during the track mode the nonlinear R_{on} introduces distortion in exactly the same way as in sub-section 5.9.1. For an SAH, typically we do not have an interference signal at the input. On the other hand, preserving the signal waveform is important. Hence low harmonic distortion is of more importance than low intermodulation distortion. Furthermore, if we assume that we have a differential SAH, HD_2 is small. Hence the harmonic distortion of importance is HD_3. The HD_3 of an SAH is the same as the HD_3 of a sampling mixer derived with V_{lo} not switching. Hence it is given by (5.55). A_{rf} becomes A_{in}, C becomes C_{sample}, and ω_{rf} becomes ω_{in}. Equation (5.55), with these modifications, is now written as:

$$HD_3 = \frac{A_{in}^2}{4} \cdot \frac{j\omega_{in}C_{sample}}{k(V_G - V_t)^3} \tag{A.2}$$

We conclude that a small HD_3 means a large k and a small C_{sample} as well.

Finally, the sampled noise in an SAH, $\sigma_{n,s}^2$, is the same as the variance of the noise of a sampling mixer with V_{lo} not switching, σ_n^2. This is given in (5.113), where C_{load} becomes C_{sample} in the present case:

$$\sigma_{n,s}^2 = kT/C_{sample} \tag{A.3}$$

5.13 Observations

Comparing between SAHs and sampling mixers, we make the following observations:

a. Structurally they are the same, both consisting of a MOS transistor followed by a capacitor.
b. Signal conditions: There are differences and similarities. One major difference is that for an SAH (as shown in Figure A.1) we have a wide baseband signal (frequency f_{in}) sampled by a clock (frequency f_s), whereas for a sampling mixer (as shown in Figure 5.10), we have a narrow baseband signal (frequency

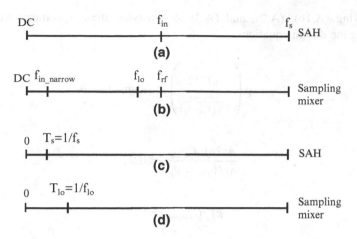

Fig. A.2   Relationship between f_{in}, f_s of a sample and hold and $f_{in\_narrow}, f_{lo}, f_{rf}$ of a sampling mixer

f_{in_narrow}) superimposed on a fast carrier (frequency f_{rf}), mixed down by an LO signal (frequency f_{lo}). What are the frequencies of interest? In SAH they are f_{in} and f_s. In a sampling mixer they are f_{rf} (not f_{in_narrow}) and f_{lo}. Now what are their relationships? Figure A.2 displays the relationship. For an SAH $f_{in} < f_s/2$ (to satisfy Nyquist criteria). This is shown in Figure A.2(a). In a sampling mixer $f_{rf} > f_{lo}$ (for downconversion mixing) but typically they are pretty close. This is shown in Figure A.2 (b). So far, we have talked about the frequency relationship within an SAH and a sampling mixer. How about the frequency relationship between the two? To make a fair comparison, we assume that the SAH is used for high-frequency application such that its f_{in} is close to f_{rf} of the sampling mixer. This is evident when comparing f_{in} in Figure A.2 (a) to f_{rf} in Figure A.2 (b). Figure A.2 (c) and (d) are just comparisons of T_s to T_{lo}.

c. Design requirements: There are differences and similarities. For an SAH the designer is primarily concerned about sampling error, whereas for a sampling mixer the designer is primarily concerned about distortion. They both have about the same requirement on noise.

5.13.1   Design Methodology

To save notations, sampled_error, HD_3, $\sigma_{n,s}^2$, G_c, IM_3, and σ_n^2 are also used to denote the required sampled_error, HD_3, $\sigma_{n,s}^2$, G_c, IM_3, σ_n^2.

We start our discussion on the design methodology by stating typical design specifications for both the SAH and the sampling mixer.

Typical design specifications (SAH): sampled_error, HD_3, $\sigma_{n,s}^2$

Now we want to develop relevant design equations for the SAH.

Applying (A.1b), (A.2), and (A.3) to satisfying these specifications means following the design equations:

$$1 - \exp\left(-\frac{T_s/2}{\frac{C_{sample}}{k(V_{gs}-V_t)}}\right) < sampled\_error \tag{A.4}$$

$$\frac{A^2 2\pi f_{in} C_{sample}}{4k(V_{GS}-V_t)^3} < HD_3 \tag{A.5}$$

$$kT/C_{sample} < \sigma_{n,s}^2 \tag{A.6}$$

Next let us turn to the sampling mixer.

Typical design specifications (sampling mixer): G_c, IM_3, $\sigma_{n,s}^2$

Again we want to develop relevant design equations for the sampling mixer. For simplification let us focus on the V_{lo} not switching case.

Applying (5.43), (5.57), and (5.113) to satisfying these specifications means following the design equations:

$$\left|\frac{\frac{1}{j\omega_{rf}C_{load}}}{\frac{1}{j\omega_{rf}C_{load}} + \frac{1}{k(V_{GS}-V_t)}}\right| > G_c \tag{A.7}$$

$$\frac{A_{interference}^2 \omega_{rf} C_{load}}{4k(V_{GS}-V_t)^3} < IM_3 \tag{A.8}$$

$$kT/C_{load} < \sigma_n^2 \tag{A.9}$$

Next, we want to look at the design parameters. We can identify the relevant design parameters for both circuits as: C_{sample}, C_{load}, k, $V_{GS} - V_t$. We then classify these design parameters according to whether they are common or distinct to designing the two circuits.

Design parameters that are designed in a common fashion to both circuits: C_{sample}, C_{load}, $V_{GS} - V_t$.

a. V_{dd} is the same for both SAH and sampling mixers. This means $V_{GS} - V_t$, fixed by V_{dd}, should be the same for both circuits.
b. Noise requirements are about the same for both SAH and sampling mixers. This means that C_{sample} and C_{load} are found by similar equations: (A.6), (A.9).

Design parameter that is designed using different approaches for two circuits: k We next focus on how this k is designed in an SAH and a sampling mixer.

For a SAH k is designed via the following steps:

Step 1: We start from (A.4). From application requirements, sampled error usually is very small. Again from application requirements, as shown in Figure A.2, (c), T_s is very small. This implies that it is highly difficult to find a k that satisfies (A.4).

Step 2: Let us go to (A.5). From application requirements, HD_3 is typically moderate in difficulty to satisfy. Again, from application requirements, as shown in Figure A.2 (a), f_{in} is high. This implies that it is moderately difficult to find a k that satisfies (A.5).

In conclusion, the strategy that we adopt is first to design k to satisfy (A.4), then (A.5) will typically be satisfied.

For a sampling mixer, k is designed via the following steps:

Step 1: We start from (A.7). From application requirements G_c is typically quite easy to satisfy. Again, from application requirements, as shown in Figure A.2(b), $\omega_{rf} = 2\pi f_{rf}$ is high. This implies that it is not that difficult to find a k that satisfies (A.7).

Step 2: Let us go to (A.8). From application requirements, IM_3 is usually moderate in difficulty to satisfy. Again, from application requirements, as shown in Figure A.2 (b), $\omega_{rf} = 2\pi f_{rf}$ is high. This implies that it is moderately difficult to find a k that satisfies (A.8).

In conclusion, the strategy that we adopt is first to design k to satisfy (A.8) and then (A.7) will typically be satisfied.

We summarize the key observation in our present design methodology. The key observation is that, as opposed to traditional sampled circuits such as an SAH, which focuses on t_{acq} in the design, for a sampling mixer, distortion requirement is more important in guiding the design process. This justifies why in this chapter we spend a substantial amount of time looking into this distortion mechanism, including the time- varying distortion and sampling distortion due to finite fall time.

5.14 Problems

5.1 In Numerical example 5.1 we have not checked if the model for high frequency is applicable. Used the same value as in the example to check if the assumptions (5.13) and (5.14) are valid for the high-frequency case.

5.2 Calculate the conversion gain in the general V_{lo} switching case of a switching mixer for DECT application using (5.29). Use parameters given in Numerical example 5.1. The W/L ratio is assumed to be the value derived in Numerical example 5.1, or 762 um/1 um. Assume that $C_{ox} = 2$ fF/um^2.

5.3 Compare the differences between the following switching mixers: mixer in Figure 5.1 and mixer in Figure 5.8. Draw the V_{if} waveform. (When you draw the V_{if} waveform, assume that the V_{rf} and V_{lo} waveforms are as given in Figure

5.10). Next repeat the comparisons between the switching mixer in Figure 5.8 and the sampling mixer in Figure 5.10. What should be the relative values of C_{ext} and C_{load}? Which of these two mixers has a poorer IM_3?

5.4 In Figure 5.13 derive the relationship between k_2 of switch M_2 and k_1 of switch M_1 to obtain equal second harmonic distortion at high frequency (assume that continuous time distortion dominates). In this case, does increasing the sampling mixer switch M_1's size always enhance the performance? At high frequency is the use of an opamp to provide virtual ground a good way of reducing the size of M_2?

5.5 The distortion formulas for a switching mixer (as shown in Figure 5.1 and Figure 5.8) or a sampling mixer (as shown in Figure 5.10), which include second-order effects such as threshold modulation, velocity saturation, and mobility degradation, have been derived at low frequency and are given by the following, where IM_3' is the modified IM_3:

a. Threshold Modulation

$$IM_3' = (1 + \theta_1)^2 \times IM_3$$

where $\theta_1 = \frac{\gamma}{2\sqrt{2\phi_f + V_s}}$
where γ = body effect parameter
$\phi_f = 2$ times the bulk potential
V_s = source voltage

b. Velocity Saturation

$IM_3' = \left(1 + \alpha L \frac{g_{ds}}{1 + 2g_{ds}R}\right) IM_3$ where $\alpha = \frac{2\mu_f}{\mu C_{ox}^2 W^2 R v_{sat}}$ and μ_f = a coefficient that helps model the absolute function of the velocity $|V|$ as a square function (i.e, $|V| \approx \mu_f V^2$), R = impedance of load (resistance if load is a resistor, impedance if load is a capacitor), $g_{ds} = 1/R_{on}$, and v_{sat} = saturation velocity.

c. Mobility Modulation from Normal Field
$IM_3' \approx [1 - \theta(V_{GS} - V_T)]IM_3$, where θ is the parameter that relates the effective mobility due to the normal field μ_{eff} to the normal mobility μ as follow: $\mu_{eff} = \frac{\mu}{1 + \theta(V_{GS} - V_T)}$.

Next assume the following parameters:

$V_{GS} - V_T = 1.45$ V
$C_{load} = 0.352\ pf\ \frac{W}{L} = \frac{110u}{0.8u}\ k' = 100\ \frac{uA}{V^2}$ or $k = k'\frac{W}{L} = 0.0137\ \frac{A}{V^2}$
$\gamma = 0.56\ \phi_f = 0.393\ V\ V_{SB} = 2.5$ V
$\mu_f = 0.5 C_{ox} = 2$ fF/um^2, $V_{sat} = 2 \times 10^5$ m/s
$\theta = 0.05$/V

The rest of the parameters are given in Numerical example 5.2.
Define the factors F_{vel}, $F_{BodyEff}$, F_{Normal}, as the ratios of IM_3' (due to second-order effects)/IM_3(no second-order effects). Calculate these factors.

5.6 Redo the IM_3 calculation part in the Numerical example 5.2, except this time include the second-order effects (i.e, the three short channel effects due to threshold modulation, velocity saturation, and mobility modulation as discussed in Problem 5.5 [and use the same formula]).

5.7 This problem recalculates NF of the sampling mixer discussed in Case 2, Numerical example 5.2 by using a more sophisticated noise model. First extrinsic noise from the LO port is included. Also, the intrinsic noise is to be calculated for the general V_{lo} switching case. To perform the calculations, we make the following assumptions. First, let us assume that all the parameters from the Numerical example 5.2 carry over. Second, we assume the following parameters for the sampling mixer: $V_{rf0} = 0$, $A_{lo} = 1$ V, $V_G - V_t = 1.45$ V, $V_G = 2.45$ V. Third, let us assume that the conversion gains, H_n, are the same for both zero and finite fall time. Furthermore as a simplification, we assume H_n, $n = 1$, 2... are all identical and are given by the following formula, which is modified from the G_c formula as given by (5.45). This modified formula incorporates the R_s and is given as

$$G_c(\omega_{out}, \omega_{lo}) = \left| G_{c0}(\omega_{out}) \left[\frac{j}{\pi} - \frac{\sin \frac{\pi}{2} \frac{\omega_{out}}{\omega_{lo}}}{\pi \frac{\omega_{out}}{\omega_{lo}}} \left(\exp^{-\frac{\pi \omega_{out}}{2 \omega_{lo}}} \right) \right] \right|$$

$$G_{c0}(\omega_{out}) = \frac{\frac{1}{\omega_{out} C_{load}}}{\frac{1}{\omega_{out} C_{load}} + R_s + R_{on}}$$

(a) Calculate NF of this 1.8 GHz sampling mixer including $\sigma_{n\_lo\_output,s}^2$ and using intrinsic noise calculated for the general V_{lo} switching case. Do the calculation for two cases. For Case 1, assumes $\sigma_{n\_lo\_input,s}^2$ in (5.120) comes from a 1kΩ resistor (you can imagine that the LO port is terminated with a 1 kΩ resistor). For Case 2, assume that $\sigma_{n\_lo\_input,s}^2$ comes from a 50 Ω resistor.

(b) Repeat (a) except this time include $\sigma_{n\_lo\_output\_dc}^2$. Again, use intrinsic noise calculated for the general V_{lo} switching case.

References

1. Cynthia D. Keys, "Low Distortion Mixer for RF Communication," Doctoral Thesis, University of California, Berkeley, 1994.
2. Wei Yu, S. Sen, and B. H. Leung, "Distortion Analysis of MOS Track & Hold Sampling Mixer using Time-Varying Volterra Series," IEEE Trans. on Circuits and Systems, II: Analog and Digital Signal Processing, Feb 1999, Vol. 46, p. 101–113.
3. Wei Yu and B. H. Leung, "Noise Analysis for Sampling Mixer using Stochastic Differential Equations," IEEE Trans. on Circuits and Systems, II: Analog and Digital Signal Processing, June 1999, Vol. 46 p. 699–704.
4. L. Breems, E. Zwan and J. Huijsing, "A 1.8 mW CMOS SD Modulator with Integrated Mixer for A/D Conversion of IF Signals," IEEE Journal of Solid State Circuits, April 2000, Vol. 35, p. 468–475.

5. http://www.cadence.com/datasheets/dat_pdf/pdistoapp.pdf, Affirma RF Simulator (SpectreRF) user guide.
6. Joel R. Phillips, "Analyzing Time-varying Noise Properties with SpectreRF", Affirma RF simulator (Spectre RF) user guide appendix I, Cadence Openbook IC product documentation, Cadence Design Systems, 1998.
7. S. Sen, B. Leung, "A 150MHz 13b 12.5mW IF Digitizer using Sampling Mixer" IEEE Custom Integrated Circuit Conference (CICC), May 1998, p.11.3.1–11.3.4.

Chapter 6
Analog-to-Digital Converters

6.1 Introduction

Traditionally in a receiver (as shown in Figure 2.1), upon mixing the input signal from RF to IF, subsequent demodulation can be performed in a couple of ways, depending on the kind of modulation used. In the case of DECT, since the input signal is phase modulated (GMSK for DE4CT), MSK demodulation for the digital encoded phase information should be performed. This can be done in a manner akin to demodulating a QPSK signal. There are three common ways to demodulate a QPSK signal: FM discriminator, IF detection, and baseband detection [9]. Each of these methods can be done entirely in the analog domain or by doing first an analog to digital (A/D) conversion at IF and then implementing these methods digitally using digital signal processing (DSP). In this sense the A/D converter becomes part of the demodulator. Using an A/D converter in a demodulator obviously is beneficial in terms of being able to integrate the post A/D conversion signal processing function on chip easily.

Instead of performing A/D conversion right before demodulation, we can perform this conversion earlier on in the receiver front end. Referring again to Figure 2.2, an A/D converter can be placed inside the front end and used to digitize the signal at the early stage of the front end. For example, if A/D conversion is performed right after the LNA, then BPF2, mixer, BPF3, and the IF amplifier can all be implemented digitally using DSP. In this sense the A/D converter becomes part of the front end and is responsible for processing (conditioning) the received signal/AWGN/interference before admitting it to the demodulator. The resulting front end architecture is shown in Figure P.2(a), Chapter 2. Using an A/D converter in the front end obviously has the same benefit as having the converter in a demodulator. In addition, since it allows the post A/D conversion signal processing function to be done digitally, it allows more flexibility in implementing the receiver front end.

B. Leung, *VLSI for Wireless Communication*, DOI 10.1007/978-1-4614-0986-1_6,
© Springer Science+Business Media, LLC 2011

In this chapter we first review the common methods of demodulation. We then discuss A/D converters most suitable to be used in the demodulator as well as in the front end. Two such A/D converters, the low-pass and bandpass sigma-delta modulators, are investigated in detail. In both cases, design procedures are developed and design examples for using them in a DECT receiver are presented.

6.2   Demodulators

In this section we review the three common methods of demodulations.

6.2.1   FM Discriminator (Incoherent)

Let us now study the first method, FM discriminator [9], which is illustrated in Figure 6.1.

This demodulation is done incoherently at passband. The FM discriminator has a phaseshift network that will introduce a delay of t_o to the carrier and a delay of t_1 to the phase. The input to this network contains a filtered (filtered by the BPF) and limited (by the limiter) version of the original QPSK modulated IF signal $S_{QPSK}(t)$. We can represent this limited and filtered version by an equivalent FM signal [9], $S_{FM}(t)$, where $S_{FM}(t) = \cos[\omega_c t + \phi(t)]$. Here ω_c is the carrier frequency (equals ω_{if}).

By definition, the phase of $S_{FM}(t)$, $\phi(t)$, is defined as $\phi(t) = 2\pi f_\Delta \int_0^t x_M(\lambda)d\lambda$, where $x_M(t)$ is the modulating signal, and f_Δ is the modulation index. Conversely we have $\dot\phi(t) = 2\pi f_\Delta x_M(t)$. Upon passing $S_{FM}(t)$ through the phaseshift network, the output will be $\cos[\omega_c t - \omega_c t_o + \phi(t-t_1)]$. Now if we design t_o such that $\omega_c t_o = 90^0$, then this output becomes $\sin[\omega_c t + \phi(t-t_1)]$. As shown in Figure 6.1, this is multiplied by the original $S_{FM}(t)$, which equals $\cos[\omega_c t + \phi(t)]$. Upon being filtered by the LPF, the output is $y_D(t) = \sin[\phi(t)-\phi(t-t_1)] \cong \phi(t)-\phi(t-t_1)$. The last approximation is valid if we assume t_1 is small enough that $|\phi(t) - \phi(t - t_1)| \ll \pi$. Under this assumption we also have $\dot\phi(t) \cong [\phi(t)-\phi(t-t_1)]/t_1$. Hence $[\phi(t)-\phi(t-t_1)] \cong \dot\phi(t) t_1 = 2\pi f_\Delta t_1 x_M(t)$. Therefore, $y_D(t) = 2\pi f_\Delta t_1 x_M(t)$ and we extract the original

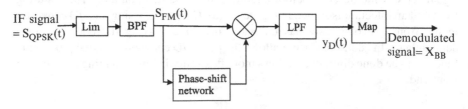

Fig. 6.1 FM discriminator, analog

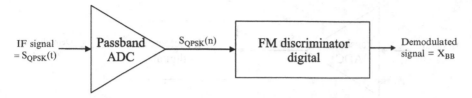

Fig. 6.2 FM discriminator, digital

FM modulating signal $x_M(t)$ and generate the demodulated FM signal $y_D(t)$ out of it. Now we can sample this demodulated FM signal and obtain $y_D(n)$, where n is the time index. This $y_D(n)$, which represents the phase shift to the proper binary bits, can be mapped to obtain the demodulated QPSK signal, $X_{BB}(n)$. Let us now represent the demodulated signal $X_{BB}(n)$ as $(x_{MI}(n)\ x_{MQ}(n))$ where $x_{MI}(n)\ x_{MQ}(n)$ are the sampled I, Q branches of the message $x_M(t)$. Hence we can map $y_D(n)$ to $(x_{MI}(n)\ x_{MQ}(n))$. For QPSK $(x_{MI}(n)\ x_{MQ}(n)) = (1\ 1)$ if $y_D(n) = \pi/4$, $(x_{MI}(n)\ x_{MQ}(n)) = (0\ 1)$ if $y_D(n) = 3\pi/4$, $(x_{MI}(n)\ x_{MQ}(n)) = (0\ 0)$ if $y_D(n) = -3\pi/4$, $(x_{MI}(n)\ x_{MQ}(n)) = (1\ 0)$ if $y_D(n) = -\pi/4$.

This FM discriminator is usually implemented in the analog domain. As an example, the IF receiver LMX2240, made by National Semiconductor, is one of the commercial realizations using this approach. This chip consists of a hard limiter at the input with an input impedance of 150 Ω and an -80 dB sensitivity for the IF signal. A Gilbert quad mixer (similar to the one covered in Chapter 4) is used as the multiplier in Figure 6.1. To get the phase-shifted signal an external tank circuit is required whose bandwidth is approximately 1% of the IF frequency and a steep phase response is required. As a result, this approach suffers from (a) the need to have external capacitors and inductors for its tank circuit, and (b) the need to have a narrowband analog filter. These analog components are expensive and are not easily integrable on a chip.

To circumvent these problems, the approach shown in Figure 6.2 can be used. In Figure 6.2 the IF signal $S_{QPSK}(t)$ is first digitized by performing an A/D conversion at the passband using a passband analog to digital converter (ADC). This generates a digital version, $S_{QPSK}(n)$. A digital version of the FM discriminator described in Figure 6.1 is then used to convert this $S_{QPSK}(n)$ to X_{BB}. This approach pushes more of the processing into the digital domain and is desirable.

6.2.2 IF Detection (Coherent)

Demodulation can also be done coherently. If performed at IF it is normally done using a matched filter [9]. As before, the demodulation can be done in both the analog and digital domains. The desirable approach, the digital-based method, consists of converting $S_{QPSK}(t)$ to $S_{QPSK}(n)$ using a passband ADC, as shown in Figure 6.3. The multiplication, low pass filter function needed in the IF detection will then be done digitally to generate the X_{BB}.

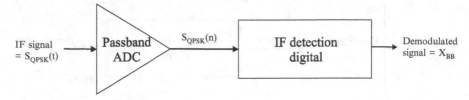

Fig. 6.3 IF detection, digital

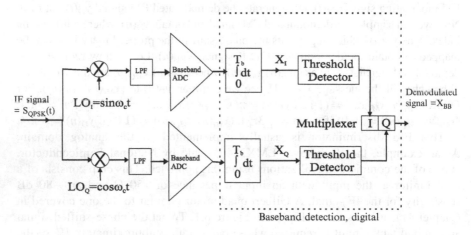

Fig. 6.4 Baseband detection, digital

6.2.3   Baseband Detection (Coherent)

If coherent detection is performed at baseband, it is typically done using a matched filter [9]. As with coherent detection at IF it can be done in both the analog and digital domains, where again the digital approach is preferable. This digital approach is shown in Figure 6.4, where the *I, Q* branches are depicted. Notice the similarity of Figure 6.4 to Figure 1.7, except that the ADCs are shown here. Referring to Figure 6.4, the incoming IF signal at $\omega_c = \omega_{if}$ is mixed by the LO signal at ω_{lo}. This ω_{lo} frequency is picked to be at the same frequency as ω_c. After analog mixing, a low-pass filter is used to eliminate the frequency component at $\omega_{if} + \omega_{lo}$, leaving behind only the baseband signal. Notice that this mixing function also performs the multiplying function necessary for detection. A baseband A/D converter is then used to convert the analog signal to its digital form, and the baseband digital signal processor performs the rest of baseband detection in the digital domain, at baseband. After the detection, equalization/decoding can also be done. Equalization has already been discussed in chapter 1. This combats ISI resulting from fading. Decoding can be further used to reduce noise.

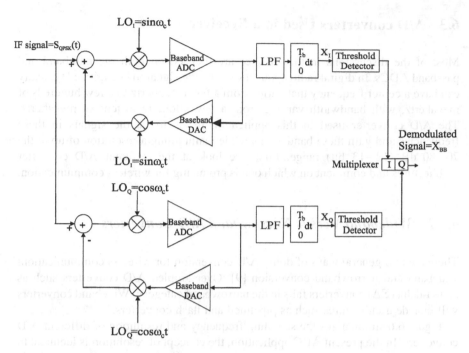

Fig. 6.5 Baseband detection, hybrid

Getting back to A/D converter, comparing the approaches in Figure 6.2 through Figure 6.4, the major difference is the requirement on these ADCs. The requirement is very high in Figure 6.2 and 6.3 since the A/D converter has to digitize directly at IF. This makes the design of the A/D converter a big challenge, especially at high IF. Therefore, usually Figure 6.4 is reserved for high IF (up to 400 MHz) and Figures 6.2 and Figure 6.3 for low IF architectures (10 MHz). This is particularly true if we observe that the performance level (especially distortion) of analog mixers at 100 MHz can be substantially better than A/D converters at the same frequency range. Recently an improvement on the baseband detection, digital approach, denoted as the hybrid approach [7], was proposed and is shown in Figure 6.5. Comparing Figures 6.4 and 6.5, everything is the same except that the mixer, baseband ADC is replaced by a feedback structure. This feedback structure does A/D conversion as usual, but then takes its output, performs D/A conversion, and subtracts the IF signal from this analog signal. In spite of its seemingly complicated structure in practical implementation a lot of simplifications can be performed, resulting in a markedly simplified final form. The important point is that because the components, most notably the analog mixers, are embedded in a feedback loop, all nonidealities are suppressed by the feedback action and thus the approach offers a significant advantage. Because of the scope of this chapter, this architecture is not covered. Interested readers are referred to the reference [7].

6.3   A/D converters Used in a Receiver

Most of the discussions in this chapter are focused on both the baseband and
passband ADCs. In digital radio applications [1], the signal to be digitized typically
can have a center frequency that varies from a few megahertz to a few hundreds of
megahertz, with bandwidth varying from a few kilohertz to tens of megahertz.
The A/D converter used in this application needs to handle signals in these
frequencies and with these bandwidths while maintaining a resolution of up to the
70−80 dB (12−13 bit) range. First, we look at the different A/D converter
architectures and comment on which one is promising for wireless communication.

6.3.1   Wideband Versus Narrowband A/D Converters

There are two general ways of doing A/D conversion for wireless communication:
wideband and narrowband conversion [9]. Oversampled A/D converters such as
sigma-delta ($\Sigma\Delta$) converters falls in the narrowband category. Wideband converters
will include architectures such as pipelined and flash converters.

Figure 6.6 summarizes the sampling frequency and resolution of different A/D
converters. In the present ADC application, the concept of resolution is identical to
dynamic range. Hence an ADC with a 80 dB resolution sweeps through a dynamic
range of 80 dB while maintaining an acceptable output SNR. Resolution can also be
specified in number of bits. In addition the concept of sampling frequency (f_s) needs
to be clarified. Sampling frequency normally is related to the bandwidth of the input
signal an ADC can digitize. Typically the sampling frequency is equal to two times
this bandwidth, as governed by the Nyquist sampling theorem. However
in oversampled ADC that performs narrowband A/D conversion, because of
oversampling, the real sampling frequency is much higher than two times the
bandwidth. To clarify, in Figure 6.6 the sampling frequency quoted for
the oversampled ADC is actually just two times the input signal bandwidth and is
different from its real sampling frequency. For the rest of the ADCs, the sampling
frequency quoted in Figure 6.6 is the real sampling frequency.

In wireless communication, there is also a stringent power requirement for
A/D converters. Figure 6.7 compares power consumption in various wideband
A/D converters. In Figure 6.6, we can see that flash A/D converters have the highest
sampling rate (input signal bandwidth) but lowest resolution. The oversampled
A/D converters have the highest resolution but the lowest input signal bandwidth.
Between them there are pipelined A/D converters, which offer a trade-off between
bandwidth and resolution. As shown in Figure 6.7, the flash A/D converters have
the highest power consumption. Because there are very few oversampled
and pipelined A/D converters operating at an 8-bit resolution, they are not presented
in this figure. In general, oversampled A/D converters have the same order of power
consumption as the algorithmic ones and pipelined A/D converters have power
consumption similar to subranging A/D converters.

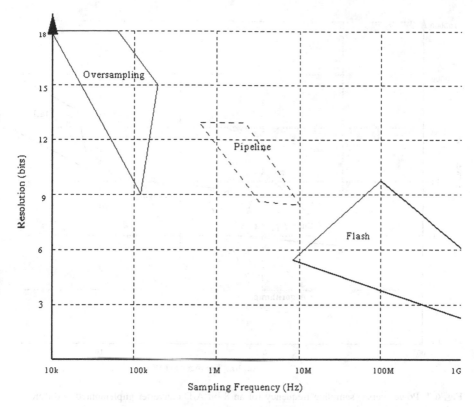

Fig. 6.6 Resolution versus sampling frequency for A/D converters implemented using different architectures

Typically, for wide signal bandwidth applications (such as CDMA, or wireless video), Nyquist rate ADCs are used. One such ADC using a pipelined architecture has been developed [8]. On the other hand, for narrow signal bandwidth applications (such as AMPS, GSM) oversampled ADCs are preferable. As mentioned previously, since power is one of the most important issues in wireless application, architectures that have high power dissipation are avoided. With this in mind we make the following observations: for wide signal bandwidth applications, the pipelined converter is appealing because its power dissipation is the lowest among high-speed Nyquist rate A/D converters. For narrow signal bandwidth applications, oversampled A/D converters (of which the sigma-delta modulator is an example) are the prime choice. In particular, low- pass and bandpass sigma-delta modulators are usually adopted in high and low IF A/D conversions, respectively. In this chapter, in order to narrow the scope we focus on narrowband applications and hence we concentrate only on sigma-delta modulators. In the following sections, the important design parameters for each architecture are explained and a design procedure is presented to help guide readers in obtaining the required design parameters. This is followed by a detail design example.

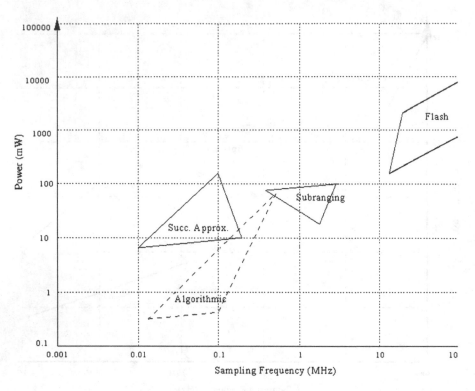

Fig. 6.7 Power versus sampling frequency for an 8-bit A/D converter implemented in different architectures

6.3.2 *Narrowband A/D Converters: General Description*

We can understand the principle of sigma-delta modulators by starting off with a flash A/D converter. A flash ADC generates quantization noise in the process of performing an analog to digital conversion. A low-pass (bandpass) sigma-delta modulator uses a low-pass (bandpass) filter (hitherto called the loop filter) in front of such a flash type ADC (hitherto called an internal quantizer) to process the quantization noise generated by this internal quantizer. This is achieved by feeding back the internal quantizer output to the input of the loop filter. By enclosing this filter/quantizer combination inside a feedback loop, we shape the quantization noise in a frequency selective way, resulting in the noise being suppressed at a frequency band of our choice (DC for low-pass and IF for bandpass $\Sigma\Delta$ modulators). For example, the lowpass $\Sigma\Delta$ modulator can be used to implement the baseband ADC in Figure 6.4 and in Figure P.2(a), Chapter 2. Similarly, the bandpass $\Sigma\Delta$ modulator can be used to implement the passband ADC in Figure 6.2 and Figure 6.3. Meanwhile, since the feedback loop broadbands the frequency response of the loop filter, the input signal is not subjected to any frequency shaping and passes unattenuated to the output.

All sigma-delta modulators (low-pass and bandpass) can be broadly categorized according to the number of bits in the internal quantizer and the order of the loop filter, which dictates the order of the modulator. In addition each of these individual modulators can be cascaded and form a multistage modulator. We review some of the basics of sigma-delta modulators, starting from a single-stage first-order modulator. Subsequently, we go through the second-order and then higher-order modulators and their variations. We focus on architectures that are of most relevance to wireless communication.

6.4   Low-Pass Sigma-Delta Modulators

An example of using a low pass sigma-delta modulator can be found in Figure 6.4 and in Figure P.2(a) Chapter 2. Here, because the baseband ADC operates on a baseband signal, a low-pass sigma-delta modulator is used. In the following, we discuss the various low-pass sigma-delta modulators, starting with the simplest one: a first-order modulator.

6.4.1   First Order Modulator

A first-order low pass (LP) sigma-delta modulator is shown in Figure 6.8. This is a baseband ADC, which takes the continuous time analog signal $x(t)$ and generates a discrete time digital signal $y(n)$, where n is the time index. This modulator implements the loop filter using an integrator with gain k and an internal quantizer using a 1-bit flash ADC. It takes the continuous time input signal $x(t)$, samples it at a sample frequency f_s, generates a discrete time (but continuous amplitude) signal $x(n)$. The sampling function is implemented by the sample and hold (SAH) described in the appendix of Chapter 5. This $x(n)$ is next discretized in amplitude by the 1-bit ADC to generate the digital code $y(n)$. This $y(n)$ consists of a bunch of 1s and 0s since the ADC is only a 1-bit ADC. The stream of 1s and 0s is a pulse density representation of the input signal $x(n)$.

We will now go through an example of how this modulator works. First we assume, for illustration purposes, that integrator has a gain $k = 1$. We also assume that $x(t)$ has a constant DC value of 0.5 [which means $x(n) = 0.5$ for all n] and that initially $w(0) = 0$. Finally the threshold of the 1-bit ADC is set such that $y(n) = 0$ for $w(n) < 0.5$ and $y(n) = 1$ for $w(n) \geq 0.5$. Since $w(0) = 0$, then $y(0) = 0$. When the time index n goes from 0 to 1, we feed the value $y(0)$ back and update $u(1)$ as follows: $u(1) = x(1) - y(0) = 0.5$. Hence $w(1) = w(0) + k \times u(1) = w(0) + u(1) = 0 + 0.5 = 0.5$. Since $w(1) = 0.5$, it exceeds the threshold and $y(1)$ becomes 1. Next n goes from $n = 1$ to $n = 2$. Then $u(2) = -0.5$, $w(2) = 0$, and $y(2) = 0$. In summary, y (n) goes from 0 to 1 and back to 0 again. If we continue on with this iteration we will find that $y(n) = 1, 0, 1, 0$, etc, and so its average value is exactly 0.5 or the same as the input value of $x(t)$. Therefore, an A/D conversion is achieved, though in an average sense.

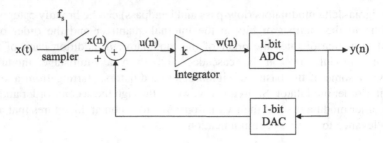

Fig. 6.8 First order modulator

From a frequency domain viewpoint, since the integrator's frequency response has infinite gain at DC, the loop gain is infinite at DC and therefore the DC component or the average of the output from the feedback 1-bit DAC will be identical to the DC component of the input signal $x(n)$. Reverting to the time domain viewpoint, this means that even though the quantization error, $e(n) = y(n) - w(n)$, at every sample is large because of the use of a 1-bit quantizer, the average of the quantized signal, and therefore the modulator output $y(n)$, tracks the signal $x(n)$. This average is computed by a digital decimation filter that is not shown here.

In general, the quantization error decreases (or the resolution of the modulator increases) when more samples are included in the averaging process or as the oversampled ratio (OSR), defined as the ratio of sampling frequency f_s to two times the signal bandwidth f_{bw}, increases. Consequently, the resolution of the modulator is a function of the OSR. The principle of operation of sigma-delta modulators relies on this fundamental trade-off between resolution and time. Since in wireless communication we typically have a narrowband signal centered at a high-frequency carrier (e.g, f_{bw} of AMPS is around 20 kHz with f_c centred at around 900 MHz), sigma-delta modulation is a natural choice for digitizing these narrowband signals.

On closer examination of the quantization process in a first-order sigma-delta modulator, we can see that when compared with Nyquist rate A/D converters, the quantization error is a differential error. In other words, the modulator tries to cancel the error by subtracting the quantization error from two adjacent samples. This principle of reducing errors by exploiting the statistics between samples can be extended to higher- order modulators, where more of past error samples are involved in the cancellation process to reduce the overall error. Viewed from the frequency domain, this difference operation acts to attenuate the quantization noise at low frequencies, thus shaping the noise.

Next let us develop a model of this first-order modulator and calculate the quantization noise. Quantization noise in this modulator depends on the input signal. For a busy signal the quantization noise is like white noise, whereas with a DC input the quantization noise is colored. To calculate the effective resolution of the $\Sigma\Delta$ modulator, it is assumed that the input signal is sufficiently busy that the quantization error of the 1-bit ADC in Figure 6.8 behaves like white noise that is uncorrelated with the input signal. Hence the 1-bit ADC in Figure 6.8 can

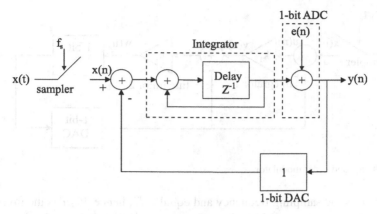

Fig. 6.9 Model of a first order modulator

be modeled as a gain block with a gain of 1 together with an additive white noise source, as shown in Figure 6.9. The 1-bit DAC is then modeled by a gain block with a gain of 1. Finally, the integrator is shown explicitly using its discrete time representation, again assuming that the integrator gain is 1. The total modulator quantization noise $e_T(n)$ [defined as $y(n) - x(n-1)$] can then be expressed as a function of the 1-bit ADC quantization noise $e(n)$ as

$$e_T(n) = e(n) - e(n-1) \qquad (6.1)$$

We next take the z-transform of (6.1) and substitute $z = e^{j\omega T}$. Here ω is the frequency of $x(t)$ in rad/s; T is period of the sampling clock and hence equals $1/f_s$. The spectral density $N(f)$ of this $e_T(n)$ can now be expressed in terms of the spectral density $E(f)$ of $e(n)$ as:

$$N(f) = E(f)|1 - \exp(-j\omega T)| = 2e_{rms}\sqrt{2T}\sin\left(\frac{\omega T}{2}\right) \qquad (6.2)$$

Here E(f), being the spectral density of $e(n)$ should be flat with an rms value equal to e_{rms}.

From (6.2) it is seen that feedback around the quantizer reduces the noise at low frequencies but increases it at high frequencies. The total noise power in the signal band, defined as $n_{bw}{}^2$, is

$$n_{bw}^2 = \int_0^{f_{bw}} |N(f)|^2 df \approx (e_{rms}^2)\frac{\pi^2}{3}(2f_{bw}T)^3 \quad f_s \gg f_{bw} \qquad (6.3)$$

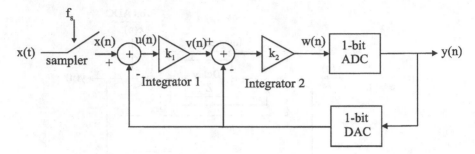

Fig. 6.10 Second order modulator

Since f_s is the sampling frequency and equals $1/T$, hence $2f_{bw}T$ is the inverse of OSR. Substituting this in (6.3) we have the rms value of n_{bw}^2, denoted as n_{rms}, given approximately by $e_{rms}\frac{\pi}{\sqrt{3}}(OSR)^{-3/2}$. Each doubling of the OSR of this circuit reduces n_{rms} by 9 dB and provides 1.5 bits of extra resolution. The improvement in the resolution requires that the modulated signal be decimated to the Nyquist rate with a sharply selective digital decimation filter, as mentioned previously. Otherwise, the high-frequency components of the noise will spoil the resolution when it is sampled at the Nyquist rate.

6.4.2 High Order Modulators

The procedure for increasing the resolution with feedback can be reiterated by replacing the 1-bit ADC and DAC in a first order modulator, as shown in Figure 6.8, by an identical first order modulator. The resulting circuit is shown in Figure 6.10. Notice that we have two integrators with two separate integrator gains: k_1 and k_2. The gains k_1, k_2 are used to optimize the output signal range of the integrators. Modulators with a second-order transfer function involve the cancellation of the two past samples and thus exhibit stronger attenuation at low frequencies. The $N(f)$ of a first- and second-order modulator are compared in Figure 6.12 [13].

Second order modulators are shown to be conditionally stable. The stability depends on the total delay in feedback loop, signal amplitude, and coefficients k_1, k_2. To avoid saturation, the maximum signal levels at the amplifier outputs should be adjusted by scaling.

The preceding analysis can be formalized to yield quantitative results for the resolution of second-order sigma-delta modulators, provided that the spectral distribution of the quantization error $e(n)$ can be assumed to be uncorrelated. The modulator can be regarded as a linear system for which the spectral density of the noise can be calculated [13]. Let us now calculate the $N(f)$ and the corresponding noise transfer function (NTF) of this second order modulator. Here NTF is defined as $\frac{N(z)}{X(z)}$, where $N(z)$ is the z-transform of the modulator noise $e_T(n)$ and $X(z)$ is the z-transform of the input $x(n)$. To simplify discussions let us assume

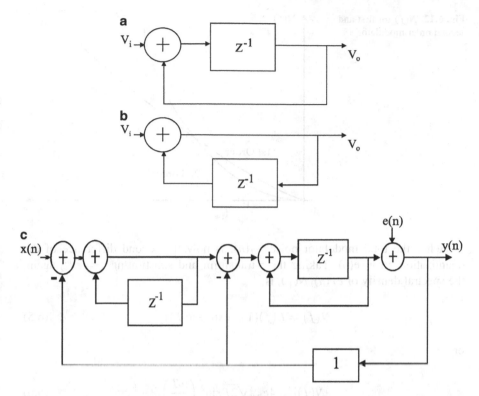

Fig. 6.11 (**a**) Integrator with forward delay. (**b**) Integrator with feedback delay. (**c**) One type of implementation of a second-order modulator

that $k_1 = k_2 = 1$. First let us go into a bit of digression on the implementations of these two integrators. It turns out that there are two types of implementation.

A block diagram representation of the first type of integrator with a gain of 1 is shown in Figure 6.11(a). Notice that this integrator has a delay element z^{-1} in the forward path. This is also the integrator that we use in Figure 6.9. Now let us turn our attention to Figure 6.11(b), where the delay has been moved to the feedback path. If we work out the equation, Figure 6.11(b) also describes an integrator.

Henceforth we label the integrator in Figure 6.11(a) type 1 and that in Figure 6.11(b) type 2. They have different transfer functions. The type 1 integrator has a transfer function given by $z^{-1}/(1 - z^{-1})$ and type 2 integrator's transfer function is $1/(1 - z^{-1})$. As it turns out, when we calculate $N(f)$ and the corresponding NTF for a second-order modulator, we can simplify the mathematics quite a bit if integrators 1 and 2 in Figure 6.10 are implemented using type 2 and type 1 integrators, respectively. The resulting block diagram is shown in Figure 6.11(c), where the output of the modulator can be expressed as

$$y(n) = x(n - 1) + (e(n) - 2e(n - 1) + e(n - 2)) \tag{6.4}$$

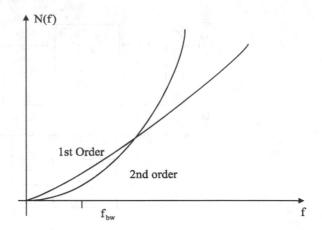

Fig. 6.12 $N(f)$ for first and second order modulator

Hence the total modulator noise, $e_T(n)$, is now the second difference of the quantization error, $e(n)$. Taking the z-transform and substituting $z = e^{j\omega T}$ again, the spectral density of $e_T(n)$, $N(f)$, is

$$N(f) = E(f)(1 - \exp(-j\omega T))^2 \tag{6.5}$$

or

$$|N(f)| = 4e_{rms}\sqrt{2T}\sin^2\left(\frac{\omega T}{2}\right) \tag{6.6}$$

and the rms noise in the signal band is given by

$$n_{rms} \approx e_{rms}\frac{\pi^2}{\sqrt{5}}(2f_oT)^{5/2} = e_{rms}\frac{\pi^2}{\sqrt{5}}OSR^{-5/2} \quad f_s \gg f_{bw} \tag{6.7}$$

This noise falls by 15 dB for every doubling of the sampling frequency, providing extra bits of resolution [13]. Figure 6.12 shows the spectral density of the quantization noise for the first- and second-order modulators. As shown, the quantization noise of the second- order modulator is smaller inside f_{bw}.

We mention in subsection 6.3.1 a general definition of dynamic range (or resolution). Let us give a more rigorous definition here. The dynamic range (DR) (or resolution) of an ADC is defined as the ratio of the input signal power for a full-scale sinusoidal input to the input signal power when the corresponding SNR is one (0 dB). Assuming that the SNR versus input amplitude curve is linear with a slope of 1, then DR also equals approximately SNR_{max}, which equals the ratio of the rms value of the largest sine wave to the rms value of the total modulator noise. Assuming that the 1-bit ADC has a step size of Δ, the full-scale sine wave that can be applied before overloading occurs has a peak value of $\Delta/2$ and its rms value is

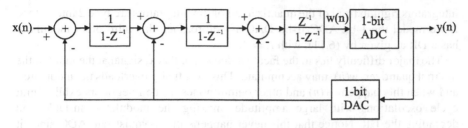

Fig. 6.13 Third-order modulator

$\Delta/(2\sqrt{2})$. Furthermore, for a quantizer of step size Δ, $e_{rms} = \frac{\Delta}{\sqrt{12}}$. Substituting this and n_{rms} from (6.7) into the DR definiton, we have

$$DR = SNR_{max} = \frac{\frac{\Delta}{2\sqrt{2}}}{n_{rms}} = \frac{\Delta}{2\sqrt{2}} \frac{\sqrt{5}}{e_{rms}\pi^2} OSR^{5/2} = \frac{\sqrt{7.5}}{\pi^2}(OSR)^{5/2} \qquad (6.8)$$

Higher-order modulators, realized by adding more feedback loop to the circuit [13], exist. In general, when a modulator has L loops and is not overloaded, it can be shown that the spectral density of the modulator noise is

$$|N(f)| = e_{rms}\sqrt{2T}\left[2\sin\left(\frac{\omega T}{2}\right)\right]^L \qquad (6.9)$$

For oversampled ratios greater than 2, the rms noise in the signal band is given approximately by

$$n_{rms} = e_{rms}\frac{\pi^L}{\sqrt{2L+1}}(2f_{bw}T)^{L+\frac{1}{2}} = e_{rms}\frac{\pi^L}{\sqrt{2L+1}}(OSR)^{-(L+\frac{1}{2})} \qquad (6.10)$$

This noise falls by $3 \times (2L+1)$ dB for every doubling of the sampling frequency, providing $(L + \frac{1}{2})$ extra bits of resolution. The extra dynamic range extended to the 1-bit ADC can now be written as [10]

$$DR = SNR_{max} = \frac{\frac{\Delta}{2\sqrt{2}}}{n_{rms}} = \frac{\Delta}{2\sqrt{2}}\frac{1}{e_{rms}\frac{\pi^L}{\sqrt{2L+1}}(OSR)^{-(L+\frac{1}{2})}}$$

$$= \sqrt{\frac{3(2L+1)}{2}}\frac{1}{\pi^L}(OSR)^{(L+\frac{1}{2})} \qquad (6.11)$$

For wireless communication, where we want to minimize power, reducing the oversampling frequency is desirable, as this would reduce the clock frequency and hence power. However, achieving this through the use of higher order modulators encounter some difficulties, particularly as we move beyond two integrators.

First let us look at one such implementation of a third-order modulator, as shown in Figure 6.13. Here again, the integrators all have gains of 1. Also, except for the integrators right in front of the quantizer, all of the integrators are type 2 integrators. If we model the 1-bit ADC, 1-bit DAC as before, then this third-order modulator has a DR as given by (6.11), with $L = 3$.

The major difficulty lies in the fact that due to feedback, signal at the input of the internal quantizer, $w(n)$, may accumulate. This eventually overloads the modulator, and when this happens $w(n)$ and other output nodes of the integrators exhibit limit cycle oscillations with large amplitude, making the modulator unstable and degrading the DR. Notice that this never happens in a Nyquist rate ADC since it does not have feedback.

To help stabilize these circuits, integrators outputs need to be clipped, resulting in DR performance much worse than predicted by (6.11). Better performance is obtained by redesigning the filter used in the feedback loop. An example is shown in the appendix, where this high-order sigma-delta modulator has been redesigned to achieve very low voltage (hence low power) operation and proves to be extremely useful as a baseband ADC.

Other stable high order modulations include

- Interpolative architecture [2]
- MASH architecture [3]

Now that we have discussed the basic theory of sigma-delta modulators, we present the implementation issues.

6.5 Implementation of Low-Pass Sigma-Delta Modulators

In our discussions of modulators, the subblocks (integrators, 1-bit ADC) are assumed to be ideal. Real life implementation of modulators falls into two broad categories, depending on how the loop filter is implemented: continuous time based and switched capacitor based. To remain focused, we assume that the sigma-delta modulators are implemented using a switch-capacitor approach. The design of switch-capacitor-based circuits can be quite involved, and complete books have been dedicated to them. For example, [11] is a good reference. Here we just review the basic switch-capacitor integrators in enough depth that we can carry out the design of sigma-delta modulators.

6.5.1 Review of Switch-Capacitor-Based Integrators

An example of a switch-capacitor-based integrator is shown in Figure 6.14 and consists of an op-amp, two capacitors, and switches. The sampling clock f_s consists of two non-oversampling phases, ϕ_1 and ϕ_2. Corresponding to these two phases, the switches configure the integrator in two configurations. In phase ϕ_1, the integrator is configured as shown. Hence input voltage V_i is sampled on capacitor C_1. In phase

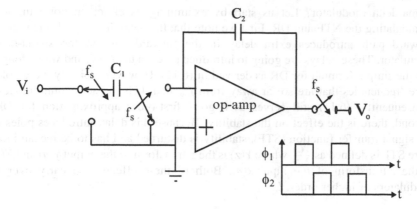

Fig. 6.14 Single-ended switched capacitor integrator

ϕ_2, the left-hand side of C_1 is switched to ground, and the right-hand side is switched to the summing node (negative terminal of the op-amp). Also, the output of the op-amp is connected to V_o. Now since this summing node is a virtual ground, the voltage across C_1 must be zero, which means the charge must be zero. From charge conservation, the charge deposited on C_1 during ϕ_1 has to go somewhere. The only place it can go to is C_2. Hence the additional charge deposited on C_2 is $V_i \times C_1$. Therefore the change in voltage across capacitor C_2, denoted as ΔV_2, is given by $(C_1/C_2) \times V_i$. Therefore, V_2 (at the end of the clock cycle) $= V_2$ (at the beginning of the clock cycle) $+ (C_1/C_2) \times V_i$. But V_2 is just the same as the output voltage of the integrator V_o and accordingly V_o (at the end of the clock cycle) $= V_o$ (at the beginning of the clock cycle) $+ (C_1/C_2) \times V_i$. This is the formula that describes an integrator with gain equal to C_1/C_2 and so the switch capacitor circuit in Figure 6.14 does realize the integrating function. As a matter of fact the circuit in Figure 6.14 realizes a type 1 integrator, as described in Figure 6.11(a), although here the gain is C_1/C_2. From a switch-capacitor implementation point of view, a type 1 integrator is more desirable. Hence, from now on we assume that all integrators are implemented as type 1, with the proper gain, of course.

6.5.2 Type 1 Switched-Capacitor-Cased Integrator

Using only a type 1 integrator has a subtle but important effect on the NTF and stability for modulators of order 2 or above. As a reminder, let us remember that for second, third-, and higher-order modulators, so far we have derived the NTF and hence DR formula using a architecture that uses type 1 integrators in only the innermost loop and type 2 integrators in the outer loops. For example, (6.4) through (6.11) have been derived with this in mind.

Since in subsection 6.5.1 we state that type 1 integrators should be used exclusively, we want to ask the following: what impact does this have on the sigma-delta modulator? Let us start by examining its effect on noise through recalculating the NTF and DR. First we note that the delay in a type 1 integrator's forward path introduces extra delay in the forward path of the sigma-delta modulator. These delays are going to introduce poles in the NTF, and we no longer has the simple formula for DR as derived in (6.11). However typically these poles have frequencies that are so far away from f_{bw} that their impact is quite modest. Consequently, (6.11) still serves as a good first-order approximation for DR. Second, there is the effect on the stability. Because this delay introduces poles in the signal transfer function (STF), stability is disturbed and has to be reexamined. Here STF is defined as $\frac{Y(z)}{X(z)}$ where $Y(z)$ is the z-transform of the output $y(n)$ and $X(z)$ is the z-transform of the input $x(n)$. Both of these effects also carry over to modulators of higher order, L.

6.5.3   Non-Ideal Integrator

Up to now we have assumed that the op-amp and switches in Figure 6.14 are ideal. For the op-amp this means it has infinite gain, speed, and zero noise and for the switch this means it has zero resistance, noise, and infinite speed. How true are these assumptions? A simple op-amp consists of a differential pair with some resistive load, and a simple switch consists of a MOS transistor. Both are not ideal due to the fact that the operational amplifier has finite DC gain, and bandwidth (and hence nonzero settling time) and the switch has similar nonidealities. In the following subsection, the impact of these nonidealities on the integrator discussed. For more complete treatment, see [11].

6.5.3.1   Finite Op-Amp Gain

Due to finite op-amp gain, the integrator's frequency response exhibits a finite DC value because the pole is no longer at DC. This movement of an integrator's pole results in the movement of the zeroes of the NTF so that the $N(f)$ changes from that as shown in Figure 6.12 to that as shown in Figure 6.15. We can see that the $N(f)$ no longer goes to zero at DC but remains finite. This introduces excess quantization noise within f_{bw} and is going to compromise the overall SNR and hence DR, as predicted by (6.11). In general, the more the leakage (the smaller the frequency response of the integrator at DC, or the smaller the frequency response of the chain of integrators at DC), the larger $N(f)$ at DC is. As a good rule of thumb, a first order modulator needs an integrator whose frequency response at DC is greater than the OSR. For a second-order modulator since the forward path consists of a cascade of two integrators, the individual integrator can have a frequency response at DC that is smaller than the OSR. This lower limit on the frequency response of the integrators at DC will put a lower limit on the gain of the op-amp that is used to realize these integrators. To see how they are related, first define $H(z)$ to be the

Fig. 6.15 N(f) for first and second order modulators with leaky integrators

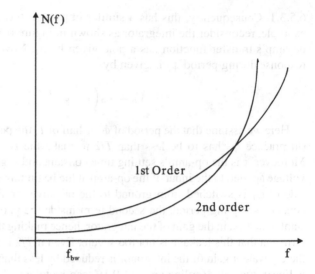

transfer function of the single-ended switched capacitor integrator shown in Figure 6.14. This transfer function, modified with finite operational amplifier gain, can be expressed as [11]

$$H(z) = \frac{(C_1/C_2)z^{-1/2}(1 - 1/A - C_1/(AC_2))}{1 - (1 - C_1/(AC_2))z^{-1}} \quad (6.12)$$

Here A is the gain of the op-amp. As an example, using our rule of thumb for a first-order modulator, the frequency response of the integrator at DC, $H(z)_{z=1}$, has to be larger than OSR. We can then substitute this in (6.12) (with $z = 1$) and find out the required A. This usually means $A >$ OSR as well.

For second- and higher-order modulators, the required $H(z)_{z=1}$ can be relaxed. The exact relaxed requirement can be worked out. However, instead of doing that, we want to simplify matters further. We can simplify matters by stating as a rule of thumb that if A, the op-amp gain for all the op-amps used in the individual integrators, is larger than OSR, the leakage is tolerable:

$$A > OSR \quad (6.13)$$

6.5.3.2 Finite Op-Amp and Switch Speed

Depending on the op-amp's settling behavior, different op-amp models can be obtained that will then be used to predict the op-amp's impact on the integrators' and hence modulators' performance. If the finite speed in the op-amp used in the integrator is modeled by representing the op-amp using a transfer function with a single pole, then incomplete settling behaves in a linear fashion. This means $H(z)$ is changed so that the integrator appears leaky again, as discussed in subsection

6.5.3.1. Consequently, this has a similar impact on the quantization noise. As an example, reconsider the integrator as shown in Figure 6.14. Let us assume that the op-amp's transfer function has a pole given by p_1. Now the integrator's transient response during period ϕ_2 is given by

$$V_o = V_a\left(1 - e^{-(T/2)}\right) \tag{6.14}$$

Here we assume that the period of ϕ_2 is half of T, the period of the sampling clock (in practice ϕ_2 has to be less than $T/2$ to make the two phases nonoverlapping). Moreover, τ is the op-amp's settling time constant and is given by $1/p_1$, and V_a is the voltage applied to the input of the op-amp at the beginning of ϕ_2, when the right-hand side of C_1 is switched from ground to the negative terminal of the op-amp. For a constant sampling period, the second term inside the parentheses represents a constant reduction in the gain of the integrator, hence making the integrator look leaky. It turns out that this leakage is not too severe. Even if $T/2$ is made to be as short as 4τ, the equivalent gain of the integrator is reduced by less than 2%. Again, if the settling is linear, typically settling error of 0.1% can be tolerated. On the other hand, if the settling is entirely dominated by slewing in the op-amp (this usually occurs at the early part of the response when a step input is applied to the op-amp, during which V_o is trying to ramp towards the applied V_a; the op-amp behaves in a nonlinear fashion during this stage), then the preceding description is not true. If the error is referred back to the input, and if every time the op-amp settles the error is made so that it always accumulated, then it can be shown that to guarantee an n-bit DR for the modulator, the settling error of the individual integrator, and hence the accompanying op-amp, should also be at the n-bit level. This is because the op-amp error is referred back to the input of the integrator and hence modulator. Thus any error it makes is identical to injecting the same error at the input of the modulator. For example, to guarantee a 100 dB DR, the op-amp needs to settle to within 0.001% of final value in $T/2$. Accordingly, slew rate limiting should be avoided and simulations should be run to ensure that the settling is essentially linear.

Finally, we have to examine settling problems due to the RC time constant of the MOS switch. This can usually be solved simply by increasing the width of the switch. Of course, there is an ultimate limit in that as the switch becomes larger in comparison to the switched capacitor, the effects due to channel charge injection and the parasitic junction capacitances also increase, as we have discussed in the appendix to Chapter 5 (where we discuss SAH).

In summary, for the integrator shown in Figure 6.14, the modified integrator transfer function due to finite op-amp bandwidth and assuming a linear settling behavior is [11]

$$H(z) = \frac{(C_1/C_2)z^{-1/2}[(1 - \epsilon) + z^{-1}\varepsilon C_2/(C_1 + C_2)]}{1 - z^{-1}} \tag{6.15}$$

where

$$e = e^{-\pi f_t T} \tag{6.16}$$

Here T is the sampling period and f_t is the unity gain bandwidth of the op-amp (in hertz). Notice that in (6.16) we assume that a feedback factor is 1, although in real life it is somewhat modified by the ratio of C_1 to C_2. We will have a chance to revisit this assumption when we talk about switch-capacitor-based resonator in subsection 6.7.2.2. For a one-pole response, $f_t = A \times 2\pi p_1$

Similarly, the modified transfer function due to nonzero switch resistance is [11]

$$H(z) = \frac{(C_1/C_2)z^{-1/2}\left(1 - 2e^{-T/4R_{on}C_1}\right)}{1 - z^{-1}} \qquad (6.17)$$

Here R_{on} is the on-resistance of the switch.

6.5.3.3   Integrator Noise

Achievable DR in a sigma-delta modulator is constrained by the available signal swing at one extreme and noise sources at the other. So far we have concentrated on quantization noise and use that to derive DR, which was given in (6.8). Noise can also arise from the power supply or substrate coupling, from clock signal feed-through, and from thermal and $1/f$ noise generation in the MOS devices.

$1/f$ noise can be a problem in the circuit but is usually taken care of by using large input devices. Thermal noise is a more fundamental problem. Because of thermal noise in the switch resistance, each voltage sampled onto a capacitance C exhibits an uncertainty of variance kT/C, where k is the Boltzmann constant and T is the absolute temperature. For an integrator circuit like that in Figure 6.14, the input capacitor C_1 sees two switch paths per clock cycle, so the noise variance increases by a factor of 2. Hence the noise sampled on it has a variance equal to $2kT/C_1$. Now this is filtered by the integrator. This integrator has a gain of C_1/C_2 and has a noise bandwidth (f_b) that is approximately half of its unity gain frequency (f_u). Hence the final output noise variance is $2kT/C_1 \times C_1/C_2 \times f_b/f_u = 2kT/C_1 \times C_1/C_2 \times 1/2 = kT/C_2$. (Actually the noise sources in the switches and op-amp interact due to bandwidth effects, but it has been established that a lower bound for the combined op-amp and switch thermal sources in this configuration is kT/C_2.) Due to over-sampling, this noise variance is reduced, effectively by the OSR to $kT/(OSR \times C_2)$, or the standard deviation is $\sqrt{\frac{kT}{OSR \times C_2}}$. Next we want to determine the maximum RMS value of a sinusoidal signal, within a full-scale voltage V_{FS}, and this turns out to be $V_{FS}/(\sqrt{2})$. Taking the ratio of the two gives a maximum signal-to-thermal-noise ratio for a sigma-delta modulator, which is also its dynamic range DR:

$$DR = SNR_{max} = V_{FS}\sqrt{OSR \times \frac{C_2}{2kT}} \qquad (6.18)$$

Notice that (6.9) gives the DR limited by quantization noise and (6.18) gives the DR limited by thermal noise.

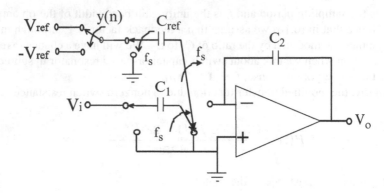

Fig. 6.16 Single-ended switched capacitor integrator with 1-bit DAC

6.5.4   1-bit ADC

A 1-bit ADC is just a single comparator. Ideally such a comparator should have zero delay, zero offset. A simple implementation of such a comparator consists of a differential pair with load (just like the op-amp), which again suffers from having finite speed and finite offset. Because of the feedback action provided by the sigma-delta modulator, this comparator offset will translate into an equivalent offset at the input of the modulator, without increasing the quantization noise of the modulator. As for the speed, the comparator has to switch and make a decision in one sampling clock period. Compared with the op-amp used to realize integrator, the comparator output does not have to settle to as high an accuracy. Moreover, the comparator operates in a open loop configuration, which is faster than the closed-loop configuration that the op-amp has to work in. Hence the speed requirement is usually met easily. The thermal noise is not important since it is noise shaped.

6.5.5   1-bit DAC

Figure 6.14 is redrawn in Figure 6.16 to show how this 1-bit D/A conversion is done. The latched comparator decision (the digital output of the 1-bit ADC) $y(n)$ is used to apply selectively a $+V_{ref}$ or $-V_{ref}$ voltage, through a separate capacitor C_{ref}, to the integrators to perform a 1-bit D/A conversion. In Figure 6.16, if $y(n)$ is 1, then V_{ref} will be applied to C_{ref} during ϕ_1. During ϕ_2, this charge will be transferred to C_2 and changes the output by $V_{ref} \times (C_{ref}/C_2)$. On the other hand if $y(n)$ is 0, then $-V_{ref}$ will be applied and at the end of ϕ_2 output changes by $-V_{ref} \times (C_{ref}/C_2)$. Consequently, a 1-bit D/A conversion plus integration is achieved. Because of the 1-bit nature, the D/A conversion is inherently linear. Of course, any noise on the V_{ref} and $-V_{ref}$ will manifest itself as input noise and must be reduced.

Now that we have introduced all the basic theory and implementation issues, we go through the design procedure.

6.5.6 Design Procedure

In this subsection we explain the steps it takes to design a low-pass sigma-delta modulator. The design procedure consists of the following steps:

1. From Chapter 2, find the specifications for ADC in terms of DR, full-scale voltage (V_{FS}), bandwidth of digitization (f_{bw}), and the IF.
2. Let us assume that we fix the order, L, of the modulator. We then assume that the modulator is implemented using type 2 integrators except in the innermost loop. Furthermore, the gains are all assumed to be 1. Then, for a given L and DR, we will determine the OSR using (6.11). Usually DR is given in log scale (i.e, in decibel). Therefore, it is more convenient to work with the logarithmic form of (6.11). The dynamic range expression in logarithmic form is

$$DR|_{dB} = 20 \log DR = 10 \log \left[\frac{3}{2} \frac{2L+1}{\pi^{2L}} (OSR)^{(2L+1)} \right] \qquad (6.19)$$

For example, for a single-stage first-order modulator the formula for DR equation in decibels becomes

$$DR|_{dB} = -3.41 + 30 \ \log(OSR) \qquad (6.20)$$

and for a second-order modulator the DR equation in decibels is

$$DR|_{dB} = -11.14 + 50 \log(OSR) \qquad (6.21)$$

After determining the OSR, the sampling frequency (f_s) can be calculated for a given bandwidth. If the calculated sampling frequency is too large for a given technology, the order must be increased. In addition, the sampling frequency should be kept below the IF.
3. The initial design from step 2 is now reiterated by changing the implementation to one based on type 1 integrators only. With the new types of integrators we perform signal scaling by changing the gains of the integrators and feedback factors so that this new modulator can achieve the original DR. Stability is then checked. If the modulator is unstable, we reduce the out-of-band gain of the NTF, again by changing the gains of the integrators. If the modulator has a large safety margin for stability, we can check and see if we can improve the DR by increasing the out-of-band gain of the NTF, also via changing the gains of the integrators. At the end of this step, the gains of individual integrators and feedback factors from the 1-bit DAC to these integrators are finalized

4. Determine the minimum requirement for each block of the sigma-delta modulator to meet the specification. To complete the design, op-amp specifications such as DC gain, and unity gain bandwidth have to be found. The important parameters can be summarized as follows:

   (a) Minimum gain for op-amp. One simple rule is to set the gain to be a few times of OSR (we arbitrarily set that to 5 here).

$$A = 5 \times OSR \qquad (6.22)$$

   (b) Unity gain bandwidth of op-amp as dictated by clock frequency and settling error. The settling behavior as explained in subsection 6.5.3.2 is dependent on the sampling frequency. To quantify this we assume linear settling behavior. First, the values for C_1/C_2, $C_2/(C_1 + C_2)$ required in (6.15) can be determined from integrator gains (determined in step 3). Also, in (6.15) $z = e^{j\omega T}$, where $\omega = 2\pi f_{bw}$ (f_{bw} has been determined in step 1) and $T = 1/f_s$ (f_s has been determined in step 2). Next we can calculate $H(z)_{no\ error}$ by setting ε to 0 in (6.15). Let us assume that we can tolerate an error of 0.1%, which is good enough for most applications. Then we find $H(z)_{error=0.1\%} = 99.9\% \times H(z)_{no\ error}$. Substituting this in (6.15) again, we can calculate ε corresponding to an error of 0.1%. Substituting this in (6.16), we can find f_t. To simplify the procedure further, we observed that for most C_1, C_2 values, the f_t so obtained is on the order of 5 to 10 times that of f_s, the sampling frequency. We can then use the following rule of thumb to find f_t:

$$f_t = 5f_s \qquad (6.23)$$

   (c) Size of capacitors of individual integrators as dictated by thermal noise consideration. The maximum dynamic range limited by thermal noise is given by (6.19) and repeated here:

$$DR = V_{FS} \times \sqrt{OSR \times \frac{C_2}{2kT}} \qquad (6.24)$$

Here V_{FS} is the full-scale input voltage of the modulator and must be less than V_{dd}, the power supply voltage. Notice that the DR obtained in (6.24) must be larger than DR in step 1 with a good safety margin. Typically for a given V_{dd}, the V_{FS} is limited. Once DR and V_{FS} are determined, the value can be substituted into (6.24) to calculate the proper value of C_2. C_1 is then calculated from C_2 and the integrator gain. C_{ref} is calculated from C_2 and the feedback factor (determined in step 3).

The following design example illustrates this procedure.

Design Example 6.1. In this design example, by using the design steps explained in the subsection 6.5.6 a complete design of a low-pass sigma-delta modulator is presented.

Step 1: We start from the example ADC given in Problem 2.9, Chapter 2. Instead of digitizing at 800 MHz, we assume that a mixer is added to translate the signal to a 20 MHz IF. Hence the ADC is digitizing a 20 MHz IF with a 200 kHz bandwidth and a 72 dB (12 bit) DR. Instead of a V_{FS} of 3.13 V, we change to a V_{FS} of 1 V.

Step 2: The order of the modulator, OSR, and the sampling frequency. First let us assume that a second-order modulator is used, again with type 2 integrator, except in the innermost loop. Let us repeat (6.20), the DR equation for a second-order modulator:

$$DR|_{dB} = -11.14 + 50 \log(OSR) \tag{6.25}$$

Substituting a DR of 72 dB into the preceding equation, we get an OSR of 46.55.

With an OSR of 46.55 and a 200 kHz bandwidth, the sampling frequency should be chosen to be at least 18.6 MHz. This is a reasonable sampling frequency for a switch-capacitor-based implementation under the current technology and no iteration on the order of modulator is needed. Since IF is 20 MHz, $f_{lo} = 20$ MHz. In general, to simplify the generation of f_{lo} and f_s, we set the two to be related by an integer ratio, with $f_s \leq f_{lo}$. In the present case this can be achieved by setting the sampling frequency to 20 MHz and hence the final OSR becomes 50.

Step 3: Determine the NTF transfer function. In step 2 we determined that a second-order modulator is sufficient to achieve the DR specification. Next we want to derive the NTF of this modulator using type 1 integrators and with proper scaling. First we take Figure 6.11(c) and change all integrators to type 1 integrators. Figure 6.11(c) now becomes Figure 6.17(a). Notice that these figures have the same NTF. Next, notice that in this new configuration integrator 2 has a gain of 2 and so the signal swing at the output of integrator 2 is large compared with the signal swing at the output of integrator 1, which only has a gain of 0.5. Hence as $x(n)$ increases in amplitude, the output of integrator 2 reaches maximum (limited by V_{dd}) sooner and limits the dynamic range of the whole modulator.

In order to maximize the dynamic range of the modulator, we want to make the internal swings all roughly equal. This can be achieved by changing the gain preceding the integrators (similar to scaling in typical filter design). In the present example, this can be done by putting a gain of less than 1 in front of each integrator. One way of performing this is shown in Figure 6.17(b). Since we have changed the gain of integrators, the NTF changes and both DR and stability must be checked. We rederive the NTF, assuming a white additive noise model:

$$NTF = \frac{4(1 - z^{-1})^2}{3z^{-2} - 6z^{-1} + 4} \tag{6.26}$$

Notice that this NTF has poles and differs from the NTF derived for the modulator in Figure 6.11(c). [the NTF derived there is restated here: $NTF = (1 - z^{-1})^2$, and has no poles]. From (6.26) we can see that there are two poles, each with a radius of 0.869 and with angles of $\pm 60°$. Hence the poles are inside the unit circle and the modulator is stable. Furthermore, $\omega_{bw}(= 2\pi f_{bw})$ is so

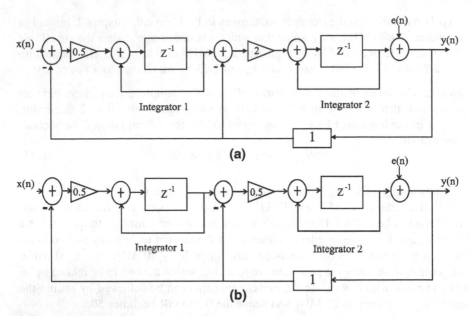

Fig. 6.17 (a) Second-order modulator using only type 1 integrator. (b) Second-order modulator using only type 1 integrator and with scaling

far away from the poles (ω_{bw}/ω_s is given approximately by the ratio $360°$/OSR or $7.2°$, which is much less than $60°$) in the present case that the poles hardly affect the NTF (hence DR) and so no iteration on the last step is needed.

Step 4: Determine the minimum requirement for each block.

(a) Op-amp gain: The minimum required op-amp gain can be calculated by satisfying (6.13). To give some safety margin we use (6.22) or $A = 250$.

(b) Op-amp unity gain bandwidth: The sampling frequency f_s is chosen to be 20 MHz. Let us assume a linear settling behavior and a tolerable error of 0.1%. This is in line with the conditions that lead to the simplifications in step 4(b) of the design procedure. Hence we can use the rule of thumb in (6.23) and set the unity gain bandwidth of the op-amp f_t to $5 \times f_s$, or 100 MHz. An op-amp with this unity gain frequency can be readily implemented in the current technology.

(c) Sampling Capacitor C_1: To find the bound on C_2 we turn (6.24) into an inequality, $DR < V_{FS} \times \sqrt{OSR \times \frac{C_2}{2kT}}$. We can then substitute the values of DR, OSR, and V_{FS} obtained from previous steps, which equal 72 dB, 50, and 1 V, respectively. Hence we have $72dB < 20 \log \left(1\ V \times \sqrt{50 \times \frac{C_2}{2kT}} \right)$ and we find that a C_2 of 0.5 pF is more than enough to meet the requirement. Next, from step 3, we have finalized the integrators' gains and feedback factors, which were shown in Figure 6.17(b). From Figure 6.17(b) we see that the gain k_1 of the first integrator is ½. Therefore, $C_1/C_2 = ½$ or $C_1 = C_2 \times ½ = 0.25$ pF. Since gain k_2 of the second integrator is also ½, we can simply copy the capacitor values for integrator 1 to those of integrator 2. In general, the size of capacitors

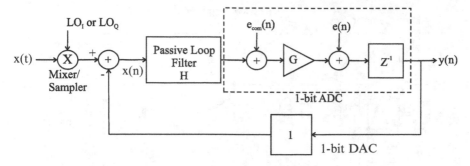

Fig. 6.18 Block diagram of IF digitizer using merged mixer/sampler and a passive loop filter

of the second stage can be made smaller, since any thermal noise from the switch associated with these capacitors is noise-shaped. Finally again from Figure 6.17(b) the feedback factor of the modulator is 1 and so $C_{ref} = C_1 = 0.25$ pF for both integrators 1 and 2.

(d) Comparator speed: The comparator speed has to be fast enough that the output code $y(n)$ is made available in less than half of the clock period, or 25 ns. As mentioned, the offset or the thermal noise of the comparator is not very important.

Now that we have gone through a design example, we discuss special low-power architectures for wireless communication.

6.5.7 Passive Low-Pass Sigma-Delta Modulator

One particular architecture of interest to wireless communication is the passive low- power sigma-delta modulator. This architecture differs from the conventional sigma-delta modulator we have described so far in two aspects. First, the mixer in front of the ADC (e.g, if we use this architecture as the baseband ADC described in Figure 6.4 it will be the mixer in front of the baseband ADC in that figure) is merged with the sampler (shown in Figure 6.8) inherent in the baseband ADC. This sampler is shown explicitly in Figure 6.8 for the case when the baseband ADC is implemented with a sigma-delta modulator. Since the mixer is implemented using a sampler-type structure, it is passive and is identical to the sampling mixer described in Chapter 5. Second, this architecture differs from a conventional sigma-delta modulator in that the integrators are replaced with a passive loop filter. In the present case the passive loop filter is implemented using R and C only, with the R being realized as a switch C. Due to the lack of gain in this passive filter, a switch-only gain-boost network provides gain without using any amplifier.

The block diagram of this architecture is shown in Figure 6.18. Since the mixer and sampler are merged, the output of this mixer/sampler is already the sampled value $x(n)$. The passive loop filter is labeled H. The 1-bit ADC is still the same.

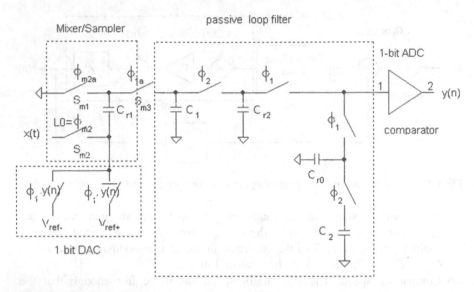

Fig. 6.19 Circuit implementation of passive sigma-delta modulator

However, as will be shown, the equivalent input noise of the comparator used in implementing this 1-bit ADC becomes very important in the present case and hence the 1-bit ADC is modeled with an explicit representation of this noise. (Remember that in subsection 6.5.2 we stated that this noise is not important.) Also notice that there is a gain block G that replaces the unity gain in the 1-bit ADC used in the conventional sigma-delta modulator.

To show the importance of the equivalent input noise of the comparator, $e_{com}(n)$, let us derive the transfer function of the sigma-delta modulator. This transfer function is:

$$Y(z) = X(z) + E(z)/GH(z) + E_{com}(z)/H(z) \tag{6.27}$$

Here $H(z)$ is the transfer function of the passive loop filter, $E(z)$ is z-transform of the quantization noise, $E_{com}(z)$ is the z-transform of the equivalent input noise of the comparator, and G is the equivalent gain provided by the 1-bit ADC. Here G can be on the order of thousands. If we keep on decreasing $E(z)$ (by increasing G, while keeping H constant), eventually the third term in (6.27) will become larger than the second term and $E_{com}(z)$ will dominate the overall noise contribution and limit the overall DR. This is one of the essential differences between the passive and conventional sigma-delta modulators.

Figure 6.19 shows the new architecture implemented with a second-order passive loop filter (i.e, H is a second-order filter). This second-order filter is implemented with 2-cascaded RC branches, with the R realized using a switched capacitor. For example, R_1 is implemented with the S_{m1}, C_{r1}, and S_{m3} combination.

This together with C_1 gives the first pole. The switches, C_{r2} and C_2, form the second pole. Finally, C_{r0} introduces a zero. ϕ_1 and ϕ_2 are the normal two-phase clocks. ϕ_{1a} is the same as ϕ_1 except that the falling edge is advanced slightly (the same as in bottom-plate sampling as discussed in Chapter 5).

Next we talk about the mixer/sampler section. This consists of ϕ_{m2} and ϕ_{m2a}, $x(t)$ and the sampling capacitor C_{r1}. Notice C_{r1} serves both to implement the loop filter and the mixer section. ϕ_{m2} and ϕ_{m2a} are the LO/sampling clock. In general, the LO signal ϕ_{m2} can have the same frequency as the sampling clock signal ϕ_2. However, as discussed in Chapter 5, the sampling distortion introduced by the mixer is dependent on the fall time of ϕ_{m2a} and ϕ_{m2}. Hence special buffers are introduced to generate these waveforms to reduce the fall time.

The G of the 1-bit ADC can be shown to be approximately $C_1 \times C_2/(C_{r1} \times C_{r2})$. By controlling the ratio of these capacitors, a large G is realized and is used to suppress quantization noise, as evident in (6.27). However, beyond a certain value, further increase in G will start to have a significant impact on H as well, so that G can no longer be increased without changing H. At this point the third term in (6.26) will start to come into the picture and should be considered to determine an optimal G to minimize the total noise contribution from the second and third terms.

In a particular implementation of this architecture, where a 10 MHz IF is digitized [5] in a 1.2 um CMOS technology, the circuit as shown in Figure 6.19 uses a capacitor C_{r0} to introduce a zero at around 750 kHz to compensate for the phase loss of the loop filter and improve the stability. C_{r2} is made small so it will not load C_1 too much. Considering the loading effect, two poles are located at 8 kHz and 34 kHz. Switches S_{m1} and S_{m2}, used for mixing/sampling, have $W/L = 20 \, um/1.2 \, um$ to reduce IM_3. IM_3 is further reduced by bottom-plate sampling and a fast rise/fall time in ϕ_{m2a} and ϕ_{m2}. The comparator is implemented with a preamplifier and a regenerative latch. It has a calculated rms value of the quantization noise, e_{rms}, of 41 uV for an equivalent noise bandwidth of 10 MHz and a G of 40 at 10 MHz. The rms input signal to the comparator is estimated to be around 147 uV.

As noted in Figure 6.19, the passive loop filter has no gain and hence the signal level at the input of the comparator in this 10 MHz IF digitizer is only on the order of 147 uV. This is quite small compared with the e_{rms} value of 41 uV. In order to increase this signal level without the use of an explicit amplifier (which consumes power), a gain-boost network implemented using switches only (no amplifiers) is used and is shown in Figure 6.20. Note that the circuit enclosed in the dotted line replaces the circuit enclosed in the box labeled mixer/sampler and the 1-bit DAC in Figure 6.19. This circuit consists of N sampling capacitors, each labeled C_{r1}.

This circuit operates as follows. During ϕ_2 (sampling phase), all N sampling capacitors C_{r1} are connected in parallel. The bottom plates are all connected to $x(t)$. Depending on whether the output code $y(n)$ is 1 or 0, the top plates are all connected to either V_{ref}^- or V_{ref}^+, respectively, thereby realizing both the 1-bit D/A conversion and subtraction. Hence all N capacitors are charged to either $x(t) - V_{ref}^-$ or x $(t) - V_{ref}^+$. Then during ϕ_1 (charge-transfer phase), the N sampling capacitors are reconfigured in a series connection with an equivalent value of $C_{r1}^* = C_{r1}/N$. This equivalent capacitor is connected between ground and the capacitor C_1 and

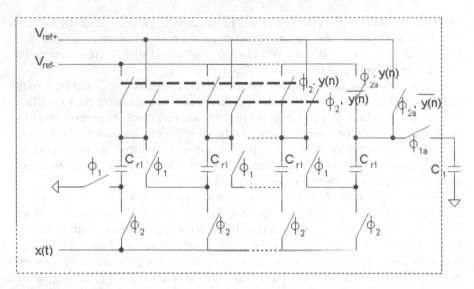

Fig. 6.20 Gain boosting network

hence the C_1 is charged to $N \times (x(t) - V_{ref}^-)$ or $N \times (x(t) - V_{ref}^+)$. Thus a voltage gain of N is achieved. Ideally the network has a transfer function:

$$\frac{\left[\frac{C_{r1}}{\frac{C_{r1}}{N}+C_1}\right]z^{-1}}{\left[1-\frac{C_1}{\left(\frac{C_{r1}}{N}+C_1\right)}z^{-1}\right]} \tag{6.28}$$

It can be shown easily that the DC gain is N and the pole is at $\frac{C_1}{\left(\frac{C_{r1}}{N}+C_1\right)}$.

Bottom-plate sampling is used in both phases to reduce charge injection error. Analysis shows that the thermal noise of the network is the same as that of a switch connected to the equivalent series capacitance. Even though it is difficult for the network to achieve high gain due to parasitic capacitance, a moderate gain of 3 is easily obtained in the design. This means that our modulator can tolerate an e_{rms} that is three times larger for the same DR. Assuming that this noise comes from thermal noise of the input differential pair of the comparator, this reduction allows a significant reduction in the bias current of the differential pair and hence power consumption of the comparator.

In summary, the aforementioned low-power modulator has been implemented in various forms, and three such examples (together with real measured performances) are summarized in Table 6.1[5,7, 12]. Notice that in example 1, 2 the modulators have the input SAH merged with the sampling mixer and therefore implement the

Table 6.1 Performance of sigma-delta modulators using merged sampler/mixer and passive loop filter.

IF	Bandwidth	Resolution	Power	Distortion (IM$_3$)	V_{dd}	Technology
10 MHz	20 kHz	13 bit	0.25 mW	≤ -70 dB @ -10 dBm	3.3 V	1.2um CMOS
150 MHz	80 kHz	13 bit	12.5 mW	≤ -65 dB @ -3 dBm	5 V	0.8u BiCMOS
400 MHz	40 kHz	12 bit	20 mW	≤ -85 dB @ -14 dBm	5 V	0.8 um BiCMOS

mixer plus baseband ADC as depicted in Figure 6.4 and in Figure P.2(a), Chapter 2. The third example [7] actually incorporates the merged mixer inside the feedback loop, as depicted in Figure 6.5, with the modulator still being passive. This example actually uses the feedback action inherent in a sigma-delta modulator (which consists of a quantized feedback loop) to suppress the distortion generated by the mixer. The details on the nature of this distortion were discussed thoroughly in Chapter 5.

6.6 Bandpass Sigma-Delta Modulators

Let us now turn our attention to bandpass sigma-delta modulators. Examples of using bandpass sigma-delta modulators are shown in Figures 6.2 and 6.3. In these figures, because the IF signal is narrowband in nature it almost always makes sense to implement the passband ADC using a sigma-delta modulator approach. In general, because a passband ADC operates at IF, we have to use a bandpass sigma-delta modulator. Since we perform the digitization at IF, the A/D conversion occurs sooner in the receiver chain. Hence we can get rid of analog narrowband IF filters (replacing them with the much easier implemented digital filters). As described previously, the quantization noise of a low-resolution quantizer can be suppressed selectively around specific ranges of frequencies by employing a combination of oversampling and feedback techniques. These techniques have so far allowed us to shape the quantization noise spectrally while maintaining a flat frequency response for a narrowband low-pass input signal. The same principle is now extended to narrowband bandpass signals, where the spectral shaping occurs around IF rather than DC. This results in improved SNR and hence DR for bandpass rather than low-pass signals.

The advantages of conventional (low-pass) modulators over Nyquist rate converters are equally applicable to bandpass modulators. Inherent linearity, and reduced antialias filter complexity are among the advantages. In addition, the design methodology of low-pass sigma-delta modulators can also be applied, with some modifications, to the bandpass case. Consequently, this section is devoted only to design aspects that are unique to a bandpass modulator.

Similar to a low-pass modulator (shown in Figure 6.8), a bandpass (BP) sigma-delta modulator can be constructed by embedding a loop filter inside a feedback loop, as shown in Figure 6.21. The major difference is that the integrator, with gain k, is now replaced by a resonator with resonating frequency ω_o. The resonator may be implemented using switched-capacitor techniques, in

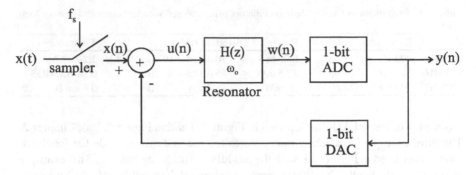

Fig. 6.21 A general noise-shaping bandpass modulator

which case it consists of two integrators connected back to back. The resonating frequency, ω_0, is then determined from the unity gain frequencies ω_u of the two integrators. Another implementation involves the use of a passive loop filter structure, similar to the passive low-pass sigma-delta modulator described in subsection 6.5.7, except the *RC* filter is replaced by an *LC* filter so that the filter can resonate. The loop filter can also be implemented by continuous time filter such as G_mC filters.

Next we show how to start from the low-pass sigma-delta modulator design procedure that we are familiar with and adapt it so we can design a bandpass modulator. Then we highlight the limitations of this architecture. Finally, we consider a design example for a complete bandpass modulator that is similar to our low-pass sigma-delta modulator design example.

6.6.1 Comparisons of Low-Pass and Bandpass Modulators

It has been shown that a low-pass sigma-delta converter of order L can be converted to a bandpass modulator of order $2L$, with a center frequency at $f_s/4$. This transformation has one major advantage: both the stability performance and the noise properties of the original low-pass prototype that we discussed previously are preserved in this new topology. Notice that the bandpass sigma-delta modulator that is used as the passband does not have a mixer in front of it. Hence the sampler in the modulator has to respond to a high frequency (IF) signal, making its design more demanding than in the case of low-pass sigma-delta converters (which are preceded by a mixer). On the other hand, in order to make a fair comparison, even in the case of a low-pass sigma delta converter, if we merge its sampler with a mixer, the sampler faces the same challenge as in the bandpass sigma-delta modulator case.

When compared with the low-pass sigma-delta modulator, we find that the high-pass NTF of a low-pass sigma-delta modulator becomes the band-reject NTF in a bandpass sigma-delta modulator. The quantization noise PSD for a second-order

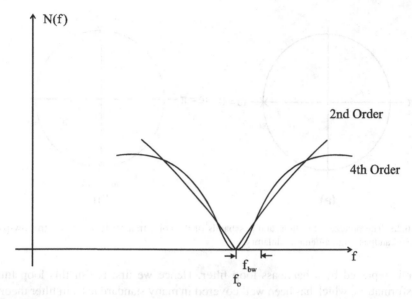

Fig. 6.22 $N(f)$ for second- and fourth-order BP modulators

and a fourth-order bandpass modulator is shown in Figure 6.22. Notice that the second-order bandpass modulator has essentially the same quantization noise PSD as a first-order low- pass modulator, except that the null frequency (frequency where the quantization noise PSD goes to zero) is moved from DC to f_0, the resonating frequency of the resonator. Hence by selecting the proper resonating frequency, we are practically selecting the quantization noise null frequency. Of course, we should choose this to be the same as f_{if}. Hence $f_0 = f_{if}$.

In this chapter we assume that the BP modulator is implemented in the sampled data domain. Hence all the poles of the resonator $H(z)$ (zeroes of the NTF) are described using the z-plane. We further assume that the poles are near or on the unit circle (but always inside to maintain stability).

Figure 6.23 compares the pole placements of $H(z)$ for a second-order low-pass modulator [Figure 6.23(a)] and a fourth-order bandpass modulator [Figure 6.23(b)]. Assume for the time being that $H(z)$ only has poles and no zeroes. Figure 6.23(a) shows that the two poles are at DC. Figure 6.23(b) shows that the four poles are at $\pm\pi/2$, or at $f_s/4$, where f_s is the sampling frequency. Hence if we set f_{if} to exactly $f_s/4$, the signal will be at the quantization noise null, as desired.

6.6.2 Low Pass Modulator to Bandpass Modulator Conversion

It seems that from an architecture point of view the bandpass modulator is essentially the same as the low-pass modulator. The only difference is that the low-pass loop

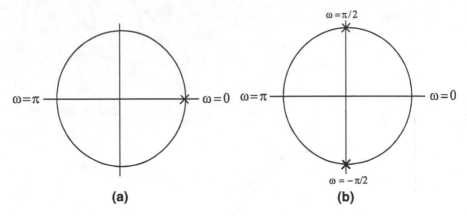

Fig. 6.23 The pole/zero locations and passbands of the noise transfer functions for (a) low-pass and (b) bandpass sigma-delta modulators

filter is replaced by a bandpass loop filter. Hence we first revisit this loop filter transformation, which has been well covered in many standard texts in filter theory.

As an example of the filter transformation, let us start with a low pass filter $H(z)$ with poles at DC and apply the transformation $z \to -z^2$. This will map the poles of $H(z)$ from DC to $\pm \pi/2$. The $z \to -z^2$ transformation is the simplest transformation. It has one major advantage: if the low-pass prototype is stable, so is the bandpass prototype. Another example of low-pass to bandpass transformations involves the use of an N-path filter.

An example of such an N-path filter, with $N = 3$, is shown in Figure 6.24(a), and its clocking waveform is shown in Figure 6.24b [11]. The clocking waveform is an N-phase clock [N = 3 in Figure 6.24(b)] with the master clock ϕ running at N times the clock of the subphases. We label its frequency as f_s. Essentially, the frequency response of the N-path filter is simply that of the LPF together with its images centered at f_s/N, $2f_s/N \ldots f_s \times (N-1)/N$. These images provide the bandpass filter response that we are interested in, with center frequency at f_s/N, $2f_s/N$, $\ldots f_s \times (N-1)/N$. Hence we just have to select the proper image that we want. The use of an N-path filter has the advantage that each path filter has more time to settle [$(N-1)T$ more time]. However, path mismatch can be a problem. If this problem is not taken care of, it may introduce a mirror image of the input signal at a mirror location centered around $f_s/4$ [6]. This path mismatch can come from mismatch in the DC gain and settling behavior of the op-amps used to realize the switch-capacitor integrators, which are in turn used to implement the separate path LPF. The impact of finite op-amp DC gain and settling error on the LPF transfer function is exactly the same as described in Section 6.5.

A final example of filter transformation consists of the use of generalized second-order low-pass-to-passband transformations. This allows us to have full control over the passband location. However, a stable low-pass prototype does not guarantee that the transformed bandpass filter is also stable.

Next we extend the quantization noise analysis for the LP modulator to the BP modulator. We can do this by repeating the analysis in subsection 6.4.2, whereby the

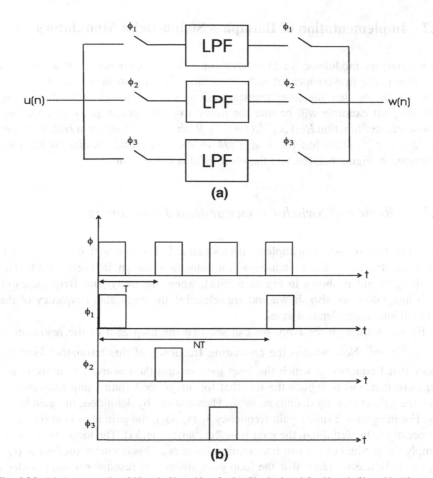

Fig. 6.24 (a) An example of N-path filter, $N = 3$. (b) Clock signal for N-path filter, $N = 3$

1-bit ADC is again replaced by an additive white noise model. Again the PSD of the quantization noise is assumed to be white with a power of $\Delta^2/12$. Using this linear analysis, the NTF has L/2 zeroes at ω_o, where L is the order of the BP modulator. The result is that for every doubling of the OSR the DR or SNR increases by $3L + 3$ dB. This is a little better than half of the improvement rate for the low pass modulator, which is $6L + 3$ dB/octave [13]. Hence we modify (6.11) by replacing the $(2L + 1)$ factor with an $L + 1$ factor and develop a corresponding DR formula for a BP modulator. The resulting formula, with DR expressed in decibels, is as follows:

$$DR|_{dB} = 10 \log \left[\frac{3}{2} \frac{L+1}{\pi^{L+\frac{1}{2}}} (OSR)^{(L+1)} \right] \tag{6.29}$$

Simplifying we have:

$$DR|_{dB} = 10 \log \left[\frac{3}{2} \frac{L+1}{\pi^{L+\frac{1}{2}}} \right] + 10(L+1) \log[OSR] \tag{6.30}$$

6.7 Implementation of Bandpass Sigma-Delta Modulators

In a bandpass modulator we also have the same components as in a low-pass modulator. The first component is the loop filter. To implement this, let us assume that we decide that the filter transformation is a z^{-1} to $-z^{-2}$ transformation. The simplest example will be that the initial low-pass prototype is an integrator whose transfer function $H(z)$ is $z^{-1}/(1-z^{-1})$. With the transformation $H(z)$ becomes $-z^{-2}/(1 + z^{-2})$. This has poles at $\pm\pi/4$ of the unit circle, similar to the case described in Figure 6.23(b), and therefore $H(z)$ is a resonator.

6.7.1 Review of Switch-Capacitor-Based Resonators

We will now discuss how to implement a resonator. Let us assume that we implement the resonator using the switch-capacitor integrator shown in Figure 6.14. The resulting circuit is shown in Figure 6.25(a), where the unity gain frequencies of each integrators are also shown and are related to the unity gain frequency of the original integrator, denoted as ω_1.

By going through the loop, we can see that the loop gain of the resonator is $\omega_1 \times \frac{\omega_o^2}{\omega_1} = \omega_o^2$. Now what is the resonating frequency of this resonator? First we show that frequency at which the loop gain around the resonator is unity is ω_o. To show that, just recognize the fact that for integrator 1 (unity gain frequency is ω_1), the gain at ω, is by definition, ω/ω_1. Hence at ω_o, by definition, the gain is ω_o/ω_1. For integrator 2 (unity gain frequency is ω_o^2/ω_1), the gain at ω is $\omega/(\omega_o^2/\omega_1)$. Hence at ω_o, by definition, the gain is $\omega_o/(\omega_o^2/\omega_1) = \omega_1/\omega_o$. The loop gain at ω_o is simply the product of the two integrator gains at ω_o. This becomes $(\omega_o/\omega_1) \times (\omega_1/\omega_o) = 1$. Hence we show that the loop gain around the resonator at ω_o is indeed unity. Second, from the definition of a resonator, the frequency at which the loop gain is 1 is its resonating frequency. Applying this definition to our first conclusion, we can now conclude that ω_o is indeed the resonating frequency of the resonator. It can further be shown that the Q (quality factor) of this resonator is ω_o/ω_1.

The aforementioned resonator works fine but has a bad component spread. This can be taken care of by scaling, so Figure 6.25(a) becomes Figure 6.25(b). Here the two integrators have identical unity gain frequency, set equal to ω_o, but each is preceded by a different attenuator. If the resonator implemented in Figure 6.25(b) (called a biquadratic structure) uses a type 1 integrator, the overall transfer function becomes

$$H(z) = Gz^{-2}/\left(1 + z^{-2}\right) \tag{6.31}$$

Here G is the gain of the resonator [6]. Notice that this function has zeroes in addition to poles. Hence if we use this to implement the BP modulator, the resulting NTF contains poles.

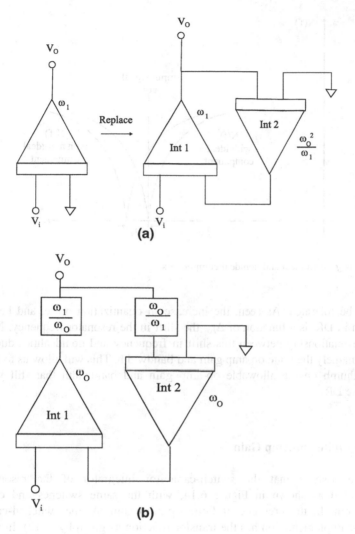

Fig. 6.25 (a) Resonator implemented using two integrators. (b) Resonator with scaled integrators. Note: Int = integrator

6.7.2 Non-Ideal Resonator

Since we implement the resonator using a biquadratic structure, the impact of op-amp settling error and finite gain will manifest as a shift in f_o. This is going to lead to excess quantization noise, as shown in Figure 6.26. The figure shows the $N(f)$ of a BP modulator using a resonator implemented with ideal components (left) and nonideal components (right). Notice that with nonideal components, the input signal is sitting at a frequency where there is substantial quantization noise and the DR is reduced. It is precisely this phenomenon that places constraints on the maximum nonidealities

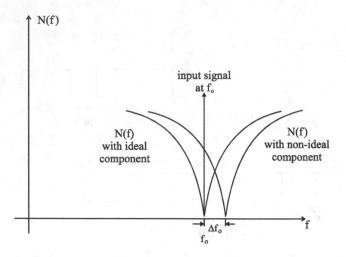

Fig. 6.26 $N(f)$ with ideal and nonideal components

that can be tolerated. As seen, the increase in quantization noise, and hence the reduction in DR, is a function of Δf_o, the shift in the resonator frequency. Next we show the relationship between this shift in frequency and nonidealities due to the op-amp, namely the finite op-amp gain and bandwidth. This will allow us to derive a rule of thumb on the allowable op-amp gain and bandwidth that still yield an acceptable DR.

6.7.2.1 Finite Op-Amp Gain

First, we assume that the switch-capacitor integrator of the resonator is implemented as shown in Figure 6.14, with the same switches and clocking arrangement. In the presence of finite op-amp gain, A, the switched-capacitor integrator implementation has the transfer function as given by (6.12). In order to separate the impact of op-amp gain on integrator gain and integrator pole location we rewrite (6.12) in the following form:

$$H(z) = \frac{C_1}{C_2}(1 - m)\frac{z^{-1/2}}{1 - (1 - p)z^{-1}} \tag{6.32}$$

Here parameters m and p highlight the separate effects on integrator gain and pole location, respectively. They are related to op-amp gain A as

$$m = \frac{1}{A}\left(1 + \frac{C_1}{C_2}\right) \tag{6.33}$$

and

$$p = \frac{1}{A}\frac{C_1}{C_2} \tag{6.34}$$

Now let us turn to the switch-capacitor resonator. It has two identical integrators. Let us relabel these integrators' capacitors as C_S and C_I. Using these new notations, (6.32) through (6.34) are rewritten as

$$H(z) = \frac{C_S}{C_I}(1-m)\frac{z^{-1/2}}{1-(1-p)z^{-1}} \tag{6.35}$$

$$m = \frac{1}{A}\left(1+\frac{C_S}{C_I}\right) \tag{6.36}$$

$$p = \frac{1}{A}\frac{C_S}{C_I} \tag{6.37}$$

Equations (6.35) through (6.37) are applicable to both integrators of the resonator. We can now find the resonator transfer function by applying (6.35) through (6.37) to each of the integrators in Figure 6.25. This expression turns out to be rather complicated. For illustration purposes we neglect all second-order terms in this expression and set $C_S/C_I = 1$. The simplified expression is

$$H(z) = G(1 - m_1 - m_2)$$
$$\times \frac{z^{-2}}{1 + (p_1 + p_2 - 2m_1 - 2m_2)z^{-1} + (1 - p_1 - p_2)z^{-2}} \tag{6.38}$$

Here G is the gain of the resonator. Variables m_1, m_2, p_1 and p_2 are defined in (6.36), (6.37) for integrators 1 and 2. Notice that if A becomes ∞, all the m's and p's become zero and (6.38) reverts back to

$$H(z) = G\frac{z^{-2}}{1+z^{-2}} \tag{6.39}$$

This is the transfer function of the resonator implemented with ideal components, as stated in (6.31). If we assume that m_1, p_1, m_2, and $p_2 \ll 1$, the fractional deviation in the center frequency, f_o, of the resonator is

$$\frac{\Delta f_o}{f_o} \approx \frac{p_1 + p_2 - 2m_1 - 2m_2}{\pi} \tag{6.40}$$

Notice that p_1, p_2, m_1, m_2 happen to cancel out one another, so the overall fractional change remains relatively small. Furthermore, it can be shown that this fractional change is not very sensitive to the ratio C_S/C_I. Hence we can use (6.40) to predict the fractional change of f_o for a general resonator.

While the tolerable decrease in DR is application dependent, a good rule of thumb is around 5 dB. The decrease in DR due to fractional shift of f_o for a BP modulator is a function of its order. The exact decrease in DR can usually be quantified by simulation only. From simulation for the DR of a fourth order BP modulators to suffer less than 5 dB reduction, f_o should shift by less than 0.2% [6].

6.7.2.2   Finite Op-Amp Speed

As is the case with low-pass sigma-delta converters, it is important that the outputs of resonators in a BP modulator settle to the desired accuracy. Now since an individual resonator is implemented using integrators as shown in Figures 6.25(a) and 6.25(b) and integrators are in turn implemented using op-amps, the finite op-amps' speed will affect the resonator's transfer function. Assuming that the switched-capacitor integrator used to implement the resonator settles linearly with a single-pole response, incomplete settling manifests itself as an integrator gain error. Hence as with finite op-amp gain error, the settling error will also result in a shift of the center frequency of the resonator. This shift of f_o will result in extra quantization noise being introduced into the passband and hence compromise the dynamic range. The effect of linear op-amp settling on the resonator can be determined by modeling each of the two switched-capacitor integrators using (6.15) again. We will simplify (6.15) by approximating the factor $1 - \epsilon + z^{-1}\epsilon C_2/(C_1 + C_2)$ in (6.15) as $1-\epsilon$, as we assume that the third term in the factor is smaller than the second term and can be neglected. Then we rename ϵ as g. Finally, we relabel these integrators capacitors as C_S and C_I. Then $H(z)$ becomes:

$$H(z) = \frac{(C_s/C_I)z^{-1/2}(1 - g)}{1 - z^{-1}} \tag{6.41}$$

Next let us re-write (6.16) using these new notations and it becomes:

$$g = e^{-T/2\tau} \tag{6.42}$$

Here T is the sampling period and we assume a two-phase clock and 50% duty cycle for each phase. τ is given by

$$\tau = \frac{1}{2\pi f_t} \frac{C_s + C_I}{C_I} \tag{6.43}$$

Here f_t is the unity gain bandwidth of the op-amp. Notice that in (6.43) we no longer assume that the feedback factor is 1 and so τ is modified from $1/2\pi f_t$ by having an extra factor $(C_s + C_I)/C_I$.

We now connect the two integrators back to back, each having the transfer function as given in (6.41). We further assume that $C_s/C_I = 1$. When second-order terms are neglected, the simplified resonator transfer function is found to be

$$H(z) = G(1 - g_1 - g_2)\frac{z^{-2}}{1 - (2g_1 + 2g_2)z^{-1} + z^{-2}} \tag{6.44}$$

Here G is the gain of the resonator, and g_1, g_2 are defined in (6.42) for integrators 1 and 2. Consequently, the fractional shift in f_o due to incomplete settling is approximately

$$\frac{\Delta f_o}{f_o} \approx \frac{2g_1 + 2g_2}{\pi} \tag{6.45}$$

6.7.2.3 Resonator Noise

The resonator noise effect on capacitor sizing is similar to the integrator noise case, with the input sampling capacitor having to be made largest. Subsequent noise sources (after the first integrator in the low pass case and after the first resonator in the bandpass case) will see their noise contribution suppressed due to noise shaping, and hence the size of the associate capacitors can be reduced substantially.

In the lowpass modulator we state that the $1/f$ noise can be a problem. In the present case this noise, which is generated mostly from gates of the MOS transistors, will lie below the band of interest of the bandpass modulators and should be of little concern.

6.7.3 Design Procedure

We now examine the design procedure of a BP modulator in a manner similar to what we did in subsection 6.5.6 for a low-pass modulator. The design steps are as follows:

1. From Chapter 2, find the specifications for ADC in terms of DR, full-scale voltage (V_{FS}), bandwidth of digitization (f_{bw}), and the IF.
2. Determine f_s from $f_s = 4 \times f_{if}$ and OSR from $OSR = \frac{f_s}{2f_{bw}}$. Use the following formula [repeated from (6.30)] to find the order L that satisfies the required DR for the given OSR: $DR|_{dB} = 10\log\left[\frac{3}{2}\frac{L+1}{\pi^{L+\frac{1}{2}}}\right] + 10(L+1)\log[OSR]$. Notice that this step is different from the low-pass case.
3. Select a low-pass modulator whose order is $L/2$ and that is stable. Determine its NTF and convert this NTF to the NTF of a bandpass modulator using the $z \to -z^2$ transformation.

4. Determine a block-level diagram implementation of the BP modulator that has the NTF determined in step 3. Each block in the diagram should consist of a resonator only.
5. Determine the minimum requirement for each block of the BP modulator. The governing equations include (6.40) and (6.45), from which the op-amp gain and bandwidth can be determined that will achieve an acceptable $\Delta f_o/f_o$ and hence acceptable DR. The resonator noise dictates the capacitor size through the use of (6.18) and the comparator has to work at four times the IF.

Design Example 6.2.

Step 1: In this design example, the same specifications are adopted as used in Design Example 6.1. Therefore, following step 1, Design Example 6.1, f_{bw} is selected to be 200 kHz and the DR is selected to be 72 dB. f_{if} is 20 MHz.

Step 2: Unlike low-pass sigma-delta modulator, f_s is larger than IF. $f_s = 4 \times f_{if} = 80$ MHz. Hence OSR is

$$OSR = \frac{f_s}{2f_{bw}} = \frac{80 \, MHz}{400 \, kHz} = 200 \tag{6.46}$$

To see if one satisfies the DR requirement, let us take (6.30), substitute in OSR = 200 and try $L = 2$. The resulting DR is

$$DR|_{dB} = 10 \log \left[\frac{3}{2} \frac{2+1}{\pi^{2+\frac{1}{2}}} \right] + 10(2+1) \log[200] = 63 dB \tag{6.47}$$

This is not sufficient to satisfy the required DR of 72 dB. Next we try $L = 4$ and we have
$DR|_{dB} = 10 \log \left[\frac{3}{2} \frac{4+1}{\pi^{4+\frac{1}{2}}} \right] + 10(4+1) \log[200] = 101 \, dB$ and the required DR is satisfied.

Step 3: From step 2, $L = 4$, hence $\frac{L}{2} = 2$. Therefore we need to select a second-order low-pass modulator that is stable. We decide to select the low-pass modulator in Design Example 6.1. The NTF for this low-pass modulator is given in (6.26) and is rewritten here:

$$NTF = \frac{4(1 - z^{-1})^2}{3z^{-2} - 6z^{-1} + 4} \tag{6.48}$$

Using the $z \rightarrow -z^2$ transformation, we obtain the NTF for the fourth-order BP modulator:

$$NTF = \frac{4(1 + z^{-2})^2}{3z^{-4} + 6z^{-2} + 4} \tag{6.49}$$

Step 4: To determine the block diagram of a BP modulator that has an NTF as described in (6.49), we start with the block diagram of an LP modulator that has an NTF as described in (6.48). This was given in Figure 6.17(b). We redraw it in Figure 6.27(a) where the integrators are redrawn with their transfer functions

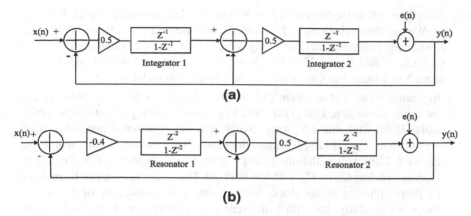

Fig. 6.27 (a) LP modulator prototype. (b) The final realization of the fourth-order BP modulator

explicitly shown. Now we replace every z in Figure 6.27(a) with $-z^2$. Hence each integrator whose transfer function is $z^{-1}/(1-z^{-1})$ becomes a block whose transfer function is $-z^{-2}/(1+z^{-2})$. This represents a resonator. Next we want to achieve proper scaling on the bandpass sigma-delta modulator while maintaining the same NTF. Therefore, the gains in front of the integrators in Figure 6.27(a) are changed. The final block diagram of the BP modulator is shown in Figure 6.27(b) [6].

Step 5: Determine requirements of each block. Let us assume that for the present application an acceptable DR reduction is 5 dB. Since from step 2 we obtain $L = 4$, then from subsection 6.7.2.1 we know that for $L = 4$ and a DR reduction of 5 dB, an acceptable $\Delta f_0/f_0 = 0.2\%$. Let us set $C_S/C_I = 1$ in all the integrators.

(a) Op-amp gain: Assume that op-amps for all the integrators have the same gain A. Substituting $C_S/C_I = 1$ in (6.36) and (6.37), we find $m = 2/A$ and $p = 1/A$. Substituting these and $\Delta f_0/f_0 = 0.2\%$ in (6.40), and we have

$$0.2\% = \frac{\frac{1}{A} + \frac{1}{A} - \frac{4}{A} - \frac{4}{A}}{\pi} \qquad (6.50)$$

Solving, we have $A = 954.9$. Rounding up we make $A = 1000$ or 60 dB.
To be exact, (6.40) is derived with C_{s1}/C_{I1} and C_{s2}/C_{I2} both set to 1, where C_{S1}, C_{S2} are the sampling capacitors for integrators 1 and 2 in the resonator described in Figure 6.15(b). Similarly, C_{I1}, C_{I2} are the integrating capacitors for integrators 1 and 2 in Figure 6.15(b). In the present case, since the gains of the resonators are -0.4 and 0.5, respectively, (6.40) may have to be modified.

(b) Op-amp bandwidth: We assume that op-amps in all integrators have the same unity gain bandwidth and hence the same settling error g. Since $C_S/C_I = 1$, (6.45) applies. Substitute $\Delta f_0/f_0 = 0.2\%$ in (6.45) we have

$$0.2\% = \frac{2g + 2g}{\pi} \qquad (6.51)$$

Solving, $g = 0.0016$. Next with $f_s = 80$ Mhz (as determined in step 2), $T = 12.5$ ns. We can then substitute g, T into (6.42) and we find $\tau = 0.97$ ns. This is the closed-loop time constant of the integrator. Substituting this value in (6.43), with $C_s = C_I$, we find f_t to be 328 MHz. Notice that f_t is about 4 times f_s. Also, this is about 3 to 4 times the f_t required for the low-pass modulator case.

(c) Resonator noise: Let us repeat (6.18): $DR = V_{FS} \times \sqrt{OSR \times \frac{C_I}{2kT}}$. Now we can substitute the required DR, OSR, and V_{FS} obtained from previous steps, which equal 72 dB, 200, and 1 V, respectively, into this equation. Upon substitution we find that a C_I of 0.125 pF is more than enough to meet the requirement. From Figure 6.27(b) we see that the gain k_1 of the first resonator is -0.4. Therefore, $C_{s1}/C_{I1} = 0.4$ or $C_{s1} = C_{I1} \times 0.4 = 0.05$ pF. The negative sign can be realized by proper phasing of the clock. We can design the capacitors of the second integrator similarly, but with a different gain. Finally from Figure 6.27(b) the feedback factor of the modulator is 1, and hence $C_{ref} = C_s$ for both resonators.

(d) Comparator requirement: in this modulator the comparator has to work at a very high speed (at least four times IF). This means that the output code $y(n)$ has to be made available in less than half of the clock period, or 6.25 ns.

6.7.4 Bandpass Versus Low-Pass Modulators

We now compare the two architectures covered in Section 6.4 and 6.6. On a system level they achieve similar goals, with the main distinction being whether mixing is performed before or after the A/D conversion. In practical terms, bandpass modulators (passband ADC in Figure 6.3) have to sample faster input signals than baseband modulators (baseband ADC in Figure 6.4 and Figure P.9(a), Chapter 2). Incidentally, the system in Figure 6.4 and Figure P.2(a), Chapter 2, is also called a zero IF system, since all the analog signal is mixed down to DC (or zero IF) before being demodulated. A/D converters for zero IF systems need to have small $1/f$ noise. The key observation is that BP modulator needs a highly linear (low distortion) input SAH whereas the zero IF systems needs a highly linear mixer.

If we want to compare the two modulators in terms of the hardware requirements, design examples 6.1 and 6.2 can give a good indication. For the op-amp gain requirement, in a bandpass modulator the NTF null frequency is not at DC and so is very sensitive to the op-amp gain. Hence the DR is very sensitive to op-amp gain. On the other hand, in a low-pass modulator the NTF null frequency is at DC, independent of op-amp gain. Finite op-amp gain only changes the noise floor of the NTF and has a much less dramatic impact on the DR. For the op-amp and comparator speed requirement, we should note that as f_{if} increases, the speed requirement for a low-pass modulator will become less than that of the bandpass modulator. The main reason is that in a bandpass modulator f_s is at a frequency higher than f_{if} (typically four times), whereas in a low-pass modulator f_s is dictated by f_{bw}, which is fixed for a given standard, and is independent of f_{if}.

Table 6.2 Comparisons of low-pass and bandpass modulators' hardware requirements

Modulator type	Op-amp gain	Op-amp bandwidth	Comparator speed
Low pass	Moderate	Moderate	Moderate
Bandpass	Moderate to high	High	High

For DECT, GSM, and other standards, once f_{if} goes into megahertz range, f_s for a bandpass modulator becomes larger than that of a low-pass modulator. This difference in f_s explains the difference in operating speed and hence requirements for op-amps and comparators in the two cases. The preceding comparisons are summarized in Table 6.2.

6.8   I/Q mismatch in Mixer and A/D Converters

6.8.1   Origin of Mismatch and Its Impact

In a wireless receiver, since the information is modulated using two paths, quadrature sampling is required to separate the data in these two paths, as shown in the IF digitizer of Figure 6.4 or Figure 6.5. Following this, digital signal processing is performed and the demodulated signal S_{DQPSK} is generated. This intermediate frequency (IF) digitizer can be adopted in a receiver using differential quadrature phase shift keying (DQPSK) (for instance, receivers for PACS, Tetra, IS-136, PDC, ETSI-DAB use DQPSK [15, 16]). Figure 6.4 or Figure 6.5 is redrawn in Figure 6.28a, with I, Q paths explicitly shown. Although ideally there are no leakage of image and crosstalk between the I and Q paths, in practice finite imbalance between the two paths does result in leakage and crosstalk [16]. This imbalance is mainly due to imperfect matching of passive and active components in the two paths, as well as LO I/Q imbalance.

Main problems with I/Q path mismatch in IF digitizers are [16, 17]:

1) Undesired image results in a leakage component that occurs in the signal channel.

   In low-IF systems, I/Q mismatch is important since the image channel is close to the frequency of the desired channel and often cannot be removed with prefiltering. Typically the image channel may be 40 dB higher than the desired channel. With I/Q mismatches this results in an image rejection of 40 dB. Thus the desired signal to undesired signal ratio is around 0 dB. If the image channel interferer looks like noise, this means the SNR is around 0 dB. For acceptable BER SNR typically should be larger than 10dB. Consequently, an image rejection circuit is needed to improve the image rejection by this additional 10dB.

   Referring to Figure 6.28a the output Y of the quadrature modulator is

$$Y = \hat{I} + j\hat{Q} \tag{6.52}$$

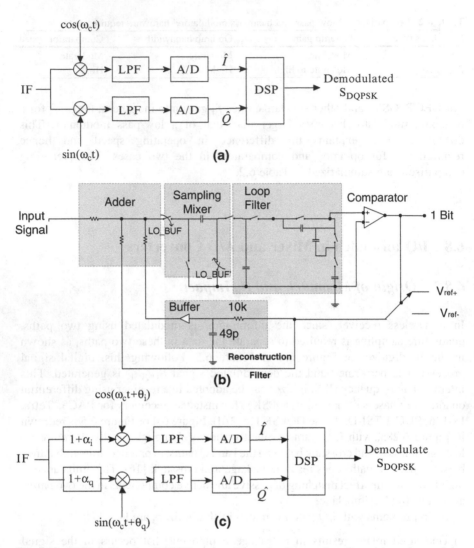

Fig. 6.28 (a) IF digitizer in a DQPSK receiver (b) one implementation of part (a), used to explain source of mismatches (c) gain and phase mismatch in part (a) explicitly shown

Sigma-delta modulators are subject to mismatch errors in the input and feedback circuits. To see this, let us now return to a specific implementation of the sigma-delta modulator architecture in Figure 6.5, which is used to realize the I, Q paths of Figure 6.28a. This implementation is shown in Figure 6.28b. Then the input and feedback circuits circuits include input resistors [7], sampling mixers as discussed in chapter 5, LO buffers to buffer VCO, which will be discussed in chapter 7, capacitors and switches in the loop filters, feedback resistors, and 1-bit feedback DACs. It is assumed that errors introduced in circuits inside the loop are suppressed by the loop gain and consequently are less important. Thus, for

Table 6.3 Notations to
Define Various I, Q
Components and Angles

Notations	Definitions
$I, Q, V, |V|, \theta$	Baseband I, Q components of transmitted symbol, vector representation and its magnitude, angle with x-axis
$\hat{I}, \hat{Q}, \hat{V}, |\hat{V}|, \hat{\theta}$	Baseband I, Q components of received symbol, vector representation and its magnitude, angle with x-axis
$\hat{I}_{2nd\_IF}(n)$, $\hat{Q}_{2nd\_IF}(n)$	I path and Q path received signals at 2nd IF (sampled and digitized); n is the sampling index
$\hat{I}_{2nd\_IF\_symbol}(n)$, $\hat{Q}_{2nd\_IF\_symbol}(n)$	A symbol period long of $\hat{I}_{2nd\_IF}(n)$, $\hat{Q}_{2nd\_IF}(n)$
$\hat{I}_{transform}, \hat{Q}_{transform}, \hat{V}_{transform}$	Baseband I, Q components of transformed received symbol and the vector representation.
$\hat{I}_{process}, \hat{Q}_{process}, \hat{V}_{process}$	Baseband I, Q components of processed received symbol and the vector representation.
$\Delta\theta, \Delta\hat{\theta}, \Delta\hat{\theta}_{scheme}$	Differential angle of transmitted dibit, received dibit, received dibit with scheme
$\hat{\theta}_{mis}$	Angle deviation in received symbol from the bisection line of the quadrant in which symbol is in, due to mismatch
$\Delta\hat{\theta}|_{impair}$, $\Delta\hat{\theta}_{scheme}|_{impair}$	Differential angle of received dibit under impairment such as noise, received dibit under impairment with scheme applied
$\Delta\hat{\theta}_{error}, \Delta\hat{\theta}_{error\_scheme}$	$\Delta\hat{\theta}|_{impair} - \Delta\theta, \Delta\hat{\theta}_{scheme}|_{impair} - \Delta\theta$

example, a mismatch between the ON-impedance of the switches in the loop filter is considered to be less important, when compared to the input switches. Matching between switches can be improved at the expense of chip area and proper layout. Problems associated with these matching techniques are parasitic delay and cross-coupling between the matched switches. The input resistors for the input adder are outside the loop. Therefore extra delay introduced does not affect stability. Thus the layout of the input resistors can be optimized for matching. On the other hand, resistors and capacitors in the reconstruction filters, which are in the feedback paths, through their parasitic, introduce extra phase shift and this could cause instability. This is especially important as sampling rate increases, although it can be alleviated somewhat through subsampling. These mismatches modify Figure 6.28a to become Figure 6.28c. The mismatch error has been modeled with two unity gain blocks that have a gain mismatch α and a phase mismatch θ, resulting in an overall mismatch of Δ. Because of the mismatch, a leakage component occurs in the image band that can be 40dB down relative to the signal carrier. In mathematical terms, the output signal can be approximated by

$$Y = (I + jQ) + \Delta(I - jQ) \tag{6.53}$$

where the term (I+jQ) represents the desired signal spectrum, and the term Δ(I−jQ) is the error spectrum due to themismatch. These two spectra are each other's mirror relative to dc. Accordingly signals and noise in the positive frequency (signal band) leak in the negative frequency (image band).

2) Crosstalk is created [16].

For ease of explanation we use QPSK signal (other signals, such as 8-phase DPSK, are explained in [27]):

$$S_{QPSK} = I \cos \omega_c t - Q \sin \omega_c t \qquad (6.54)$$

where I and Q are the baseband I, Q components of the transmitted symbol, with vector representation V and magnitude V.

Table 6.3 summarizes the notations in this section. ω_c is the carrier frequency. As discussed above, mismatches are explicitly shown in Figure 6.28c, whereby, upon quadrature mixing, and following Figure 6.28c, the resulting I, Q path signals are given as $(1 + \alpha_i)[S_{QPSK} \cos(\omega_c t + \theta_i)]$, $(1 + \alpha_q)[S_{QPSK} \sin(\omega_c t + \theta_q)]$, respectively. Collecting only the baseband terms we have:

$$\hat{I} = (1 + \alpha_i)[I \cos \theta_i + Q \sin \theta_i] \qquad (6.55)$$

$$\hat{Q} = (1 + \alpha_q)[-I \sin \theta_q + Q \cos \theta_q] \qquad (6.56)$$

The symbol "∧" in $\hat{I}, \hat{Q}$, $\hat{V}$, $|\hat{V}|$ designate that they are received symbols. Since Q now appears in the $\hat{I}$ expression and I now appears in the $\hat{Q}$ expression, there is crosstalk. Similar reasoning explains how crosstalk occurs with a DQPSK signal, which is expressed as:

$$S_{DQPSK} = \cos(\omega_c t + \theta_k + \pi/4) \qquad (6.57)$$

Here $\theta_k = \theta_{k-1} + \Delta\theta$; k (=2) is the symbol index and $\Delta\theta$ is the differential angle of the transmitted dibit. $\Delta\theta$ changes the phase shift of the carrier.

6.8.2 Techniques to Combat Problems Due to Mismatch

Past techniques have been proposed to combat both problem (1): image leakage and problem (2): crosstalk, as well as (1) or (2) alone.

In [17] a DEM based approach has been proposed to combat mainly problem (1). Hardware has been built to operate the scheme under practical impairments. It works by swapping the two paths, depending on the relative sign of the I, Q sigma delta modulators. Specifically the two 1-bit output of the sigma-delta modulator are compared. Depending on whether they are same or not, swapping is performed. With a 1-bit output, comparison is easily performed using an EXOR gate. Swapping is done on a per bit basis (bit timing). The DEM scheme is simple in implementation but is applicable to IF digitizer using sigma-delta modulator with 1-bit quantizer only.

Compensation based approaches have been proposed to combat problem (1) and (2) in the IF digitizer part of the receiver as well as the entire receiver itself ([18]–[25]). These typically involve estimating the gain and phase error, with a calibration loop. The circuitry in the receiver are made programmable in gain and phase, and they are adjusted accordingly. This is an adaptive process, and some kind of adaptive signal processing, like least mean square method (LMS) is adopted so that the gain and phase are adjusted to compensate for the mismatch. Because it is adaptive in nature, compensation is not instantaneous, but a certain convergence time is required. In general, they are complex and involve doing up to eight full multiplications per sample. In [18] only simulations (but no hardware) were used to test the operation of the scheme for QPSK, GMSK and 16-QAM modulation. [18] measures imbalances using a self-generated test-tone and corrects them at start-up. It requires the generation of a signal with a frequency of four times the local oscillator frequency and possibly an extra filter to suppress spurious tones. [19]-[21] uses LMS based adaptive filter to correct for imbalances for 16-QAM and QPSK modulation. [19] uses an LMS based adaptive filter to compensate the mismatch in the IF digitizer part of the receiver. In [20] and [21] simulations and hardware for compensation were used to test the operation of the scheme. [20] and [19] are similar but [20] compensates the mismatch in the receiver (of which the IF digitizer is a part of) and the implementation for compensation is complicated, involving a 32-bit floating point signal processor operating at 33MHz. [21] uses an analog-digital adaptive image-reject technique to compensate the mismatches. It compensates the mismatch in the receiver and the implementation for compensation is complicated, involving D/A converter and digital multiplier. However both [20], [21] do not need training signal and on-line correction is possible. [22] reports its own calibration has some restriction. [23]'s calibration needs an external image tone. [24] needs training signal. [25] focuses on the IF digitizer part, and uses sign-bit only in the compensation algorithm, which can potentially be unstable. This is summarized in the Table 6.4.

Finally dynamic quadrant swapping approach [26, 27] has been proposed to combat mainly problem (2). Like dynamic quadrant swapping reported in [17], this also performs swapping on a dynamic basis. However unlike [17], it swaps the received vector, rather than the 1-bit output from the sigma-delta modulator. Thus it swaps on a per symbol, rather than per bit basis. Secondly since the whole symbol is swapped, no assumption is made about the ADC architecture. Therefore it can be applied to ADC other than sigma-delta modulator, for example, pipelined ADC as reported in [8]. Here the receiver architecture is assumed such that for the IF under consideration the preceding channel filter suppresses the image sufficiently so that problem (1) is no longer the main problem, although the technique also corrects for some of the image leakage associated with problem (1) too. [26, 27] proposes a

Table 6.4 Comparison of I/Q mismatch reduction schemes, "—" means "not reported"

| Class | ref | f_{in} | Complexity of approach | | | Algorithm/ Implement | Source Of I/Q mismatch | IRR (dB) | Time to converge | Modulation/ Access/ IEEE standards |
			area (mm²)	CMOS (um)	Components					
Dynamic swap	[27]	10–160M	9×10^{-4}	0.09	6 12-bit registers, 1 6-bit counter	Dynamic quadrant swap/ digital	IF digitizer	65@ 10M	none	DQPSK/TDMA/ PACS, Tetra, IS-136, PDC
	[17]	10 MHz	1×10^{-3}	0.35	1 EXOR, 1 INV, 2 DAC, 2 Choppers	Dynamic element matching/ digital	IF digitizer	63	none	Applicable to sigma-delta modulator only
Compensation	[21]	6 MHz	0.45	0.18/FPGA	DAC, digital multiplier	mixed	receiver	54	~ 600 Symbol periods	CATV/16-PAM
	[22]	1.8GHz	3.6	0.35	IF multiplier, variable delay-gain cell	analog	receiver front end	59	—	—
	[23]	2.1GHz	1.2	0.25	DAC, variable-delay circuit	mixed	receiver front end	57	~ 15000 symbol periods	TDMA
	[24]	1 GHz	5×10^{-2}	0.09	Full digital multiplier	SAD, digital	receiver	55	~ ms	GSM
	[25]	1 MHz	0.21	0.18	Trim caps, sign detection circuits	Sign-bit only/ Mixed	IF digitizer	62–65	2^{20} samples	256-QAM
	[18]	13 MHz	—	—	Full multiplier	software	—	—	~ 20 Symbol periods	GSM
	[19]	—	—	—	Full multiplier	software	IF digitizer	—	~ 30000 s amples	GMSK
	[20]	1.9 GHz	—	—	Full multiplier	software	receiver	—	~1500 symbol periods	GMSK

scheme that is simple in implementation. For ease of explanation, let us follow 6.8.1, and use DQPSK as an example. The flowchart in Figure 6.29 explains the algorithm part. The notations are explained in Table 6.3.

Let us start by considering the case when dibit=01. With dibit=01 (I_1, Q_1) lies in the first quadrant and (I_2, Q_2) lies in the second quadrant. Applying this information to (6.55), (6.56) and we have the following equations for the two received $\hat{I}(t)$, $\hat{Q}(t)$:

$$\hat{I}_1 = (1 + \alpha_i)[\cos \theta_i + \sin \theta_i] \tag{6.58}$$

$$\hat{Q}_1 = (1 + \alpha_q)[\cos \theta_q - \sin \theta_q] \tag{6.59}$$

$$\hat{I}_2 = (1 + \alpha_i)[-\cos \theta_i + \sin \theta_i] \tag{6.60}$$

$$\hat{Q}_2 = (1 + \alpha_q)[\cos \theta_q + \sin \theta_q] \tag{6.61}$$

When $(\hat{I}_1, \hat{Q}_1)$, $(\hat{I}_{1\_2nd\_IF}, \hat{Q}_{1\_2nd\_IF})$ are applied to the flowchart, they correspond to the $(\hat{I}, \hat{Q})$, $(\hat{I}_{2nd\_IF}, \hat{Q}_{2nd\_IF})$ input to the quadrant detection box. The quadrant detection box takes $(\hat{I}_1, \hat{Q}_1)$, as represented by (6.58), (6.59), and detects the quadrant to be one. The quadrant detection box then states that we do not perform processing and just pass $(\hat{I}_1, \hat{Q}_1)$ as is. Therefore in this case $(\hat{I}_1, \hat{Q}_1)$ corresponds to the $(\hat{I}, \hat{Q})$ input to the $\Delta\theta$ detection box in Figure 6.29. Next, let us consider received symbol 2. When $(\hat{I}_2, \hat{Q}_2)$, $(\hat{I}_{2\_2nd\_IF}, \hat{Q}_{2\_2nd\_IF})$ are applied to the flowchart in Figure 6.29, they now correspond to the $(\hat{I}, \hat{Q})$, $(\hat{I}_{2nd\_IF}, \hat{Q}_{2nd\_IF})$ input to the quadrant detection box. The quadrant detection box takes $(\hat{I}_2, \hat{Q}_2)$, as represented by (6.60), (6.61), and detects the quadrant to be two. The quadrant detection box then states that we do perform processing. This is done by passing $(\hat{I}_{2\_2nd\_IF}, \hat{Q}_{2\_2nd\_IF})$ to the dotted box, to be processed. We henceforth denote the received symbol that comes out of the dotted box as the *processed symbol*, or $(\hat{I}_{process}, \hat{Q}_{process})$. $(\hat{I}_{process}, \hat{Q}_{process})$ is at baseband. In the present case $(\hat{I}_{process}, \hat{Q}_{process})$ correspond to. $(\hat{I}_{2process}, \hat{Q}_{2process})$. $\hat{V}_{2process}$ is the vector formed by $(\hat{I}_{2process}, \hat{Q}_{2process})$. Thus the input to the $\Delta\theta$ detection box is $(\hat{I}_{2process}, \hat{Q}_{2process})$, or $\hat{V}_{2process}$. Examining the dotted box, it is seen that processing consists of transforming followed by swapping. In the present case $(\hat{I}_{2nd\_IF}, \hat{Q}_{2nd\_IF})$ corresponds to $(\hat{I}_{2\_2nd\_IF}, \hat{Q}_{2\_2nd\_IF})$ and $(\hat{I}_{transform}, \hat{Q}_{transform})$ corresponds to $(\hat{I}_{2\,transform}, \hat{Q}_{2\,transform})$. Let us consider for the time being $\hat{I}_{2\_2nd\_IF}$. Transformation is done by first shifting $\hat{I}_{2\_2nd\_IF}$ by 90° (i.e. ¼ of the period of the second IF signal, denoted as T_{2nd_IF}). We then mix the shifted signal with a 2nd LO, at frequency ω_{c2} (since the signal is already in digital domain, mixing is done digitally), and then collect baseband terms. The result is denoted as $\hat{I}_{2\,transform}$. Physically the 90° shifting can be explained as follows. We first write $\hat{I}_{2\_2nd\_IF}$ explicitly as $\hat{I}_{2\_2nd\_IF}(n)$, where n, the sampling index, highlights

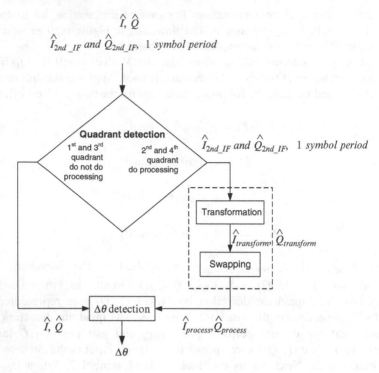

Fig. 6.29 Flowchart of algorithm

that $\hat{I}_{2\_2nd\_IF}$ is a sampled (and digitized) signal. Its sampling frequency is denoted as f_s i.e. its sampling period is T_s. 90° shifting means shifting n by $(1/4) \times (T_{2nd\_IF}/T_s)$ and the result is $\hat{I}_{2\_2nd\_IF}\left(n - \frac{1}{4}\frac{T_{2nd\_IF}}{T_s}\right)$. Next we consider $\hat{Q}_{2\_2nd\_IF}$. The above is repeated, except we shift by -90°. The result is denoted as $\hat{Q}_{2\ transform}$.

To see the effect of the 90° shifting, let us consider its impact on the I and Q path signals, starting from the 1st IF signal. In the present case the 1st IF signal corresponds to S_{2QPSK}. It is more convenient to represent the signal along the IF digitizer in continuous time domain i.e. keep treating the A/D converter as a gain block with gain one. Thus, without shifting, the I-path signal can mathematically be written as $(1 + \alpha_i)\left[S_{2QPSK}\cos(\omega_c t + \theta_i)\right]$ and the Q-path signal as $(1 + \alpha_q)$ $\left[S_{2QPSK}\sin(\omega_c t + \theta_i)\right]$. With 90° shifting they are modified to $(1 + \alpha_i)\left[S_{2QPSK}\cos(\omega_c t + \theta_i - 90°)\right]$, $(1 + \alpha_q)\left[S_{2QPSK}\sin(\omega_c t + \theta_q - 90°)\right]$, respectively.

To derive $\hat{I}_{2\ transform}$ and $\hat{Q}_{2\ transform}$ mathematically, we start from (6.54). With I (t) = I_2= -1 and Q(t) = Q_2 = 1, we have $S_{2QPSK}(t) = (-\cos \omega_c t) - \sin \omega_c t$. We substitute this for S_{2QPSK} in the above modified mathematical representations. We then mix and collect only the baseband terms, and we obtain:

$$\hat{I}_{2transfprm} = -(1 + \alpha_i)[\cos \theta_i + \sin \theta_i] \qquad (6.62)$$

$$\hat{Q}_{2transform} = (1 + \alpha_q)[\cos\theta_q - \sin\theta_q] \tag{6.63}$$

Next we take the negative of each of $\hat{I}_{2\,transform}$ and $\hat{Q}_{2\,transform}$, swap them:

$$\hat{I}_{2process} = -\hat{Q}_{2transfprm} = -(1 + \alpha q)[\cos\theta_q - \sin\theta_q] \tag{6.64}$$

$$\hat{Q}_{2process} = -\hat{I}_{2transform} = (1 + \alpha_i)[\cos\theta_i + \sin\theta_i] \tag{6.65}$$

Remember in this case $(\hat{I}_{2process}, \hat{Q}_{2process})$ corresponds to the $(\hat{I}_{process}, \hat{Q}_{process})$ input to the $\Delta\theta$ detection box in Figure 6.29. With $(\hat{I}_1, \hat{Q}_1)$ and $(\hat{I}_{2process}, \hat{Q}_{2process})$ applied to the $\Delta\theta$ detection box, $\Delta\theta$, at the output of this box, has its cosine given by:

$$\cos(\Delta\theta) = \frac{\hat{I}_1 \cdot \hat{I}_{2\,process} + \hat{Q}_1 \cdot \hat{Q}_{2\,process}}{|\hat{V}_1||\hat{V}_{2\,process}|} \tag{6.66}$$

Substituting (6.58), (6.59), (6.64), (6.65) in (6.66), the numerator of (6.66) becomes:

$$\begin{aligned} numerator &= -(1 + \alpha_i)(1 + \alpha_q)(\cos\theta_i + \sin\theta_i)(\cos\cos\theta_q - \sin\theta_q) \\ &+ (1 + \alpha_i)(1 + \alpha_q)(\cos\theta_i + \sin\theta_i)(\cos\theta_q - \sin\theta_q) \\ &= 0 \end{aligned} \tag{6.67}$$

Thus $\cos(\Delta\theta)$ is identically zero for arbitrary θ_i and θ_q. A $\Delta\theta$ of $90°$, the correct value, is detected for all mismatch values.

The application of the algorithm to the case for other dibits can be explained similarly.

6.9 Appendix: Low-Voltage Low-Pass Modulator

So far in this chapter we have achieved low-power dissipation in a sigma-delta modulator via the use of a passive loop filter. Another approach is to minimize the power supply voltage V_{dd}. Various low voltage designs have been proposed [28, 29]. Among them, the one discussed in [29] achieves low voltage ($V_{dd} = 1.95$ V) operation by using a new architecture called a local feedback loop. Here a high order modulator is stabilized by a local feedback loop around each integrator. Unlike multistage architecture, this architecture becomes very tolerant of the modest gain from low-voltage op-amps. We spend the rest of this appendix discussing this approach.

In stabilizing typical high-order, single-stage, single-bit modulators by decreasing integrator gains, an undesirable effect is that their dynamic range is reduced. The stable multistage (MASH) modulator with first- or second-order individual

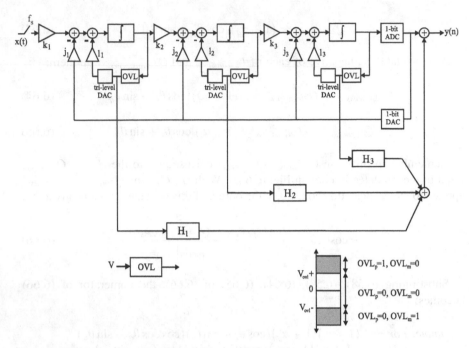

Fig. A.1 Block diagram of low-voltage sigma-delta modulator

stages requires high gain op-amps to achieve the necessary quantization noise cancellation. High-gain cascode op-amps are not attractive choices at low supply voltage because of the further loss in output swing. Although high-gain op-amps in low supply voltage can be realized with cascaded stages, they tend to consume more power due to the larger number of current branches. With this new architecture, it can tolerate low gain op-amps, which are readily realized at low supply voltage and hence achieve low power dissipation.

Figure A.1 shows a third-order example of this low-voltage modulator scheme [4]. Apart from the three local feedback loops (each consisting of an overload detector [OVL], a local tri-level DAC, and scale factors [l_1 or l_2 or l_3]) and the three digital compensators H_1, H_2, H_3, the basic circuit is a third-order single-stage modulator (i.e. the same as in Figure 6.13, with the same stability problem). When the integrators operate within their normal range (-V_{ovl}, V_{ovl}), the OVLs output 0 (Figure A.1) and therefore both the local feedback loops and the digital compensators are not activated. In essence, the modulator operates just like an ordinary third-order single-stage modulator that is very tolerant of modest op-amp gain and component mismatch but is prone to instability, particularly at high input levels. If an exceedingly large integrator level is detected by any of the

Table A.1 Performance of local feedback low voltage modulator

Bandwidth	Resolution	Power	Distortion (HD$_3$)	V_{dd}	Technology
8 kHz	12 bit	0.34 mW	$< -95\text{dB@}-5\text{dBm}$	1.95 V	1.2 um CMOS

OVLs, the corresponding integrator is forced back to the normal operating range ($-V_{ovl}$, V_{ovl}) via the tri-level DAC. Note that by activating this local feedback loop we have injected an extra signal into the basic third-order loop, thereby degrading the SNR. Therefore, an associated digital compensator (H_1, H_2, H_3) is introduced and designed such that its transfer function matches that of the transfer function corresponding to the path from this tri-level DAC to $y(n)$. By doing so, the extra signal injected into the basic third- order loop is canceled digitally at the final output and the original SNR of the basic modulator is restored.

A tri-level DAC can be realized with good linearity in a differential architecture. The digital compensators can also be approximated in the baseband by the simple FIR filters:

$$H_1(z) = \frac{l_1}{j_1} z^{-3} \tag{A.1}$$

$$H_2(z) = \frac{l_2}{j_1 k_2} z^{-2} \left(1 - z^{-1}\right) \tag{A.2}$$

$$H_3(z) = \frac{l_3}{j_1 k_2 k_3} z^{-1} \left(1 - z^{-1}\right)^{-2} \tag{A.3}$$

The accuracy of this digital compensation does depend on the matching between the analog modulator and the digital compensators as in a MASH architecture. However, since in reality the OVLs are invoked relatively infrequently, the requirement on the matching is relaxed. Therefore, simple op-amps (without cascode devices or cascaded stages) having only modest DC gain, but readily realized even at low supply voltage, can be used as in ordinary single-stage modulators without the inherent stability problem.

In general, the modulator k, j, and l coefficients have to satisfy certain relationships that are sufficient (but not necessary) conditions for stability. Practically, however, some of these criteria can be relaxed to accommodate easier implementation and better SNR performance without jeopardizing the modulator stability. One of the key issues is to select the coefficients such that most of the overloadings for high input level would occur at OVL$_2$ and OVL$_3$ rather than at OVL$_1$. Such a modulator has been implemented with measured performance [4] as summarized in Table A.1.

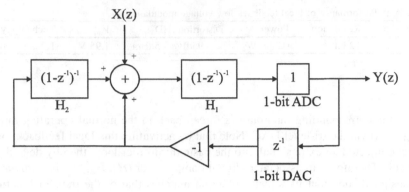

Fig. P.6.1 Modified second order sigma-delta modulator

6.10   Problems

6.1 In this problem we investigate the time domain response of a sigma delta modulator. In the following cases, assume that the input V_{in} consists of a ramp that goes from 0 to V_{FS} in 50 clock cycles (assume $V_{FS} = 1$ V). Plot V_{out} for these 50 clock cycles in the following cases:

(a) A first-order sigma-delta modulator with a 2-bit internal quantizer
(b) A second-order sigma-delta modulator with a 1-bit internal quantizer
(c) A second-order sigma-delta modulator with a 2-bit internal quantizer

6.2 Find the signal transfer function (STF) and noise transfer function (NTF) of a modified second-order sigma-delta modulator. This modulator has a loop filter in the forward and feedback path (shown in Figure P.6.1). Note that $X(z)$ is the input and $Y(z)$ is the output.

6.3 Show that the total quantization noise power in the band of interest $(-f_N/2, f_N/2)$ of an L th-order low-pass sigma-delta modulator, assuming a white noise model, is given by

$$n_{bw}^2 \approx \frac{\pi^{2L}}{2L+1} \left(\frac{f_N}{f_s}\right)^{2L+1} e_{rms}^2$$

where L is the order of the sigma-delta modulator. Here f_N is the Nyquist frequency, e_{rms}^2 is the quantization noise power of the quantizer inside the modulator, and the oversampling ratio (OSR) $= f_s/f_N$ is assumed to be much larger than 1.

6.4 The block diagram of a cascaded modulator, in which a second-order sigma-delta modulator is cascaded with a first-order modulator, is shown in Figure P.6.2. The blocks labeled z^{-1} are delay blocks. Assume that the quantization noise of the internal ADCs can be modeled as additive white noise and that the

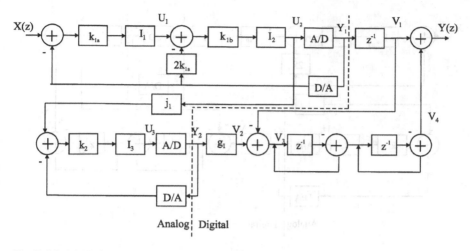

Fig. P.6.2 Multi-Stage sigma-delta A/D converter

integrators have a transfer function of $\frac{z^{-1}}{1-z^{-1}}$. Assume D/A is modelled with gain 1 (and no delay), just like in the chapter. Assume NTF and STF for the individual modulators are the same as in the chapter (i.e. $k_{1a}k_{1b}\text{gain}_{A/D}=1$).

(a) Derive an expression for the output $Y(z)$ in terms of the input $X(z)$ and the quantization noise of the two ADCs, $E_1(z)$ and $E_2(z)$. Simplify the final result for the case when $g_1j_1k_{1a}k_{1b} = 1$ (perfect matching).

(b) What is the increase in the quantization noise of the output due to the presence of a mismatch characterized by the factor $\delta = 1 - g_1j_1k_{1a}k_{1b}$? Express your answer in terms of the following parameters: $\sigma_\delta, g_1, OSR, \sigma_{Q1}, \sigma_{Q2}$, where σ_δ is the standard deviation of δ and σ_{Q1}^2, σ_{Q2}^2 are the quantization noise power of the two internal quantizers, respectively.

6.5 For the architecture shown in Figure P.6.2, evaluate an expression for the increase in quantization noise due to the finite gain of the first integrator's op-amp. The transfer function of the integrator with finite op-amp gain is

$$I_1(z) = \frac{z^{-1}}{(1 - z^{-1})(1 + \mu) + \alpha\mu}$$

where $\mu = \frac{1}{A}$, A is the gain of the op-amp, and α is the ratio of the sampling capacitor to the integrating capacitor.

6.6 For the third-order interpolative sigma-delta modulator shown in Figure P6.3 the transfer function of the integrators is $I(z) = (z - 1)^{-1}$. Derive the STF and NTF.

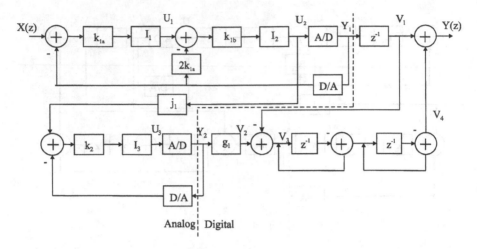

Fig. P.6.3 Third-order interpolative sigma-delta modulator

6.7 Repeat the design of the low pass sigma-delta modulator given in Design Example 6.1, except this time we have the following specs: $f_{if} = 20$ MHz, $f_{bw} = 4$ kHz, DR $= 78$ dB, power supply $= 5$ V. $V_{FS} = 2$V.

6.8 In Figure 6.27 of Design Example 6.2 we have indicated a block-level diagram implementation of a resonator in a BP modulator. The resonator has transfer function $Gz^{-2}/(1-z^{-2})$, where G is the gain of the resonator, using two integrators whose transfer functions both equal $z^{-1}/(1-z^{-1})$.

   (a) Using integrators whose transfer function are $z^{-1/2}/(1-z^{-1})$, show a block-level diagram implementation of the same resonator.
   (b) Modify the integrator as shown in Figure 6.14 so that it has the new transfer function $z^{-1/2}/(1-z^{-1})$ and show the new circuit schematic. Show clearly the clock phasing in the circuit schematic.
   (c) Translate the block level diagram obtained in (a) into a circuit schematic. The circuit should use the integrator obtained in (b). Show clearly the clock phasing in the circuit schematic.

6.9 Repeat the design of the BP sigma-delta modulator as given in Design Example 6.2, except this time use the following specs: $f_{if} = 20$MHz, $f_{bw} = 4$ kHz, DR $= 78$ dB, power supply $= 5$ V, $V_{FS} = 2$ V, tolerable reduction in DR $= 5$ dB. In section 6.7.2.1 we stated that for $L = 4$, in order to achieve a DR reduction of less than 5 dB, the fractional change in f_o ($\Delta f_o/f_o$) must be less than 0.2%. In the present problem (which may involve a different L), what is the tolerable fractional change in f_o? (*Hint*: Instead of doing simulation, you may find this by comparing the NTFs of the two BP modulators.) This new fractional change should be used in the problem for proper calculation.

References

1. Y. Manoii, "Reconfigurable ADCs for 3.xG and 4G," ISSCC 2009 Forum F3.
2. K. C. H. Chao, S. Nadeem, W. L. Lee, and C. G. Sodini, "A Higher Order Topology for Interpolative Modulators for Oversampled A/D Conversion," *IEEE Trans. Circuits Sys.*, vol. CAS-, pp. 309–318, March. 1990.
3. W. Chou, P. W. Wong, and R. M. Gray, "Multistage Sigma-Delta Modulation," *IEEE Trans. Inform. Theory*, vol. IT-35, pp. 784–796, July 1989.
4. S. Au and B. Leung, "A 1.95-V, 0.34-mV, 12-b Sigma-Delta Modulator Stabilized by Local Feedback Loops," *IEEE Journal of Solid-State Circuits*, vol. JSSC-32, pp. 321–328, March 1997.
5. F. Chen and B. Leung, "A 0.25 mW Low Pass Passive Sigma-Delta Modulator with Built-In Mixer for a 10-MHz IF Input," *IEEE Journal of Solid-State Circuits*, vol. JSSC-33, pp. 774–782, June 1997.
6. Ong and B. Wooley, "A 2-Path Band Pass Sigma-Delta Modulator for Digital IF extraction at 20 MHz," *IEEE Journal of Solid-State Circuits*, vol. 32, pp. 1920–1934, December 1997.
7. Namdar, B. Leung, "A 400 MHz 12-Bit 18 mW IF Digitizer with Mixer Inside a Sigma-Delta Modulator Loop," *IEEE Journal of Solid State Circuits*, Vol. 34, pp. 1765–1777, December 1999.
8. Y. Ren, B. Leung, and Y.M. Len, "A Mistmatch Independent DNL Pipelined Analog to Digital Convernter," *IEEE Transactions on Circuits and System II*, Analog and Digital Signal Processing, vol. 46, no. 6, pp. 699–704, June 1999.
9. T. S. Rappaport, *Wireless communications : principles and practice*, Prentice Hall, 1996, problem 5.23.
10. Leung, "Oversampled A/D Converter," Chapter 10 (pp. 467–505), *Analog VLSI; Signal and Information Processing*,. M. Ismail and T. Fiez, McGraw-Hill, 1993.
11. Roubik Gregorian, and Gabor C. Temes, *Analog MOS Integrated Circuits for Signal Processing*, Wiley, 1986.
12. S. Sen and B. Leung, "A 150 MHz 13b 12.5 mW IF digitizer," *Proceedings of the IEEE 1998 Custom Integrated Circuits Conference*, 1998, pp. 233–236.
13. James C. Candy and Gabor C. Temes, Oversampled Delta-Sigma Data Converters: Theory, Design and Simulation. IEEE Press, 1992.
14. J. Dunlop, D. Girma and J. Irvine, "Digital mobile communications and the Tetra system", Wiley 1999.
15. T. Rappaport, "Wireless communications, Principles and Practice", Prentice Hall, 2002, Ch. 11.
16. L. Liu, "Impacts of I/Q imbalance on QPSK-OFDM-QAM detection," IEEE Transactions on Consumer Electronics, p. 984–989, August 1998.
17. Lucien J. Breems, E. Carel Dijkmans, and Johan H. Huijsing, "A quadrature data-dependent DEM Algorithm to improve image rejection of a complex sigma delta modulator", IEEE Journal of Solid State Circuits, vol 36, p. 1879–1887, Dec, 2001.
18. J. Glas, "Digital I/Q imbalance compensation in a low-IF receiver," IEEE Global Telecomm., Conf., Nov. 1998, p. 1461–1466.
19. K. Pun, J. Franca, C. Azeredo-Leme, "The correction of frequency-dependent I/Q mismatches in quadrature receivers by adaptive signal separation", [ASIC], 2001, pp. 424–427.
20. L. Yu, W. Snelgrove, "A Novel Adaptive Mismatch Cancellation System for Quadrature IF Radio Receivers", IEEE Trans. Circuits Syst. II, pp. 789–801, June 1999.
21. M. Hajirostam, K. Martin, "On-chip image rejection in a low-IF CMOS receiver", IEEE international solid state circuit conference [ISSCC], 2006, pp. 457–458.
22. M. Elmala, S. Embabi, "Calibration of Phase and Gain Mismatches in Weaver Image-Reject Receiver", IEEE Journal of Solid State Circuits, p. 283–289, February 2004.
23. L. Der, B. Razavi, "A 2-GHz CMOS image-reject receiver with LMS calibration," IEEE Journal of Solid State Circuits, p. 167–175, Feb., 2003.

24. Elahi, K. Muhammad, P. Balsara, "I/Q mismatch compensation using adaptive decorrelation in a low-IF receiver in 90-nm CMOS process" IEEE Journal of Solid State Circuits, p. 395–404, Feb., 2006.
25. S. Lertavessin, B. Song, "A complex image rejection circuit with sign detection only", IEEE international solid state circuit conference [ISSCC], 2006, pp. 454–456.
26. Y. Lai, N. Lam, B. Leung, "A 10MHz IF Digitizer Using a Novel Quadrant Based Swapping Scheme for I,Q Mismatch Elimination That Achieves an Equivalent 65dB Image Rejection Ratio", European Solid-State Circuits Conference, paper 32.5, p.591–594, September 2002.
27. N. Lam, B. Leung, "Dynamic Quadrant Swapping Scheme Implemented in a Post conversion Block for I,Q Mismatch Reduction in a DQPSK receiver", *IEEE Journal of Solid State Circuits*, Vol. 45, pp. 322–337, February, 2010.
28. Grilo, E. MacRobbie, R. Halim and G. Temes, "A 1.8 V 94 dB Dynamic Range Sigma Delta Modulator for Voice Applications", 1996 IEEE International Solid-State Circuits Conference, *Digest of Technical Papers*, 1996, pp. 230–231, 451.
29. S. Au and B. Leung, "A 1.95 V, 0.34 mW 12-Bit Sigma-Delta Modulator Stabilized by Local Feedback Loops," *Proceedings of the IEEE 1996 Custom Integrated Circuits Conference*, 1996, pp. 411–414.

Chapter 7
Frequency Synthesizer: Phase/Frequency Processing Components

7.1 Introduction

As discussed in Chapter 2, a receiver front end consists of two blocks: a signal conditioning block and a signal controlling block. We have discussed the components needed for the signal conditioning block in the last few chapters: how the LNA amplifies the small RF input signals, the mixer mixes down the high-frequency RF input signal, and the A/D converter digitizes the mixed-down signal. In addition, we have assumed that someone is going to generate the LO (local oscillator) signal for the mixer. It turns out that the requirement on this signal is not trivial, as discussed in the Chapters 4 and 5. This is because any impairment on this LO signal's integrity, such as noise or finite fall time of the clock edge, will adversely affect the integrity of the mixed-down signal. Accordingly, a well-controlled oscillator is required to attain this requirement. Henceforth this subcomponent of the receiver front end is denoted as the frequency synthesizer. In addition, since there are multiple users in the system, we need a way of allowing each user access to the same shared transmission medium (air in this case). This multiple access control can be implemented through frequency, time, or code division multiplexing. In the case of frequency division multiplexing, the frequency synthesizer also carries out the access control function.

Traditionally there are two major ways of generating frequencies: direct and indirect approaches. The direct approach consists of generating a sine wave digitally (by storing the sine wave in a read only memory (ROM), for example) followed by digital to analog (D/A) conversion. Even though this approach enjoys the advantage of being mostly digital and is therefore easy to design, manufacture, and test, it needs a high-speed D/A converter, which can be the major bottleneck. In addition, it tends to consume high power and is therefore not considered further in this book. Instead we look at the indirect way of generating frequencies: the use of a controlled oscillator. The general idea is to control the frequency of oscillation via some external control input (usually voltage), and hence the resulting oscillator is called a voltage controlled oscillator (VCO). Since the noise property and frequency

B. Leung, *VLSI for Wireless Communication*, DOI 10.1007/978-1-4614-0986-1_7,
© Springer Science+Business Media, LLC 2011

stability of such a VCO is usually rather poor, it is typically put in a feedback loop to enhance these properties. Since the loop tracks the phase (and hence frequency), it is called a phase locked loop (PLL) and so the approach is denoted as a PLL-based frequency synthesizer. This approach is the main focus in this book.

For low system cost and portability, frequency synthesizers need to be integrated with an on-chip VCO. The difficulty of integrated VCO lies in its poor frequency stability compared with external high-Q resonator-based VCO. The phase noise of the integrated VCO also contributes directly to reducing the signal-to-noise ratio (SNR) of the receiver front end. The phase noise (which we define more precisely later) problem can be alleviated partially by broadbanding the PLL. Essentially, the feedback action allows the PLL to suppress the inband VCO phase noise by high-pass filtering it.

This chapter begins with a discussion of a typical integrated frequency synthesizer, followed by some of its subcomponents: namely, the phase detector (PD), divider, and VCO. These subcomponents are new in this book in the sense that they allow us to process the frequency/phase as opposed to the amplitude (either voltage/current) of a signal. We pay special attention to familiarizing readers with the components possessing this property. The nonidealities of these components will have an impact on the synthesizer. Again, we focus on these nonidealities' impact on the frequency/phase properties of the output signals from these subcomponents (e.g., spur from PD, phase noise from VCO). The loop filter, which can be looked at as the controller of the synthesizer, will be discussed in Chapter 8 as its role is to coordinate these other subcomponents to achieve the goal of frequency synthesizing with the least impairment.

7.2   PLL-Based Frequency Synthesizer

Most books on PLL [13,14] focus on aspects that cover the use of PLL in general applications, such as data acquisition and carrier recovery. This chapter concentrates on and highlights those aspects of PLL that are unique to frequency synthesizer application. For example, the use of phase frequency detector (PFD), although not suitable for bit extraction, is universally used in the frequency synthesizer, and is emphasized here, as is the use of a charge pump. Again, aspects that are unique to frequency synthesizer applications, but less relevant to other PLL-based applications, such as the divider, are given extensive treatment. Even in discussion of subcomponents usually found in a PLL book, such as VCO, the unique environment facing its design in the present application (namely, the large tuning range) is given special attention.

A typical PLL-based frequency synthesizer is shown in Figure 7.1 and contains a reference source oscillating at frequency f_r and a VCO oscillating at frequency f_o. The reference frequency is divided by an integer N and the VCO frequency is divided by M; the two divided waves are then compared in a phase detector. When the two phases are equal (phase locking), then $\frac{f_r}{N} = \frac{f_o}{M}$. This also means that the output frequency is locked to a rational fraction of the reference frequency. In essence, the synthesizer is

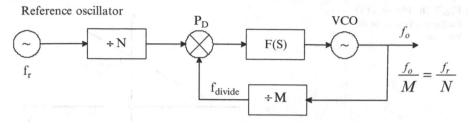

Fig. 7.1 PLL based frequency synthesizer

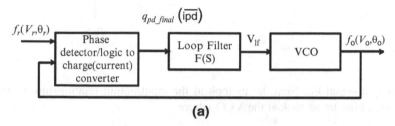

(a)

Fig. 7.2a Block diagram of a PLL

capable of generating a large number of highly accurate output frequencies. Frequency selection is achieved by changing the divider ratios M and N.

Since the frequency synthesizer is PLL based, a background discussion of the PLL is in order.

Figure 7.1 is now redrawn in Figure 7.2a to facilitate the explanation of the operating principle of a PLL. Notice that the dividers are omitted, but this does not have any major impact on our explanation.

In general, a PLL synchronizes an output signal (usually generated by an oscillator) with a reference signal in terms of its frequency and phase. In the locked state, the phase error between the output and reference signal is very small, as ensured by the negative feedback principle. When there is change in the input, output, or the internal parameters of the synthesizer itself (e.g., changing the divider ratio, if there is a divider), then a phase error is introduced. Means are devised whereby this error is converted to some form that can control and direct the oscillator output frequency and phase. Together with negative feedback this frequency and phase change occurs in such a way as to reduce the error. As shown, the PLL consists of the following subcomponents:

1. A phase detector (PD) and logic to charge converter
2. A loop filter (LF)
3. A voltage control oscillator (VCO)

The signals are as follows: reference signal V_r with frequency f_r and phase θ_r, output signal V_o (which also equals the VCO output signal) with frequency f_o, phase θ_o, phase detector output signal q_{pd_final} (a charge variable), phase error θ_e, and loop

Fig. 7.2b Plot of VCO frequency versus tuning voltage

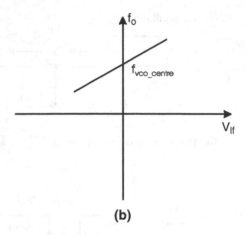

(b)

filter output signal V_{lf}. Next, let us look at the input/output relationships of these subblocks. First let us look at the VCO where

$$f_o = \frac{d\theta_o}{dt} = f_{vco-center} + K_{vco} \times V_{lf} \tag{7.1}$$

Here $f_{vco\text{-}center}$ is the center frequency of the VCO and K_{vco} is the VCO gain in units of deg/s/V. f_o is plotted as a function of V_{lf} in Figure 7.2b. Second, the phase detector compares the phase of the output signal, θ_o, with the phase of the reference signal, θ_r, and develops θ_e. θ_e turns on the logic to charge (current) converter and generates a current i_{pd} and starts to deliver charge q_{pd} to the capacitor in the loop filter $F(s)$. At the end of a reference signal cycle, the charge delivered, denoted as q_{pd_final}, is designed to be linearly proportional to phase error θ_e, as shown in Figure 7.2c and expressed as follows:

$$q_{pd\_final} = K'_{pd} \times (\theta_r - \theta_o) = K'_{pd} \times \theta_e \tag{7.2a}$$

Here K_{pd}' is the phase detector gain with units in coulomb/deg. The q_{pd_final}, when integrated on the capacitor of the loop filter, will generate a voltage V_{lf}, whose value is the integrated (sum) value of all the θ_e changes. The instantaneous value of V_{lf} is not proportional to the instantaneous value of θ_e, but rather the integrated value of θ_e. If the loop filter is a simple capacitor C, then

$$V_{lf}(n) = q_c(n)/C \tag{7.2b}$$

where n is the nth cycle of the reference clock. Here $V_{lf}(n)$, $q_c(n)$ are the voltage and charge on capacitor C at reference cycle n. Also,

$$q_c(n) = q_c(n-1) + q_{pd\_final}(n) \tag{7.3}$$

Fig. 7.2c Plot of charge
delivered and average current
in a phase detector versus
phase error

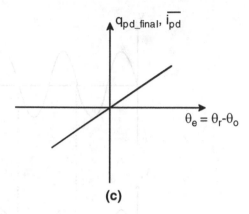

(c)

Alternatively, we can adopt another variable $\overline{i_{pd}}$, defined to be the average of i_{pd} over one reference clock cycle. This $\overline{i_{pd}}$ is also designed to be linearly proportional to θ_e, as shown in Figure 7.2c:

$$\overline{i_{pd}} = K_{pd}\theta_e \qquad (7.4a)$$

K_{pd} is the phase detector gain with units A/deg. (Note the difference between K'_{pd} and K_{pd}.) Again this current will flow into the loop filter and create a V_{lf}. Then we can write

$$V_{lf}(n) = \frac{i_{pd}}{C}\Delta T + V_{lf}(n-1) \qquad (7.4b)$$

Here ΔT is the period when the phase detector is on. Also, we have

$$V_{lf}(n) = \frac{\overline{i_{pd}}}{C}T + V_{lf}(n-1) \qquad (7.4c)$$

Here T is the period of V_r and equals $\frac{1}{f_r}$.

We are now in a position to explain qualitatively the operation of the PLL. Let us assume that initially f_r is equal to f_{vco_center} and the VCO also oscillates at f_{vco_center}. Therefore, θ_e is zero and q_{pd_final} ($\overline{i_{pd}}$) is zero. Assuming initially that V_{lf} is zero, this means that V_{lf} stays at zero and everything is in equilibrium. If θ_e is nonzero initially, then q_{pd_final} and hence V_{lf} would be nonzero after a delay. f_o will then deviate, forcing θ_e to settle back to zero. This is depicted in Figure 7.3 before time t_o. What happens when there is a change of frequency now? Let us assume that the reference signal V_r's frequency f_r is arbitrarily increased by Δf at t_o. As shown in Figure 7.3b, a higher frequency means a shorter period, and therefore the phase of V_r starts leading the phase of V_o. This nonzero phase error θ_e increases with time. After some delay, this increasing θ_e will lead to an increase in q_{pd_final} ($\overline{i_{pd}}$). This causes V_{lf} to increase, resulting in a change to f_o, as described in Figure 7.3c

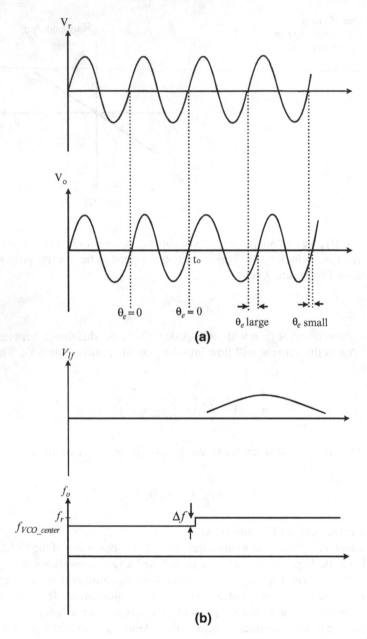

Fig. 7.3 Plot of internal voltage and frequency changes in a PLL when reference frequency changes. (**a**) Reference voltage. (**b**) VCO voltage. (**c**) Loop filter output voltage. (**d**) VCO frequency

and Figure 7.3d. Due to negative feedback, this leads to a decrease of θ_e. Eventually the output frequency f_o will be the same as f_r. Notice that in contrast to the change in the input phase case, depending on the order of the filter $F(s)$, the final θ_e may go back to zero or a finite value.

7.3   Phase Detector/Charge Pump

The purpose of the phase detector is to produce a signal that is proportional to the difference in phase between two signals. There are two main categories of phase detectors: analog and digital. In frequency synthesizers, almost all phase detectors are digital. The key characteristics are as follows:

1. Gain (sometimes associated with the output or charge pump stage)
2. Linearity
3. Steering characteristics

There are many different types. Of special interest are the following:

1. Tristate phase/frquency detector (PFD)
2. JK flip-flop based PD
3. EXOR gate based PD

For most PLL-based frequency synthesizers, the PFD is the preferred choice. In other applications, other types of PD, such as EXOR and JK based PD, can also be used. In this chapter we will concentrate on PFD but we will briefly compare it with another PD.

7.3.1   Phase Frequency Detector

Let us first look at an example of a PFD [14]. Figure 7.4 shows a PFD based on D-FF. (It should be noted that another realization also exists. A JK-FF based PFD is very popular and potentially dissipates less power.) The PFD differs greatly from the other type of phase detector. The major difference is the existence of a third state, which will be shown to lead to major advantage over other types of PDs. The PFD's output signal depends not only on phase error θ_e but also on frequency error $\Delta f = f_r - f_o$ before locking is acquired. The present PFD is built from two D-FFs, whose outputs are denoted Up and Down and whose state variable is denoted as d_{pd}. Since we have two storage elements, d_{pd} has four states:

1. Up $= 0$, Down $= 0$
2. Up $= 1$, Down $= 0$

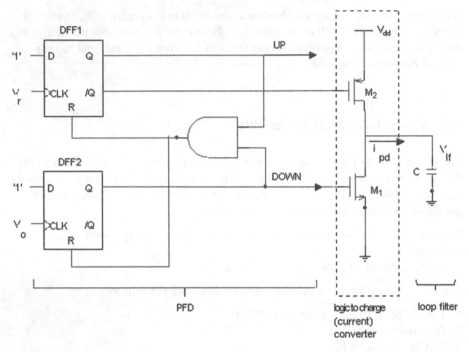

Fig. 7.4 Phase frequency detector based on D-FF

3. Up = 0, Down = 1
4. Up = 1, Down = 1

Since we only need a tri-state device, one of the states is unused. In this case we arbitrarily select that to be the fourth state. However, we want to ensure that if we accidentally end up in this unused state, it will not get stuck there (called lockout). This is guaranteed by ensuring that when the fourth state is entered the PFD will exit into one of the used states (arbitrarily picked to be the first state).

In actual hardware, this is achieved so that whenever the fourth state is detected, the AND gate in Figure 7.4 is activated and its output resets both flipflops back to the first state. Notice that this AND gate is also instrumental in determining the state transition. The state and d_{pd} are now defined as follows:

1. Up = 0, Down = 1, state = d_{pd} = −1
2. Up = 0, Down = 0, state = d_{pd} = 0
3. Up = 1, Down = 0, state = d_{pd} = 1
4. Up = 1, Down = 1, state = d_{pd} = inhibited

The actual state of the PFD is determined by the signals V_r, V_o, the two trigger (both positive edge trigger) signals applied to the two D-FFs. Figure 7.5 is the state diagram that describes the operation of the circuit as depicted in Figure 7.4. First let us go through how the circuit in Figure 7.4 implements the state diagram in

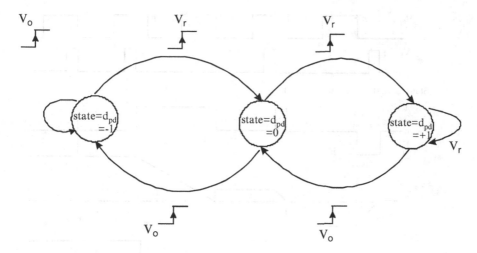

Fig. 7.5 State diagram of phase frequency detector

Figure 7.5 by going through a state transition sequence. Assume from Figure 7.5 that we are initially in the −1 state (i.e., Up = 0, Down = 1). Now from Figure 7.4, a positive transition of V_r will trigger the DFF1 [i.e., gating the D input (=1) of DFF1 to the output Q, which is also the Up signal]. Since $D = 1$, Q becomes 1 or Up changes from 0 to 1. Meanwhile, since V_o does not move DFF2's output of course remains the same as before (i.e., $Q = 1$ or Down stays at 1). In summary, the (Up, Down) pair moves from (0,1) to (1,1).

Notice that this (1,1) is the fourth or inhibited state. From Figure 7.4, since both Up and Down = 1, the AND gate is enabled and the reset (R) signal is high. Therefore, DFF1 and DFF2 are both resetted and (Up, Down) goes to (0,0) or d_{pd} becomes 0, just as depicted in Figure 7.5.

Next let us look at how the transition from state 0 to 1 occurs in Figure 7.5. Since V_r's edge rises again, Up (= Q of DFF1) goes from 0 (the reset state) to 1. Again, since V_o does not change, Down (= Q of DFF2) stays at 0. Therefore the state, or d_{pd} goes from 0 to 1, as shown in Figure 7.5. From symmetry the exact opposite will happen when we start from the 1 state. Here repeated activation of V_o will eventually take the PFD to the −1 state. To summarize, a positive edge of V_r forces the PFD to go into its next higher state, unless it is already in the 1 state. Similarly, a positive edge of V_o results in the PFD going into its next lower state, unless it is already in the −1 state.

We now go through an example and see how this circuit actually detects phase. Referring to Figure 7.6a, where events are labelled from 1 to 8, right next to the positive edge of V_r and V_o, imagine that originally the PFD is in the 0 state. When the V_o edge goes up (event 1), according to Figure 7.5, the state changes from 0 to −1. This is shown in the d_{pd} trace of Figure 7.6a. Then when the V_r edge rises (event 2), according to Figure 7.5 the state changes from −1 to 0. In Figure 7.6a this means that d_{pd} goes from −1 to 0. Hence d_{pd} goes from 0 to −1 and back to 0 again. Notice

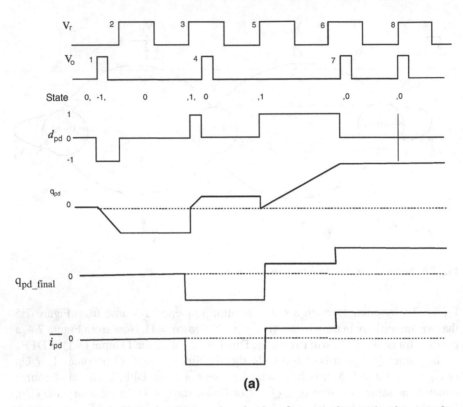

(a)

Fig. 7.6a An example illustrating the operation of a phase frequency detector (q_{pd} is assumed to be resetted at the beginning of each phase comparison; similarly q_{pd_final} is assumed not to accumulate between phase comparison)

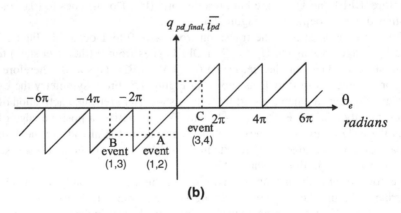

(b)

Fig. 7.6b Plot of charge delivered and average current in a phase frequency detector versus phase error

that the duration when d_{pd} stays in −1 measures the phase difference between V_o and V_r and therefore its value is linearly proportional to the phase difference θ_e between V_r and V_o. This is evident in that part of the d_{pd} trace that corresponds to event (1,2).

Now we will see how to generate q_{pd_final} $(\overline{i_{pd}})$ for event (1,2) and show that q_{pd_final} $(\overline{i_{pd}})$ is proportional to the duration when d_{pd} is in the '-1' state. First we reiterate that the signal d_{pd} is a logical variable that represents the PFD state and has three states. Most logical circuits used today generate binary signals $(1,-1)$; however, the third state $(d_{pd} = 0)$ can be substituted by a high-impedance state. The circuitry in Figure 7.4 within the dashed box shows how the three states are generated and how logic to charge (current) conversion is achieved. When the Up signal is high, the PMOS transistor M_2 conducts, so the phase detector current, i_{pd}, is positive (pushing) and the charge delivered by the phase detector to the capacitor C, q_{pd}, is positive. When the Down signal is high, the NMOS transistor M_1 conducts, i_{pd} is negative (sinking) and q_{pd} is negative. When neither signal is high, i_{pd} is zero (high impedance) and q_{pd} is zero. Hence the sign of q_{pd} and i_{pd} follows the same sign as d_{pd}. A trace of q_{pd} is shown in Figure 7.6a. Now let us assume that when M_1 or M_2 is conducting, i_{pd} is constant and whose value is denoted as I_{charge_pump}. Then if we look at the final charge delivered to C, or q_{pd_final}, it is obviously proportional to the time when M_1 or M_2 is on. Hence q_{pd_final} is proportional to the duration when the phase detector is on, which is denoted in (7.4b) as ΔT. For event (1,2), since q_{pd} is negative, q_{pd_final} is also negative. We can then write

$$q_{pd\_final} = -I_{charge\_pump}\Delta T \tag{7.5}$$

Next we can replace ΔT by $-\theta_e T/2\pi$, where T is the period of the reference signal V_r (a constant) and θ_e is the phase difference to be detected. There is a negative sign in front of θ_e because ΔT is always positive while θ_e for event (1,2) is negative. With this replacement (7.5) becomes

$$q_{pd\_final} = I_{charge\_pump}\frac{\theta_e T}{2\pi} \tag{7.6}$$

Looking at (7.6) again, we finally show that q_{pd_final} is proportional to θ_e, as we have stipulated in (7.2a), with

$$K'_{pd} = \frac{I_{charge\_pump}T}{2\pi} \tag{7.7}$$

Similarly, from definition

$$\overline{i_{pd}} = \frac{1}{T}\int_0^T i_{pd}dt$$

$$= \frac{1}{T}I_{charge\_pump}\frac{\theta_e T}{2\pi}$$

$$= I_{charge\_pump}\frac{\theta_e}{2\pi} \tag{7.8}$$

$\overline{i_{pd}}$ is shown in Figure 7.6a. From (7.8) it is seen that this $\overline{i_{pd}}$ is also proportional to θ_e. Comparing (7.8) with (7.4a) we have

$$K_{pd} = \frac{I_{charge\_pump}}{2\pi} \tag{7.9}$$

Note that

$$K_{pd} = K'_{pd}/T \tag{7.10}$$

The current sources M_1, M_2 are collectively denoted as logic to charge (current) converter, more popularly known as the charge pump. It serves to convert the logic variable d_{pd} to a signal variable $\overline{i_{pd}}$. Capacitor C (the simplest example of the loop filter) then converts $\overline{i_{pd}}$ to q_{pd_final}. Both the charge pump and the loop filter are discussed in further detail later.

Having explained how the charge pump generates q_{pd_final} ($\overline{i_{pd}}$) for the event (1,2), it is helpful to go back to Figure 7.6 to illustrate that it works equally well for other events. Now if we progress to event 3, V_r comes along first this time. According to Figure 7.5, the state goes from 0 to 1. On event 4 when V_o comes along, the state goes back to 0. Consequently, d_{pd} goes up and then down. Again it is seen that q_{pd_final} assumes a value that is proportional to the phase difference. Notice that in the present case, with a positive phase difference (defined as phase of V_r - phase of V_o), the PFD outputs a positive d_{pd}. Therefore, this PFD measures both the magnitude and the sign of the phase difference. This is highlighted in Figure 7.6b, which plots q_{pd_final} and $\overline{i_{pd}}$ as a function of θ_e. From Figure 7.6b it is seen that q_{pd_final} becomes largest when the phase error is positive and approaches 360°. Beyond 2π what happens? The PFD behaves as if the phase error recycled at zero. Why does this happen? An easy answer is that the state-machine whose state diagram is described in Figure 7.5 cannot distinguish between θ_e and $\theta_e + 360°$. An example of this is shown in Figure 7.6a, when we trace the value of d_{pd} during the events (1,2) and events (1,3). If we take a look at the phase error corresponding to these two events [as shown in Figure 7.6a], θ_e (1,3) = θ_e (1,2) + 360°. This is because edge 3 is exactly one period delayed from edge 2. Now if we look at d_{pd} in the same figure, we see that d_{pd} does not change as we move from edge 2 to edge 3. It stays at 0. Hence q_{pd_final} for event (1,2) = q_{pd} for event (1,3). If we look at Figure 7.6b, this corresponds to points A and B on the transfer curve. Hence θ_e differs by 360° but has the same q_{pd_final}. Basically, the PFD slips a cycle. Therefore, the q_{pd_final} versus θ_e curve as described in Figure 7.2c repeats itself (like a sawtooth) and the relationship is no longer linear (it is periodic). This is shown in Figure 7.6b. This nonlinear behaviour of PFD is denoted as cycle slipping, and hence the PLL is a nonlinear system. In this chapter we assume that we always work within the 360° region (which is true for a frequency synthesizer) and therefore we assume that the PLL is linear.

Other phase relationship possibilities are depicted in events 5-8, which are self-explanatory if we go through Figure 7.5. One interesting scenario is event 8, when V_r and V_o have exactly the same phase. This is an ill-defined situation, and what happens really depends on the internal delay of the logic. There will be some transients but d_{pd} will always settle back to state 0. When in lock, $V_r \approx V_o$. Imagine for the time being that V_r has the same frequency as V_o but is slightly leading in phase, d_{pd} therefore consists of a narrow pulse train at frequency f_r. This pulse train generates spurious components (spurs) and will have serious impact on the final spectral purity of the frequency synthesizer.

Finally, let us point out that the PFD is operating as an asynchronous (or fundamental) digital circuit and therefore all the problems in the asynchronous circuits will appear (like hazards) and will be a problem. Also, the setup time and the hold time need to be further investigated to see (1) if they will be violated and (2) if they have been violated, what impact they will have on the operation of the PFD. Fortunately, since the PFD for a frequency synthesizer typically works at rather low frequency, the preceding constraints do not usually pose a problem.

To highlight the advantages of the PFD, we next review another phase detector based on the combinational circuit: the EXOR gate.

7.3.2   EXOR Phase Detector

Figure 7.7a shows the EXOR gate-based PD [14], and Figure 7.7b depicts the waveform of the EXOR PD for different θ_e. Let us show how the EXOR PD differs from the PFD. First it should be noted that d_{pd} only has two states: -1 and 1, as opposed to the three states available in a PFD. Now from the waveform it can be seen initially that V_o's phase leads V_r's. Since V_r is at a higher frequency than V_o, eventually V_r's phase leads V_o. Notice that upon comparing events (1,2) and (3,4), d_{pd} is the same, which means that the EXOR gate cannot tell which edge is leading. Also from Figure 7.7c, q_{pd_final} is exactly 0 when θ_e is exactly $90°$ (it does not matter which wave is leading) rather than when $\theta_e = 0$. Comparing Figure 7.6a and Figure 7.7a, it is seen that d_{pd} changes twice as fast here. Therefore, following the same argument on PFD, when the PLL is in lock, the present PD will generate a spur that is at twice the frequency of f_r. The duty cycle of d_{pd} is exactly 50%.

Having highlighted the difference between the two detectors, let us go through some detailed explanations of the EXOR circuit by following through Figure 7.7b. Initially V_r is lagging V_o [event (1,2)], θ_e is negative, and d_{pd} stays in the -1 state more often than in the 1 state. Hence M_2 is on more often than M_1. Therefore, more charge is sunk from C through M_2 to ground than is sourced to C through M_1 from V_{dd}. Accordingly, q_{pd_final} is negative. As time progresses, V_r starts to lead V_o (event (3,4)). We may expect q_{pd_final} to be positive. However, if we focus on d_{pd} in Figure 7.7b during event (3,4) one quickly finds out that q_{pd_final} is still negative. This is because, as discussed previously, q_{pd_final} is exactly 0 when the phase

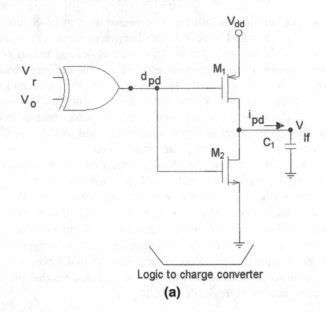

Fig. 7.7a Phase detector based on EXOR gate

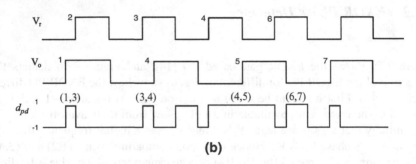

Fig. 7.7b An example illustrating the operation of a phase detector based on EXOR gate

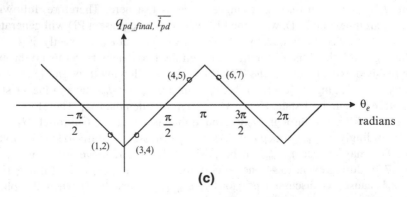

Fig. 7.7c Plot of charge delivered and average current in a phase detector based on EXOR gate versus phase error

difference is exactly 90° and so there is an offset (note that in the PFD case there is no such offset). In the present case, since V_r does not lead V_o quite by 90°, therefore $q_{\text{pd\_final}}$ is still negative, as shown in Figure 7.6b. Eventually θ_e is positive and larger than 90° [event (4,5)]. We can see that in this case $q_{\text{pd\_final}}$ becomes positive, as shown in Figure 7.7c. Consequently, we can see that the segment from $\theta_e = 0$ to 180° is the same as in the PFD case except that it is offset both in the θ_e (by 90°) and $q_{\text{pd\_final}}$ (by $\frac{q_{\text{pd\_final\_max}}}{2}$) axis. What happens when θ_e increases beyond π radians? As shown in event (6,7) of Figure 7.7b, d_{pd}'s duty cycle starts to decrease and hence $q_{\text{pd\_final}}$ also starts to decrease, as drawn in Figure 7.7c. This continues until $q_{\text{pd\_final}}$ decreases to its minimum value. Note that this is different from Figure 7.6b. In Figure 7.7c, we have a triangular plot as opposed to the sawtooth plot in Figure 7.6b. Next, what happens as θ_e increases beyond 2π radians? Exactly the same phenomenon as in the PFD case will happen. Since the PD cannot distinguish between θ_e and $\theta_e + 360°$, the PD will slip a cycle. That is, the transfer characteristic will repeat itself. Notice that the same repetition occurs in the negative θ_e axis (that is, if θ_e goes below 0°). On the other hand, PFD behaves differently with a negative θ_e.

Despite all of these differences, it seems that EXOR is not fundamentally inferior to PFD. Its major disadvantage, however, is that it is very sensitive to waveform symmetry. It is sensitive to both the duty cycle and the finite rise/fall time of V_r and V_o. From event (6,7) in Figure 7.7b, it is obvious that d_{pd} depends not only on the rising edge but on the falling edge of V_r and V_o as well (not so in PFD); that is, on the duty cycle. Figure 7.8a, and Figure 7.8b further highlight this dependency on finite rise, fall time. In Figure 7.8a,b, we have two identical inputs (V_r, V_o) to the EXOR PD, except that in Figure 7.8a, the V_r waveform has a finite rise and fall time. Right next to each edge we attach an edge number. Comparing d_{pd} in Figure 7.8a and Figure 7.8b they can be seen to be different because of this finite rise/fall time of V_r. Specifically, because of finite rise and fall time in Figure 7.8a, the d_{pd}'s edge 1 is delayed compared with the d_{pd}'s edge 1 in Figure 7.8b (edge 1 comes along only when V_r in Figure 7.8a crosses the logic threshold). Therefore, in Figure 7.8a, d_{pd}'s (1, 2) pulse's on duration is shorter than its counterpart in Figure 7.8b. Similarly, because of finite fall time in Figure 7.8a, the d_{pd}'s edge 3 happens before the d_{pd}'s edge 3 in Figure 7.8b. Therefore, in Figure 7.8a, d_{pd}'s (3 4) pulse's on duration is longer than its counterpart in Figure 7.8b.

This dependency on the actual waveform (not just edge) runs contrary to what a true phase detector is supposed to respond to: only to edge timing. On closer look we can see why. In an EXOR PD, and for that matter any combinational logic-based PD, the device is really not a phase detector but a duty-cycle detector. It so happens that if the rise/fall time of the waveform is zero, then there is a linear relationship between the duty cycle of a waveform and its phase, making this circuit useful in detecting phase.

Exactly the opposite is true for sequential logic-based PD (of which PFD is one type; JK-FF based PD is another popular one). Here since the PD responds to an edge, there is no waveform dependency. The difference of responding either to duty cycle or edge timing also helps to explain one of the differences observed earlier in the $q_{\text{pd\_final}}$ versus θ_e curve for these two types of phase detector. It is

Fig. 7.8 Sensitivity of
EXOR gate based phase
detector to finite rise, fall
time. (**a**) d_{pd} with finite rise,
fall time in V_r (**b**) d_{pd} with
zero rise, fall time in V_r

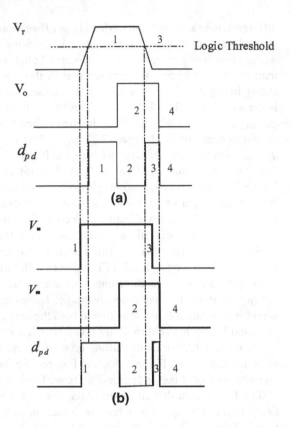

obvious that for the combinational logic-based PD since the duty cycle changes continuously at the cycle boundaries [e.g., at $\theta_e = 2\pi$ in Figure 7.7c], $q_{\text{pd\_final}}$ also changes continuously with θ_e at these cycle boundaries. Because the duty cycle changes continuously, only the slope (or $\frac{dq_{\text{pd\_final}}}{d\theta_e}$) changes abruptly at these boundaries. The same cannot be said for a sequential logic-based detector. For sequential logic-based PD at the boundaries [e.g., at $\theta_e = 2\pi$ in Figure 7.6b], $q_{\text{pd\_final}}$ also changes abruptly because once you miss the edge, you miss the whole cycle and you start all over again. This is exactly what was observed before as a difference between Figure 7.7c and Figure 7.6b at these boundaries.

7.3.3 Charge Pump

As shown previously, the PD (both PFD and EXOR based) needs a logic to charge (current) converter or a charge pump. This charge pump does one more thing: it realizes integration without an op-amp (or any other active device). We now

Fig. 7.9 Generic model of a
charge pump

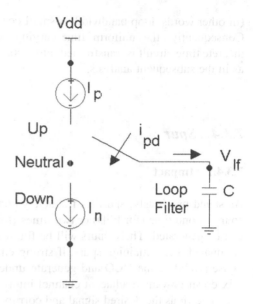

consider some unique features of this charge pump. To do so it is helpful to develop a
generic model of the charge pump, which is shown in Figure 7.9 [22]. In general, the
C can be a more complex loop filter. Most of the time the three output states are
used with a PFD, but it is also possible to have combinatorial phase detectors with
three-state logic outputs. These types of PD, however, do not have the frequency-
detection property of the PFD. One unique feature (and major advantage) of having a
three-state output, like the one in the charge pump, is the existence of an idle state,
which occurs when the PLL is in lock. This idling means that output is zero and ideally
spur does not exist. Therefore, the VCO is not modulated with spur and V_o does not
have sidetones due to frequency modulation. In practice, there will be voltage ripple,
which causes spurs to occur at frequency f_r. This comes about because there is always
a small amount of charge that needs to be supplied to compensate for charge leaking
away. This charge leakage is due to mismatch between current sources I_p and I_n.

Another feature of any charge pump is its time-varying properties. Because of
the switching inherent in the charge pump, the PLL is a time-varying network and is
best analyzed as a discrete time circuit. One complication about analyzing the
charge pump circuit as a discrete time circuit is the fact that the sampling operation
in a PFD/charge pump is nonuniform. This is because the V_o's edge can arrive at the
input to the PFD any time (nonuniformly), in particular during the acquisition
phase. When the system is locked, it becomes uniform sampling only when the
reference frequency is constant, which fortunately is the case in the present
application. Hence, we can treat the PLL as a discrete time circuit with uniform
sampling once it is in lock. Further simplification in analysis is possible if we
observe that V_{lf} varies by a very small amount on each cycle of the reference signal

(in other words, loop bandwidth is small compared with the reference frequency). Consequently, the uniform time-varying operation can be averaged and the discrete time circuit is transformed into a continuous time (time invariant) network, as in the subsequent analysis.

7.3.4   Spur

7.3.4.1   Impact

As stated previously, spurious frequency components, which we have denoted as spurs, at one time (for PFD) or two times (for EXOR PD) the reference frequency can be generated. These spurs will be filtered by the loop filter. However, though attenuated, the remaining spurs, if strong enough, when applied to the VCO will phase modulate the VCO and generate undesired sidetones. These sidetones can mix down unwanted adjacent channel interference (or other unwanted signals) to the same IF as the desired signal and corrupt the desired signal.

7.3.4.2   Origin and Calculation

Let us refer to Figure 7.4 for the time being and assume that we have a fixed steady-state phase error θ_e that is positive. This means that V_r is always on for a fixed period of time before V_o is on. One such example is event (3, 4) in Figure 7.6a. Here it is seen that d_{pd} is 1 for the $\frac{\theta_e}{2\pi}T$ period and 0 for the rest of the period. This part of the d_{pd} trace is redrawn in Figure 7.10a. The corresponding i_{pd} waveform is shown in Figure 7.10b, where it is seen that the charge pump delivers a current $i_{pd} = I_{charge\_pump}$ for $\frac{\theta_e}{2\pi}$ fraction of a period T (where $T = \frac{1}{f_r}$, f_r being the reference frequency). For the rest of the period $i_{pd} = 0$. Now where does this steady-state phase error θ_e come from? The answer lies in the mismatch between the top and bottom transistors M_2, M_1 in Figure 7.4. Referring again to Figure 7.10a and Figure 7.10b, we assume that $i_{pd}(t)$ has a duty cycle $d = \frac{\theta_e}{2\pi}$. For typical technology and transistor sizing, d turns out to be on the order of 0.1%. If we take a Fourier series expansion of $i_{pd}(t)$, it follows that the spectrum of the waveform, denoted as $i_{pd}(f)$, consists of discrete frequency components spaced f_r apart, as shown in Figure 7.11. These are the spurs. The envelope of these frequency components follows a $\sin(f)/f$ function. The dc component is given by dc $= \overline{i_{pd}} = d \times I_{charge\_pump}$. The spur is represented by the frequency component closest to dc, as that component is most damaging. This is denoted as i_{spur} and is given by $A_{spur}\sin\omega_r t$. Since f_r is close to dc, then $A_{spur} \cong i_{pd}(f)_{f=0}$ and so $A_{spur} \cong \overline{i_{pd}}$. Consequently,

$$i_{spur} \cong \overline{i_{pd}}\ \sin\omega_r t = d \times I_{charge\_pump}\sin\omega_r t \qquad (7.11)$$

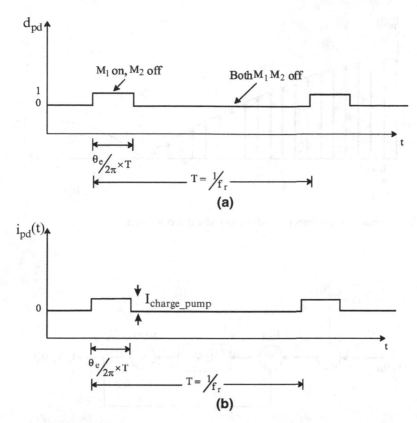

Fig. 7.10 (a) Plot of steady-state d_{pd} as a function of time. (b) Plot of steady-state phase detector current as a function of time

We may choose to represent spur in the phase rather than in the amplitude domain. Here we denote spur in the phase domain as $\theta_{spur}(t)$, and this is given by

$$\theta_{spur}(t) = d \times 2\pi \times \sin \omega_r t \qquad (7.12)$$

To see the impact of such a sinusoidal perturbation of i_{spur} or θ_{spur} on the output frequency, we start with Figure 7.2a, inject the spur, and redraw the resulting diagram in the S-domain as shown in Figure 7.12. Now why is the spur represented by $i_{spur}(s)/K_{pd}$ in Figure 7.12? Dividing (7.11) by (7.12), we have

$$\frac{i_{spur}}{\theta_{spur}} = \frac{I_{charge\_pump}}{2\pi} \qquad (7.13)$$

Substituting (7.9) into (7.13), we have

$$i_{spur}(t) = K_{pd}\theta_{spur}(t)$$

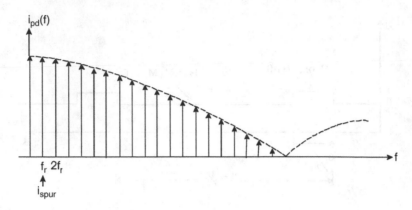

Fig. 7.11 Frequency spectrum of steady-state phase detector current

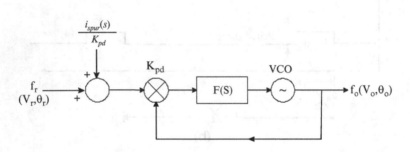

Fig. 7.12 Block diagram of a PLL with spur sources highlighted

Taking the Laplace transform and rearranging, we have:

$$\theta_{\text{spur}}(s) = \frac{i_{\text{spur}}(s)}{K_{\text{pd}}} \tag{7.14}$$

Hence the spur, $\theta_{\text{spur}}(s)$, is indeed represented by $i_{\text{spur}}(s)/K_{\text{pd}}$, which explains the representation in Figure 7.12.

Continuing with our discussion on the impact of of i_{spur}, let us now apply a sinusoidal i_{spur} at the phase detector. This would result in a sinusoidal phase change at the output that equals θ_o.

In the S-domain θ_o is given by

$$\theta_o = \frac{\theta_o}{\theta_{\text{spur}}}\theta_{\text{spur}} = \frac{\theta_o}{\theta_{\text{spur}}}\frac{i_{\text{spur}}}{K_{\text{pd}}} = H_{\text{pd}}\frac{i_{\text{spur}}}{K_{\text{pd}}} = H_{\text{ref}}\frac{i_{\text{spur}}}{K_{\text{pd}}} \tag{7.15}$$

Here H_{pd}, H_{ref} are the transfer functions from the PD and reference source to the synthesizer output, respectively. Notice that since the PD is located at the same node as the reference source, $H_{pd}=H_{ref}$, as indicated in the last equality. From now on we will use H_{ref} to represent H_{pd}.

The θ_o would phase modulate the VCO. Let us assume that the VCO output before any spur is injected is given as $V_o=(A_c/2)\sin\omega_o t$. Here $A_c/2$ is the amplitude of the carrier. If we assume that the phase modulation is narrowband modulation (with the modulation index being small, which is almost always the case as θ_{spur}, i_{spur} is small), then V_o will consist of the original frequency component at f_o and two new sidetones at $f_o \pm f_r$ [15]. Mathematically, V_o is given as

$$V_o = \frac{A_c}{2}\sin(\omega_o t) + \frac{A_c\theta_o}{2}\sin((\omega_o + \omega_r)t) + \frac{A_c\theta_o}{2}\sin((\omega_o - \omega_r)t) \qquad (7.16)$$

The two new sidetones constitute spurs at the output. The ratio of the amplitudes of these spurs to the amplitude of the carrier is obtained from (7.16) as

$$\frac{\text{spur\_amplitude}}{\text{carrier\_amplitude}} = \frac{\frac{A_c\theta_o}{2}}{\frac{A_c}{2}} = \theta_o \qquad (7.17a)$$

and

$$\frac{\text{spur\_power}}{\text{carrier\_power}} = \theta_o{}^2 \qquad (7.17b)$$

where θ_o has the dimension of rad and is a ratio. If we look at $\theta_o{}^2$ and take the log of it, then the ratio has the dimension of dBc.

From (7.17) we can see that to find spur amplitude/power we need θ_o. In turn, from (7.15), we need the reference transfer function H_{ref}. This transfer function depends on the loop filter transfer function $F(s)$. Hence we will postpone any further discussion on how to calculate spur amplitude/power until after the loop filter section in Chapter 8.

7.3.5 K_{pd}

We have explained the operation of the PD/charge pump using both q_{pd_final} (K_{pd}') and $\overline{i_{pd}}$ (K_{pd}). For design purposes, it is easier to deal with current rather than charge and so from now on we will characterize the PD/charge pump with $\overline{i_{pd}}$ and K_{pd}.

Here are the considerations for determining K_{pd}:

1. K_{pd} should not be so large that the transistors M_1, M_2 in Figure 7.4 or Figure 7.7a become excessively large.
2. Phase noise considerations. We will jump ahead of ourselves a little bit and extend the definition of the synthesizer output phase noise due to VCO phase

noise, as defined in (7.119), to the definition of the synthesizer output phase noise due to PD phase noise, $S_{\theta o\_pd}$:

$$S_{\theta o\_pd} = |H_{ref}|^2 S_{\theta\_pd} \qquad (7.18)$$

Here S_{θ_pd} is the phase noise of the PD. If we assume that the PD phase noise comes from the charge pump, then S_{θ_pd} can be related to the PSD of the equivalent input current noise of the charge pump, as given by $\frac{\overline{i_{nd}^2}}{\Delta f}$. The relationship can be derived by the same procedure that leads to (7.14) and is given as

$$S_{\theta\_pd} = \frac{\frac{\overline{i_{nd}^2}}{\Delta f}}{K_{pd}^2} \qquad (7.19)$$

Substituting (7.19) into (7.18), we have

$$S_{\theta o\_pd} = |H_{ref}|^2 \frac{\overline{i_{nd}^2}/\Delta f}{K_{pd}^2} \qquad (7.20)$$

The input here is $\overline{i_{nd}^2}$, which has the dimension of A^2. $S_{\theta o\_pd}$ has the dimension of dBc/Hz.

As will be shown in Chapter 8, $\overline{i_{nd}^2}$ is only weakly dependent on I_{charge_pump}. From (7.9) this means that $\overline{i_{nd}^2}$ is also only weakly dependent on K_{pd}. Hence from (7.20) $S_{\theta o\_pd}$ depends roughly on the inverse of K_{pd}^2. Accordingly, K_{pd} should be selected large enough to make any output phase noise due to charge pump insignificant when compared with other sources. The exact value, of course, depends on H_{ref}. H_{ref} again depends on the loop filter. We will postpone further discussion until the loop filter section in Chapter 8. After that discussion in Chapter 8, we will have a chance to use (7.18) through (7.20) to determine K_{pd}. Once K_{pd}'s value is picked, we can use the following equation to design I_{charge_pump}

$$K_{pd} = \frac{I_{charge\_pump}}{2\pi} \qquad (7.21)$$

Then we can determine the size of M_1 and M_2.

7.4 Dividers

7.4.1 Survey of Different Types of Divider

We begin our discussion by showing some examples of dividers in terms of increasing complexity (Figure 7.13) [19]. Referring back to Figure 7.1, we note that the input to the divider is denoted as $f_o(V_o)$ and output is denoted as $f_{divide}(V_{divide})$. This notation is

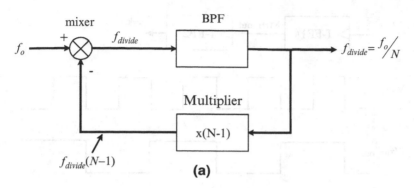

Fig. 7.13a Indirect divider

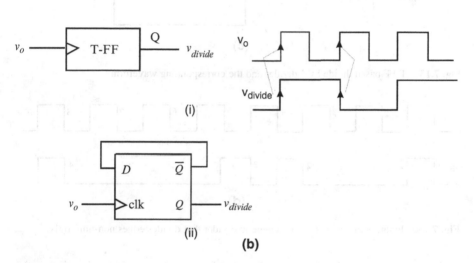

Fig. 7.13b (i) T-FF based divide by 2 divider. (ii) T-FF implemented using D-FF

carried over to Figure 7.13. Figure 7.13a shows an indirect divider where the waveform is assumed to be sinusoidal. Let us assume that the output of the mixer has a frequency component f_{divide}. Hence the output of the BPF has a frequency at f_{divide} as well. When multiplied, this becomes $f_{\text{divide}} \times (N-1)$. If we apply this signal to the mixer, the mixer output has a frequency component at $f_o - f_{\text{divide}} \times (N-1)$, which we assumed previously is f_{divide}. Equating these two expressions, we can solve for f_{divide}, which is given as $f_{\text{divide}} = \frac{f_o}{N}$, and division is achieved as desired. Notice that in the present case, division is performed by using a multiplier (hence the term INDIRECT DIVISION). Due to feedback, this divider is slow.

Figure 7.13b shows a faster divider that is based on digital circuits. This time the input is assumed to be a square wave. Let us assume that the T-FF (toggle FF) is a positive edge trigger. Accordingly, this FF has its output level toggled every time a

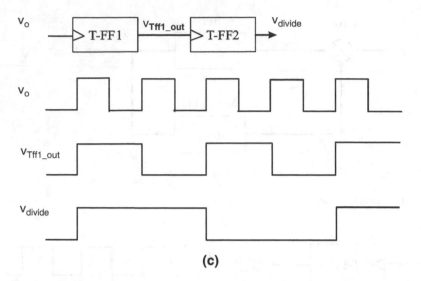

(c)

Fig. 7.13c T-FF based divide by 4 divider and the corresponding waveform

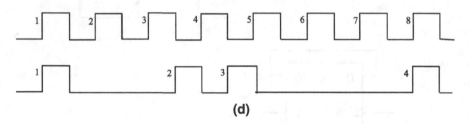

(d)

Fig. 7.13d Input, output waveform of a generic divider that divides edges non-uniformly

positive edge arrives. Then it is easy to see that for every two positive edge changes in the input waveform, there is one positive edge change in the output waveform, and hence the number of positive edges is reduced by 2 (divide by 2). Figure 7.13b(ii) shows how this T-FF can be easily implemented by connecting the $\overline{Q}$ output of a D-FF back to itself. Figure 7.13c has two of these cascaded. In general, we can also employ feedback to obtain a non-binary division ratio.

Let us now do some comparisons among Figure 7.13a, b, and c. It is seen that the divider of Figure 7.13a divides the input frequency by two, but that in Figure 7.13b it divides the input edges by 2. In general, for a frequency synthesizer the divider only needs to divide down the number of input edges, and so the scheme in Figure 7.13b is more general. Taking this one step further, theoretically we do not even have to divide down the input edges uniformly. For example, a divider that divides according to Figure 7.13d will also work. This divider essentially produces four output positive edges for eight positive input edges, and therefore achieves a divide by 2 operation, although in an average sense. Along this line of thinking, it is necessary to revisit the concept of division and classify the dividers accordingly, as will be done in the following section. Finally, to have a variable division ratio, the

Fig. 7.13e A programmable divider

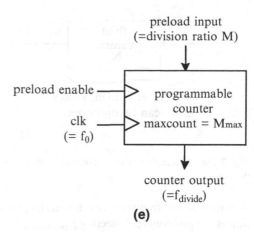

preload input
(=division ratio M)

preload enable ──▷

programmable counter
maxcount = M$_{max}$

clk
(= f$_0$) ──▷

counter output
(=f$_{divide}$)

(e)

divider can be implemented as shown in Figure 7.13e, as a programmable counter [20]. This divider is essentially a counter that counts from the preloaded input M to 0, thereby achieving a division of M. Here $M \leq M_{max}$, the modulus of the counter, and so the maximum division ratio is M_{max}. This counter can be implemented either as a ripple or a synchronous counter. As a side note, unless otherwise specified, from now on the terms divider, prescaler, and counter are used synonymously [20].

The preceding discussion gives examples of various types of dividers. To be more rigorous and generic, let us do the following classifications: fixed divider and programmable divider [20].

A fixed divider means that the divisor (or dividing ratio) is fixed and cannot be changed. Examples are shown in Figure 7.13b, c. This divider has the inherent advantage that since the configuration is fixed, the critical path is fixed and we can optimize speed along this critical path. This is because for the early stage that carries out division on the most significant bit (MSB), the input signal is faster and therefore the size and current (hence power) should be designed to make it run faster. Alternately, we can see that the critical path from the stage for the MSB back to the stage for the least significant bit (LSB) is fixed and thus can be optimized.

As the name suggests, a programmable divider has a programmable divisor. As stated previously, this can be achieved by using a programmable counter, as shown in Figure 7.13e. For a given speed requirement a programmable divider is less power optimized. This is so because with programmability the critical path is dependent on the loaded value. In the example of Figure 7.13e, the critical path is dependent on the value M and hence the designer cannot optimize the transistor sizing a priori.

In general, a frequency synthesizer does need a programmable division ratio while maintaining low power. To meet this challenge, we can use the architecture as shown in Figure 7.14a. Here the output frequency f_o is divided by a fixed counter at high frequency. Since the first counter is operating at high frequency, it consumes a lot of power. Since it is a fixed counter, its power can be optimized. To make the entire division ratio programmable, the second counter must be programmable. Since it is programmable, it cannot be optimized for power. However, since it is

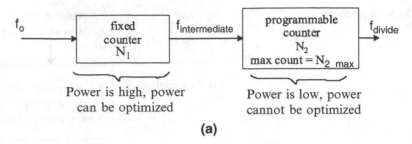

(a)

Fig. 7.14a A complete divider consisting of a fixed divider cascaded with a programmable divider

operating at low frequency, even without power optimization, the power consumption is not prohibitive. Selecting $f_{\text{intermediate}}$ involves a proper trade-off between the relative power consumption between the first and second counter.

Even though the aforementioned divider achieves better overall power consumption than a single programmable counter, it does reduce the resolution when compared with the single programmable counter approach. Assume that the overall division ratio is

$$M = N_1 \times N_2 \qquad (7.22)$$

Here M is the complete divisor needed, N_1 is the divisor of the fixed divider, and N_2 is the divisor of the programmable divider. The overall resolution is N_1. This is because changing counter N_2 by 1 results in the overall division ratio changing by N_1. In general, we have

$$\text{resolutionn}_{\text{complete\_divider}} = \text{resolution}_{\text{programmable\_divider}}$$
$$\times \text{division\_ratio}_{\text{fixed\_divider}} \qquad (7.23)$$

To improve the resolution of the complete divider, one approach is to make the fixed divider also programmable. Since full programmability consumes too much power, we would like to restrict its programmability, which is done by restricting the number of moduli it can divide. The simplest case is to make it a dual modulus counter. A dual modulus counter can further be viewed as a fixed counter with a pulse swallower, which has the flexibility of counting one more pulse before generating an output or, equivalently, inserting one more pulse before generating an output. This concept is illustrated by an example as shown in Figure 7.14b through Figure 7.14e. First, back in Figure 7.14a we pick $N_{2\ max} = 4$. This will be used in Figure 7.14b through Figure 7.14e. In Figure 7.14b, we set $N_1 = 3$, $N_2 = 4$ and $f_{\text{divide}} = \frac{f_o}{12}$.

Next, we want to switch to another frequency, say, by making $f_{\text{divide}} = \frac{f_o}{11}$. How do we achieve this? The simplest way is if $M = N_1 \times N_2$ is programmable by 1 and we change M from 12 to 11. This is not possible here. Since the only

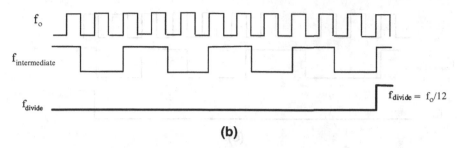

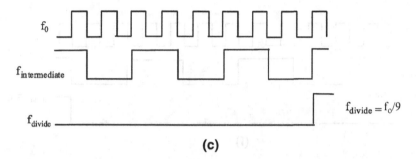

(b)

Fig. 7.14b Waveform of the complete divider, $N_1=3$, $N_2=4$

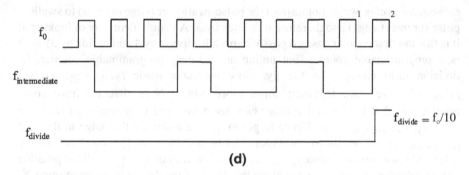

(c)

Fig. 7.14c Waveform of the complete divider, $N_1=3$, $N_2=3$

(d)

Fig. 7.14d Waveform of the complete divider, with the fixed divider replaced by a dual modulus divider. $N_1=3/4$, $N_2=3$. Here $N_1=4$ for 1 N_2 cycle

programmable counter is the second counter, let us first decrease N_2 by 1 with N_1 remaining at 3. The result is shown in Figure 7.14c, where there are only nine f_o edges for 1 f_{divide} edge and so $f_{\text{divide}} = \frac{f_o}{9}$. Next, to make up for the lost edges, we modify the first counter so that it can swallow pulse. The first counter is now called a pulse swallower or edge inserter. It works by swallowing one extra pulse (or inserting one extra edge) before we generate the output. Hence if we turn on the pulse swallower (edge inserter) function of the first counter, the results will be as shown in Figure 7.14d. Here the first counter, in the first cycle, counts four rather than three f_o edges before it generates a $f_{\text{intermediate}}$ edge. Let us assume that in the

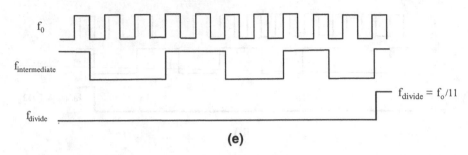

(e)

Fig. 7.14e Waveform of the complete divider, $N_1=3/4$, $N_2=3$. Here $N_1=4$ for two N_2 cycles

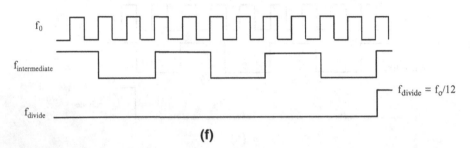

(f)

Fig. 7.14f Waveform of the complete divider, $N_1=4/5$, $N_2=3$. Here $N_1=4$ for three N_2 cycles

subsequent cycles, we do not instruct the pulse swallower (edge inserter) to swallow pulse (or insert edge) and therefore it divides by 3. An equivalent way of looking at it is that the first counter has two possible moduli (divide by 4 and divide by 3), so it is a programmable counter, but unlike an ordinary programmable counter, its division ratio changes on the fly. Now we have made $f_{\text{divide}} = \frac{f_o}{10}$. To get $f_{\text{divide}} = \frac{f_o}{11}$, we need to insert some more edges. Note that we have some restrictions: that is, the first counter can insert one and only one edge (not two edges) per $f_{\text{intermedite}}$ cycle. Hence to get to $\frac{f_o}{11}$, we insert another edge in the next $f_{\text{intermediate}}$ cycle, as shown in Figure 7.14e and $f_{\text{divide}} = \frac{f_o}{11}$. We have finally achieved resolution $= 1$ since f_{divide} can assume values of $\frac{f_o}{10}$, $\frac{f_o}{11}$ or all the possible values between $\frac{f_o}{12}$ and $\frac{f_o}{9}$. We achieve this level of resolution by programming N_1 (telling it which modulus to use) and also how many times it uses that particular modulus (how many times it insert edges).

Finally, note that we have an overall restrictions in choosing $N_{2\,\text{max}}$ and N_1. The restriction is

$$N_{2\,\text{max}} \geq N_1 \qquad (7.24)$$

Let us violate this inequality deliberately and see what happens. For Figure 7.14f through Figure 7.14h, let us go back to the divider in Figure 7.14a and pick $N_{2\,\text{max}} = 3$. Initially, in Figure 7.14f, we set $N_1=4$, $N_2 = N_{2\,\text{max}} = 3$, and so f_{divide} is still $\frac{f_o}{12}$. To get to $\frac{f_o}{11}$, we first decrease N_2 to 2 with N_1 remaining at 4. As shown in Figure 7.14g,

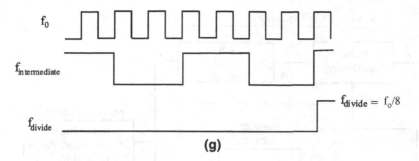

(g)

Fig. 7.14g Waveform of the complete divider, $N_1=4/5$, $N_2=2$. Here $N_1=4$ for two N_2 cycles

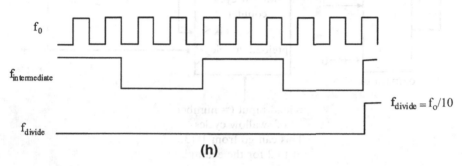

(h)

Fig. 7.14h Waveform of the complete divider, $N_1=4/5$, $N_2=2$. Here $N_1=5$ for two N_2 cycles

$f_{\text{divide}} = \frac{f_o}{8}$. Next, in Figure 7.14h, we try to make up for the lost edges by inserting extra f_o edges in all the $f_{\text{intermediate}}$ cycles and the maximum we can get is only 10 edges (i.e., $f_o = \frac{f_{\text{divide}}}{10}$). Since we have the restrictions that only 1 f_o edge can be added per $f_{\text{intermediate}}$ cycle, we have run out of $f_{\text{intermediate}}$ cycles to add any more edges and $f_o = \frac{f_{\text{divide}}}{11}$ cannot be generated. This is the consequence of violating (7.24).

7.4.2   Example of a Complete Divider (DECT Application)

Let us now apply this dual modulus counter concept to our DECT example [20]. To do this we start from Figure 7.14a, replace the fixed counter N_1 by a dual modulus counter N_1 and keep the second-stage programmable counter N_2. The resulting divider is shown in Figure 7.15 [19]. For a DECT standard, the division ratio is

$$M = 1089 - 1098 \tag{7.25}$$

First we need to decide on $f_{\text{intermediate}}$. Now $f_{\text{intermediate}}$ can be chosen subject to power consumption and the inequality (7.24). One choice involves making

$$N_{2\,\text{max}} \cong 16N_1 \tag{7.26}$$

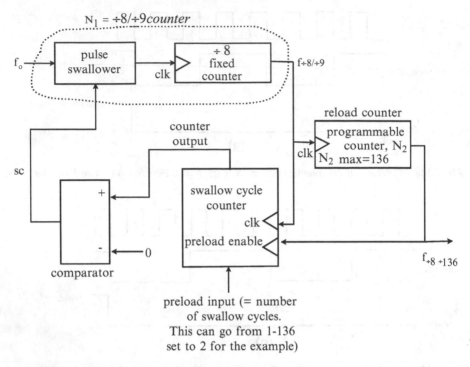

Fig. 7.15 A complete divider for DECT application

Substituting (7.26) and (7.25) into (7.22), we see that we can satisfy these equations by assigning $N_1 = 8/9$, $N_{2\ max} = 136$. Hence the divider has a dual modulus divide by 8/9 first stage counter followed by a programmable divide by 136 second stage. The $f_{intermediate}$ becomes $f_{\div 8/\div 9}$. Note that for DECT, there are only 10 channels, so the second divider, strictly speaking, does not need to be fully programmable. We will leave it as a fully programmable divider so that the divider architecture presented can be applied in all general cases.

Let us examine how the N_1 counter shown in Figure 7.15 is implemented. This is shown in Figure 7.16 [19]. Figure 7.16a shows the state table of the $\div 8/9$ counter whose implementation is shown in Figure 7.16b. Let us see how it works. From Figure 7.16a notice that when the swallow control signal (SC) = 1, there are two counts of 0 state. Normally as far as the output of the counter is concerned, it expects one output pulse for every eight input pulses. This time it is still getting the same one output pulse without knowing that the input has counted one extra pulse. This is the principle of pulse swallowing.

Let us look at Figure 7.17 for another example of a dual modulus counter (this time it is a $\div 4$, $\div 5$ counter) [21]. Figure 7.17a shows the state diagram and Figure 7.17b shows the implementation. From Figure 7.17a, the state diagram is more interesting than the state table of Figure 7.16a. Here the control signal is the SC signal (labeled along the paths), where an X means "don't care". Notice that there

Count sequence close to swallow states

d_0	d_1	d_2	$d_3(=fps)$	
1	0	1	1	
0	1	1	1	
1	1	1	1	divide by 8, SC = 0:
0	0	0	1	only 1 pulse for '0' state,
1	0	0	1	no swallow
.	.	.	.	
.	.	.	.	
.	.	.	.	
1	0	1	1	
0	1	1	1	
1	1	1	1	divide by 9, SC = 1:
0	0	0	0	2 pulses for '0' states,
0	0	0	1	pulse swallow
1	0	0	1	

(a)

Fig. 7.16a State table of a ÷8/÷9 dual modulus counter (dual modulus counter whose state table has one path)

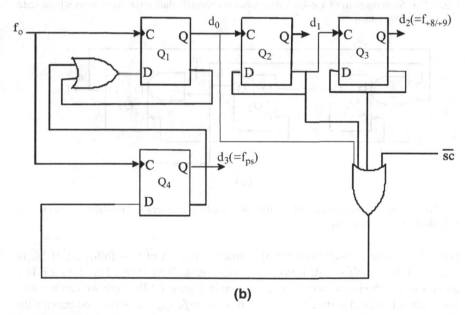

(b)

Fig. 7.16b Logic implementation of a ÷8/÷9 dual modulus counter (dual modulus counter whose state table has one path)

are two possible state paths: A and B, each consisting of two sequences, a ÷ 4 and a ÷ 5 sequence. Assuming the notation of d_2, d_1, d_0, then in path A, the ÷5 sequence is from 000, 001, 011, 010, 100, then back to 000. The ÷4 sequence is 000, 001, 011, 010, 000. For path B, the ÷5 sequence is 000, 001, 011, 110, 100, 000 and the ÷4 sequence is 000, 001, 011, 110, 000. Notice the SC signal decides

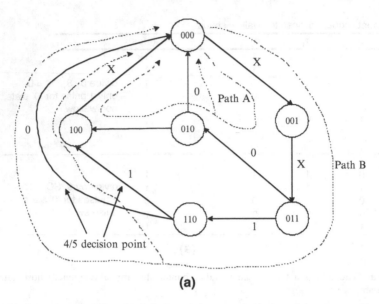

(a)

Fig. 7.17a State diagram of a ÷4/÷5 dual modulus counter (dual modulus counter whose state diagram has two paths)

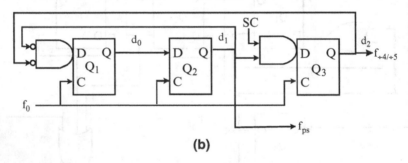

(b)

Fig. 7.17b Logic implementation of a ÷4/÷5 dual modulus counter (dual modulus counter whose state diagram has two paths)

not only if pulse is swallowed but also whether path A or B is followed. If SC is enabled at 011, path B will follow, and vice versa. Why is this being done? The answer lies in the implementation, as shown in Figure 7.17b. Here we can see that less logic is involved in the critical path from SC to $f_{intermediate}$, when compared with Figure 7.16b, making the divider faster.

Next let us go back to Figure 7.15 and finish our discussion on how the complete divider works. We have explained in Figure 7.16 how the ÷ 8/9 dual modulus counter operates. According to Figure 7.14d, in a dual modulus divider a mechanism is needed to control the number of N_2 cycles in which the counter N_1 insert edges (swallow pulses). This is implemented by the blocks labelled "swallow cycle counter" and "comparator" in Figure 7.15. The swallow cycle counter is preloaded with the number

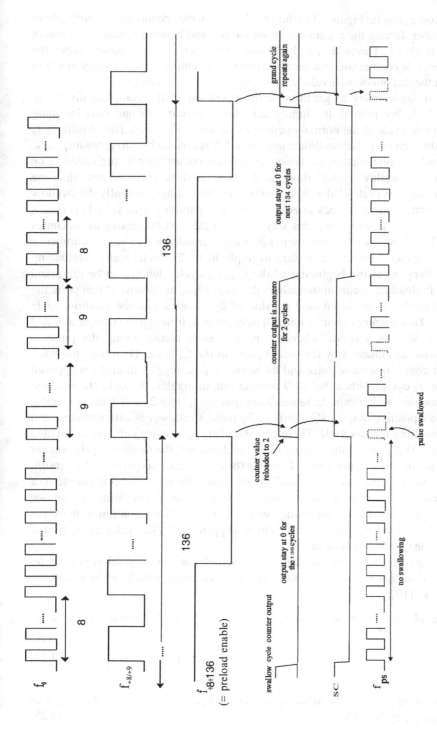

Fig. 7.18 Timing diagram showing the operation of the complete divider for DECT application

of swallow cycles [in Figure 7.14d this should be 1]. It then counts down, starting from this number. During the countdown its output is 1 and hence the comparator output (=SC) is also 1. Therefore, the dual modulus divider swallows pulses. When the swallow cycle counter counts to zero, the comparator output (=SC) changes from 1 to 0. Then the dual modulus divider will no longer swallow pulses [19].

We are now ready to go through the operation of the complete divider in Figure 7.15. We refer to the timing diagram in Figure 7.18 and start by going through one cycle of the normal operation (i.e., no swallowing). Hence initially f_o passes through the pulse swallower unchanged. It is divided by 8 to generate $f_{\div 8/\div 9}$. The swallow cycle counter is a down counter that counts from its preloaded value "number of swallow cycles" down to 0 and stays there until the next time the counter output is preloaded with a non 0 value. Accordingly, initially the swallow cycle counter output is stuck at zero and the comparator output is 0. This means that SC is zero, as expected. This stays on until the next interesting event comes along. This happens when the counter is being preloaded with a value "number of swallow cycles", set in the present example to be 2. (Incidentally, preloading always happens at the beginning of the $f_{\div 8/\div 136}$ cycle, denoted to be the grand cycle.) Preloading occurs at the rising edge of $f_{\div 8/\div 136}$, as shown. At that point the preload enable is activated, and a value of 2 is loaded into the swallow cycle counter. This counter output is immediately set to 2, firing up the comparator and enabling SC. As a result, when the next f_o edge comes along, the pulse is swallowed, as indicated by the dotted pulse in the f_{ps} trace. Therefore, the $\div 8/9$ counter counts one more pulse and its output, $f_{\div 8/\div 9}$, toggles in nine (as opposed to eight) f_o cycles. When the $\div 8/9$ counter output toggles, it clocks the swallow cycle counter, which then decreases its output by 1, from 2 to 1. The comparator still has a positive output, SC is still 1, the pulse swallower is still activated, and the same thing repeats itself. Therefore, the next $f_{\div 8/\div 9}$ cycle contains again nine f_o cycles. At the end of the next $f_{\div 8/\div 9}$ cycle, however, the swallow cycle counter output counts down from 1 to 0. This sets the comparator output to 0. As a result, SC is 0 and the pulse swallower is disabled. Since the swallow cycle counter is a down counter, once its output reaches 0, it stays there. Hence from now on one $f_{\div 8/\div 9}$ cycle reverts to containing eight f_o cycles. This continues until the counter is reloaded with 2 again, which happens 134 $f_{\div 8}$ cycles later, by the output edge of the N_2 counter.

To show that the complete divider can actually achieve a resolution of 1, let us look at the number of input pulses in a $f_{\div 8/\div 136}$ cycle, which can be calculated as follows [19]:

Number of input pulses to the N_1 divider in a $f_{\div 8/\div 136}$ cycle = total f_o pulses in a

$$f_{\div 8/\div 136}\text{cycle} \;=\; 2 \times 9 \;+\; 134 \times 8 \qquad (7.27)$$

Number of output pulses from the N_1 divider in a $f_{\div 8/\div 136}$ cycle = total $f_{\div 8/\div 9}$ pulses in a $f_{\div 8/\div 136}$ cycle =136   (7.28)

$$\text{Division ratio of } N_1 = \frac{\text{number of input pulses}}{\text{number of output pulses}} = \frac{\text{number of } f_o \text{ pulses}}{\text{number of } f_{\div 8/\div 9} \text{ pulses}}$$

$$= \frac{(2 \times 9 + 134 \times 8)}{136} = 8.0147 \tag{7.29}$$

Therefore, we achieve a fractional division of 0.0147

What is the smallest fraction we can achieve? This will be achieved obviously when number of swallow cycles $=1$. Then

$$\text{division ratio of } N_1 = \frac{(1 \times 9 + 135 \times 8)}{136} = 8.00735$$

Hence the smallest fractional part is 0.00735 and the resolution of N_1 is 0.00375. Accordingly, we have

$$\text{Resolution}_{\text{complete\_divider}} = 0.00735 \times 136 = 1 \tag{7.30}$$

To go through some more examples, let us set "number of swallow cycles"$=0$ and we observe that the division ratio of N_1 becomes 8. On the other hand, if "number of swallow cycles"$=135$, it becomes 8.99265 and if "number of swallow cycles"$=136$ (maximum number of swallow cycles allowable), it becomes 9. So we can see that the division ratio of N_1 can be programmed to cover all 0.00725 increments, spanning the entire interval from 8 to 9. In other words the complete division ratio M varies from 8×136 to 9×136, or 1088 to 1224. This spans the division ratio used for DECT (1088–1098). What happens if the modulus of counter N_2 changes? In that case the division ratio spans the rest of the range (8 through 1088), which is of no interest to DECT.

7.5 VCO: Introduction

7.5.1 Categorization

VCOs can be categorized by method of oscillation into resonator-based oscillators versus waveform-based oscillators, as shown in Figure 7.19. Primary examples of each category are the LC oscillator and the ring oscillator, respectively. Each type has different ways of doing frequency tuning [which is mathematically descrbed by (7.1)]: current steering for ring oscillators and variable capacitor (or varactor) for LC oscillators. Because of the need for integrability, ring oscillator is very desirable

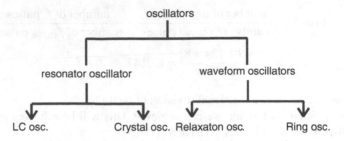

Fig. 7.19 Categorization of VCOs

in a VLSI environment, and most of our discussion focused on this type of oscillator. Relaxation VCO, another waveform based VCO, is usually not a good choice for the present application due to the huge amount of phase noise introduced as a result of positive feedback. A more thorough organization and description of some of these various structures follows.

We start by focusing our attention on resonator based oscillator. First, we review the underlying principles that analyze the operation of resonator based oscillator. There are two approaches to the analysis of resonator based oscillators:

1. Feedback approach
2. Negative resistance approach

The advantage of adopting the feedback approach is that it makes use of negative feedback amplifier theory, which is familiar to most circuit designers. Hence we adopt this approach. In Chapter 3 we adopted feedback theory in analyzing the distortion, frequency response, and other behaviour of an amplifier, but in the context that the feedback action is negative. Since an oscillator works under the principle of positive feedback, we review feedback theory and see how it applies to the present situation.

As a side note, we should note that we can apply small signal analysis to resonator-based oscillators. Accordingly, these oscillators can be described by a linear time invariant (LTI) system using S-domain representation. Thus feedback theory for the LTI circuit applies. On the other hand, for waveform-based oscillators, small-signal analysis is not applicable. Hence the feedback theory reviewed in subsection 7.5.2 is only applicable to resonator-based oscillators.

7.5.2   Review of Positive Feedback Theory

We first review positive feedback theory for an LTI circuit [23]. Let us represent a feedback circuit in the S-domain as shown in Figure 7.20.

Fig. 7.20 Block diagram of a feedback circuit

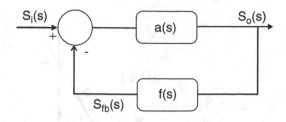

Here $\dfrac{S_0}{S_i} = \dfrac{a(s)}{1 + a(s)f(s)}$ (7.31)

As an example, let us assume that

$$a(s) = \frac{a}{(1 + s\tau)^3} \qquad (7.32)$$

We further assume that $f(s)$ is negative (so overall feedback is positive), resistive, and constant. Let us now close the feedback loop. Then

$$\frac{S_0}{S_i} = \frac{a(s)}{1 + a(s)f(s)} = \frac{k}{\left(1 - \frac{s}{s_1}\right)\left(1 - \frac{s}{s_2}\right)\left(1 - \frac{s}{s_3}\right)} \qquad (7.33)$$

We see that the transfer function consists of three poles, where the poles are roots of the equation.

$$1 + a(s)f(s) = 0 \qquad (7.34a)$$

$$1 + \frac{af}{(1 + s\tau)^3} = 0 \qquad (7.34b)$$

$$(1 + s\tau)^3 = -af \qquad (7.34c)$$

$$1 + s\tau = \sqrt[3]{-af} \qquad (7.34d)$$

To take care of the −1 inside the cube root we observe that

$$-1 = 1\angle 180° + n360° \; for \; n = 1, 2, 3 \qquad (7.35a)$$

$$\sqrt[3]{-1} = 1\angle 60° + n120° \qquad (7.35b)$$

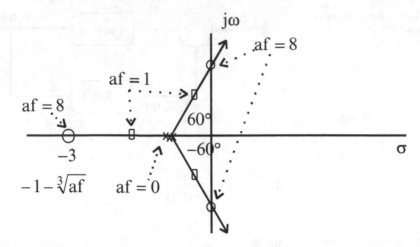

Fig. 7.21 Root locus of the example feedback circuit

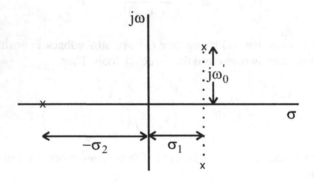

Fig. 7.22 Pole/zero diagram of the example feedback circuit

$$= -1, 1\angle 60°, 1\angle - 60° \qquad (7.35c)$$

Substituting the cube root of -1 and normalizing $\tau = 1$, we find the poles of the transfer function as

$$s_1 = -1 - \sqrt[3]{af}$$
$$s_1 = -1 + \sqrt[3]{af}\angle 60°$$
$$s_3 = -1 + \sqrt[3]{af}\angle - 60° \qquad (7.36)$$

where af is the low-frequency loop gain.

Let us plot root locus as a function of af in Figure 7.21.

Notice that when $af = 8$, two of the poles have just reached the imaginary axis. For $af > 8$ we have right half plane (RHP) poles, shown in Figure 7.22.

Eventually oscillation reaches steady state. Notice that our analysis so far only shows that oscillation will start. It needs nonlinear analysis, which is outside the scope of this book, to shows that a steady state is reached.

To get a more complete picture of the behaviour of this transfer function, let us plot the Nyquist plot, which is a plot of $af(j\omega)$ in magnitude and phase on a polar plot using ω as the variable. In the present case, $af(j\omega) = \frac{af}{(1+j\omega)^3}$, with $\tau = 1$.

As a check as $\omega \to \infty (1 + j\omega)^3 \to (j\omega)^3$ so angle is $-270\,°$, which agrees with the plot. As a condition of oscillation, we are interested in the frequency where $\angle af(j\omega) = -180°$.

Let us try

$$\omega = \sqrt{3}, \tau = 1 \tag{7.36a}$$

Upon substitution, we get

$$af(j\omega) = -\frac{af}{8} \tag{7.36b}$$

Hence we can make two observations:

1. Feedback in this circuit is negative at $\omega = 0$.
2. Feedback is positive at $\omega = \sqrt{3}$

To oscillate, we need one more condition:

$$\text{loop gain} \geq 1 \tag{7.37}$$

We now derive what this condition translates into. Mathematically,

$$\frac{s_o}{s_i}(j\omega) = \frac{a(j\omega)}{1 + af(j\omega)}$$

Now we put $\omega = \sqrt{3}$.

$$af(j\omega) = -\frac{af}{8}$$

To set loop gain $= 1$, we put $af = 8$. Transfer function with feedback applied is

$$\frac{s_o}{s_i}(j\omega) = \frac{a(j\omega)}{1 + af(j\omega)} = \frac{a(j\omega)}{1 - 1} = \infty \tag{7.38}$$

This indicates oscillation and hence a loop gain at low frequency $= 8$ causes oscillation.

As a check, referring to Figure 7.23, oscillation should occur when the Nyquist plot passes through $(-1,0)$, and this indeed happens at $af = 8$. As a final check,

Fig. 7.23 Nyquist plot of the
example feedback circuit.
This diagram scales directly
and linearly with the low
frequency loop gain (*af*)

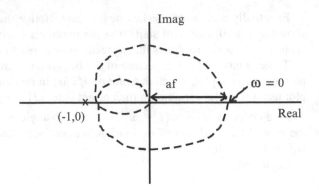

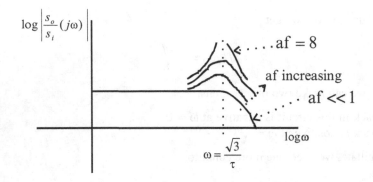

Fig. 7.24 Frequency response of the example feedback circuit

referring to Figure 7.21, S-plane analysis tells us that oscillation should occur when
the root locus pass through the imaginary axis, and the plot shows that this does
happens at *af*=8.

   One natural question to ask is, what happens if loop gain > 8? It turns out that
this is not obvious from the Nyquist plot. Mathematically, the number of times the
Nyquist plot encircles $(-1,0)$ = the number of RHP poles but does not give us any
more information. To answer that question, let us plot overall gain, $\left|\frac{s_o}{s_i}(j\omega)\right|$, as *af*
increases in Figure 7.24. As *af* > 8, gain goes to ∞. This indicates the presence of
RHP poles.

   Complete response of a circuit with RHP poles to any input is

$$S_o = f(input) + K_1 e^{-\sigma_2 t} + K_2 e^{\sigma_1 t} \cos \omega_0' t \qquad (7.39)$$

Here *f*(input) is the forced response (amplifier response),$K_1 e^{-\sigma_2 t}$ is the
decaying exponential natural response, $K_2 e^{\sigma_1 t} \cos \omega_0' t$ is the oscillation, which is
shown in the Figure 7.25.

   Notice that in Figure 7.25, linear analysis is valid in the first part of the waveform
and hence the zero crossing frequency equals ω_0' and is set by RHP pole location.

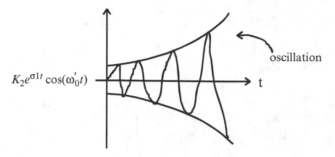

$$K_2 e^{\sigma_1 t} \cos(\omega_0' t)$$

oscillation

t

Fig. 7.25 Transient response of the example feedback circuit

In the second part of the waveform the effect of nonlinearity in active devices eventually dampens the growing exponential so that it has a steady-state waveform with constant amplitude. Here the zero crossing frequency is not necessarily ω_0'. Since we want the frequency of oscillation to be stable, the circuit is not often used for oscillation.

To summarize, for oscillation, we want positive feedback so the phase shift of the loop gain is 180° and the magnitude of loop gain is equal to or greater than 1 while maintaining a stable and predictable frequency of oscillation. To do this we must establish accurately the frequency where the total loop phase shift is 0°. This is key to all oscillator designs. Our first attempt toward maintaining a stable frequency of oscillation involves the use of a *LC* tank in an oscillator. This oscillator further uses a transformer to do phase inversion, which is necessary for positive feedback. This architecture has the advantage that one can accurately define the frequency of zero loop phase shift.

7.6 *LC* Oscillators

7.6.1 *Basic* **LC** *Oscillators*

A resonator-based oscillator that uses a *LC* tank as a resonator is shown in Figure 7.26. The transformer is used to do phase inversion, making $f(s)$ in (7.31) negative and thus achieving positive feedback. Transistor M_1 is, of course, used to implement $a(s)$ in (7.31). Here C_1, C_2 are large bypass capacitors and R_1, R_2, R_E set bias.

L, C resonates at

$$\omega_0 = \frac{1}{\sqrt{LC}}. \tag{7.40}$$

In (7.40), C is not simply C_L, as the transformer reflects impedance from the input.

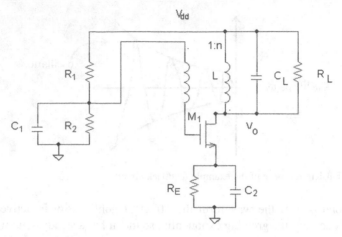

Fig. 7.26 Transformer based *LC* oscillator

A stable and well-specified frequency in this circuit is ω_0, the *LC* resonance frequency. We will try to make this the frequency of oscillator. Hence we want 360° phase shift at

$$\omega_0 = \frac{1}{\sqrt{LC}} \tag{7.41}$$

and

$$af \geq 1 \text{ at } \omega_o \tag{7.42}$$

for oscillation.

To analyze, let us eliminate the bypass capacitors and bias resistors from Figure 7.26 and redraw Figure 7.26 in Figure 7.27. V_{dd} becomes ac ground as well. At resonance, impedance of a LC tank circuit becomes infinite. The transformer's action is represented by setting $v_i = v_o/n$, where n is the turn ratio. Hence at resonance Figure 7.27 becomes Figure 7.28.

From Figure 7.28, loop gain *af* at ω_o is

$$\text{Loop gain} = \frac{g_m R}{n} \tag{7.43}$$

In (7.43), g_m is the transconductance of transistor M_1, but R is not simply R_L, as the transformer reflects impedance from input.

To find R, let us form the small-signal equivalent of the complete ac circuit as described in Figure 7.27 and analyze using Laplace transforms. This small-signal circuit is shown in Figure 7.29, where

$$f = 1/n \tag{7.44}$$

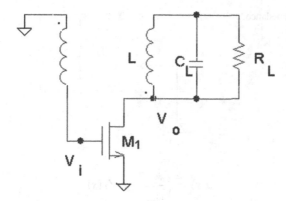

Fig. 7.27 Transformer-based *LC* oscillator with biasing removed

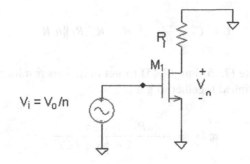

Fig. 7.28 Transformer-based *LC* oscillator at resonance

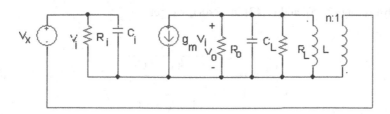

Fig. 7.29 Small-signal circuit of transformer-based *LC* oscillator

Fig. 7.30 Total impedance, including reflected impedance

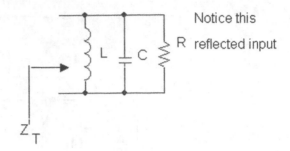

Notice this

R reflected input

$$a(s) = \frac{V_o}{V_i} = -g_m Z_T(s) \tag{7.45}$$

Z_T is described in Figure 7.30. R_i and C_i in Figure 7.29 are the input resistance and capacitance of M_1. R_i and C_i are reflected through the transformer to affect R and C in Figure 7.30. We can refer to Figure 7.29 to find R, C:

$$C = C_L + \frac{C_i}{n^2} \qquad R = R_o \| R_L \| n^2 R_i \tag{7.46}$$

Next we substitute (7.46) into (7.45) to get $a(s)$. This resulting expression and (7.44) are then multiplied together to get

$$af(s) = -\frac{g_m R}{n} \frac{\frac{L}{R} s}{1 + \frac{L}{R} s + LCs^2} \tag{7.47}$$

To start oscillation, we insert a small noise source V_x as shown in Figure 7.29. This has the transfer function

$$\frac{V_0}{V_x} = \frac{a(s)}{1 + af(s)} \tag{7.48}$$

Substituting (7.45), and (7.47) in (7.48), we get

$$\frac{V_0}{V_x} = \frac{-g_m R \dfrac{\frac{L}{R} s}{1 + \frac{L}{R} s + LCs^2}}{1 - \dfrac{g_m R}{n} \dfrac{\frac{L}{R} s}{1 + \frac{Ls}{R} + LCs^2}} \tag{7.49a}$$

$$\frac{V_0}{V_x} = -\frac{g_m R \dfrac{L}{R} s}{1 + \dfrac{L}{R} s + LCs^2 - \dfrac{g_m R}{n} \dfrac{L}{R} s} \tag{7.49b}$$

This transfer function has one zero and two poles and so can be represented as

$$\frac{V_o}{V_x} = \frac{Ks}{\left(1 - \frac{s}{s_1}\right)\left(1 - \frac{s}{s_2}\right)} \tag{7.50}$$

We now apply feedback by making n a finite number. Specifically, we set

$$\frac{g_m R}{n} = 1 \tag{7.51}$$

Substituting this in (7.49b), we find that the denominator of the transfer function becomes

$$1 + LCs^2 = \left(1 + js\sqrt{LC}\right)\left(1 - js\sqrt{LC}\right) \tag{7.52}$$

This has solutions:

$$s_1 = \frac{j}{\sqrt{LC}}, s_2 = -\frac{j}{\sqrt{LC}}$$

Hence the poles/zeroes of (7.49b) are as shown in Figure 7.31.

Consequently, if we plot $\left|\frac{V_o}{V_x}(\omega)\right|$ as a function of ω, it will peak up at ω_0, as shown in Figure 7.32.

To plot the root locus, let us revisit the denominator of the transfer function. From (7.49b) we have

$$\text{denominator of transfer function} = 1 + as + bs^2 = b\left(s^2 + \frac{a}{b}s + \frac{1}{b}\right) \tag{7.53}$$

Setting the denominator to zero, we have the following solutions:

$$s_1, s_2 = \frac{\left(-\frac{a}{b} \pm \sqrt{\left(\frac{a}{b}\right)^2 - \frac{4}{b}}\right)}{2} \tag{7.54}$$

Simplifying, we have

$$s_1, s_2 = -\frac{a}{2b} \pm \frac{a}{2b}\sqrt{1 - \frac{4b}{a^2}} \tag{7.55}$$

If the quality factor (Q) of the *RLC* tank is high, s_1, s_2 are complex and (7.55) can be rewritten as

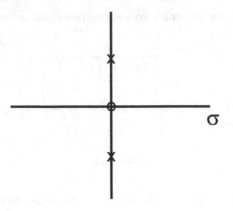

Fig. 7.31 Pole/zero diagram of transformer-based LC oscillator

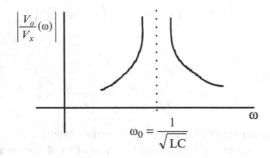

Fig. 7.32 Frequency response of transformer-based LC oscillator

$$s_1, s_2 = -\frac{a}{2b} \pm j \frac{a}{2b} \sqrt{\frac{4b}{a^2} - 1} \qquad (7.56)$$

Let $s = x \pm jy$; then from (7.56)

$$|s| = \sqrt{x^2 + y^2} = \sqrt{\frac{a^2}{4b^2} + \frac{a^2}{4b^2}\left(\frac{4b}{a^2} - 1\right)} = \frac{1}{\sqrt{b}} = \frac{1}{\sqrt{LC}} = \omega_0 \qquad (7.57)$$

From (7.57) we infer that the poles always have constant magnitudes. This is true for any loop gain and so the root locus is a circle. This is shown in Figure 7.33.

7.6.2   Alternate LC Oscillator Topologies

7.6.2.1   Common gate oscillator

The oscillator in Figure 7.26 can be redrawn with a common gate configuration, as shown in Figure 7.34. Since there is no phase shift in the transformer, we do not need transformers and we can use capacitors (use capacitive transformers). The result is the Colpitts oscillator shown in Figure 7.35.

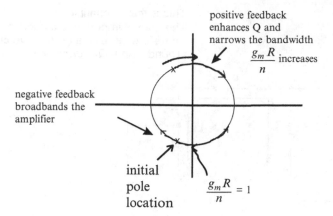

Fig. 7.33 Root locus of transformer-based *LC* oscillator

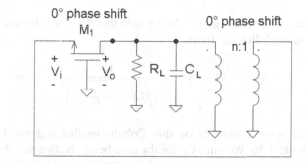

Fig. 7.34 Common gate *LC* oscillator

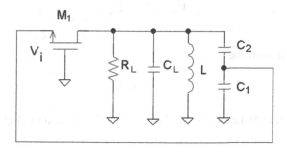

Fig. 7.35 Colpitts oscillator

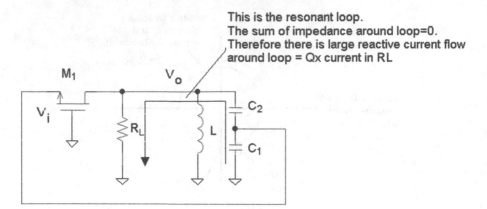

Fig. 7.36 Colpitts oscillator, resonant loop highlighted

7.6.2.2 Colpitts Oscillator

Comparing Figure 7.35 with Figure 7.26 we note that the n, ω_0 derived for Figure 7.26 now assume the following form:

$$n = \frac{C_2}{C_1 + C_2}, \omega_0 = \frac{1}{\sqrt{L\left(C_L + \frac{C_1 C_2}{C_1 + C_2}\right)}} \tag{7.58}$$

We now offer some comments on this Colpitts oscillator. First, Figure 7.35 is redrawn in Figure 7.36. We omit C_L for the time being. Notice that the L, C_1, C_2 loop is the resonant loop. Hence, the sum of impedance around the loop is 0 at ω_0. Therefore, there is a large reactive current flowing around this loop, whose value is given by Q × current in R_L.

Now we can analyze Figure 7.36 as follow. From the capacitive divider formula,

$$\frac{V_i}{V_o} = \frac{C_1}{C_1 + C_2} \tag{7.59}$$

From the gain formula,

$$V_o = g_m R_L V_i \tag{7.60}$$

Getting f from (7.59) and a from (7.60) and multiplying them together, we have

$$af = g_m R_L \frac{C_1}{C_1 + C_2} \tag{7.61}$$

Oscillation frequency is

$$\omega_0 = \frac{1}{\sqrt{L\frac{C_1 C_2}{C_1 + C_2}}} \tag{7.62}$$

Since at resonance the sum of impedance around the loop $= 0$,

$$\therefore \frac{1}{j\omega L} = -j\omega \frac{C_1 C_2}{C_1 + C_2} \quad \text{at oscillation frequency} \tag{7.63}$$

Therefore

$$\frac{1}{j\omega L} + \frac{1}{j\omega C_1} + \frac{1}{j\omega C_2} = 0 \tag{7.64}$$

$$\therefore jX_L + jX_1 + jX_2 = 0 \tag{7.65}$$

where

$$X_1 = -\frac{1}{\omega C_1} \qquad X_2 = -\frac{1}{\omega C_2}$$

or

$$\frac{1}{j\omega L} = -\frac{j\omega C_1 C_2}{C_1 + C_2} \tag{7.66}$$

Hence we see that at oscillation we create a negative impedance from C_1, C_2 whose value is equal and opposite to the impedance of L. Accordingly, the condition of oscillation can also be explained using this negative impedance.

Next let us redraw Figure 7.36 in Figure 7.37, where there is a change in location of ground. Now it becomes the common source (CS) Colpitts oscillator.

Furthermore, let us redraw Figure 7.37 in Figure 7.38. It is now relatively easy to see why this is an oscillator.

First, from Figure 7.38 we can see that the reverse transmission transfer function is given by

$$\frac{V_i}{V_o} = \frac{\frac{1}{j\omega C_2}}{j\omega L + \frac{1}{j\omega C_2}} = \frac{1}{1 - \omega^2 L C_2} \tag{7.67}$$

At oscillation frequency, ω becomes

$$\omega_0 = \frac{1}{\sqrt{L\frac{C_1 C_2}{C_1 + C_2}}} \tag{7.68}$$

Fig. 7.37 Common source
Colpitts oscillator

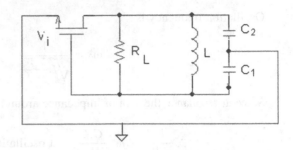

Fig. 7.38 Common source
Colpitt oscillator, alternate
orientation

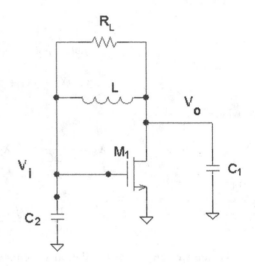

Setting $\omega = \omega_0$ in (7.67) and substituting (7.68) in (7.67), we have

$$\frac{V_i}{V_o} = -\frac{C_1}{C_2} \tag{7.69}$$

Hence, at ω_0 there is exactly 180° phase shift in this network, making the total phase
shift 360°, or satisfying the condition for oscillation.

7.6.3 Tuning, K_{vco}

How do we perform tuning? One way is to use so-called varactor tuning. This is
achieved by adding a variable capacitor in series with the original capacitor.
For example C_1 in Figure 7.38 is replaced by a series combination of C_1 and C_v,
as shown in Figure 7.39. C_v is the capacitance from a varactor.

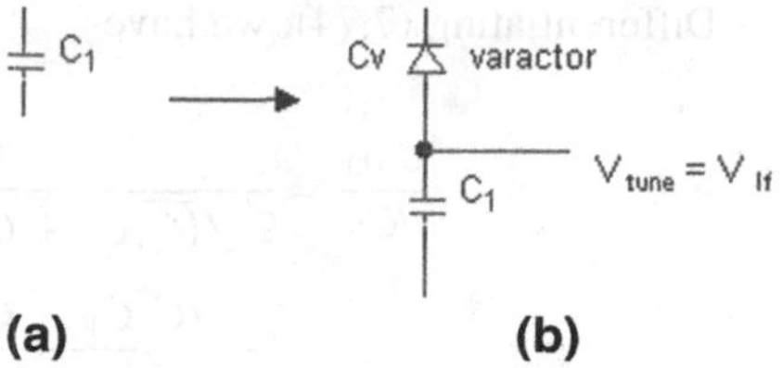

Fig. 7.39 Tuning circuit for
LC oscillator

    (a)   (b)

    The varactor is basically a reverse biased *pn* junction diode whose depletion capacitance is a function of voltage. Even though popular, this type of tuning does not have a wide tuning range and is also sensitive to temperature variation. We now quantify the K_{vco} of such an arrangement. We start from (7.1) and rewrite the definition of K_{vco} with units in rad/s:

$$K_{vco} = \frac{d\omega_o}{dV_{lf}} = \frac{d\omega_o}{dV_{tune}} = 2\pi \frac{df_o}{dV_{tune}} \tag{7.70}$$

Whether K_{vco} is in units of deg/s/V or rad/s/V should be evident from the context.
    Now from pn junction theory we have,

$$C_V = \frac{C_{V0}}{\sqrt{1 - \frac{V_{tune}}{\phi_0}}} \tag{7.71}$$

Here C_{V0} is the capacitance at zero bias and V_{tune}, the tuning voltage, is the input to the VCO. In a PLL this also equals V_{lf}, the loop filter voltage. ϕ_0 is a constant that is around 0.6 V.
    Differentiating (7.71), we have

$$\frac{dC_V}{dV_{tune}} = \frac{C_{V0}}{2\phi_0 \left(\sqrt{1 - \frac{V_{tune}}{\phi_0}}\right)^3} \tag{7.72}$$

According to Figure 7.39, C_1 in Figure 7.38 becomes:

$$C_1 \rightarrow \frac{C_1 C_V}{C_1 + C_V} \tag{7.73}$$

Substituting (7.73) in (7.68), we have

$$\omega_0 = \frac{1}{\sqrt{L \frac{C_1 C_2 C_V}{C_1 C_2 + C_1 C_V + C_2 C_V}}} \tag{7.74}$$

Differentiating (7.74), we have

$$
\frac{d\omega_0}{dC_V} = \frac{C_1 + C_2}{2\sqrt{(C_1C_2 + C_1C_V + C_2C_V)(C_1C_2C_VL)}}
$$
$$
- \frac{\sqrt{C_1C_V + C_2C_V + C_1C_2}}{2\left(\sqrt{LC_1C_2}\right)^3}
\tag{7.75}
$$

Finally, expanding (7.70) and substituting (7.75) and (7.72) in the expansion, we have

$$
K_{\text{vco}} = \frac{d\omega_0}{dV_{\text{tune}}} = \frac{d\omega_0}{dC_V}\frac{dC_V}{dV_{\text{tune}}}
$$
$$
= \frac{C_{V0}}{2\phi_0\left(\sqrt{1 - \frac{V_{\text{tune}}}{\phi_0}}\right)^3}
$$
$$
\times \left(\frac{C_1 + C_2}{2\sqrt{(C_1C_2 + C_1C_V + C_2C_V)(C_1C_2C_VL)}} - \frac{\sqrt{C_1C_V + C_2C_V + C_1C_2}}{2\left(\sqrt{LC_1C_2}\right)^3}\right)
\tag{7.76}
$$

What are the design considerations of K_{vco}? K_{vco} in general should be designed to be as small as possible. However, it must be large enough that f_o can span the whole frequency range with a tuning voltage, V_{1f}, that is within the power supply. K_{VCO} should be made small because the varactor is connected to the LC tank via a small fixed capacitor. If K_{vco} is large, then this coupling capacitor is large and the varactor has a large influence on the resonant frequency of the LC tank. In addition, the varactor itself has a low Q factor, in particular when compared with the inductor or capacitor in the oscillator. This is due to the resistance in the varactor itself (on the order of 1 Ω) and also due to packaging. A large varactor influence (due to a large K_{vco}) and a low Q varactor mean that the varactor resistance is translated across to the tank circuit, and this would reduce the Q of the tank significantly.

7.7 Ring Oscillators

We now turn our attention to the second oscillator category: waveform-based oscillators. To illustrate its nature we focus our discussion on one specific type: ring oscillators. A ring oscillator is shown in Figure 7.40. Notice that this structure also employs positive feedback to achieve oscillation. However, because switching is involved, it can no longer be treated as an LTI system and all the results

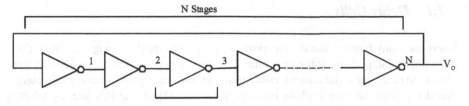

Fig. 7.40 Block diagram of ring oscillator

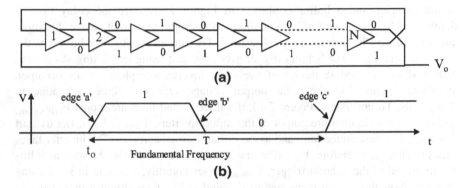

Fig. 7.41 (a) Differential implementation of ring oscillator. (b) Corresponding waveform

developed in sub-section 7.5.2 are no longer applicable. We now study this using large-signal time domain analysis.

Let us start with Figure 7.41a, which depicts an N-stage ring oscillator realized using differential cells (which have complementary outputs). A source coupled pair (SCL) inverter will be a typical implementation. Assume that at time t_o the output of stage 1 changes to logic 1 (denoted by edge a), as shown in Figure 7.41b. When this logic 1 propagates to the end, it creates a logic 1 at the N_{th} stage, which, when fed back to the input of the first stage, creates a logic 0 in the first stage output. This is edge b. When this logic 0 is propagated through the chain again, it toggles the output of stage 1 and triggers edge c. Notice that it takes two passes through the chain to complete a period (each pass generates an edge transition, and we need an up transition followed by a down transition to complete a period). Denoting t_p as the propagation delay through each stage, then period $T = 2Nt_p$. For a single-ended output cell, N has to be odd, but for a differential cell N can be odd/even, to start an oscillation. There is, of course, a minimum N. The minimum N depends on the ω_0 and t_p available from a given technology. Assuming that 180° of phase shift is provided by the chain of N stages, each stage must provide $\frac{180^\circ}{N}$ of phase shift and sufficient gain (so that the overall gain is greater than 1) at ω_0. This usually means that the gain of each individual stage has to be greater than 1 as well.

7.7.1 Delay Cells

There are many features that differentiate the delay cells used in a ring oscillator. The most important, perhaps, is the slew time. Contrary to popular belief, it is actually the slew time (a dynamic parameter), rather than the interstage gain (a static parameter), that determines the overall phase noise performance [4]. Along this line, therefore, we categorize the delay cells of a ring oscillator into three different types [16]. The first one is a fast-slewing saturated delay cell [17]. This delay cell has fast rise time and fall time. It also performs full switching and therefore belongs to the saturated class of delay cell. An example that consists of one voltage-based inverter plus a Schmitt trigger and a buffer is shown in Figure 7.42. Here the delay time is determined by the amount of current supplied through the current source, the input capacitance, and the Schmitt trigger threshold. The current supplied is adjusted by the tuning voltage V_{tune}. The Schmitt trigger gives you fast rising and falling. As shown, each PMOS and NMOS device of the input inverter completely turns off upon switching. Figure 7.43 shows the output voltage swings of three such adjacent delay cells. To understand Figure 7.43, the three dotted lines are inputs $V_{in1} \ldots V_{in3}$ and the three solid lines are output of the input inverters, $V_{inv1} \ldots V_{inv3}$. Let us start from V_{in1} (the first dotted line) and assume that it drops abruptly. The inverter takes times to change. Therefore, V_{inv1} (the first solid line) rises slowly. As soon as it hits the threshold of the Schmitt trigger, $V_{schmitt1}$ rises abruptly, resulting in V_{out1} rising abruptly. Now this V_{out1} is connected to V_{in2} and so V_{in2} rises abruptly again (second dotted line). V_{in2} has the same steepness as V_{in1} but is delayed from V_{in1} by a certain period. The process repeats itself, as V_{inv2} (second solid line) responds to V_{in2} and drops slowly. Therefore, V_{inv2} is delayed from V_{inv1} by the same delay as V_{in2} is from V_{in1}. Finally, the process is repeated for the third stage, whose input is V_{in3} (third dotted line) and whose inverter output is V_{inv3} (third solid line). In general, we can make the following observations:

1. V_{inv} changes slowly, whereas V_{in} changes abruptly
2. When we go from one stage to the other, V_{in} and V_{inv} change in the opposite direction.
3. V_{inv} starts to change when V_{out} from the previous stage hits the voltage rail (either V_{dd} or ground).
4. $V_{schmitt1}$ and hence V_{out} (and the next stage V_{in}) start to change when V_{inv} crosses $V_{threshold}$, the threshold voltage of the Schmitt trigger. Usually the Schmitt trigger has two thresholds. However, to simplify the present explanation, only one threshold is assumed.

The present design is a fast-slewing delay cell because of the rapid switching of V_{out}, made possible by the use of Schmitt trigger. Another variation of this type of delay cell consists of only one inverter without buffer.

The second type of delay cell is a slow-slewing saturated delay cell. An example of this type is shown in Figure 7.44. Here the inverter consists of a source coupled pair (SCP) and hence this is a current-based inverter. In this case, full switching also occurs.

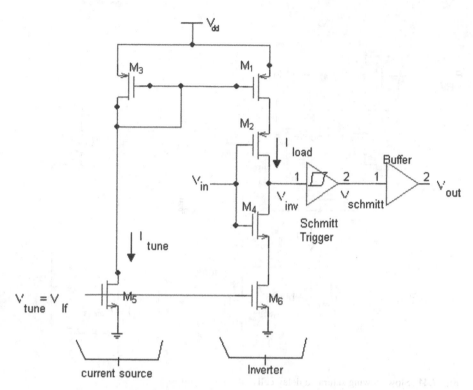

Fig. 7.42 Fast slewing saturated delay cell

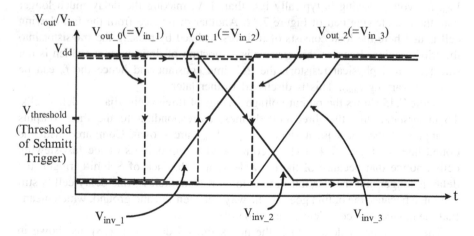

Fig. 7.43 Waveform of three cascaded fast-slewing saturated delay cells

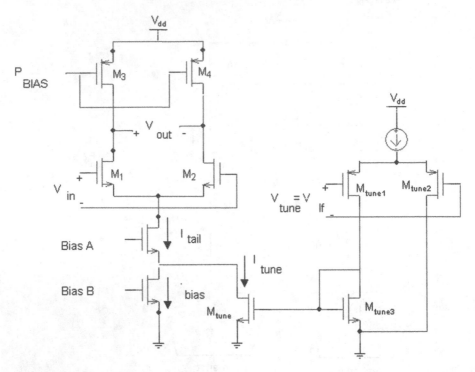

Fig. 7.44 Slow-slewing saturated delay cell

Therefore, this delay cell belongs to the saturated class of delay cell, just like the one described in Figure 7.42. On the other hand, it is called slow slewing because it has a longer gate delay. Because of the square law drain current characteristics, the delay is inversely proportional to the voltage swing. For a low supply voltage V_{dd}, the voltage swing is typically less than 1 V, making the delay much longer than the fast-slewing cell of Figure 7.42. Another difference from the fast-slewing cell is that here the load consists of resistors realized by biasing the transistor into the triode region. Because this resistor is a function of drain current (which is not true for a real physical resistor), the RC time constant and hence the f_o can be tuned by varying I_{tune}. This is discussed further later.

Figure 7.45 shows the output voltage swings of three such adjacent delay cells. To understand this, the three dotted lines, corresponding to the three outputs (= inputs of the subsequent stages) $V_{out_0} ... V_{out_2}$ are shown. Compared with the dotted lines of Figure 7.43 (which correspond to the outputs of the fast-slewing cells), notice that because of the extra delay and the lack of Schmitt trigger, the delay is longer and the rise/fall time is longer. Nonetheless, this delay cell is still saturated because the output goes all the way between V_{dd} and ground, which means that the devices are completely turned on/off.

The third type of delay cell is the non-saturated delay cell [18], as shown in Figure 7.46. This is also a voltage inverter based delay cell. Compared with

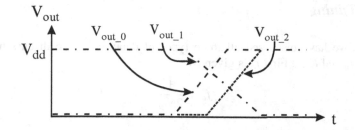

Fig. 7.45 Waveform of three cascaded slow-slewing saturated delay cells

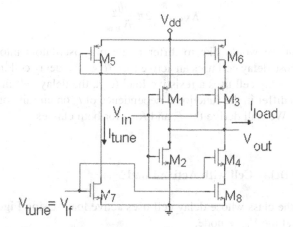

Fig. 7.46 Non-saturated delay cell

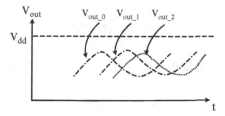

Fig. 7.47 Waveform of 3 cascaded non-saturated delay cells

Figure 7.42, it does not have a Schmitt trigger to enable fast turning on/off. Compared with both Figure 7.42, and Figure 7.44, the transistors M_1, M_2, M_3, M_4, M_6, M_8 never fully turn on/off. As a result, as shown in Figure 7.47, the output waveforms $V_{out0} \ldots V_{out2}$ never reach V_{dd} or ground which is why this type of delay cell is called non-saturated.

7.7.2   Tuning

Now that we have discussed all three types of delay cells, we discuss how we calculate f_o and K_{vco}. First f_o is given by

$$f_o = \frac{1}{2Nt_p} \tag{7.77}$$

where N = number of stages and t_p = delay of each stage.

Next we want to calculate K_{VCO}. First let us repeat its definition from (7.70):

$$K_{VCO} = 2\pi \frac{df_o}{dV_{tune}} \tag{7.78}$$

At this point we would like to differentiate ring oscillators into two classes, namely one whose delay cell uses an active load (e.g., the delay cell in Figure 7.42) and one whose delay cell uses a resistive load (e.g., the delay cell in Figure 7.44). The delay cells differ in the functional dependence of t_p on circuit components and tuning voltage. We will derive the f_o and K_{vco} for both classes.

7.7.2.1 K_{vco}: Delay Cell with Active Load

We start with the class whose delay cell uses active load. Using Figure 7.42 as an example, here at the V_{inv}'s node,

$$I_{load} = C_l \frac{V_{swing}}{t_p} \tag{7.79}$$

Here C_l is the load capacitor at V_{inv}, and V_{swing} is V_{inv}'s voltage swing. Substituting (7.79) into (7.77), we have

$$f_o = \frac{I_{load}}{2C_l V_{swing} N} \tag{7.80}$$

Taking differentials on both sides, we have

$$df_o = \frac{dI_{load}}{2C_l V_{swing} N} \tag{7.81}$$

From Figure 7.42, $I_{load} = I_{tune}$ and so $dI_{load} = dI_{tune}$. Hence

$$df_o = \frac{dI_{tune}}{2C_l V_{swing} N} \tag{7.82}$$

Fig. 7.48 Simplified slow-slewing saturated delay cell circuit schematic

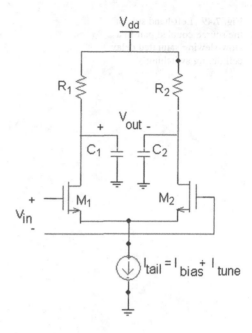

Finally, substituting (7.82) into (7.78),

$$K_{VCO} = 2\pi \frac{df_o}{dV_{tune}} = 2\pi \frac{dI_{tune}}{2C_l V_{swing} N dV_{tune}} = \frac{2\pi G_m}{2C_l V_{swing} N} \qquad (7.83)$$

The last equality comes about because definition $G_m = dI_{tune}/dV_{tune}$, where G_m is the transconductance of tuning transistor M_5, M_6.

7.7.2.2 K_{vco}: Delay Cell with Resistive Load

Next we turn to the class whose delay cell uses resistive load. This is the class used for the rest of this book. We use Figure 7.44 as an example to facilitate discussion. We redraw Figure 7.44 in Figure 7.48, where loads M_3, M_4 are represented by real resistors R_1, R_2. C_1, C_2 are load capacitors (from device, parasitic capacitance) hanging at V_{out}^+, V_{out}^-.

First let us derive f_o. To derive this we know from (7.77) that we need t_p. Hence we would now derive t_p, which is the time between the zero crossings of V_{out}. When do zero crossings occur? As an example, when V_{in} is positive, V_{in} switches I_{tail} from M_2 to M_1. Hence

Fig. 7.49 Left-hand side of the source coupled pair in a slow-slewing saturated delay cell during switching

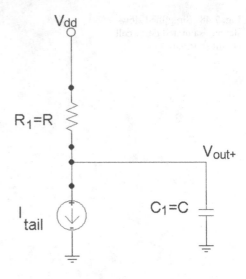

$$V_{out}{}^+ \text{goes from } V_{dd} \text{ to } V_{dd} - V_{swing} \qquad (7.84)$$

$$V_{out}{}^- \text{ goes from } V_{dd} - V_{swing} \text{ to } V_{dd} \qquad (7.85)$$

The zero crossing occurs when

$$V_{out}{}^+ = V_{out}{}^- = V_{dd} - (V_{swing}/2) \qquad (7.86)$$

Let us calculate the time between the zero crossings. First we note that for any first-order RC circuit with output voltage V_{out}, we have

$$V_{out}(t) = V_{out}(\text{final}) + [V_{out}(\text{initial}) - V_{out}(\text{final})] \exp(-t/RC) \qquad (7.87)$$

Let us now redraw left-hand side of the circuit in Figure 7.48 in Figure 7.49. Here we assume that V_{in} is positive so that all of I_{tail} is switched to M_1. Hence we replace M_1 by a current source with value I_{tail}. Next we assume that $R_1=R_2=R$ and $C_1=C_2=C$.

Focusing on Figure 7.49, we note that it is a first-order RC circuit and we can use (7.87) to find $V_{out}{}^+$. To apply (7.87) we need to find $V_{out}{}^+(\text{initial})$ and $V_{out}{}^+(\text{final})$. From (7.84) we know that

$$V_{out}{}^+(\text{initial}) = V_{dd} \qquad (7.88)$$

Fig. 7.50 Right-hand side of the source coupled pair in a slow-slewing saturated delay cell during switching

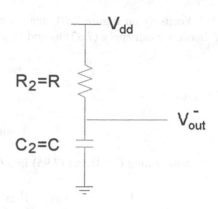

$$V_{out}{}^+(final) \ = V_{dd} - V_{swing} = V_{dd} - I_{tail} R \qquad (7.89)$$

Substituting (7.88) and (7.89) into (7.87), we have

$$
\begin{aligned}
V_{out}{}^+(t) &= V_{dd} - V_{swing} + (V_{dd} - V_{dd} + V_{swing})\exp(-t/RC) \\
&= V_{dd} - V_{swing}(1 - \exp(-t/RC))
\end{aligned}
$$

$$
\begin{aligned}
V_{out}{}^+(t) &= V_{dd} - V_{swing} + (V_{dd} - V_{dd} + V_{swing})e^{\left(\frac{-t}{RC}\right)} \\
&= V_{dd} - V_{swing}(1 - e^{\left(\frac{-t}{RC}\right)})
\end{aligned} \qquad (7.90)
$$

From (7.86), zero crossing occurs when

$$V_{out}(t) \ reaches V_{dd} - V_{swing}/2 \qquad (7.91)$$

Substituting (7.91) into (7.90), we have

$$V_{dd} - V_{swing}/2 \ = V_{dd} - V_{swing}\left(1 - \exp(-t_p/RC)\right) \qquad (7.92)$$

Solving, we have

$$t_p = RC\ln 2 \qquad (7.93)$$

To be consistent, let us rederive this by using the right-hand side of the circuit. We redraw the right hand side of the circuit in Figure 7.48 in Figure 7.50. Remember that we have assumed that V_{in} is positive and so M_2 is off. Since M_2 is off, it is not shown in Figure 7.50.

Focusing on Figure 7.50, again we note that it is a first-order RC circuit and hence we can apply (7.87) to find V_{out}^-. From (7.85) we have

$$V_{out(initial)}^- = V_{dd} - V_{swing} \tag{7.94}$$

and

$$V_{out(final)}^- = V_{dd} \tag{7.95}$$

Substituting (7.94) and (7.95) into (7.87) we have

$$\begin{aligned} V_{out}^-(t) &= V_{dd} + \left(V_{dd} - V_{swing} - V_{dd}\right) \exp\left(-t/RC\right) \\ &= V_{dd} - V_{swing}\exp\left(-t/RC\right) \end{aligned} \tag{7.96}$$

From (7.86), zero crossing occurs when

$$V_{out}(t) \text{ reaches } V_{dd} - V_{swing}/2 \tag{7.97}$$

Substituing (7.97) into (7.96), we have

$$V_{dd} - V_{swing}/2 = V_{dd} - V_{swing}\left(1 - \exp(-t_p/RC)\right) \tag{7.98}$$

Solving again we have:

$$t_p = RC\ln 2 \tag{7.99}$$

Now that we have derived t_p, let us go back and substitute (7.93) or (7.99) into (7.77) to get f_o:

$$f_o = 1/(2NRC\ln 2) \tag{7.100}$$

Next we want to derive K_{vco}. To find K_{vco}, take differentials on both sides of (7.100):

$$df_o = \frac{1}{2NC\ln 2}\frac{-1}{R^2}dR \tag{7.101}$$

To find f_o's dependence on I_{tune}, we express dR in terms of dI_{tune}.

$$df_o = -\frac{1}{2R^2NC\ln 2}\frac{dR}{dI_{tune}}dI_{tune} = \frac{-f_o}{R}\frac{dR}{dI_{tune}}dI_{tune} \tag{7.102a}$$

Alternately, (7.102a) can be expressed in terms of dV_{tune}:

$$df_o = \frac{1}{2R^2NC\ln 2}\frac{dR}{dI_{tune}}G_m dV_{tune} = \frac{-f_o}{R}\frac{dR}{dI_{tune}}G_m dV_{tune} \tag{7.102b}$$

Here G_m is the transconductance of the V_{tune}-I_{tune} converter. This V_{tune}-I_{tune} converter was shown in Figure 7.44. Referring to Figure 7.44, G_m is simply the g_m of transistor M_{tune1} in Figure 7.44, denoted as $g_{mMtune1}$ (we assume the current mirror M_{tune}-M_{tune3} has a gain of 1). How about $\frac{dR}{dI_{tune}}$? To find this again we refer to Figure 7.44. There are two cases: the first is when P_{bias} is a constant, and the second is when P_{bias} varies in a way that maintains a constant V_{swing} across M_3 and M_4.

Case 1: $P_{bias} = $ constant
Since M_3, M_4 are biased in the triode region, we know that

$$R = \frac{1}{k_p(W/L)_3(V_{gs3} - V_t - V_{ds3})} \tag{7.103}$$

Differentiating (7.103) with respect to I_{tune} and noting that $dV_{ds}/dI_{tune} = R$, we have

$$\frac{dR}{dI_{tune}} = \frac{R^2}{V_{gs3} - V_t - V_{ds3}} \tag{7.104}$$

Substituting (7.104) into (7.102a),

$$\frac{df_o}{dI_{tune}} = \frac{-f_o R}{V_{gs3} - V_t - V_{ds3}} \tag{7.105a}$$

Substituting (7.104) into (7.102b),

$$\frac{df_o}{dV_{tune}} = \frac{-f_o R g m_{Mtune1}}{V_{gs3} - V_t - V_{ds3}} \tag{7.105b}$$

Substituting (7.105b) into (7.78), we have,

$$K_{vco} = \frac{-2\pi f_o R g m_{Mtune1}}{V_{gs3} - V_t - V_{ds3}} \approx \frac{-2\pi f_o R g m_{Mtune1}}{V_{GS3} - V_t - V_{DS3}} \tag{7.106}$$

The last equality occurs because we assume that V_{gs}, V_{ds} do not change much. Substituting (7.103) in (7.106),

$$K_{vco} = \frac{-2\pi f_o g m_{Mtune1}}{k_p(W/L)_3(V_{GS3} - V_t - V_{DS3})^2} \tag{7.107}$$

Case 2: V_{swing} is a constant.
V_{swing} is set constant by a circuit called the replica bias. The idea is that we use a feedback loop that sets f_o to be constant, irrespective of temperature, power supply, and process variation. We call this coarse tuning. This can also be used to do fine tuning: that is, changing f_o as a function of V_{tune}. In practice, this poses some

additional problems because it will put the replica bias circuit inside the PLL. Hence extra poles introduced by this circuit will compromise the dynamics of PLL. If we do indeed choose to use the replica bias circuit to perform fine tuning, then we can also derive K_{vco} with this method.

First we note that, by design,

$$V_{swing} = RI_{tail} = \text{constant} \tag{7.108}$$

Differentiating (7.108) with respect to I_{tune},

$$\frac{dR}{dI_{tune}} = -\frac{R}{I_{tune}} \tag{7.109}$$

Substituting (7.109) into (7.102b) and simplifying,

$$\frac{df_o}{dV_{tune}} = \frac{f_o}{I_{tune}} g_{m_{Mtune1}} \tag{7.110}$$

Substitute (7.110) into (7.78), we finally have

$$K_{vco} = \frac{2\pi f_o}{I_{tune}} g_{m_{Mtune1}} \tag{7.111}$$

As with K_{vco} of an LC oscillator, K_{vco} of a ring oscillator should also be designed as small as possible. Of course, it must also be large enough that f_o can span the whole frequency range. Why is a small K_{vco} also desirable in the case of a ring oscillator? In this case since the propagation delay is a function of C_l, V_{swing}, and I_{bias}, we can tune using C_l (as in the varactor case) or using V_{swing} (by changing the load resistance) or using I_{bias}. It turns out that tuning using I_{bias} is the more convenient choice. In the case of tuning by varying I_{bias}, the thermal noise in I_{bias} will degrade the phase noise. The degradation becomes worse as K_{vco} becomes larger; hence it is also desirable to keep K_{vco} small.

7.8   Introduction to Phase Noise

7.8.1   Definition

What is phase noise? We briefly mentioned that in Chapter 2 and compared it to amplitude noise. Since we are familiar with amplitude noise, we start by repeating its definition. From Chapter 2, we know that we can describe voltage noise (a type of amplitude noise) by

$$\overline{V^2} = S_v(f)\Delta f \tag{7.112}$$

Here $S_v(f)$ is the power spectral density in V^2/Hz and $\overline{V^2}$ is the average of the square of noise voltage. The unit of measurement, for $\overline{V^2}$ is volt^2. When we extend this definition to the definition of phase noise, we have

$$\overline{\theta^2} = S_\theta(f)\Delta f \tag{7.113}$$

Here $S_\theta(f)$ is the power spectral density (PSD) of phase at modulation frequency f, and $\overline{\theta^2}$ is the average of the square of phase deviation for frequencies from $f - \frac{1}{2}\Delta f$ to $f + \frac{1}{2}\Delta f$. The unit for $\overline{\theta^2}$ is rad^2 or cycles^2 and for S_θ is rad^2/Hz or cycles^2/Hz. For example, if S_θ is 8 cycles^2/Hz, the average of the square of phase deviation in a 2-Hz band is 16 cycles. If we pass such a signal through a phase detector, it will produce a current power spectrum from a phase power spectrum and phase noise will be converted into current noise.

7.8.2   Interpretation

In this sub-section we clear up a subtle point that relates phase noise and amplitude (voltage or current) noise. As a start, let us assume that the oscillator has a sinusoidal output $S_o(t)$ (like the one given in Figure 7.20) that is now corrupted by phase noise. Hence the form becomes

$$S_o(t) = \cos(\omega_c t + \Delta\theta(t)) \tag{7.114}$$

Here ω_c is the oscillation or carrier frequency, and $\Delta\theta(t)$ is the random phase fluctuation due to noise inside the oscillator. We assume that the amplitude of $S_o(t)$ is 1. Let us further assume that $\Delta\theta(t)$ has a spectral density, denoted as $\Delta\theta(f)$, as shown in Figure 7.51. ($|\Delta\theta(f)|^2$ is, of course, the PSD of the phase noise.) This is a baseband spectrum and is plotting the phase variable, $\Delta\theta(f)$, as a function of frequency. Next we will represent this spectral density by individual frequency impulses. As an example, let us refer to Figure 7.52, where two pairs of the impulses are shown, at frequencies $\pm f_{m1}$, $\pm f_{m2}$. Using the principle of superposition, let us apply each of these individual pairs of frequency impulses to (7.114). We start with the pair $\Delta\theta(f_{m1})$, $-\Delta\theta(f_{m1})$, with frequency impulses at $\pm f_{m1}$. They are represented in the time domain as:

$$\Delta\theta_{fm_1}(t) = A_{\Delta\theta}(f_{m1})\cos(\omega_{m1}t) \tag{7.115}$$

We then substitute (7.115) in (7.114):

$$S_o(t) = \cos\left(\omega_c t + A_{\Delta\theta}(f_{m1})\cos(\omega_{m1}t)\right) \tag{7.116}$$

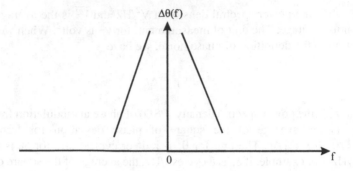

Fig. 7.51 Oscillator phase noise spectral density, phase representation

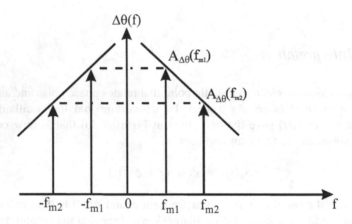

Fig. 7.52 Equivalent representation of oscillator phase noise spectral density, phase representation, using frequency impulses at $\pm f_{m1}$, $\pm f_{m2}$

We can see that (7.116) represents phase modulation. Next we assume that the phase fluctuation is small and so $\Delta\theta_{fm1}(t)$ is small, which means that $A_{\Delta\theta}(f_{m1})$ is small as well. Hence this phase modulation is a narrowband phase modulation, which can be approximated by amplitude modulation. Accordingly, $S_o(t)$ in (7.116) consists of $A_{\Delta\theta}(f_{m1})\cos((\omega_c\pm\omega_{m1})t)$ and $\cos(\omega_c t)$ [15]. This is shown in Figure 7.53. We can see that they become bandpass signals, centered at f_c. Moreover, the phase variable $\Delta\theta(f)$ is transformed to an amplitude variable, $S_o(f)$. Looking at Figure 7.53, we can see that $S_o(f)$ consists of the carrier at f_c and two frequency impulses spaced $\pm f_{m1}$ from f_c, with the amplitudes being $A_{\Delta\theta}(f_{m1})$.

Next let us repeat the procedure for the pair $\Delta\theta(f_{m2})$, $\Delta\theta(-f_{m2})$ the frequency impulses at $\pm f_{m2}$, as shown in Figure 7.52. Then (7.114) becomes

$$S_o(t) = \cos(\omega_c t + A_{\Delta\theta}(f_{m2})\cos(\omega_{m2}t)) \qquad (7.117)$$

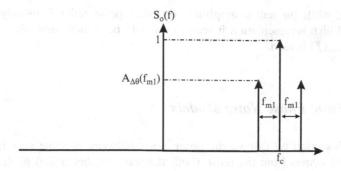

Fig. 7.53 Frequency impulses at the oscillator output due to frequency impulses injected at $\pm f_{m1}$

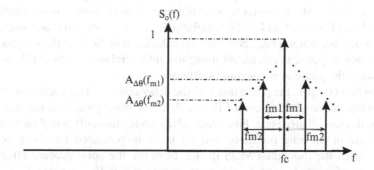

Fig. 7.54 Frequency impulses at the oscillator output due to frequency impulses injected at $\pm f_{m1}$, $\pm f_{m2}$. The envelope formed by connecting the peaks of these frequency impulses at the oscillator output becomes the oscillator phase noise spectral density, amplitude representation

This is again narrowband phase modulation and hence $S_o(t)$ consists of the carrier at f_c and two frequency impulses spaced $\pm f_{m2}$ away from f_c, with the amplitudes being $A_{\Delta\,\theta}(f_{m2})$. If we add these new frequency impulses to Figure 7.53, we have Figure 7.54. Furthermore, let us connect the peaks of these frequency impulses by two dotted lines, as shown in Figure 7.54. These dotted lines are called the envelope. Now let us compare the envelope in Figure 7.54 with the spectral density in Figure 7.52. We note that they have the same shape. The difference, of course, is that Figure 7.52 is a baseband plot and is plotting the spectral density of the phase variable, whereas Figure 7.54 is a bandpass plot (or RF) and is plotting the spectral density of the amplitude variable. Strictly speaking, Figure 7.52 is the spectral density of the true phase noise. However, because the plot in Figure 7.54 has the same shape as that in Figure 7.52, it can be used to represent indirectly the spectral density of the phase noise as well. Moreover, Figure 7.54 represents what we can actually measure (amplitude). From now on, we will use either representation as suited and we will use $S_{\theta\_vco}(f)$ to denote the PSD of both representations of the phase noise in a VCO. The unit in phase domain representation is usually given in rad^2/Hz, while the unit in amplitude domain representation is usually given in dBc/Hz. Which representation it refers to should be evident from the context in which $S_{\theta\_vco}(f)$ is used.

7.8.3 Basic Phase Noise Models

In this sub-section we discuss the phase noise behavior of some specific VCOs. Phase noise comes from the noise (both thermal and shot noise) of the internal transistors and resistors. Rough first order estimate of phase noise of a resonator-based VCO, such as the *LC* tank VCO, can be carried out by continuing to assume the VCO to be a LTI system. We will have a chance to do that in the Problem 7.7. More sophisticated treatment of phase noise calls for treating the resonator based VCO as a LPTV system, similar to what has been done in mixer's noise analysis in Chapter 4 and Chapter 5 and in the simulator *SpectreRF*. For waveform oscillators such as a ring oscillator, phase noise is best calculated in the time domain and then its frequency response is calculated using the autocorrelation method [4]. We will now discuss the phase noise of ring oscillators.

Let us first compare the phase noise of the delay cells in a ring oscillator. For the fast-slewing saturated delay cell of Figure 7.42, the noise process is not stationary since the devices change state. First, each MOS device turns off completely once in every cycle of the ring oscillator and no noise is generated for those periods. Second, when the individual MOS device turns on, the noise process affects the delay time of each cell up to the time instant when V_{out} crosses $V_{threshold}$ of the next delay cell. Again that is because once V_{out} crosses $V_{threshold}$, the inverter of the next delay cell switches state.

For the slow-slewing saturated delay cell of Figure 7.44, full switching also occurs and the noise process is still non-stationary. However, the noise process affects the delay long after the time instant when V_{out} crosses $V_{threshold}$ [4,16].

Finally, for the non-saturated delay cell of Figure 7.46, because transistors are never turned off, the noise process can be approximated as a stationary process. For calculation of noise, each delay cell can be modeled as a linear amplifier. Comparing the three classes of ring oscillators based on the three types of delay cells, the one in Figure 7.46 has the worst phase noise performance. In general, in terms of phase noise performance the fast-saturated type is better than the slow-saturated type, which in turn is better than the non-saturated type [16].

In this section let us concentrate on the ring oscillator based on the slow saturated type. We assume that the noise is dominated by thermal noise in transistor M_1, M_2 in this type of delay cell (shown in Figure 7.44). We first present the phase noise PSD of this ring oscillator, which is given as [4]

$$S_{\theta\_vco}(f_i) = \left(\frac{f_o}{f_i}\right)^2 \frac{kTa_v\eta^2}{I_{bias}(V_{GS1} - V_t)} \tag{7.118}$$

Here N is the number of stages, f_i is the frequency of interest (as offset from f_o), and f_o is the oscillator frequency. For the rest of the parameters let us refer to Figure 7.44. a_v is the voltage gain of the differential pair M_1, M_2. η is a constant that depends on the relative amount of time that transistor M_1 is in saturation and triode region. I_{bias} is the bias current, and $V_{GS_1} - V_t$ is the overdrive voltage of M_1. Notice that $S_{\theta\_vco}(f_i)$ has bandpass characteristics around $\omega_0 = 2\pi \times f_o$. This shape is called Lorentizian and falls off as $\frac{1}{f_i^2}$ from the center frequency.

From (7.118) notice that as power dissipated in the delay cell goes up I_{bias} times, $V_{GS_1} - V_t$ goes up and $S_{\theta\_vco}(f_i)$ goes down. Hence there is a trade-off between power and phase noise.

Now this oscillator is embedded inside a synthesizer, so to calculate the PSD of the synthesizer output phase noise due to VCO phase noise, denoted as $S_{\theta o\_vco}(f_i)$, we need to know the transfer function from the VCO to the synthesizer output, which we denote as H_{vco}. Using H_{vco}, $S_{\theta o\_vco}(f_i)$ can be expressed as

$$S_{\theta o\_vco} = |H_{vco}|^2 S_{\theta\_vco} \tag{7.119}$$

H_{vco} depends on the loop filter. The loop filter will be discussed in Chapter 8.

7.9 Advanced Phase Noise Model of Ring Oscillators

Let us return to Figure 7.40, which shows an N-stage ring oscillator consisting of N identical delay cells. Elaborating on section 7.7.1, and also following [2, 3] ring oscillators can be categorized as having saturated and unsaturated outputs. Ring oscillators with saturated outputs are those whose delay cells' outputs reach the power supply V_{dd} within the period of oscillation. Ring oscillators with unsaturated outputs are those whose delay cells' outputs never reach V_{dd} within the period of oscillation. Typically in ring oscillators with saturated outputs the transistors in each delay cell are fully switched. However in ring oscillators with unsaturated outputs, depending on the delay cells' design, either all the transistors in each delay cell remain on all the time and never turn off, or in each delay cell some of the transistors never turn off while some of the transistors are fully switched. In the case when all the transistors remain on all the time the ring oscillator is denoted as a non-saturated ring oscillator. In the case when some of the transistors never turn off while some transistors are fully switched the ring oscillator was denoted, in this paper, as an unsaturated oscillator. This categorization is shown in Figure 7.55.

7.9.1 Saturated Ring Oscillators

In the past ring oscillators with saturated outputs have been reported to have better phase noise characteristics than ring oscillators with unsaturated outputs [3]. This is because once output reaches V_{dd} in a cycle, noise is resetted and there is no

Fig. 7.55 Categorization of
ring oscillators

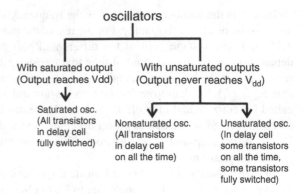

correlation in noise to the next cycle. To make oscillators saturated, one way is to increase the number of stages, N, in a ring oscillator. As N increases, the oscillator outputs eventually become saturated. This is because once the output crosses the threshold and triggers the next stage, it has time to ramp up to V_{dd} before the change propagates through the chain of delay cell, feeds itself back to the input of the delay cell, and makes the output ramp down again. Increasing N, however, increases hardware and/or decreases the oscillation frequency.

To look at the phase noise of saturated ring oscillator we start with section 7.8 above. There we present some basic phase noise theory. Assumptions were made to simplify the analysis. These assumptions affect the accuracy of the model as well as design insights.

To develop a more advanced phase noise model, we state explicitly these assumptions. For illustration we use the single ended oscillator, as shown above in Figure 7.40 whose delay cells are simple inverters, each consists of transistors, M_1 and M_2. We focus on the rising transition of V_{out}, where an example simulation is shown in Figure 7.57. V_{out} around the moment of switching is zoomed in and noise is visible. t_d is divided into two regions: t_1 (M_1 in triode) and t_2 (M_1 in saturation). An expression for phase noise of this single ended oscillator, with simple inverter delay cell, was derived with the following assumptions [12]:

1. When V_{out} crosses the threshold, it crosses the threshold only one time.
2. V_{in} switches instantly to 0V and stays constant during the rising transition of V_{out}, so that M_1 is replaced by a current source (Figure 7.56a), whose mean (deterministic part) is constant. The noise part of the current source has constant power spectral density (psd).
3. Referring to transistor M_1, M_2, during the rising transition, it is assumed that M_1 is on and in saturation throughout, while M_2 is off, as shown in Figure 7.56a i.e. it essentially ignores the fact that M_1 typically enters triode.

The consequences of these assumptions are now discussed.

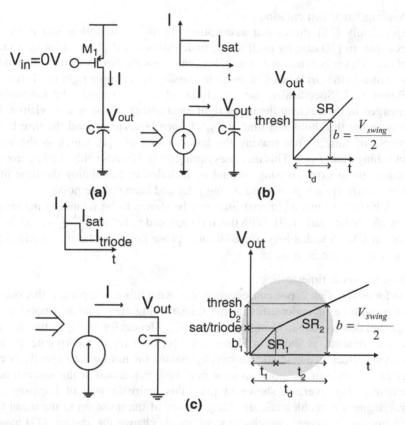

Fig. 7.56 Representation for simple inverter based ring oscillator

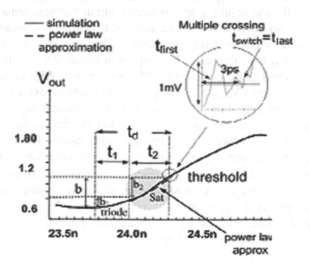

Fig. 7.57 Noise simulation (include device cap) of an example differential pair based ring oscillator with physical resistor load

A) Assumption 1: last crossing

Specifically [28] shows that assumption (1) (this assumption was made, for example, in [7]) can, by itself alone, underestimate the phase noise by a factor of two. This is because, with noise, V_{out} can cross the threshold more than once, as shown in the circle labelled multiple crossing at the upper right hand corner of Figure 7.57. Specifically, the waveform when V_{out} crosses the threshold is enlarged in the circle at the upper right hand corner. Noise is now visible. V_{out} crosses the threshold first time at t_{first}. It then bounces around the threshold a couple of time before making the last crossing at t_{last}, which is the actual switching time, t_{switch}. This increases timing jitter. Because only the last crossing counts, these extra crossings should be included in calculating the time jitter. These extra crossings increase timing jitter and hence phase noise.

A bound for this additional jitter can be shown to be related to the current noise/slew rate ratio [29]. With this ratio reduced to below $\frac{1\,pA/\sqrt{Hz}}{1\,V/ns}$, and for the typical CMOS technology, the additional phase noise has been minimized to a value that it can be ignored.

B) Assumption 2: time scaling

As for assumption 2, one consequence of making this assumption is that one can over-estimate or under-estimate the time varying slew rate and noise power spectral density, due to changing V_{in} and V_{out}. Hence, for example, the slew rate is not constant, as shown in Figure 7.57. [6] does try to incorporate the time varying effect, in a heuristic manner, by making the timing jitter (per delay cell) equal the ratio of the noise to slew rate, both calculated at the mean time of crossing. However, as shown in [29], this heuristic way of handling time varyingness is problematic since any history of the noise up to the mean time of crossing is ignored, which can significantly change the answer. [27] handles assumption (3) by the concept of time scaling.

Let us review the concept of time scaling by using the region t_2 in Figure 7.57 as an example. The resulting approximated output voltage, denoted as power law approx, is shown in as dotted line in region t_2, with time dependency written as t^{2p}. All parameters are subscripted with "2" because M_1 is in region t_2.

Let us now redraw the power law approx in Figure 7.58a. The time dependency is assumed to be $\sqrt{t}$ (i.e. p=0.25). Focusing on SR_2, because it is not constant, the mean of V_{out2}, $\overline{V_{out2}}$ ramps up with a non-constant slope. To show this we now label the time axis, t, as 0, 1, 2, 3. Notice the time spacing is *uniform*, i.e. change in time from $t = 0$ to $t = 1$ is identical to the change in time from $t = 1$ to $t = 2$, which in turn equals the time change from $t = 2$ to $t = 3$, where the time spacings are all 1 time unit. On the other hand, the change in $\overline{V_{out2}}$, over each corresponding time spacing, is increasing. To see this, note by definition, $\frac{dV_{out2}}{dt} = SR_2(t)$. As stated above, $SR_2(t)$ is proportional to $\sqrt{t}$. For illustration, we set the proportionality constant to 1. Then $SR_2(t) = \sqrt{t}$. This means the slew rate in region 2 is increasing with time (from $SR_2(1)$, at $t = 1$, to $SR_2(3)$, at $t = 3$), as shown in Figure 7.58a.

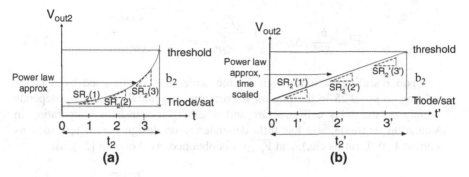

Fig. 7.58 Time scaling concept

Next we discretize the equation, $\frac{dV_{out2}}{dt} = SR_2(t)$, as $\frac{\Delta V_{out2}(n\Delta t)}{\Delta t} = SR_2(n\Delta t)$, $n = 1, 2 \ldots$, where n is the time index. Then as the time marches from $t = 0$ to $t = 1$ to $t = 2 \ldots$, we calculate $\Delta \overline{V_{out2}}(n\Delta t)$ by setting $\Delta t = 1$, and thus $\Delta \overline{V_{out2}}(n\Delta t) = \Delta \overline{V_{out2}}(n) = SR_2(n) \times 1$.   Then   at   n=1, $\Delta \overline{Vout2}$ is $SR_2(1) \times 1 = \sqrt{1} \times 1 = 1$ voltage unit.   Next,   at   n=2, $\Delta \overline{V_{out2}}$ is $SR_2(2) \times 1 = \sqrt{2} = 1.41$ voltage unit, and from $t = 2$ to $t = 3$ is 1.73 voltage units and so on. Thus as time marches along the time index n, $\overline{V_{out2}}$ ramps up with a non-constant slope.

Next we scale the original time t to a new time t', as shown in Figure 7.58b. The labels for the time axis are now superscripted with primes and become 1', 2', 3'. In the new time, the time spacing is *increasing*. Thus change in time from $t' = 0'$ to $t' = 1'$, equals 1 time unit, and change in time from $t' = 1'$ to $t' = 2'$, equals 1.41 time units. In turn, change in time from $t' = 2'$ to $t' = 3'$, equals 1.73 time units. This is shown in Figure 7.58b. *This increase exactly matches the change in $\overline{V_{out2}}$ in the original time.* This also means the slew rate SR_2' (primed to indicate time scaled) is now constant (i.e. $SR_2'(1')$, at $t' = 1'$, is the same as $SR_2'(3')$, at $t' = 3'$). This continues until M_1 hits threshold at time t_2'. t_2' has mean $\overline{t_2'}$ and jitter $\Delta t_2'$.

So far we have concentrated on $SR_2(t)$. Similar argument applies to the current noise power spectral density, $S_{I_{n2}}$, which is responsible for the noise part of V_{out2}. Working out the mathematics, it was shown in [27] that, in the new time scale, t', $S_{I_{n2}}'(t')$ (primed to indicate time scaled), are also constant.

In summary, whereas in time t, slew rate and noise psd were functions of time, in scaled time t', they become constants. Thus, in scaled time, assumption 2 becomes valid. Therefore $\overline{t_2'}$, $\Delta t_2'$ can be calculated by simply dividing the barrier and accumulated noise by the slew rate, respectively. If we assume the noise comes from thermal noise, then from [12, 27], the accumulated noise is $\frac{1}{\sqrt{2}} \frac{\sqrt{\overline{t_2'}S_{I_{n2}}'}}{C} \cdot \overline{t_2'}$, $\Delta t_2'$ are:

$$\overline{t_2'} = \frac{b_2}{SR_2'}, \Delta t_2' = \frac{1}{\sqrt{2}} \frac{\frac{\sqrt{\overline{t_2'}S_{In2}'}}{C}}{SR_2'} = \frac{1}{\sqrt{2}} \frac{\frac{\sqrt{\frac{b_2}{SR_2'} \times S_{In2}'}}{C}}{SR_2'} \qquad (7.120)$$

Upon rescaling back to original time scale, $\Delta t_2'$ is stretched by a factor relating to $p(=0.25$ in the present illustration; in general, value of p depends mainly on the delay cell structure and whether the transistor is operating in saturation or triode, and has little dependency on design parameters such as current I, W/L ratios etc.), and $\overline{V_{out2}}(t)$ is obtained, as shown in [27], as:

$$\Delta t_2 = \left(\frac{b_2}{SR_2'}\right)^{\left[\frac{1}{2p+1}-1\right]} \times (2p+1)^{\left[\frac{1}{2p+1}-1\right]} \times \frac{1}{\sqrt{2}} \frac{\frac{\sqrt{\frac{b_2}{SR_2'} \times S_{In2}'}}{C}}{SR_2'} \qquad (7.121)$$

C) Assumption 3: multiple threshold crossing

The consequence of assumption 3 is that with technology scaling, power supply is reduced. Hence the voltage headroom is reduced and M_1 can get into triode for a portion of the operation. In triode, the transistor slew rate is reduced and phase noise deteriorates. [26] handles assumption 3 by proposing using replica bias design to set V_{swing} so as to maintain M_1 remains in saturation, if possible. If that is not possible [26] gives analytical equation that predicts the phase noise, including the effect of M_1 entering triode.

Let us elaborate on this.

Returning to Figure 7.56a, let us take a look at M_1. V_{in} is at 0V. Initially M_1 is in saturation, and $\overline{I} = I_{sat}$, as shown in the I versus t diagram in Figure 7.56c. V_{out} increases with a constant slew rate, denoted as SR_1 in the V_{out} versus t diagram in Figure 7.56c (SR_1 in Figure 7.56c is the same as SR in Figure 7.56b). As V_{out} increases and crosses the sat/triode value, M_1 enters the triode region. However since $|V_{ds}|$ of M_1 equals $V_{dd} - V_{out}$, with M_1 in triode, I and noise change as V_{out} change. For assumption 2 to remain valid, we set $\overline{I}$ and noise power spectral density (psd) equal their respective average values over time (note the ensemble average is different from the average referred to here, which means time average. The ensemble average still varies with time). The time average value of $\overline{I}$ in the triode region is denoted as I_{triode}, as shown in the I versus t diagram. I_{sat} is larger than I_{triode}. Thus as M_1 changes from saturation to triode, $\overline{I}$ drops from I_{sat} to I_{triode}. As $\overline{I}$ decreases, the slew rate decreases, so when M_1 switches from saturation to triode, the slope of V_{out} decreases from SR_1 to SR_2, as shown in the V_{out} versus t diagram. Similarly S_{I_n} also changes when M_1 moves from saturation to triode.

To reiterate, in Figure 7.56c, physically when M_1 goes from saturation to triode, there is a change in current and noise. The change from saturation to triode happens at $V_{out} - V_{in} = V_t$, where V_t is the transistor threshold voltage. Since V_{in} equals 0, this happens at $V_{out} = V_t$. Mathematically V_{out} can be viewed as the state of the transition, which starts from 0, and increases towards to V_{dd}.

Thus the change from saturation to triode region can be interpreted as a change in state. A change in state, in turn, mathematically is governed by the Markov property [28]. This essentially means that as far as the transition is concerned, what happens after this change in state is captured by what happens at only the instant when $V_{out} = V_t$ (or sat/triode). To elaborate, at the beginning of a region, the inverter's voltages and currents are known. They specify the initial condition (initial voltages and currents). Since the only noise source taken into account is white, the uncertainty of the voltages and current in that region is independent of what happens before. Thus the circuit's voltages and currents in a region depend only on the initial condition of the circuit at the beginning of that region and not before. Therefore they possess Markov property.

Because of noise V_{out} reaches the value of V_t at a random time, in the same way V_{out} reaches threshold of the next stage inverter at a random time. The randomness in time for V_{out} to reach V_t is characterized by a probability distribution (which possesses a standard deviation), in much the same way the randomness in time for V_{out} to reach the threshold is also characterized by a (different) probability distribution and standard deviation (which we have denoted as the timing jitter). To repeat, whereas in the case where V_{out} crosses the threshold at $V_{dd}/2$, the following delay cell is switched, in the case where V_{out} crosses the value V_t, M_1 is switched from saturation to triode. Therefore V_{out} crossing V_t can also be viewed as crossing a threshold. To differentiate we call this an internal threshold. Just like the random time to threshold, which has a mean as well as a standard deviation/timing jitter, the random time to this internal threshold also has its own mean as well as its own standard deviation/ timing jitter. The mean time to the internal threshold is denoted as $\overline{t_1}$, the barrier is b_1 and the timing jitter is Δt_1. We further label the transition of V_{out} from 0 to the internal threshold as the first leg of the transition.

Returning back to the transition, after V_{out} reaches the internal threshold it continues to rise and eventually crosses the threshold. We label this as the second leg of the transition. Like the first leg, which starts from a known state ($V_{out} = 0$) and reaches another known state ($V_{out} = V_t$), the second leg also starts from a known state ($V_{out} = V_t$) and reaches another known state ($V_{out} = V_{dd}/2$). Then the noise injected in the second leg impacts its crossing of the threshold in much the same way the noise injected in the first leg impacts its crossing of the internal threshold. This means, like the first leg, the time associated with the second leg (a random time) also possesses a mean (denoted as $\overline{t_2}$), a barrier (denoted as b_2) and a timing jitter (denoted as Δt_2).

Because of Markov property, the influence of the first leg on the second leg is captured by what happens only at $V_{out} = V_t$, and is otherwise independent of what has happened earlier in the first leg. What happens only at $V_{out} = V_t$, is captured by Δt_1. Because Δt_1 and Δt_2 are independent, Δt_1 can be added to Δt_2 (in the r.m.s sense) to obtain Δt_d. Thus:

$$\Delta t_d = \sqrt{\Delta t_1 + \Delta t_2} \qquad (7.122)$$

For completeness we also have:

$$\bar{t}_d = \bar{t}_1 + \bar{t}_2 \tag{7.123}$$

Now referring to Figure 7.56c, within each of leg 1 and leg 2, M_1 stays in only one region. Thus slew rate and S_{I_n} are constant. Also in the present situation it is assumed that not only does V_{out} cross the threshold only one time, it also crosses the internal threshold (sat/triode) only one time. Thus assumptions (1)-(3) are satisfied and so we can easily find each of Δt_1 and Δt_2. Thus:

$$\Delta t_1 = \frac{1}{\sqrt{2}} \frac{\frac{\sqrt{\bar{t}_1 S_{I_{n1}}}}{C}}{SR_1} \text{ and } \Delta t_2 = \frac{1}{\sqrt{2}} \frac{\frac{\sqrt{\bar{t}_2 S_{I_{n2}}}}{C}}{SR_2} \tag{7.124}$$

Substituting (7.124) in (7.122):

$$\Delta t_d = \frac{1}{2} \sqrt{\frac{\bar{t}_1 S_{I_{n1}}/C^2}{SR_1^2} + \frac{\bar{t}_2 S_{I_{n2}}/C^2}{SR_2^2}} \tag{7.125}$$

From Figure 7.56c, $SR_1 = I_{sat}/C$ and $SR_2 = I_{triode}/C$. Since $I_{sat} > I_{triode}$, $SR_1 > SR_2$.

As before, we have:

$$\bar{t}_1 = \frac{b_1}{SR_1} \text{ and } \bar{t}_2 = \frac{b_2}{SR_2} \tag{7.126}$$

$b_1 = V_t$ and $b_2 = V_{dd}/2 - V_t$

Finally we have:

$$b = b_1 + b_2 = V_{swing}/2 \tag{7.127}$$

The final equation is:

$$psd(f_{offset}) = 10 \log \left(f_o \left(\frac{\sqrt{2N} \times \Delta t_d}{T_o} \right)^2 \frac{1}{f_{offset}^2} \right) \tag{7.128}$$

$\Delta t_d = \sqrt{\Delta t_1 + \Delta t_2}$. p=0.25, $b_1 = V_{dd}/2 - (V_{dd} - V_t)$, $b_2 = V_{dd} - V_t$

If in addition to assumption 3, assumption 2 is also addressed, (7.124) can be further refined by using time scaling, and we have:

$$\Delta t_1 = \left(\frac{b_1}{SR\_const_1} \right)^{\left[\frac{1}{2p+1} - 1 \right]} \times (2p+1)^{\left[\frac{1}{2p+1} - 1 \right]}$$

$$\times \frac{1}{\sqrt{2}} \sqrt{\frac{\left(\left(\frac{b_1}{SR\_const_1} \right) \times Noise\_const_1^2 \right)}{SR\_const_1^2}} \tag{7.129}$$

$$\Delta t_2 = \left(\frac{b_2}{SR\_const_2}\right)^{\left[\frac{1}{2p+1}-1\right]} \times (2p+1)^{\left[\frac{1}{2p+1}-1\right]}$$

$$\times \frac{1}{\sqrt{2}} \sqrt{\left(\frac{\left(\frac{b_2}{SR\_const_2}\right) \times Noise\_const_2^2}{SR\_const_2^2}\right)} \tag{7.130}$$

$$SR\_const_1 = \frac{1}{\overline{t_1}^{2p}} \times \frac{\overline{I_1}(\overline{t_1})}{C}, \; Noise\_const_1 = \frac{1}{\overline{t_1}^{p}} \times \frac{\sigma_{I_{n1}}(\overline{t_1})}{C}$$

$$\sigma_{I_{n1}}(\overline{t_1}) = \sqrt{S_{I_{n1}}(\overline{t_1})} = \sqrt{4kT\left(\gamma_{triode}(\overline{t_1})g_{ds\_0_{M_1}} + \frac{1}{R3}\right)} \tag{7.131}$$

$$SR_{const2} = \frac{1}{\overline{t_2}^{2p}} \times \frac{\overline{I_2}(\overline{t_2})}{C}, \; Noise_{const2} = \frac{1}{\overline{t_2}^{p}} \times \frac{\sigma_{I_{n2}}(\overline{t_2})}{C}$$

$$\sigma_{I_{n2}}(\overline{t_2}) = \sqrt{S_{I_{n2}}(\overline{t_2})} = \sqrt{4kT\left(\gamma g_{m_{M_1}}(\overline{t_2}) + \frac{1}{R3}\right)} \tag{7.132}$$

7.9.2 Unsaturated Ring Oscillators

To reduce hardware and/or increase oscillation frequency, the number of stages, N, in a ring oscillator can be reduced. As N decreases, the oscillator outputs eventually become unsaturated. In this section we assume N=2. In the past ring oscillators with unsaturated outputs have been reported to have poorer phase noise characteristics than ring oscillators with saturated outputs [3] (all other things being equal). This is due to cycle to cycle correlation. Thus unsaturated ring oscillator has higher oscillation frequency but also higher phase noise.

To look at the phase noise of unsaturated ring oscillator, we employ all the methods that handle assumptions 1-3 in 7.9.1, which apply to saturated ring oscillator. In addition we incorporate the effect of cycle to cycle correlation, which appears only in unsaturated ring oscillator. Similar to past phase noise models on unsaturated ring oscillator we focus our analysis on thermal noise, as a first attempt to investigate the impact of cycle to cycle correlation. The 1/f noise can likewise be treated, as presented in [25]. For illustration the design whose delay cell is shown in Figure 7.59 is adopted. Transistors M_3, M_4 are biased in the triode region by the bias voltage P_{bias}. They are always on (never turn off) since V_{out}^+ and V_{out}^- never reach V_{dd}. Transistor M_5, M_6 are configured as a cross-coupled pair. M_3–M_6 together form the load. An explicit capacitor is used at the output. This,

Fig. 7.59 Delay cell of the unsaturated ring oscillator

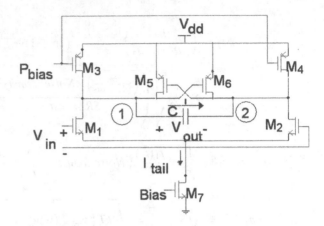

together with any parasitic capacitance coming from the transistors' device capacitance, is lumped together and denoted as C. The use of an explicit capacitor as an additional mean to control delay was reported, for example in [4] (see fig.7 of [4]). Transistor M_1, M_2 are alternately fully switched on and off and C is charged and discharged.

This delay cell and some existing designs (e.g. [24]), which are gaining in popularity, are similar. Specifically when comparing this delay cell, and these existing designs, with the more conventional differential ring oscillators ([12], [27], [28]), this delay cell and these existing designs all have an additional cross-coupled pair (M_5, M_6). This cross-coupled pair is supposed to speed up the slew rate and thus reduces the timing jitter. In some of these existing designs (see fig. 5c of [24]) there is no explicit capacitor at the output to control the delay. Instead the parasitic capacitances hanging at the output nodes are used to control the delay. This existing design also has an extra pair of diode connected PMOS transistors (one of which was denoted in fig. 5c of [24] as M_2) that are connected in parallel to M_3, M_4 and which form part of the load.

With N=2, we denote the outputs of the 1st and 2nd stage as V_1, V_2, respectively. Next let us concentrate on the 1st stage. Since we are using the delay cell in Figure 7.59, hence V_{out} now is V_1. Because this is a 2 stage fully differential ring oscillator, from Figure 7.41, the output of the 2nd stage, V_2, is flipped and fedback to become the input of the 1st stage. Hence V_{in} now is $-V_2$. We will now look at the typical voltages.

First let us take a look at the waveforms of the delay cell. Some typical waveforms of the differential outputs are shown in Figure 7.60a with noise exaggerated again for illustration purpose. From Figure 7.60a we can see V_1, V_2 are identical except shifted in the time axis by a quarter of the period. It is seen that V_1 starts at 0 and its mean ramps up. Meanwhile V_2 starts at A_0 at t=0 and independently its mean ramps down with V_2 eventually crossing 0 at the quarter period τ_1. At the same time V_1 reaches A_1. Its mean then turns around and starts to ramp down. Hence V_1 is never given a chance to saturate before it starts to ramp down. Notice the amplitude of nth quarter period is denoted as A_n. It is defined as

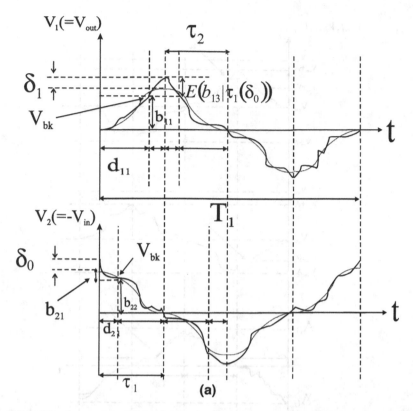

Fig. 7.60a Waveforms of the differential outputs V_1, V_2. In times, there are 3 scales: d (durations), τ(quarter period) and T(period). In amplitude there is A, A_{mean}, δ (noise at the end of the quarter period, which equals $A-A_{mean}$, V_{bk} (threshold), b (barrier) $E(b_{13}|\tau_1(\delta_0))$ is the conditional expected barrier, conditioned on τ

the magnitude of the non-zero crossing voltage process at the end of the nth quarter period (the zero crossing voltage process has the deterministic value zero at the end of the nth quarter period) e.g. we have just stated that at τ_1, V_2 crosses zero and so it is the zero-crossing voltage process for the 1st quarter period, τ_1. The non-zero crossing voltage process for τ_1 is then V_1, and we have just stated that its magnitude at the end of τ_1 is A_1. Hence the amplitude of τ_1 is $A_1=|V_1(\tau_1)|$. Similarly A_2 is the amplitude of τ_2 and $A_2=|V_2(\tau_2)|$, $A_3=|V_1(\tau_3)|$ etc.

Now as V_1, V_2 are differential output voltages, we show in Figure 7.60b some of the waveforms of the single-ended output voltages. For example waveforms for the two single-ended output voltages corresponding to $V_1=V_{out}$, denoted as V_{out}^+, V_{out}^- are shown. Also shown is V_{in}. Since V_{in}, V_{out} are identical except shifted in the time axis by a quarter of the period hence V_{in}, V_{out}^+, V_{out}^- are also shifted in the time axis by a quarter of a period. Let us first look at V_{out}^+ where it is seen to center at V_{cm}, the common mode voltage. It swings symmetrically to its amplitude,

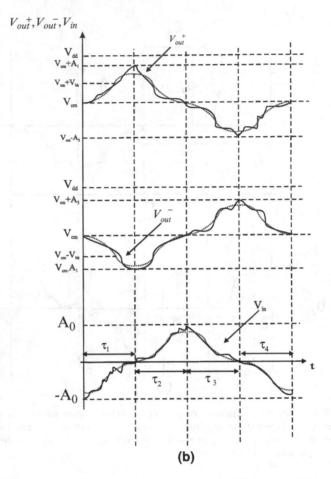

Fig. 7.60b Typical voltages of a two-stage unsaturated ring oscillator using delay cell in Figure 7.59. Noise is exaggerated and superimposed on the mean

above and below V_{cm}, but never reaches V_{dd}. Essentially V_{out}^{+} ramps up towards V_{dd}. However before V_{out}^{+} reaches V_{dd}, V_{in} crosses zero and then goes above zero, turning off M_2, turning on M_1 and C starts to get discharged. Hence V_{out}^{+} turns around and starts to ramp down i.e. it is never given a chance to reach V_{dd} before it starts to ramp down. Hence the ring oscillator has an unsaturated output. Finally let us look at the waveform V_{in}. Following [4] we assume that the switching threshold of the differential pair M_1-M_2 is zero. Hence referring to Figure 7.59 this means that during time τ_1 (see Figure 7.60b) when V_{in} is below zero M_1 is off and M_2 is on. On the other hand during time $\tau_2 V_{in}$ is above zero and so M_1 is on and M_2 is off. This means M_1, M_2 are fully switched. Meanwhile at all times V_{out}^{+}, V_{out}^{-} never reach V_{dd}. This means M_3, M_4 are always on. Therefore according to Figure 7.55 this is an example of an unsaturated oscillator.

Let us use the model to explain qualitatively where the cycle to cycle correlation arises from. First referring to Figure 7.60a, δ_0 is the randomness in V_2, at t=0. This, together with the noise injected in stage 2 during τ_1, is the reason why, τ_1, the time it takes V_2 to ramp down and cross the threshold at zero, is random. This randomness is the timing jitter of τ_1. Meanwhile V_1, (a random process with noise injected in stage 1 during τ_1), starts from zero at t=0 and ramps up. As V_2 crosses zero at t=τ_1, sign(V_2) changes sign and hence slew rate changes sign, or V_1 finishes ramping up, turns around and starts to ramp down. Hence $V_1(\tau_1)$ is the voltage that V_1 assumes at the moment when the drift of V_1 turns around. $V_1(\tau_1)$ is random and it differs from the amplitude mean, A_{mean}, by δ_1, i.e. $\delta_1 = A_1 - A_{mean} = |V_1(\tau_1)| - A_{mean}$. Thus physically δ_1 is the randomness in V_1 at the end of τ_1 and it involves two contributions:

a) during τ_1, V_1, a random process, has noise continuously injected to it
b) the random process also runs for a random time τ_1

Because of (b), δ_1 is a function of τ_1 and hence of δ_0.

Next let us look at V_1 during τ_2, when V_1 ramps down and crosses the threshold at zero at time $\tau_1 + \tau_2$. Just like the case of V_2 previously, now V_1 also starts with randomness, this time δ_1. Again together with the noise injected in stage 1 during τ_2, is the reason why τ_2, the time it takes V_1 to ramp down and crosses the threshold at zero, is random. Again τ_2 depends on the initial randomness δ_1. However since we have already established that δ_1 depends on τ_1, therefore τ_2 in turn depends on τ_1 and so there is correlation between these two consecutive quarter periods.

The final equation of the phase noise, incorporating cycle to cycle correlation and all the methods that handle assumptions 1-3 in 8.1, can now be derived. It is however, quite complicated. For highlighting the effect of cycle to cycle correlation we now present the phase noise equation with this effect alone. Following [25], this is:

$$psd(\omega_{offset}) = 10 \log \left(\frac{1}{T_{mean}} \left(\frac{4\left(\frac{\sigma_\varepsilon^2}{(1-\theta)^2} \right)}{T_{mean}^2} \right) \frac{1}{\omega_{offset}^2} \right) \text{dBc/Hz} \qquad (7.133)$$

Here θ, the correlation coefficient, summarized the amount of cycle to cycle to correlation, and is between 0 and 1. As it increases phase noise increases. It can be analytically related to the circuit parameters, such as amplitude of oscillation, A_{mean}, slew rate, characterisitics of the load etc. [25]. σ_ε, is the noise source and is analytically related to the circuit parameters, such as tail current value, I_{tail_mean}, etc. [25]. For thermal noise, σ_ε decreases as I_{tail_mean} decreases, since thermal noise (current) depends on g_m, transconductance of the devices, which decreases with decreasing g_m. [25] shows that to improve phase noise, one way is to improve the characteristics of the load so that correlation is reduced. Shown in

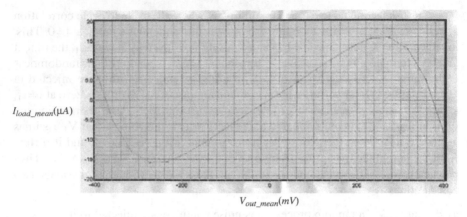

$$I_{load\_mean}(\mu A)$$

$$V_{out\_mean}(mV)$$

Fig. 7.61 I-V characteristics of load, with V_{in} set to be negative; impedance is negative of slope

Figure 7.61 is an example I vs V characteristics across the load, used to characterize the load impedance. The slope at the center, s_c, is positive, because of the negative impedance presented by the cross coupled pair is designed to be more than the positive impedance presented by the triode based load, thus having an overall negative impedance (hence positive slope). The slope at the side, s_s, is negative, because at those voltages, the cross coupled pair is turned off, leaving only the positive impedance presented by the triode based load (hence negative slope). An I-V characteristics of this form (i.e. with the center portion of the slope positive) is not unexpected for a 2-stage ring oscillator, as a stable limit cycle can be achieved. To improve phase noise we can increase $|s_s|$, $|s_c|$ (defined in Figure 7.61), while keeping I_{tail_mean} constant. Heuristically, by doing this, the resulting waveform is less "linear", and thus perturbation of the amplitude in stage 1's output voltage, due to noise, is going to have less perturbation on the change in time at which it crosses the threshold. Stage 1's output voltage, being the same as stage 2's input voltage, trigger stage 2's output to start ramping up/down upon this crossing of the threshold. With less perturbation in the crossing time, stage 2's output's amplitude uncertainity is also reduced. Thus perturbation in the amplitude of stage 1's output affects less the amplitude in stage 2's output and subsequently stage 1's output in the subsequent cycle (hence less cycle to cycle correlation). It should be remembered though, this propagation of perturbation of amplitude from 1 cycle to the next comes not just from the noise injected at the peak of stage 1's output, but also from noise injected into the output voltage during the ramp (up/down). Thus all the noise should be included in computing the correlation. This way of increasing $|s_s|$, $|s_c|$, while keeping I_{tail_mean} constant, can be done by decreasing P_{bias}, $\frac{W_3}{L_3}$, and $\frac{W_4}{L_4}$, as well as increasing $\frac{W_5}{L_5}$, and $\frac{W_6}{L_6}$. Since I_{tail_mean}, N, V_{dd} remain constant, this means power remains constant.

7.10 Problems

7.1 A new PFD implemented with RS-FF is shown.

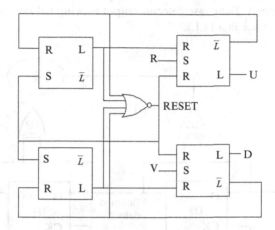

a) Given the following R, V waveforms (they have the same frequency but different phases), show the U, D, reset waveforms and explain how they perform phase detection.

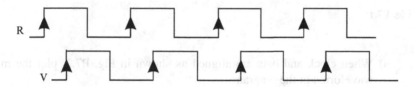

b) The R-V waveforms are changed and are as follows, where they have different frequencies.

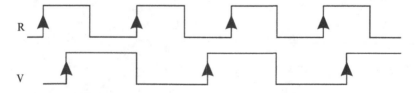

(i) Show the U, D, reset waveforms.

(ii) Let us define $\alpha = \frac{f_V}{f_R}$, $\beta = \frac{f_R - f_V}{f_V} = \frac{1}{\alpha} - 1$, $\gamma = \frac{f_V - f_R}{f_R}$ and $\overline{(U - D)}_{ave}$ as the time average of the different outputs. Show that $\overline{(U - D)}_{ave} = \frac{\gamma + 0.5}{\gamma + 1.0}$. (Hint: Start by observing that the probability of a single V transition in $[t, t+T_R]$ $(=\alpha)$. Here T_R is the time interval such that R has two successive transitions at time t and $t+T_R$. Also assume the corresponding probability density function within that interval is uniformly distributed). Plot $\overline{(U-D)_{ave}}$ versus β and explain how it achieves frequency detection.

7.2 Shown in Figure P7.1 is a new type of phase detector (the D-FF are positive edge triggered). The phase detector output is being fed into an integrator (used as the loop filter in a PLL).

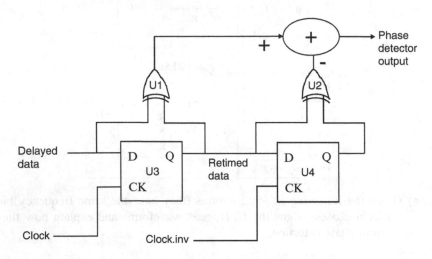

Fig. P7.1

a) When clock and data are aligned as shown in Fig. P7.2, plot the missing waveforms in the diagram.

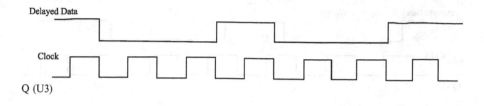

Delayed Data

Clock

Q (U3)

Q (U4)

Output U1

Output U2

Phase detector
output1

Loop integrator
Output

Fig. P7.2

b) When the data is ahead of clock as shown in Fig. P7.3, plot the missing
waveforms in the diagram.

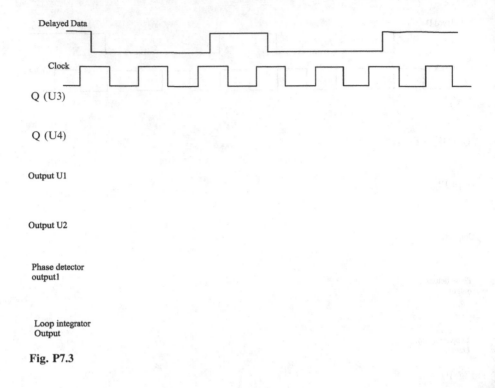

Fig. P7.3

c) Plot the transfer characteristics of this phase detector for phase error varying from -2π to 2π. You can assume maximum data transition density.

7.3 Let us design another phase detector (still based on D-FF). The resulting waveform is shown in Fig. P7.4.

a) Design the phase detector that outputs this phase detector output waveform (and hence the corresponding loop integrator output waveform).

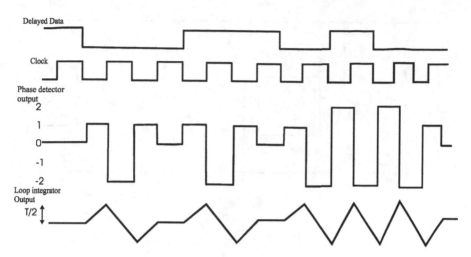

Delayed Data

Clock

Phase detector output

Loop integrator Output

Fig. P7.4

b) Plot the waveforms at the internal nodes of this phase detector (you should include at least output of all the D-FF).

7.4 Figure P7.5 shows a divider circuit. The input is labelled clk, and the output is labelled Out. There is a control input labelled MC. Assume that clk is a 50% square wave clocking at 1 GHz.

a) When $MC=0$, draw the waveform along the signal path [i.e. clk(L_1), $Q(L_1)$, $D(L_2)$, $\overline{Q}(L_2)$, clk(FF_1), $\overline{Q}(FF_1)$, out, MC_1]. The waveform should be long enough to show the divider action. What is the total division ratio?

b) Repeat the same procedure when $MC=1$. Now what is the total division ratio?

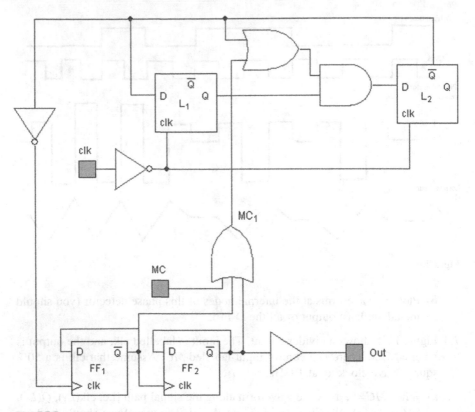

Fig. P7.5

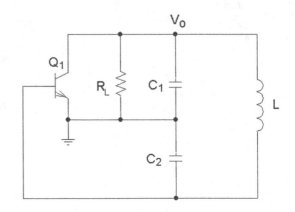

Fig. P7.6

7.5 Using a BJT biased at $I_C = 1$ mA, design a Colpitts oscillator as shown in Fig. P7.6 to operate at $\omega_0 = 1.9 \times 10^9 rad/s$. Use C_1=5.26 pF and assume that the C_1 available has a Q of 100 (this can be represented by a resistance in parallel with C_1 whose value is given by $Q/(\omega_0 C_1)$). Also assume that R_L=2kΩ and that for the BJT, $r_0 = 100$ kΩ. Find C_2 and L.

7.6 Figure P7.7 is a LC based oscillator. The block-level diagram is given in Figure P7.7a and the transistor-level diagram is given in Figure P7.7b.

a) Explain the operation of the circuit in Figure P7.7b by explaining how it oscillates. In the circuit in Figure P7.7b there is a positive feedback path and a negative feedback path. Identify the paths.

b) Now we want to tune the circuit using a varactor diode. Show how we can modify the circuit in Figure P7.7b to achieve that. Show the complete modified circuit. Derive the K_{vco} for this modified circuit.

7.7 In this question we will investigate the phase noise's PSD of the LC oscillator as described in Figure P7.6. Since this oscillator is a resonance based oscillator, its phase noise PSD will be derived by the simplifying assumption that the oscillator is an LTI system (a gross approximation).

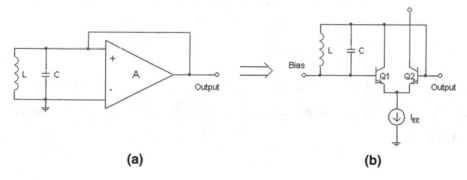

(a)   **(b)**

Fig. P7.7

To derive the phase noise's PSD of this Colpitts oscillator, it is represented in the block diagram in Figure P7.8.

Using this block diagram, one can express noise at the output as a product of noise injected at the summing node and the corresponding transfer function. The detail has been worked out at [9] with the results given below:

$$\frac{\frac{v_o^2(f_i)}{\Delta f}}{\frac{v_n^2(f_i)}{\Delta f}} = \frac{|a|^2}{(f_i)^2 \left|\frac{d(af)}{d\omega}\right|^2} \tag{P7.1}$$

Here $\frac{\overline{v_o^2(f_i)}}{\Delta f}$ = PSD of noise at the output (S_o) of the VCO, at an offset frequency f_i from the center frequency f_o.

Fig. P7.8

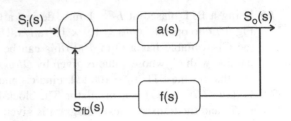

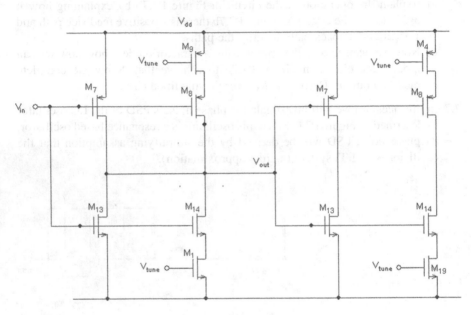

Fig. P7.9

$\frac{\overline{v_n^2(f_i)}}{\Delta f}$ = PSD of noise injected at the summing node of the VCO (just before the a(s) block), at an offset frequency f_i from the center frequency f_o.

a=forward gain

f=feedback factor

a) Derive expressions for a(s) and a(s)f(s) in Figure P7.6 in terms of circuit elements R_L, L, C_1, C_2, g_m of Q_1. (Hint: a good starting point is the transfer function described in (7.47), derived for the transformer coupled LC oscil-

lator). Then substitute these into (P7.1) and find an expression for $\frac{\frac{\overline{v_o^2(f_i)}}{\Delta f}}{\frac{\overline{v_n^2(f_i)}}{\Delta f}}$ (you

can assume the only noise source is lumped as an equivalent input noise source placed at the base of transistor Q_1)

b) Assume the noise injected at the summing node of the oscillator, which is represented by an equivalent voltage noise source $\overline{v_n^2}$, has an equivalent noise resistance $= \frac{1}{2g_m}$ where g_m is the transconductance of Q1. Write down the expression for $\frac{\overline{v_n^2(f_i)}}{\Delta f}$ and hence derive the expression for $\frac{\overline{v_o^2(f_i)}}{\Delta f}$.

c) Derive an expression for $v_{carrier}^2$, the mean square voltage of the oscillator output voltage when no noise is injected. This is also called the carrier signal. (Hint: a good reference is [24]). Using the same values from problem 7.5, calculate the numerical value of $v_{carrier}^2$. Then use this value and the expression of $\frac{\overline{v_o^2(f_i)}}{\Delta f}$ derived in (b) to find an expression for $S_{\theta\_vco}(f_i)$.

7.8 Fig. P7.9 shows the circuit schematic of a non-saturated delay cell. Two identical cells are shown, one driving the other. Assume that a method is devised such that the cell can switch fast so that its switching waveform resembles that of a fast-slewing saturated delay cell. Derive an expression for the K_{vco} of the ring oscillator built using these non-saturated delay cells. The expression should be in terms of g_m, C_{gs}, C_{gd} and other device parameters of specific transistors, $N(=$number of stages), and V_{swing}. You can assume V_{dd}-V_{tune}' equals V_{tune}.

References

1. W.R. Robins. "Phase Noise in Signal Sources (Theory and Applications) Peter Peregrinus Ltd.
2. Z. Shu, K. Lee, B. Leung, "A 2.4GHz ring oscillator based CMOS frequency synthesizer with a fractional divider dual PLL architecture", IEEE Journal of Solid State Circuits, Vol. 39, No.3, pp.452–462, March 2004
3. C. Park, B. Kim "A low-noise 900MHz VCO in 0.6um CMOS", IEEE Journal of Solid State Circuits, Vol. 34, No. 5, pp. 586–591, May 1999.
4. T.C. Weigandt, B. Kim, P.R. Gray, "Analysis of Timing Jitter in CMOS Ring Oscillators" in Proc ISCAS June 1994, pp.4.31–4.35.
5. McNeil, John A. "Jitter in Ring Oscillators" IEEE JSSC Vol32 No.6 p870 June 1997.
6. R. Navid. T. Lee, Dutton, "Minimum Achievable Phase Noise of RC oscillators," IEEE Journal of Solid-State Circuits, pp. 630–637, March 2005
7. A. Hajimiri, S. Limotyrakis, T. H. Lee, "Jitter and phase noise in ring oscillators", IEEE Journal of Solid State Circuits, pp. 790–804, June 1999
8. Craninckx, J. M. Steyaert "Low-Noise Voltage Controlled Oscillator Using Enhanced LC-Tanks" IEEE TC&S II Vol 42 No 12 p 794 Dec 1995.
9. B. Razavi "A Study of Phase Noise in CMOS Oscillators" IEEE JSSC Vol31. Vol 3 p 331 Mar 1996
10. N.M. Nguyen, R.G. Meyer. "A 1.8GHz Monolithic LC Voltage Controlled Oscillator" IEEE JSSC Vol27 No.3 p. 444 March 1992.
11. S.K. Enam A.A. Abidi "A 300-MHz CMOS Voltage Controlled Ring Oscillator" IEEE JSSC Vol25 No1 p. 312 February 1990
12. A. Abidi, "Phase Noise and Jitter in Ring Oscillators", IEEE Journal of Solid State Circuits, pp. 1803–1816, August 2006.

13. F. Gardner, "Phaselock Techniques", 2nd edition, John Wiley
14. R. Best, "Phase-Locked Loops, Theory, Design and Applications", 2nd edition, McGraw Hill
15. A. Carlson, "Communication Systems", 3rd edition, McGraw Hill
16. B. Kim, "High Speed Clock Recovery in VLSI using Hybrid Analog/Digital Techniques", Ph.D. Thesis, U.C. Berkeley, 1992
17. D. Jeong, G. Borrielle, D. Hodges, R. Katz, "Design of PLL-Based Clock Generation Circuits," IEEE JSSC vol. sc22, no.2, April 1987
18. K. Ware, H. Lee, C. Sodini, "A 200MHz CMOS PLL with Dual Phase Detectors", ISSCC, vol32, pp 192–193, Feb, 1989
19. W. Egan, "Frequency Synthesis by Phase Lock", Wiley and Sons.
20. S. Mehta, "Design of Gigahertz CMOS Prescalers", MSc. Thesis, U.C. Berkeley, 1997
21. F. Martin, "Frequency Synthesizers in RF Wireless Communications", Course Notes, 1994, Motorola, Plantation, Florida F. Gardner, "Charge Pump Phase Lock Loops", IEEE Transactions on Communications, Vol 28, No. 11, pp1849–1858, November 1980
22. F. Gardner, "Charge Pump Phase Lock Loops", IEEE Transactions on Communications, Vol 28, No. 11, PP. 1849–1858, November 1980
23. R. Meyer "Non-linear integrated circuits", course notes, 1986, U.C. Berkeley
24. R. Zamora, T. Lee "CMOS VCOs for frequency synthesis in wirless biotelemetry", Proceedings of International Symposium on Low Power Electronics, pp. 91–94, August 1998
25. B. Leung and D. Mcleish, " Phase Noise of a Class of Ring Oscillators Having Unsaturated Outputs with Focus on Cycle to Cycle Correlation", IEEE Transactions on Circuits and Systems I, Vol 56, No. 9, pp. 1689–1707, August 2009
26. B. Leung, "A Switching Based Phase Noise Model for CMOS Ring Oscillators Based on Multiple Thresholds Crossing", IEEE Trans on Circuits and Systems I, pp.2858–2869 Nov 2010
27. B. Leung, D. Mcleish "Investigation of Phase Noise of Ring Oscillators with Time Varying Current and Noise Sources", IEEE Trans on Circuits and Systems I, pp.1926–1939 Oct 2004
28. B. Leung "A Novel Model on Phase Noise in Ring Oscillator Based on Last Passage Time", IEEE Transactions on Circuits and Systems I, pp. 471–482 March 2004
29. B. Leung "Comparison of Phase Noise Models on a Class of Ring Oscillators Using Low Voltage Swing Fully Differential Delay Cells", Analog Integrated Circuits and Signal Processing, pp. 129–147 November 2009

Chapter 8
Frequency Synthesizer: Loop Filter and System Design

8.1 Introduction

In Chapter 7 we discussed the phase/frequency processing elements part of a frequency synthesizer. We noted that due to noise, mismatch in these components, spurs, and phase noise are generated. In this chapter we investigate how spurs and phase noise from individual components affect the spurs and phase noise of the complete synthesizer. To discuss these issues, we need to know the transfer functions from these components to the synthesizer output. These transfer functions depend on the loop filter of the synthesizer. Hence we start by discussing the loop filter. We analyze the loop filter from the view point of using it as a key to trade off spurs and phase noise while maintaining stability. Then we provide a design flow chart of the synthesizer and a detailed design example of a synthesizer that is used in a DECT receiver front end.

8.2 Loop Filter: General Description

In this section we discuss the loop filter, the single most important component in the synthesizer with regard to providing trade-off of various performances (e.g., spur, phase noise, capture, and lock range).

8.2.1 Basic Equations and Definitions

We start by introducing the proper definition of the following terms [5]: filter transfer function $F(s)$, open-loop transfer function $G(s)$, and closed-loop transfer function $H(s)$ of the synthesizer. They will be used again later in the design of the complete synthesizer.

B. Leung, *VLSI for Wireless Communication*, DOI 10.1007/978-1-4614-0986-1_8,
© Springer Science+Business Media, LLC 2011

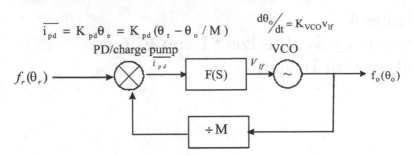

Fig. 8.1 Block diagram of synthesizer

In Figure 8.1 we redraw the block diagram of the PLL in Figure 7.2a. Note that we have reinserted the divider M. As stated in sub-section 7.3.3 (Chapter 7) the PD/charge pump is now represented by an LTI circuit with gain K_{pd}. Also we stated in sub-section 7.3.1 (Chapter 7) that we assume that no cycle slipping occurs. Hence, the whole synthesizer is an LTI system and can be described using the familiar Laplace transform.

We define everything in phase, use S-domain representation, and generalize the filter from a simple capacitor to a general network with transfer function $F(s)$. Hence we generalize (7.1), (7.4a), (7.4c) to the following equations:

$$\theta_o(s) = \frac{K_{vco}v_{lf}(s)}{s} \tag{8.1}$$

$$\overline{i_{pd}(s)} = K_{pd}\theta_e(s) = K_{pd}[\theta_r(s) - \theta_o(s)/M] \tag{8.2}$$

$$v_{lf}(s) = F(s)\overline{i_{pd}(s)} \tag{8.3}$$

From the preceding equations we can derive the transfer function between any two variables. The most important one is the closed-loop input/output transfer function:

$$H(s) = \frac{\theta_o(s)}{\theta_r(s)} = \frac{K_{vco}K_{pd}F(s)}{s + \frac{K_{vco}K_{pd}F(s)}{M}} \tag{8.4}$$

Another transfer function of interest is the error transfer function, which can be related and expressed in H(s):

$$\frac{\theta_e(s)}{\theta_r(s)} = K_{pd}\left[1 - \frac{H(s)}{M}\right] \tag{8.5}$$

The open-loop transfer function $G(s)$ is simply defined as

$$G(s) = \frac{K_{pd}K_{vco}}{sM}F(s) \tag{8.6}$$

As can be seen, all the transfer functions depend on the loop filter $F(s)$, and so a review on the basics of filter design is in order.

Fig. 8.2 First-order loop
filter

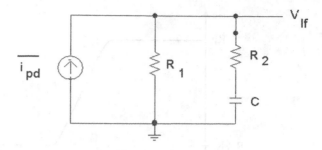

8.2.2   First-Order Filter

The simplest loop filter consists of a single capacitor C (this is driven by current as opposed to voltage source), as shown in Figure.7.4. Such a simple first-order filter (which is simply an integrator) has very limited flexibility on optimizing the performance, as will be shown later. As a result there is a need for more complex filters. A general first-order filter is shown in Figure 8.2.

For this first-order filter we have

$$F_1(s) = \frac{R_1(sR_2C + 1)}{1 + s(R_2 + R_1)C} \tag{8.7}$$

Figure 8.3a and Figure 8.3b show the frequency responses of $F_1(s)$ and $G_1(s)$. Substituting (8.7) into (8.6) the loop transfer function becomes

$$G_1(s) = \frac{K_{pd}K_{vco}}{Ms} F_1(s) = \frac{K_{pd}K_{vco}(s + z_1)}{Ms(s + p_2)} \tag{8.8}$$

Notice from Figure 8.3b that the unity gain frequency of the loop transfer function is ω_u.

Now (8.8) can be rewritten as

$$G_1(s) = \frac{K_{vco}K_{pd}}{Ms} \frac{R_1(1 + s\tau_2)}{(1 + s\tau_1)} = \frac{K_{pd}K_{vco}R_1}{M} \frac{1}{s} \frac{s + \frac{1}{\tau_2}}{s + \frac{1}{\tau_1}} \frac{\tau_2}{\tau_1}$$

$$= \frac{K_{pd}K_{vco}}{sM}(R_1 \| R_2) \frac{s + z_1}{s + p_2} \tag{8.9a}$$

Here

$$z_1 = \frac{1}{\tau_2} = R_2C \tag{8.9b}$$

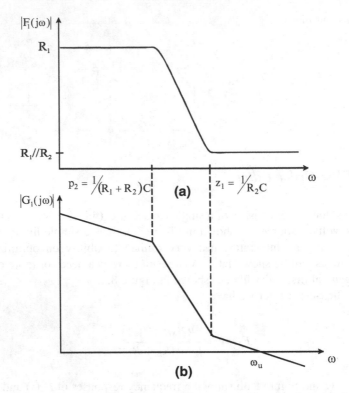

Fig. 8.3 Frequency responses of first-order loop filter and resulting open-loop transfer function

$$p_2 = \frac{1}{\tau_1} = \frac{1}{(R_1 + R_2)C} \tag{8.9c}$$

How about $H_1(s)$? Substituting (8.9b) (8.9c) into (8.7) and then the resulting $F_1(s)$ in (8.4), we have

$$H_1(s) = \frac{K_{vco}K_{pd}R_1(s\tau_2 + 1)}{\tau_1 s^2 + (1 + \frac{K_{VCO}K_{pd}\tau_2}{M})s + \frac{K_{vco}K_{pd}}{M}R_1} \tag{8.10}$$

The $H_1(s)$'s frequency response is shown in Figure 8.4, where ω_n is the natural frequency of the frequency response and ξ is the damping factor.

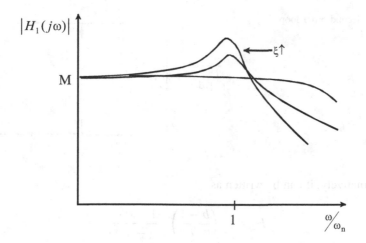

Fig. 8.4 Frequency response of closed-loop transfer function with a first-order loop filter

8.2.3 *Second-Order Filter*

A second-order filter, whose transfer function is denoted as $F_2(s)$, is shown in Figure 8.5. This is one of the most popular filters [4] for frequency synthesizer applications. When compared with the first-order filter in Figure 8.2, the present filter has one fewer resistor but one more capacitor. The elimination of the resistor means one pole is moved to the origin. The extra capacitor makes the filter a second-order filter and the synthesizer a third-order loop. The extra capacitor C_3 is just put there for second-order consideration (mainly to smooth off voltage jumps across R_2 due to current switching), and the extra pole it introduces is far away from the main pole. Hence for most purposes (like stability), the filter behaves like a first-order filter and the synthesizer behaves like a second-order synthesizer with similar transfer functions. The transfer function of this filter is

$$F_2(s) = \frac{s + z_1}{sC_3(s + p_3)} \tag{8.11a}$$

where

$$z_1 = \frac{1}{R_2 C} \tag{8.11b}$$

and

$$p_3 = \frac{1}{R_2\left(\frac{CC_3}{C+C_3}\right)} \tag{8.11c}$$

Fig. 8.5 Second-order loop
filter

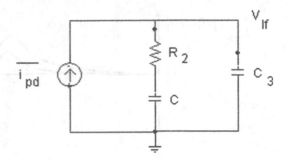

Alternatively, it can be written as

$$F_2(s) = \left(\frac{b-1}{b}\right) \frac{1 + s\tau_2}{sC\left(\frac{s\tau_2}{b} + 1\right)} \tag{8.12}$$

where

$$b = 1 + \frac{C}{C_3} \tag{8.12b}$$

and

$$\tau_2 = R_2 C \tag{8.12c}$$

Equation (8.12a) shows that $F_2(s)$ has the dimension of $1/\Omega$ because it takes a current input $\overline{i_{pd}}$ (from the charge pump) and generates a voltage output, v_{lf}. As was shown in [4], typically b is large and $F_2(s)$ becomes

$$F_2(s) = \frac{1 + s\tau_2}{sC} = \frac{R_2(s + z_1)}{s} \tag{8.13}$$

where $z_1 = 1/R_2 C$.

Now again assuming that b is large, $F_2(s)$ can also be written as

$$F_2(s) = \frac{(b-1)}{C} \frac{(s + z_1)}{s(s + p_3)} \tag{8.14}$$

Here

$$z_1 = \frac{1}{\tau_2} \tag{8.14a}$$

$$p_3 = \frac{b}{\tau_2} \tag{8.14b}$$

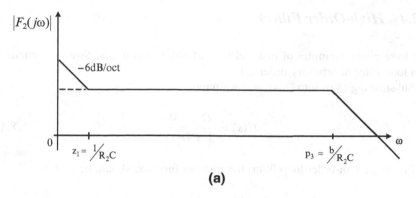

Fig. 8.6a Frequency response of second-order loop filter

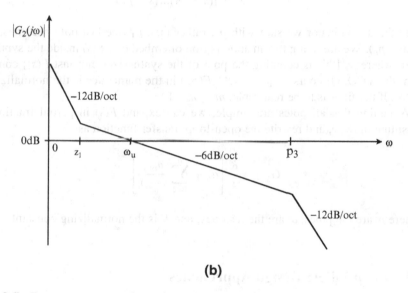

Fig. 8.6b Frequency response of open-loop transfer function with a second-order loop filter

Substituting (8.12a), (8.12b) into (8.6), we get $G_2(s)$.

$$G_2(s) = \frac{K_{pd}K_{vco}}{sM} \cdot \frac{s + Z_1}{s^2 C_3 (s + p_3)} \tag{8.15}$$

Alternately, using the approximated form for $F_2(s)$ as given in (8.12a) and substituting that into (8.6) we get

$$G_2(s) = \frac{K_{pd}K_{vco}}{sM} \cdot \frac{R_2(s + Z_1)}{s} \tag{8.16}$$

$F_2(s)$ and $G_2(s)$ are plotted in Figure 8.6a and Figure 8.6b respectively. Finally, substituting (8.12a) into (8.4), we get $H_2(s)$.

8.2.4 High-Order Filters

We have given examples of first and second-order loop filters. Now we generalize to a loop filter of arbitrary order [5].

Substituting (8.6) into (8.4), and we have

$$H(s) = \frac{G(s)/M}{1 + G(s)} \tag{8.17}$$

For an (n-1)th-order loop filter, the transfer function should be

$$F(s) = \frac{F(\infty)(s - z_1)(s - z_2)...(s - z_m)}{(s - p_2)(s - p_3)...(s - p_n)} \tag{8.18}$$

In the denominator we start with p_2, rather than p_1 (the last pole becomes, of course, p_n). We are doing this in anticipation of embedding $F(s)$ inside the synthesizer, where p_1 [this is counting the pole of the synthesizer, not just $F(s)$] comes from the VCO. Of course, p_1 is at DC. $F(\infty)$ in the numerator is the normalizing factor. If the filter is to be realizable, $m \leq n - 1$.

Assuming that all poles are simple, we can expand $F(s)$ in partial fractions, substitute in (8.6), and rewrite the open-loop transfer function as

$$G(s) = \frac{K}{s}\left[a_1 + \sum_{i=2}^{n} \frac{a_{i+1}}{s - p_i}\right] \tag{8.19}$$

where p_i are the poles, a_i are the residues, and K is the normalizing constant.

8.3 Loop Filter: Design Approaches

In sub-section 7.3.4, 7.3.5 and 7.8.3, we stated that to calculate the output spur and phase noise due to the PD($=\theta_o$) and the output phase noise due to the VCO ($=S_{\theta o\_vco}$), we need the transfer functions (H_{ref} or H_{n_vco}) from the particular source of disturbance (PD or VCO) to the synthesizer output. Similarly output phase noise from the reference also depends on the transfer function H_{ref}. These transfer functions depend on $F(s)$. To generalize, spur or noise from internal nodes affects the synthesizer output depending on the transfer functions from these internal nodes to the output. Hence we conclude that the loop filter affects the performance of the synthesizer through its impact on these transfer functions. In addition F(s) also affects the stability of the synthesizer. Therefore the design of the loop filter is guided by achieving an optimal compromise between these impacts. There are two approaches: phase noise based and spur based approaches.

8.3.1 Phase Noise Based Approach

In this approach we design $F(s)$ based on primarily phase noise and stability requirements. The resulting synthesizer can then be checked and see if it meets the spur requirement. Following this philosophy, it turns out that phase noise has a more direct (one to one) dependency on $G(s)$. Hence, we would first design $G(s)$ based on phase noise requirement. Then from $G(s)$ we can in turn determine $F(s)$.

8.3.1.1 Phase Noise Requirement

First we describe qualitatively the impact of the synthesizer's phase noise on the receiver front end's performance. These impacts are then quantified into specific requirements on the synthesizer [10], [11].

We assume that the output of the synthesizer from Figure 8.1 consists of a sinusoidal waveform with frequency f_o. When corrupted with phase noise, the synthesizer output can be written as

$$V_o = \cos(\omega_o t + \Delta\theta(t)) \tag{8.20}$$

Here $\Delta\theta(t)$ is the random phase fluctuation due to noise inside the synthesizer. If we compare (8.20) to (7.114), we find that they have similar form. Of course, $\Delta\theta(t)$ in (8.20) refers to phase noise of the synthesizer whereas $\Delta\theta(t)$ in (7.114) refers to phase noise of the oscillator alone. They would have different PSDs. However, the interpretation of phase noise, as outlined in Section 7.8, applies in both cases. In other words, the PSD of $\Delta\theta(t)$, denoted as $S_{\theta\_syn}(f)$, can be interpreted both in the phase and amplitude domain.

Now the synthesizer's phase noise has two independent impacts on the receiver front end's SNR and hence BER. The first impairment is called phase impairment, and is best explained by interpreting the $\Delta\theta(t)$ term in (8.20) in the phase domain. To facilitate explanation let us assume a heterodyne architecture is used in which the frequency synthesizer mixes with the received signal and downconverts it to an IF signal. This IF signal is then demodulated coherently using a carrier recovery PLL loop. (This carrier recovery PLL loop is separate from the frequency synthesizer.). The carrier recovery PLL loop output is assumed to demodulate a phase encoded signal, like the BPSK encoded signal as described in Chapter 1. The carrier recovery PLL loop output has phase impairment similar to a frequency synthesizer and hence its output is also given by (8.20). Therefore the original equations (1.3) and (1.4), rewritten as

$$\text{logic } 0: \quad s_0(t) = \int_0^{T_b} (A\cos\omega_0 t)(A\cos\omega_0 t)\,dt = \frac{A^2 T_b}{2} \tag{8.21}$$

$$\text{logic 1}: \quad s_1(t) = \int_0^{T_b} (-A \cos \omega_0 t)(A \cos \omega_0 t)dt = -\frac{A^2 T_b}{2} \qquad (8.22)$$

become

$$\text{logic 0}: \quad s_0(t) = \int_0^{T_b} (A \cos \omega_0 t)(A \cos(\omega_0 t + \Delta\theta(t)))dt \neq \frac{A^2 T_b}{2} \qquad (8.23)$$

$$\text{logic 1}: \quad s_1(t) = \int_0^{T_b} (-A \cos \omega_0 t)(A \cos(\omega_0 t + \Delta\theta(t)))dt \neq -\frac{A^2 T_b}{2} \qquad (8.24)$$

We can see that because of $\Delta\theta(t)$ in (8.23), (8.24), upon demodulation the result is no longer simply $\pm A^2 T_b/2$. Again, referring to Section 1.3, the E_b is no longer simply $A^2 T_b/2$. This will degrade the P_e, or BER. Let us return to our original discussion and ask what the impact of the $\Delta\theta(t)$ in the synthesizer output, which is given by (8.20), has on the demodulated signal. Remember the synthesizer output is used to mix with the received signal and downconvert it before the IF signal is being demodulated using the carrier recovery PLL loop output. However in the mixing operation, this $\Delta\theta(t)$ in the synthesizer would have corrupted the phase of the downconverted signal. Hence the first $\cos(\omega_0 t)$ terms in (8.20), (8.21) (which represent the IF signal) will become $\cos(\omega_0 t + \Delta\theta(t))$ (Contrast this with the previous case when we state the only $\Delta\theta(t)$ comes from the carrier recovery PLL loop and hence the second $\cos(\omega_0 t)$ terms become $\cos(\omega_0 t + \Delta\theta(t))$). From (8.20), (8.21) it does not matter whether $\Delta\theta(t)$ is introduced in the first $\cos(\omega_0 t)$ terms or second cos $(\omega_0 t)$ terms. (8.23), (8.24) still follow. Accordingly in the demodulation process, the $\Delta\theta(t)$ introduced by the synthesizer degrades the final P_e in exactly the same way that the $\Delta\theta(t)$ in the carrier recovery PLL loop does. Of course if we have $\Delta\theta(t)$ in both the synthesizer and carrier recovery PLL loop then both the first and second $\cos(\omega_0 t)$ terms in (8.20), (8.21) will be corrupted.

The second impact, called reciprocal mix, comes about because of the presence of adjacent channel interference. This is best understood by viewing the $S_{\theta\_syn}(f)$ associated with the $\Delta\theta(t)$ term in (8.20) in the amplitude domain. $S_{\theta\_syn}(f)$ for a typical synthesizer has the shape as shown in Figure 8.7.

Hence we can view the phase noise as making the output tone not pure. If strong enough it will mix the strong adjacent channel interference down to the desired IF. The frequency domain explanation of how this happens is given in Figure 8.8 and Figure 8.9. To demonstrate this mixing of adjacent channel phenomeon, Figure 8.8 takes Figure 8.7 and represents all the relevant phase noise energy on each side of f_o by two frequency impulses. These impulses are situated at frequencies denoted as $f_o \pm f_{\text{noise}}$. We assume that this phase noise energy is situated one channel away from f_o and hence we have $f_{\text{noise}} = f_{\text{channel}}$. Next we go to Figure 8.9, where for

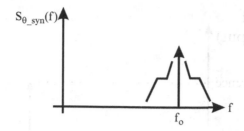

Fig. 8.7 Power spectral density of synthesizer phase noise

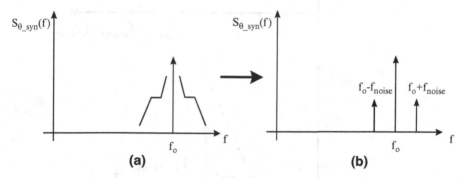

Fig. 8.8 (a) Power spectral density of synthesizer phase noise. (b) Equivalent representation using frequency impulses

simplicity only the $f_o + f_{noise}$ impulse is drawn. As shown in Figure 8.9, this undesired sideband energy from the synthesizer (phase noise) "reciprocal mixes " with an adjacent channel interference at $f_{desire} + f_{channel}$ (blocker) and manifests itself as noise on top of the desired signal. The final effect is pretty much the same as that due to intermodulation between two adjacent channel interferences as described in subsection 2.5.2; that is, a reduction of SNR and hence larger BER.

Notice that the phase impairment effect is always there, with and without interference, whereas the reciprocal mix effect is there only when there is adjacent channel interference. Which impact is more dominant depends, among other factors, on the applications.

We assume that the frequency synthesizer is used to generate the LO signal for the first mixer in the heterodyne architecture of Chapter 2.

Criterion 1: For phase impairment the final degradation in BER depends, in a complicated fashion, on the modulation scheme and statistics of $\Delta\theta(t)$. The relationship between this integrated phase noise and the resulting BER is best obtained via simulation [10]. This will then give us a requirement on the tolerable integrated phase noise. As an example, in [10] for QPSK modulation, a SNR of 9 dB, a BER of 10^{-3} (these are close to the requirements for DECT) the syntheszier integrated phase noise should be less than -15 dBc. Hence we build in some safety margin and specify a requirement of -20 dBc for DECT.

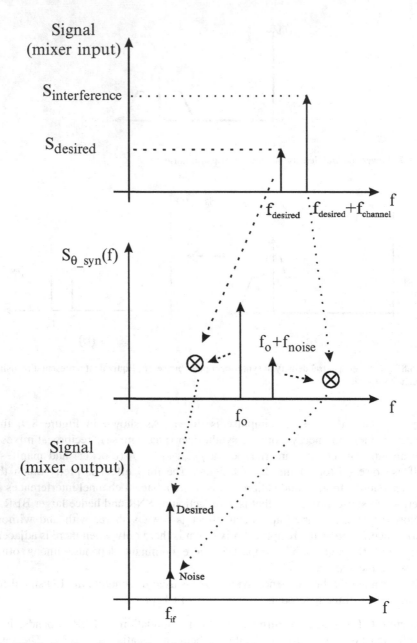

Fig. 8.9 Explanation of reciprocal mixing phenomenon

Criterion 2: Next we quantify the reciprocal mix effect [11]. Here the phase noise must be minimized such that under the worst-case blocking scenario, the power of the reciprocal mixed interferer is below the desired signal level by the minimum SNR the demodulator required, denoted in Chapter 2 as SNR_{demod_in}.

In determining the phase noise figure, the following simplifying assumptions were made:

1. The receiver channel and front end are assumed to be noise free.
2. Phase noise is assumed to be flat across the band of interest at a certain offset from the carrier.
3. The only interference produced within the signal band moving through the receiver front end is due to the phase noise reciprocal mixing with out-of-signal band blockers.

In general, the calculation of allowable phase noise of a frequency synthesizer for the first-stage mixer can be summarized by the following formula [11]:

$$10 \log S_{\theta\_allowable\_syn}(f_i)(Hz) = 10 \log S_{interference}(f_i)(dBm/dBV)$$
$$+ 10 \log S_{desired}(f_i)(dBm/dBV)$$
$$- SNR_{demod\_in}(dB) - 10 \log(BW)) \qquad (8.25)$$

Here $10 \log S_{\theta\_allowable\_syn}(f_i)$ is the PSD of allowable phase noise f_i away from carrier in dBc/Hz; $10 \log S_{interference}$ is the power of the interference in dBm or dBV at the mixer input; $10 \log S_{desired}$ is the power of the desired signal in dBm or dBV at the mixer input; SNR_{demod_in} is the required signal-to-noise ratio at the demodulator input; BW is bandwidth of the desired signal.

We illustrate the calculation by using the DECT standard and heterodyne architecture given in Chapter 2. Considering the DECT standard, the maximum $S_{interference}$ can be at -62 dBm (worst case) compared with a minimum $S_{desired}$ at -77 dBm (worst case). First we note that if BPF1, LNA provide the same gain to interference and desired signal, then their difference in power level remains the same when we go from antenna to mixer. Hence $-10 \log S_{interference} + 10 \log S_{desired}$ $= -(-62$ dBm$) + (-77$ dbm$)$. Second, we conclude from assumptions 1 and 3 and Chapter 2 that $10 \log SNR_{demod\_in} = 25$dB. BW of DECT is 1.728 MHz. Substitute all these in (8.25), we have:

$$10 \log S_{\theta\_allowable\_syn}(f_{channel})(dBc/Hz) = -(-62dBm) + (-77dBm) - 25dB$$
$$- 10 \log(1.728 \, MHz)$$
$$= -102 \, dBc/Hz$$

$$(8.26)$$

Let us go back and use these requirements to guide the design $G(s)$. We first repeat (8.6) here:

$$G(s) = \frac{F(s)K_{pd}K_{vco}}{sM} \qquad (8.27)$$

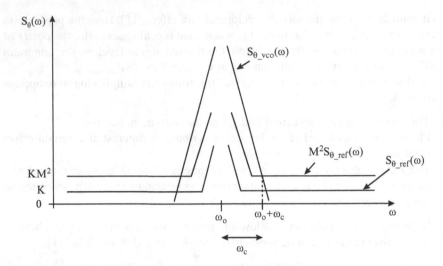

Fig. 8.10 Power spectral density of phase noise sources

How do we determine $G(s)$? The first thing we want to find out about $G(s)$ is its unity gain frequency, ω_u. We note that ω_u of $G(s)$ is directly related to the synthesizer phase noise requirement. Hence we need to find S_{θ_syn}'s dependency on ω_u of $G(s)$.

8.3.1.2 Power Spectral Density of Synthesizer Phase Noise (S_{θ_syn})

To find S_{θ_syn}'s dependency on ω_u, we need to obtain the following information [8]:

1. PSD of phase noise sources
2. Transfer function from phase noise sources to synthesizer output

To help clarify notation we define ω_i (f_i) as the frequency offset from the oscillation frequency ω_o (f_o) in rad/s (Hz). This is to be distinguished from ω (or f), the absolute frequency under consideration. Of course, $\omega_i = \omega - \omega_o$ ($f_i = f - f_o$).

PSD of phase noise sources

1. Figure 8.10 shows $S_{\theta\_vco}(\omega)$ from the VCO. In the ring oscillator case we discussed that its phase noise has $1/(\omega-\omega_o)^2$ characteristics, and we can general-ize that by saying that in general oscillator's phase noise PSD has a bandpass characteristics around ω_o [8], the oscillation frequency, as shown in Figure 8.10.
2. There are other sources of phase noise (other than from oscillators). As a matter of fact, any noise sources in the synthesizer (from divider, phase detector, charge pump, reference, noise coupling from substrate and supply) can result in phase

noise. For the time being, we assume that aside from VCO noise source, reference noise is the other major contributor to phase noise [9]. We further know that PSD of reference phase noise, denoted as $S_{\theta\_ref}(\omega)$, is relatively white around ω_o [8], as shown in Figure 8.10 (in practice it will start to rise as frequency becomes close enough to ω_o [9]). This flat PSD is then taken to have a value equal to K dBc/rad. When referred to the output, this PSD is multiplied by M^2, the square of the division ratio.

As shown in Figure 8.10, $S_{\theta\_vco}(\omega)$ falls off from infinity as we move from ω_o. Since $M^2 S_{\theta\_ref}(\omega)$ is flat, eventually it will become larger than $S_{\theta\_vco}(\omega)$. The distance, between ω_o and when these two PSDs interesect, is defined as ω_c, the crossover frequency. Mathematically it is defined as:

$$S_{\theta\_vco}(\omega)|_{@\omega_o+\omega_c} = M^2 S_{\theta\_ref}(\omega)|_{@\omega_o+\omega_c} = KM^2$$

In terms of offset frequency we have:

$$S_{\theta\_vco}(\omega_i)|_{@\omega_c} = M^2 S_{\theta\_ref}(\omega_i)|_{@\omega_c} = KM^2 \qquad (8.28)$$

Transfer Functions for Phase Noise Sources

Figure 8.11 shows the diagram redrawn from Figure 8.1, but with phase noise from reference source, $\overline{\theta_{i\_ref}^2}$, and VCO, $\overline{\theta_{i\_vco}^2}$, explicitily shown. Let us also redraw Figure 8.10 in Figure 8.12, using offset frequency ω_i. In (7.119), we denoted the transfer function from the VCO node to the synthesizer output as H_{vco}. Similarly, in Chapter 7 we also denoted the transfer function from the reference source node to the synthesizer output as H_{ref}. We have already used this H_{ref} to relate the following disturbances injected at the reference source node to the synthesizer output: spur from the PD [see (7.15)] and phase noise from the PD [see (7.18)]. We now use H_{ref} to relate the phase noise from the reference source node to the synthesizer output.

For the frequency synthesizer in Figure 8.11 let us denote the forward gain $A(s)$ and the feedback factor B. Hence

$$G(s) = A(s)B \qquad (8.29)$$

and

$$B = 1/M \qquad (8.30)$$

From classical feedback theory we know that if the synthesizer is modeled as an LTI system, then

$$H_{vco} = \frac{1}{1 + A(s)B} = \frac{1}{1 + G(s)} \qquad (8.31)$$

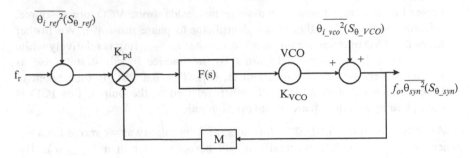

Fig. 8.11 Block diagram of synthesizer with phase noise sources

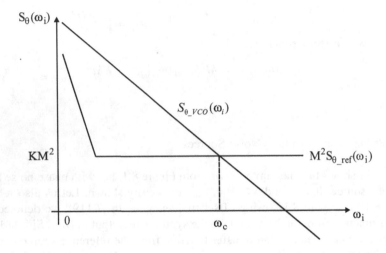

Fig. 8.12 Power spectral density of phase noise sources referenced using offset frequency (ω_i)

and

$$H_{\text{ref}} = \frac{A(s)}{1 + A(s)B} = \frac{A(s)}{1 + G(s)} \tag{8.32}$$

Now what would these transfer functions look like? From (8.18) and (8.19), $F(s)$ and $G(s)$ are low-pass functions. If we substitute (8.18), and (8.19) into (8.31) and (8.32), we find that H_{vco} is a high-pass function and H_{ref} is a low-pass function [8]. We should emphasize that terms like low-pass and high-pass are used with the understanding that the underlying frequency variable is ω_i (not ω). First let us apply the approximation: when $\omega_i < \omega_u$, $|G(s)| \gg 1$; when $\omega_i > \omega_u$, $|G(s)| \ll 1$ [basically replace $G(s)$ by 0], then we can approximate H_{vco} as shown in Figure 8.13. Let us apply the same $G(s)$ approximation to H_{ref}. Then when $\omega_i < \omega_u$, $H_{\text{ref}} \cong M$ and when:

$$\omega_i > \omega_u, H_{\text{ref}} \cong A(s) \tag{8.33}$$

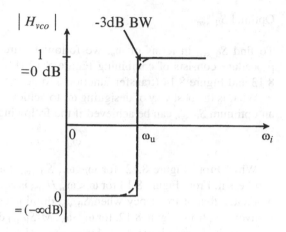

Fig. 8.13 Frequency response of VCO transfer function

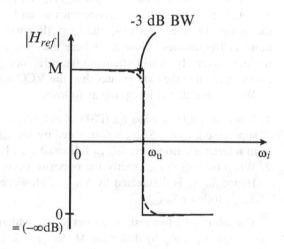

Fig. 8.14 Frequency response of reference transfer function

We now make one more approximation: when $\omega_i > \omega_u$, $|A(s)|$ becomes 0. Substituting this approximation into (8.33), we have

$$H_{\text{ref}} = 0 \text{ for } \omega_i > \omega_u \qquad (8.34)$$

This is called the first-order approximation to H_{ref} at $\omega_i > \omega_u$. Both H_{vco} and H_{ref} are shown in Figure 8.13 and Figure 8.14, where the exact responses are shown in dotted lines and the idealized responses are shown in solid lines.

It has been shown in [8] that for a PLL with an arbitrary loop filter $F(s)$, the unity gain frequency of the resulting open-loop transfer function $G(s)$ becomes the -3 dB bandwidth of the closed-loop transfer function. Hence we can conclude that ω_u of $G(s)$ is also the -3 dB bandwidth for H_{ref}. By the same argument, we can conclude that ω_u is also the -3 dB bandwidth for H_{vco}, except that this time H_{vco} is a high-pass function. In summary, ω_u of $G(s)$ is the -3 dB bandwidth for H_{vco}, H_{ref}.

Optimal S_{θ_syn}

To find S_{θ_syn} in terms of ω_u, we follow Figure 4.36 and Fig.4.37 of [8]. The procedure consists of combining Figure 8.12 (PSD of input phase noise), Figure 8.13 and Figure 8.14 (transfer functions) to derive S_{θ_syn}.

What is the best way of designing ω_u to achieve the optimal S_{θ_syn}? It seems that an optimum S_{θ_syn} can be achieved if the following is satisfied [9]:

$$\omega_u = \omega_c \qquad (8.35)$$

Why? From Figure 8.12, for $\omega_i < \omega_c$, S_{θ_vco} dominates, and hence we want to suppress it. From Figure 8.13 for $\omega_i < \omega_u$, H_{vco} is zero. Hence by making $\omega_c = \omega_u$, we guarantee that at frequency when S_{θ_vco} dominates, it is suppressed at the output. Conversely, from Figure 8.12, for $\omega_i > \omega_c$, $M^2 S_{\theta\_ref}$ dominates, and hence we want to suppress it. From Figure 8.14, for $\omega_i > \omega_u$, H_{ref} is zero. Hence again by making $\omega_c = \omega_u$, we guarantee that at frequencies when S_{θ_ref} dominates, it is suppressed at the output. In other words [9], the loop filter will attenuate the VCO phase noise more at frequencies below ω_c, where they are greater than the reference phase noise. Conversely, it will attenuate the reference phase noise more at frequencies above ω_c, where they are greater than the VCO noise.

We can restate the foregoing as follows:

1. When $\omega_i < \omega_u (=\omega_c)$ So_{θ_vco} (PSD of output phase noise due to VCO) is heavily suppressed. Hence S_{θ_syn} is dominated by So_{θ_ref} (PSD of output phase noise due to reference). However, So_{θ_ref} follows S_{θ_ref}. Hence S_{θ_syn} follows S_{θ_ref}.
2. When $\omega_i > \omega_u (=\omega_c)$, exactly the opposite occurs. So_{θ_ref} is heavily suppressed. Hence S_{θ_syn} is dominated by So_{θ_vco}. However, So_{θ_vco} follows S_{θ_vco}. Hence S_{θ_syn} follows S_{θ_vco}.

Combining the two results, we get S_{θ_syn}, which is shown in Figure 8.15. Notice since $\omega_u = \omega_c$, at ω_u, by definition $M^2 S_{\theta\_ref} = S_{\theta\_vco}$. This is also shown in Figure 8.15. Hence the optimal S_{θ_syn}'s dependency on $G(s)$ is established.

So far we have established the general criteria for selecting ω_u. This does not allow us to determine its specific value. For example, according to (8.28), ω_c can be set arbitrarily by varying K, and hence ω_u's specific value can also be set arbitrarily. To determine the specific value of ω_u we need the phase noise requirement as explained in criteria 1, and 2.

8.3.1.3 Crossover Frequency (ω_c), Unity Gain Frequency (ω_u)

From criterion 1, subsection 8.3.1.1, we are interested in the integrated phase noise. Referring to Figure 8.15, integrated phase noise can be interpreted as the area underneath the S_{θ_syn} curve. Specifically, if we neglect close-in phase noise (i.e., that part of PSD that is very close to $\omega_i = 0$) and assume that noise beyond $\omega_c (=\omega_u)$ is negligible, then

$$\text{integrated phase noise} \cong 2KM^2\omega_c \qquad (8.36)$$

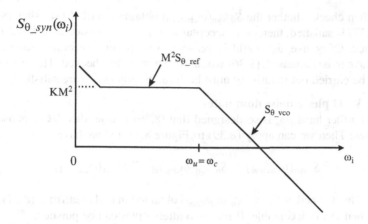

Fig. 8.15 Power spectral density of synthesizer phase noise referenced using offset frequency (ω_i)

A tolerable value on this integrated phase noise, as explained under criterion 1, sub-section 8.3.1.1, is best obtained from simulation. Substituting this value in (8.36) we have one equation and two unknowns: K and ω_c. Combining this equation with (8.28), we have two equations and two unknowns and we can solve for K and ω_c. Applying this value of ω_c to (8.35), ω_u is determined.

Next we want to check whether the ω_u we have obtained will result in an S_{θ_syn} that satisfies criterion 2. If this is acceptable, then the ω_u we have selected has led to a design with acceptable phase noise performance. Otherwise we have to modify ω_u. To carry out this check, let us start with (8.25), which gives a value on the allowable S_{θ_syn}, $S_{\theta\_allowable\_syn}(\omega_i)$. Criterion 2 is satisfied if

$$S_{\theta\_syn}(\omega_i = \omega_{channel}) < S_{\theta\_allowable\_syn}(\omega_i = \omega_{channel}) \qquad (8.37)$$

To obtain the proper value of S_{θ_syn} let us consider Figure 8.15. We note depending on the ω_u of $G(s)$

$$\omega_{channel} < \omega_u \qquad (8.38)$$

or

$$\omega_{channel} > \omega_u \qquad (8.39)$$

Case 1: reference phase noise dominates

If ω_u is so designed that (8.38) is true, then reference phase noise dominates. Then we can apply (8.38) to Figure 8.15 and we have

$$S_{\theta\_syn}(\omega_{channel}) \cong M^2 S_{\theta\_ref}(\omega_{channel}) = KM^2 \qquad (8.40)$$

We then check whether the $S_{\theta\_syn}(\omega_{channel})$ obtained in (8.40) satisfies (8.37).

If (8.37) is satisfied, then ω_u is acceptable. If not, we can decrease K until (8.37) is satisfied. Of course, ω_c would increase as a result and, according to (8.35), ω_u would have to be increased. (8.36) would then have to be checked. The process may have to be carried out iteratively until both (8.36) and (8.37) are satisfied.

Case 2: VCO phase noise dominates

If, on the other hand, ω_u is so designed that (8.39) is true, then VCO phase noise dominates. Then we can apply (8.39) to Figure 8.15 and we have

$$S_{\theta\,syn}(\omega_{channel}) \cong S_{\theta o\,vco}(\omega_{channel}) = S_{\theta vco}(\omega_{channel}) \tag{8.41}$$

We then check whether the $S_{\theta\_syn}(\omega_{channel})$ obtained in (8.41) satisfies (8.37). If this is true, then ω_u, is acceptable. If not, two alternatives can be pursued.

In the first alternative we can redesign ω_u such that reference phase noise becomes dominant again. We can do this by increasing ω_u of $G(s)$ until (8.38) is satisfied. Then we can follow the same procedure as described in "**Case 1: Reference Phase Noise Dominates**" to ensure that (8.37) is satisfied. In the second alternative we can redesign VCO and ω_u together. We first redesign VCO such that $S_{\theta\_vco}(\omega_{channel})$ is lowered until (8.37) is satisfied. Then this new S_{θ_vco} is used in (8.28) to calculate the new ω_c. Of course, the new ω_c is smaller, and according to (8.35), ω_u would have to be decreased.

In summary, in both cases we modify ω_u of $G(s)$, if necessary, until (8.37) and (8.36) are satisfied.

8.3.1.4 Other Poles, and Zeroes

ω_u is designed in the above section for F(s) of any order. Other poles and zeroes design depend on the exact order of F(s).

We omit the other poles and zeroes design of $F_1(s)$ because it is relatively simple. For $F_2(s)$ as shown in (8.14) we have already made

$$p_2 = 0 \tag{8.42}$$

This will be shown to help reduce spur. Next we note that ω_u of $G_2(s)$ [given in (8.15)] should be smaller than p_3. This ensures that when $\omega=p_3$, which means $\angle G_2(j\omega) = -135$ degrees, $|G_2(j\omega)|$ is less than 1. This guarantees that the synthesizer is stable. Furthermore, ω_u should be greater than z_1 because $G_2(j\omega)$ starts off with two poles, p_1 and p_2, at origin, giving a 180-degree phase shift. Hence if ω_u is smaller than z_1, then at ω_u, $G_2(j\omega)$ has close to 180 degrees phase shift and the phase margin of the synthesizer is very poor [6]. This will result in a sharp peak in | $G_2(j\omega)|$ [6]. Hence, to avoid this, ω_u is set to be greater than z_1. Accordingly, ω_u is between z_1 and p_3, as depicted in Figure 8.6b.

To eliminate the peaking, we need to set z_1 precisely. How? We note that since p_3 is far away, the PLL behaves like a second-order LTI system, with a damping factor denoted as ξ. To eliminate peaking we want to set $\xi = 1$. This means that

$$z_1 = \frac{1}{4}\omega_u \qquad (8.43)$$

We now want to find p_3 by making use of the stability requirement. We note that

$$
\begin{aligned}
\text{phase margin} &= 360° - 180° + \angle G_2(j\omega_u) \\
&= 360° - 180° - 180° + \tan^{-1}\frac{\omega_u}{z_1} - \tan^{-1}\frac{\omega_u}{p_3}
\end{aligned}
\qquad (8.44)
$$

Substituting (8.43) into (8.44),

$$\text{phase margin} = \tan^{-1} 4 - \tan^{-1}\frac{\omega_u}{p_3} \qquad (8.45)$$

From the specifications, phase margin is usually given and this determines p_3.

For a higher-order filter with a transfer function $F(s)$ [as described in (8.18)], we first assume for the same reasons as in the $F_2(s)$ case that (8.42) holds. We further assume that the higher-order poles, and zeroes $(p_4 \ldots p_n, z_2 \ldots z_m)$ are so far away that (8.43), and (8.44) still hold. We can therefore determine z_1, p_3 of $F(s)$ using the same procedure in $F_2(s)$'s case. The higher-order poles and zeroes are just for fine-tuning of the synthesizer's performance, and their assignments (for typical applications) can be rather loose as long as they are far enough from p_3.

We have now come up with an initial design of $F(s)$ based on phase noise and stability. We will have a chance to do a complete design using this approach in section 8.4, when we would also check if the spur requirement is satisfied.

8.3.2 Spur-Based Approach

In this approach we design $F(s)$ based on primarily spur and stability requirements. The resulting synthesizer can then be checked and see if it meets the phase noise requirement. We again denote ω_u as the unity gain frequency of the resulting $G(s)$. First we omit the design of $F_1(s)$ because it is rather simple. For second and higher-order filters, we follow sub-section 8.3.1.4 and determine p_2, according to (8.42). We then relate z_1, p_3 to ω_u via (8.43) and (8.45).

Next, to determine ω_u, we again bring in the spur consideration. To save space, we skip the discussion of spur requirements. We assume that spur requirements have been derived in a procedure similar to the derivation of phase noise requirements due to reciprocal mix effect in sub-section 8.3.1.1. Remember that

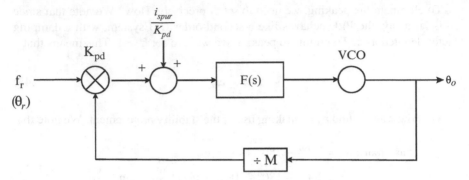

Fig. 8.16 Block diagram of synthesizer with spur sources

[from (7.17a)] spur depends on θ_o, which according to (7.15), depends on H_{ref}. H_{ref} in turn depends on $F(s)$. Hence our goal is to find the dependency of spur on $F(s)$.

First, Figure 8.16 is redrawn from Figure 7.12, with divider M in the feedback path. Following (8.29) and (8.30), we call the forward gain $A(s)$, where

$$A(s) = \frac{K_{pd}F(s)K_{vco}}{s} \tag{8.46}$$

and the feedback factor B, where

$$B = \frac{1}{M} \tag{8.47}$$

Then from feedback theory we can immediately write

$$H_{ref} = \frac{A(s)}{1 + A(s)B} \tag{8.48}$$

If we follow the first-order approximation to H_{ref} at $\omega_i > \omega_u$ as given in (8.34), this creates a problem because the spur frequency, ω_{spur}, typically is larger than ω_u. Hence this approximation will make $H_{ref} = 0$. Substituting this in (7.15), predicts a θ_o of 0, or zero output spur, which is clearly an underestimation and will not help us in the design process. We can overcome this problem by going back to (8.33), which is repeated here:

$$H_{ref} \cong A(s), \omega_i > \omega_u \tag{8.49}$$

We will find another approximation to $A(s)$. Referring to (8.46) we can do that by finding another approximation to $F(s)$, as long as it does not lead to (8.34). To do this we start from the loop filter function $F(s)$ in (8.18) and approximate it by the following form [12]:

$$F(s) = \frac{(s + z_1)}{(s + p_2)...(s + p_n)} \tag{8.50}$$

This will make

$$H_{\text{ref}} = \frac{K_{\text{pd}}K_{\text{vco}}}{s}\left(\frac{(s + z_1)}{(s + p_2)...(s + p_n)}\right) \text{ for } \omega_i > \omega_u \tag{8.51}$$

This is called the second-order approximation to H_{ref} at $\omega_i > \omega_u$.

Turning back to $F(s)$, we note that the approximation consists of assuming that $F(s)$ has only one zero and that $F(\infty)$ is set to 1. Then we substitute (8.50) into (8.46) to get $A(s)$. Following that, we substitute the resulting $A(s)$ in (8.48) and we get the complete H_{ref} as:

$$H_{\text{ref}} = \frac{\frac{K_{\text{pd}}F(s)K_{\text{vco}}}{s}}{1 + \frac{K_{\text{pd}}F(s)K_{\text{vco}}}{sM}} \tag{8.52a}$$

$$= \frac{K_{\text{pd}}\frac{(s+z_1)}{s(s+p_2)...(s+p_n)}K_{\text{vco}}}{1 + K_{\text{pd}}\frac{1}{sM}\frac{(s+z_1)}{(s+p_2)...(s+p_n)}K_{\text{vco}}} \tag{8.52b}$$

We make three comments about this transfer function:

1. z_1 typically is close to origin and so the pole due to VCO is canceled by this zero.
2. Even though $F(s)$ is approximated differently, H_{ref} still has its −3 dB frequency at ω_u. This means that H_{ref} has a factor $\frac{K}{s+\omega_u}$.
3. When we put this simplified $F(s)$ inside the synthesizer loop, we assume that since $p_3,.. p_n$ of $F(s)$ are so far away from ω_u that these poles do not move much.

From comments 1 through 3, we can further simplify (8.52b) as

$$H_{\text{ref}} = \frac{K}{(s + \omega_u)(s + p_3)\cdots(s + p_n)} \tag{8.53}$$

Only K now remains an unknown. At low frequency: that is, when $\omega_i \ll \omega_u$, we still have $|G(s)| = |A(s)B| \gg 1$. Applying this approximation, to (8.48) and it becomes

$$H_{\text{ref}}|_{\omega_i<<\omega_u} = \frac{A(s)}{1 + A(s)B}|_{\omega_i<<\omega_u}$$

$$\approx \frac{1}{B} \tag{8.54}$$

Substituting (8.47) into (8.54),

$$H_{\text{ref}}\big|_{\omega_i \ll \omega_u} = M \tag{8.55}$$

Again at low frequency, $\omega_i \ll \omega_u$, (8.53) becomes

$$H_{\text{ref}}\big|_{\omega_i << \omega_u} = \frac{K}{\omega_u p_3 \cdots p_n} \tag{8.56}$$

(This is because at low frequency $s \ll \omega_u$, $\ll p_3 \ldots \ll p_n$) Equating (8.55) and (8.56),

$$M = \frac{K}{\omega_u p_3 \cdots p_n}$$

or

$$K = \omega_u p_3 \cdots p_n M \tag{8.57}$$

Substituting (8.57) back into (8.53), we finally determine the complete H_{ref} expression:

$$H_{\text{ref}} = \frac{\omega_u p_3 \cdots p_n M}{(s + \omega_u)(s + p_3)..(s + p_n)} \tag{8.58}$$

Notice that since comment 3 assumes that the higher order poles do not move upon closing the loop, hence (8.58) does not account for peaking of the closed-loop response near ω_u. Remember that from (7.17b), spur is related to $\theta_o{}^2$ as

$$\frac{\text{spur\_power}}{\text{carrier\_power}} = \left(\frac{\frac{A_c \theta_o}{2}}{\frac{A_c}{2}}\right)^2 = \theta_o^2 \tag{8.59}$$

To calculate θ_o we substitute (8.58) into (7.15), and we get:

$$\theta_o = \frac{\omega_u p_3 .. p_n M}{(s + \omega_u)(s + p_3)..(s + p_n)} \frac{1}{K_{\text{pd}}} i_{\text{spur}} \tag{8.60}$$

If we replace θ_o, i_{spur} by its amplitude and express everything in decibels we have

$$\text{spur}\big|_{\text{dBc}} = 10 \log \frac{\text{spur\_power}}{\text{carrier\_power}} = 10 \log \theta_o^2 = 20 \log \theta_o$$

$$= 20 \log \left(\left| \frac{\omega_u p_3 \cdots p_n M}{(s + \omega_u)(s + p_3) \cdots (s + p_n)} \frac{1}{K_{\text{pd}}} i_{\text{spur}}(\text{amplitude}) \right| \right) \tag{8.61}$$

Here amplitude of θ_o is denoted by the notation spur, which actually represents the spur ratio. We use spur to denote both the spur component and the spur ratio. Its actual meaning should be clear from the context. Obviously, spur depends on the spur frequency, ω_{spur}. We want the loop filter to filter out the spur at ω_{spur}. Therefore, $\omega_{spur} \gg \omega_u$ (usually on the order of at least 10 or more). Typically, $|j\omega_{spur}| \gg |p_3|...|p_n|$ as well and therefore in the denominator of (8.61) we have

$$s + p_3 \cong s + p_4 \cong \cdots \cong s + p_n \cong j\omega_{spur} \qquad (8.62)$$

We now substitute (8.62) into (8.61). Then (8.61) becomes

$$spur|_{dBc} = 20\log\left(\left|\frac{\omega_u p_3 \cdots p_n M}{\omega_{spur}^{n-1}} \frac{1}{K_{pd}} i_{spur}(amplitude)\right|\right) \qquad (8.63a)$$

$$= 20\log\left(\left|\frac{\omega_u p_3 \cdots p_n M}{\omega_{spur}^{n-1}} \frac{2\pi}{I_{charge\_pump}} i_{spur}(amplitude)\right|\right) \qquad (8.63b)$$

Note that going from (8.63a) to (8.63b) we have used the definition $K_{pd} = I_{charge\_pump}/2\pi$, obtained from (7.9). What is $i_{spur}(amplitude)$? From (7.7),

$$amplitude\ of\ i_{spur} = d \times I_{charge\_pump} \qquad (8.64)$$

Substituting this in (8.63b), we have

$$spur|_{dBc} = 20\log\left(\left|\frac{\omega_u p_3 \cdots p_n M}{\omega_{spur}^{n-1}} \frac{2\pi}{I_{charge\_pump}} d \times I_{charge\_pump}\right|\right) \qquad (8.65a)$$

$$= 20\log\left(\left|\frac{\omega_u p_3 \cdots p_n M}{\omega_{spur}^{n-1}} 2\pi \times d\right|\right) \qquad (8.65b)$$

Now that we have established the dependency of spur on $F(s)$, how do we finish our design of the loop filter? For a second order filter, we start from (8.50) and retain only z_1, p_2, p_n. $F(s)$ then becomes

$$F_2(s) = \frac{s + z_1}{(s + p_2)(s + p_3)} \qquad (8.66)$$

Previously, we made $p_2 = 0$ to get some preliminary spur reduction. Then $F_2(s)$ becomes

$$F_2(s) = \frac{s + z_1}{s(s + p_3)} \qquad (8.67)$$

With this $F_2(s)$, (8.65b) becomes

$$spur|_{dBc} = 20\log\left(\left|\frac{\omega_u p_3}{\omega_{spur}^2} M_2 \pi d\right|\right) \qquad (8.68)$$

It should be noted that $\text{spur}|_{dBc}$ and ω_{spur} in (8.68) are given in the spur specification and d is a mismatch and is given for a given technology. As stated in sub-section 7.3.4.2, d typically is around 0.1%. Hence (8.68) gives an equation that relates the two unknowns: p_3, ω_u. We have stated that p_3 is related to ω_u through (8.45). Hence when we look at (8.45) and (8.68), we find that we have two equations and two unknowns: ω_u, p_3. With two equations and two unknowns we can solve for ω_u, p_3, and then we can go back to (8.43) and solve for z_1. $F_2(s)$ is then completely determined. For higher-order filters, we will follow the same argument as given at the end of sub-section 8.3.1.4 and extend the results from $F_2(s)$ to $F(s)$. In summary this section shows us how to come up with an initial design on $F(s)$ based on spur and stability requirement.

We have now finished our discussions of synthesizer subblocks and have given design procedures for all of them. Next we put the pieces together and see how they fit with each other.

8.4 A Complete Synthesizer Design Example (DECT Application)

Figure 8.17 is a flowchart that shows how to design a PLL-based frequency synthesizer. First, in the flowchart we choose to design the loop filter based on phase noise; that is, we follow subsection 8.3.1. If we had chosen to design a loop filter based on spur requirements (i.e., sub-section 8.3.2), the flowchart will be different.

Second, the key parameters to design are K_{vco}, K_{pd}, $F(s)$. Of course, to determine K_{vco}, we need to do a detailed design of the VCO, and to determine K_{pd}, we need to do a detail design of the PD/charge pump. In both cases, the W/L ratio and bias currents need to be calculated. As for the loop filter, the R, C have to be designed. Notice that as opposed to some traditional design methodologies [7] used for discrete circuit design, the present methodology is more suitable for integrating the entire synthesizer on a chip. The major difference is that here we do not assume that we can have individual blocks (such as PD, VCO) available from off-the-shelf components, whose design values for K_{vco}, K_{pd} span a large range (due to the fact you can mix different technologies, such as GaAs, bipolar, MOS). Instead, we are designing everything in one technology (CMOS) and so the range of K_{vco} and K_{pd} is more restricted. Consequently, some of the steps should follow a specific order (e.g., VCO should be designed before loop filter).

Third, as will be shown, during the design, if we are not careful, passive components such as resistors and capacitors can have design values too large to be realizable in integrated form. It is our philosophy to avoid this. In addition, we make sure that the design for transistors is reasonable [e.g., we do not end up with a $\left(\frac{W}{L}\right)$ ratio of thousands or much less than one, and that bias current is on the order of

Fig. 8.17 Flowchart of
synthesizer design procedure

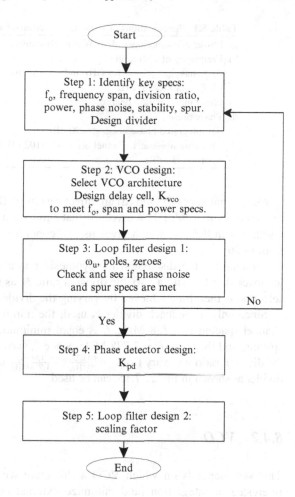

milliampere). This will give readers the feeling that they are doing a realistic design
and allow them to map the design to their application.

We now illustrate this methodology by going through an example step by step,
using realistic numbers for a typical application. Consistent with previous chapters,
we do the design to satisfy the DECT standard. We assume that the technology is a
0.6 μm CMOS process.

8.4.1   Specifications

For DECT application, we assume that a direct conversion architecture is used.
The key specifications relevant to the frequency synthesizer are summarized in
Table 8.1.

Table 8.1 Specifications of frequency synthesizer for DECT application

a) Power consumption < 33 mW (mostly consumed by VCO) [2]

b) Frequency of operation

    i) f_o has $f_{\text{vco\_center}} = 1.9$ GHz and sweeps from 1.884 GHz to 1.9 GHz.

    ii) $f_{\text{channel}} = 1.728$ MHz

c) Phase margin > 55 degrees

d) Phase noise

    i) Integrated phase noise < -20 dBc

    ii) Phase noise at 1 channel offset < -102 dBc/Hz

e) Spur < -30 dBc

Notice that specs (b) and (d) are direct results of DECT standards. For example, (b) comes from Table 1.1. We assume that simulation is performed according to the discussion in 8.3.1.1, which gives us (d), condition (i); (d), condition (ii), comes from (8.26).

To simplify the discussion the synthesizer is assumed to consist of only the feedback divider with a variable division ratio M as shown in Figure 8.1. Channel selection is therefore achieved by varying the divider ratio M.

Since only a feedback divider is used, the minimum resolution is 1. Since the channel spacing is 1.728 MHz, f_r is equal minimum resolution times the channel spacing, and therefore $f_r = 1.728$ MHz. Since f_o varies from 1.884 GHz to 1.9 GHz, the division ratio will vary from $\frac{1.884\text{GHz}}{1.728\text{MHz}}$ to $\frac{1.9\text{GHz}}{1.728\text{MHz}}$ or $M = 1090$ to 1099. Hence the divider as shown in Figure 7.15, can be used.

8.4.2 VCO

This step depends on which VCO architecture we have selected. Since we are interested in integration (and minimize external components such as L, C) we choose the ring oscillator [2] as shown in Figure 7.40. What is N? In Section 7.7, we noted that this depends on ω_o and technology. In this example we have arbitrarily picked this to be

$$N = 4 \tag{8.69}$$

To carry on the analysis, we have to make some assumptions on the internal details of the ring oscillators. Three types of ring oscillators have been discussed in sub-section 7.7.1 (each with both single-ended and differential implementations). For the present example we choose the one whose delay cell was shown in Figure 7.44. This is redrawn in Figure 8.18, with M_{tune} replaced by M_{10} and M_{11}. The P_{bias} for M_3, M_4 is also shown explicitly and is set to be a constant. Finally, the loading from the next stage delay cell is shown. We are now going to design a VCO that oscillates fast enough to achieve f_o and so that its K_{vco} is sufficient to attain all channels while staying within the power specification [basically satisfying specs

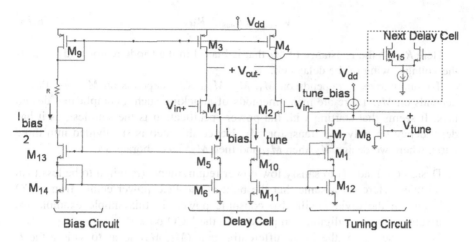

Fig. 8.18 Slow-slewing saturated delay cell

(a), and (b) of Table 8.1]. We should of course check whether this VCO generates excessive phase noise at the synthesizer output. This check will be postponed until after the loop filter is designed in the next section (section 8.4.3). If the phase noise generated is excessive we may have to come back to this section and redesign the VCO (perhaps by relaxing the power specification).

First let us rewrite the formula for f_o from (7.100), where C is relabelled C_l:

$$f_o = \frac{1}{2NRC_l \ln 2} \tag{8.70}$$

As shown in (8.70), f_o depends on C_l. Now C_l is the loading capacitance on either $V_{out}{}^+$ or $V_{out}{}^-$. Let us consider C_l on $V_{out}{}^+$ only. From Figure 8.18, C_l depends on M_{15}, M_1, M_3.

Next let us repeat the formula for K_{vco}. Since the bias for M_3, M_4 (=P_{bias} in Figure 7.44) is set to a constant, we can repeat the K_{vco} formula from (7.106).

$$K_{vco} = \frac{-2\pi f_o R G_{m7-8}}{|V_{GS3}| - |V_t| - |V_{DS3}|} \tag{8.71}$$

Here M_7-M_8 are used for tuning and hence G_{m7-8} replaces $g_{m_{M_{tune1}}}$. Finally, let us write the formula for the power specification. In Figure 8.18, $I_{bias} \gg I_{tune}$. Hence power in the tuning circuitry can be neglected. Furthermore, the bias circuit is shared among all delay cells and hence its power consumed per delay cell can be neglected. Consequently, power for a ring oscillator with $N=4$ can be approximated as

$$power = 4I_{bias} \times V_{dd} \tag{8.72}$$

This I_{bias} is, in turn, related to the design of M_3 by the following formula:

$$V_{\text{swing}} = I_{\text{bias}} \times R_{M3} \tag{8.73}$$

Here R_{M3} is the resistance of M_3 that is biased in the triode region and V_{swing} is the voltage swing of the delay cell.

To summarize, f_o depends on M_1, M_3, M_{15}. K_{vco} depends on M_3, M_7. Power depends on M_3. There are many methods of attacking such a complicated design task. It seems that starting with the power specification is the simplest, as it has dependency on only one transistor, M_3. Hence this step is subdivided into three parts, where we design M_3, then M_1, and then M_7. To elaborate, we

1. Design the load M_3 to satisfy low power requirements (required to be less than 33 mW). Here we assume that the bulk of the PLL power comes from VCO (although the divider will also consume power, in this simple example we assume it to be negligible compared with the VCO power).
2. Design the G_m of the input differential pair (M_1, M_2) so as to satisfy the f_o requirement. This makes the VCO oscillate at a DECT center frequency of 1.9 GHz and allows us to determine C_1.
3. Design the G_m of the tuning circuit (M_7, M_8) and hence the K_{VCO} to attain the frequency sweep from 1.884 GHz to 1.9 GHz.

8.4.2.1 Design for Power Specification

Step 1: Determine I$_{\text{bias}}$. First, from the power specification, set power consumption to be 33 mW. Assume that all power is consumed in the VCO. Substituting this in (8.72) we have

$$4 \times I_{\text{bias}} \times 3.3 \text{ V} = 33 \text{ mW} \tag{8.74}$$

Solving, we have

$$I_{\text{bias}} = 2.5 \text{ mA} \tag{8.74a 8.74b}$$

Step 2: Calculate $\left(\frac{W}{L}\right)_3$. Next we want to determine the W/L ratio of M_3. This can be determined from the resistance of M_3.

$$R_{M_3} = \frac{V_{\text{swing}}}{I_{m3\_\text{max}}} \tag{8.75}$$

From Figure 4.2, the maximum current flowing through M_3, $I_{M_{3\_\text{max}}}$ is

$$I_{M_{3\,\text{max}}} = I_{\text{bias}} = 2.5\text{mA} \tag{8.76}$$

This happens when all I_{bias} is steered to the branch containing M_3.

According to [2] V_{swing} should be picked to ensure that the PMOS transistor is deep enough in triode so that R_{M_3} is sufficiently linear. We follow the value selected in [2] and choose

$$V_{\text{swing}} = 1 \text{ V} \tag{8.77}$$

Substituting (8.76) and (8.77) in (8.75),

$$R_{M_3} = \frac{V_{\text{swing}}}{I_{\text{Bias}}} = \frac{1V}{2.5mA} = 400\Omega \tag{8.78}$$

If we set R_{M_3} equal to the small-signal resistance of M_3 in the triode region, then we have

$$R_{M_3} = \frac{1}{k_p\left(\frac{W}{L}\right)_3(|V_{GS_3}| - |V_t| - |V_{DS_3}|)} \tag{8.79}$$

V_{GS} is made high to favour a high f_o [2]. The upper limit is the power supply voltage. Hence we choose

$$|V_{GS3}| = 3.3 \text{ V} \tag{8.80}$$

Notice that to make the small signal resistance of M_3 as given by (8.79) truly represent the resistance R_{M_3}, we should bias M_3 at the midpoint of the V_{ds3} vs I_{d3} curve[2]. This is the point when $|V_{DS3}| = V_{\text{swing}}/2$. Hence we set

$$|V_{DS3}| = V_{\text{swing}}/2 = 0.5 \text{ V} \tag{8.81}$$

Substituting (8.78), (8.80), and (8.81) in (8.79),

$$400\Omega = \frac{1}{k_p\left(\frac{W}{L}\right)_3(3.3V - V_t - 0.5V)} \tag{8.82}$$

Let us assume that our 0.6 μm CMOS process has k_p=50μA/V^2, V_t=1V. Substituting these in (8.82), we find

$$\left(\frac{W}{L}\right)_3 = \frac{17\mu m}{0.6 \ \mu m} \tag{8.83}$$

8.4.2.2 Design for f_o

Next we want to determine $\left(\frac{W}{L}\right)_1$. To do this we first have to calculate C_l, the load capacitance. Let us substitute condition b(i) in Table 8.1, (8.69), and (8.78) into (8.70)

$$1.9 \text{ GHz} = \frac{1}{2 \times 4 \times 400\Omega \times C_l \ln 2} \tag{8.84}$$

Solving,

$$C_l = 0.24 \text{ pF} \tag{8.85}$$

From Figure 8.18, we see that C_{gs} (M_{15}) dominates the contribution to C_1. Hence

$$C_l = C_{gs}(M_{15}) \tag{8.86}$$

Therefore,

$$C_{gs}(M_{15}) = 0.24 pF \tag{8.87}$$

Also, since each delay cell in the oscillator is assumed to be identical, we assume that

$$\left(\frac{W}{L}\right)_{15} = \left(\frac{W}{L}\right)_1 \text{ and so} \tag{8.88}$$

$$C_{gs}(M_{15}) = W_{15} \times L_{15} \times C_{ox} = W_1 \times L_1 \times C_{ox} = W_1 \times 0.6 \ \mu m \times C_{ox} \tag{8.88a}$$

For this 0.6 μm process, we assume that

$$C_{ox} = 2.5 \text{ fF}/\mu m^2 \tag{8.88b}$$

Substituting (8.87), and (8.88b) into (8.88a) we have

$$0.24 pf = W_1 \times 0.6u \times 2.5 fF/\mu m^2 \tag{8.89}$$

Solving we have

$$\left(\frac{W}{L}\right)_1 = \frac{160\mu m}{0.6 \ \mu m} \tag{8.90}$$

8.4.2.3 Design for f_{sweep}, Determine K_{vco}

Substituting condition b(i), Table 8.1, in the definition of K_{vco} as given by (7.70)

$$K_{\text{vco}}\big|_{\text{in Hz/V}} = \frac{df_o}{dV_{\text{tune}}} = \frac{f_{\max} - f_{\min}}{V_{\text{tune\_min}} - V_{\text{tune\_max}}} = \frac{1.9 \text{GHz} - 1.884 \text{GHz}}{V_{\text{tune\_min}} - V_{\text{tune\_max}}} \tag{8.91}$$

The tuning range, which equals $V_{\text{tune\_max}} - V_{\text{tune\_min}}$, is upper bounded by V_{dd} and lower bounded by noise. Following [2], we choose

$$V_{\text{tune\_max}} = 1.3 \text{ V} \tag{8.92}$$

and

$$V_{\text{tune\_min}} = 0 \text{ V} \tag{8.93}$$

Substituting (8.92) and (8.93) in (8.91), we have

$$K_{\text{vco}} = -0.016 \text{ GHz}/1.3 \text{ V} = -12.3 \text{ MHz/V} \tag{8.94}$$

Let us build a safety margin by setting

$$K_{\text{vco}} = -15 \text{ MHz/V} \tag{8.95}$$

Substituting (8.95), (8.80), (8.81), and (8.78) in (8.71) and rearranging, we have

$$
\begin{aligned}
Gm_{7-8} &= \frac{-15\text{MHz/V} \times (|V_{GS3}| - |V_t| - |V_{DS3}|)}{-f_o R_{M_3}} \\
&= \frac{-15\text{MHz/V} \times (3.3\text{V} - 1\text{V} - 0.5\text{V})}{-1.9 \text{ GHz} \times 400\Omega} \\
&= 28\mu\text{A/V}
\end{aligned}
\tag{8.96}
$$

Referring again to the tuning circuit in Figure 8.18, from the definition of G_m we have

$$I_{\text{tune\_max}} - I_{\text{tune\_min}} = G_{m7-8}(V_{\text{tune\_max}} - V_{\text{tune\_min}}) \tag{8.97}$$

$$\text{Now } I_{\text{tune\_min}} = 0 \tag{8.98}$$

Substituting (8.92), (8.93), (8.96) and (8.98) into (8.97) and solving,

$$I_{\text{tune\_max}} = 28\mu\text{A/V} \times 1.3 \text{ V} = 36\mu\text{A} \tag{8.99}$$

Since $I_{\text{tune\_bias}}$ has to be larger than $I_{\text{tune\_max}}$, let us set

$$I_{\text{tune\_bias}} = 40\mu\text{A} \tag{8.100}$$

Next, for any differential pair, its G_m is related to the (W/L) and bias current as:

$$Gm_{7-8} = \sqrt{I_{\text{tune\_bias}} \times k_n \times (W/L)_7} \tag{8.101}$$

Given $k_n = 125 \ \mu\text{A/V}^2$ for the present process, substitute this value, (8.96), (8.100) into (8.101). Then

$$28 \ \mu\text{A/V} = \sqrt{40 \ \mu A \times 125 \ \mu\text{A/V}^2 \times (W/L)_7} \tag{8.102}$$

Solving,

$$(W/L)_7 = 0.16 \tag{8.103}$$

This $(W/L)_7$ may be small. It can be circumvented by making the gain of current mirror M_{11}-M_{12} to be smaller. We assume that is being done in the present example and we set a designed $(W/L)_7$ of

$$(W/L)_7 = 1 \tag{8.104}$$

We have now completed our design of VCO.

8.4.3   Loop Filter: Part 1

For our present application, experience shows us that a second-order filter is a good choice, in particular to smooth out voltage jumps during switching transients of the charge pump circuit [4]. Therefore, we use the filter $F_2(s)$ described in (8.11). How do we determine all the parameters of $F_2(s)$? We choose to follow sub-section 8.3.1 to design $F_2(s)$.

8.4.3.1   Determine Crossover Frequency (ω_c) Using Integrated Phase Noise Specification, and Unity Gain Frequency (ω_u)

From condition d(i), Table 8.1 we set integrated phase noise as –20 dBc. We substitute this into (8.36), rewrite the formula in hertz, and we have

$$-20\mathrm{dBc} = 2KM^2 f_c \tag{8.105}$$

In addition, using (8.28), we have another equation

$$S_{\theta\_vco}(\omega_i)|_{@\omega_c} = M^2 S_{\theta\_ref}(\omega_i)|_{@\omega_c} = KM^2 \tag{8.106}$$

In our present example, we are interested in the ring oscillator, so S_{θ_vco} is given by (7.118). Expressing (8.106) in hertz and substituting (7.118), we have

$$\left(\frac{f_o}{f_c}\right)^2 \frac{kTa_v\zeta^2}{I_{\mathrm{bias}}(V_{GS_1} - V_t)} = KM^2 \tag{8.107}$$

To obtain the parameters needed to solve (8.107), we do the following: first we calculate V_{GS1}-V_t by substituting $\left(\frac{W}{L}\right)_1$ from (8.90) and I_{bias} from (8.74b) in the I-V equation of a SCP:

$$\frac{I_{bias}}{2} = \frac{k_n}{2} \left(\frac{W}{L}\right)_1 (V_{GS_1} - V_t)^2 \tag{8.108a}$$

which becomes

$$\frac{2.5mA}{2} = \frac{125\mu A/V^2}{2} \left(\frac{160\mu m}{0.6\ \mu}\right)(V_{GS_1} - V_t)^2 \tag{8.108b}$$

Solving, V_{GS1}-$V_t = 0.27$ V.

Next, from [1], there is a range for a_v and ζ. For example a_v, the gain, has to be at least larger than 1 to ensure that the loop gain is larger than 1 to sustain oscillation. We set the values on the high side. Therefore, we set a_v to be 3, $\zeta=2.8$. kT is always 4.14×10^{-21}J.

Finally, from condition b(i), Table 8.1 $f_o = 1.9$ GHz, and from (8.74b), $I_{bias} = 2.5$mA.

Substituting all these values in (8.107), we have

$$\frac{258}{f_c^2}\ dBc/Hz = KM^2 \tag{8.109}$$

Now we have two equations, (8.105) and (8.109) and two unknowns, K and f_c. We can then solve for f_c:

$$f_c = 51.6\ KHz \tag{8.110}$$

Also, from sub-section 8.4.1 we obtain the maximum value of M:

$$M = 1099 \tag{8.111}$$

Substituting (8.110) and (8.111) into (8.105), we solve for K:

$$K = -131\ dBc/Hz \tag{8.112}$$

Finally, we substitute (8.110) in (8.35) and we have

$$f_u = 51.6\ kHz\ (\omega_u = 324\ krad/s) \tag{8.113}$$

8.4.3.2 Check Phase Noise Specification Imposed by Reciprocal Mixing

Now that we have obtained ω_u of $G_2(s)$, have designed a VCO and have determined all the parameters in (8.107), we want to check and see whether the $G_2(s)$ so designed does lead to a synthesizer that satisfies condition d(ii), Table 8.1. To do this we follow the procedure in sub-section 8.3.1.3. First we substitute (8.113) and condition b(ii), Table 8.1, in (8.39) and we note that (8.39) is satisfied. Hence we should follow the procedure in case 2 of sub-section 8.3.1.3.

Next we repeat (7.118)

$$S_{\theta\_vco}(f_i) = \left(\frac{f_o}{f_i}\right)^2 \frac{kTa_v\eta^2}{I_{bias}(V_{GS_1} - V_t)} \tag{8.114}$$

With parameters we have chosen before, this becomes

$$S_{\theta\_vco}(f_i) = \frac{258}{f_i^2} \text{ dBc/Hz} \tag{8.115}$$

From condition b(ii), Table 8.1 we know that $f_{channel} = 1.7$ MHz. Substituting this in (8.115) and we have

$$S_{\theta\_vco}(f_{channel}) = -100.5 \text{ dBc/Hz} \tag{8.116}$$

Substituting this in (8.41), we have

$$S_{\theta\_syn}(f_{channel}) = -100.5 \text{ dBc/Hz} \tag{8.117}$$

Equation (8.117) just meets condition d(ii), Table 8.1. Hence there is no need to go back to sub-section 8.4.2 to redesign the VCO.

8.4.3.3 Determine Other Poles, and Zeros

We next determine z_1 from ω_u and use condition (c) in Table 8.1 to find p_3. This can be done by following the same procedure as described in sub-section 8.3.1.4. First, rewriting (8.43), we have

$$\frac{\omega_u}{z_1} = 4 \tag{8.118}$$

Next let us go back to condition (c) in Table 8.1, and set the phase margin to 58°. Substituting this in (8.45) and solving, we have

$$\frac{p_3}{\omega_u} = 3 \tag{8.119}$$

Substituting (8.113) into (8.118), we have

$$z_1 = 81 \text{ Krad/s} \tag{8.120}$$

Substituting (8.113) into (8.119), we have

$$p_3 = 972 \text{ Krad/s} \tag{8.121}$$

As a side note dividing (8.14b) by (8.14a), we get

$$b = \frac{p_3}{z_1} \tag{8.122}$$

Substituting (8.120) and (8.121) in (8.122), we have

$$b = 12 \tag{8.123}$$

8.4.3.4   Check Spurs Specification

Now that we have obtained ω_u and all poles and zeroes of $G_2(s)$, we want to go back and check if our design satisfies condition (e), Table 8.1. To do this we make use of the discussion offered in sub-section 8.3.2 to find the output spurs. As discussed in sub-section 8.3.2, we have assumed that the single most important spurious source is from the PD. We jump ahead a bit to sub-section 8.4.4 and learn that we will be using a PFD for our PD. Hence $\omega_{spur} = \omega_r = 2\pi \times 1.728$ Mrad/s.

To find the output spur for our present design, we use (8.68), repeated here:

$$\text{spur}|_{dBc} = 20 \log \left(\left| \frac{\omega_u p_3}{\omega_{spur}^2} M_2 \pi d \right| \right) \tag{8.124}$$

For the present technology we assume that we are given a mismatch that leads to

$$d = 0.001 \tag{8.125}$$

Substituting (8.113), (8.111), (8.121), (8.125), and $\omega_{spur} = 2\pi \times 1.728$ Mrad/s into (8.124), we have

$$\text{spur}|_{dBc} = 20 \log \left(\left| \frac{324 \text{Krad/s} \times 972 \text{Krad/s} \times 1099 \times 2\pi \times 0.001}{(10.2 \text{Mrad/s})^2} \right| \right)$$
$$= -35 \text{ dBc} \tag{8.126}$$

This satisfies condition (e) of Table 8.1.

At this point in the design we have finished picking all the poles and zeroes as well as the ω_u of $G_2(s)$ that allows the synthesizer to meet the specifications. How does this information allow us to determine $F_2(s)$? Referring to (8.11a) to determine F_2 we need to find z_1, p_3, C_3. To obtain these values we recognize that p_3, z_1, for F_2 are the same as z_1, p_3 of G_2, which have already been determined.

We ask ourselves, have we finished designing the loop filter $F_2(s)$ since we have determined all its poles and zeroes? The answer is no. This is because $F_2(s)$ takes current as the input and generates voltage as the output. Hence, besides poles and zeroes, this transfer function needs a term that transforms current to voltage. When $F_2(s)$ is expressed in the form given in (8.13), it can be seen that R_2 is this term, which we denote as the scaling factor. When $F_2(s)$ is expressed in the form used in this section, namely the form given in (8.11a), C_3, instead of R_2, becomes the scaling factor.

To determine this C_3 we recognize that C_3 in the F_2 expression is the same as the C_3 in the G_2 expression as given in (8.15). To determine C_3 in the G_2 expression we must first find all the other unknowns in G_2. It turns out that the only unknown is K_{pd}

and hence we first determine K_{pd}. Then we set $s=j\omega_u$ in (8.15). From the definition, | $G_2(j\omega_u)| = 1$. Hence we have one equation and one unknown, C_3, and it can be determined. This is what we do next.

8.4.4   Phase Detector: Determine K_{pd}

In this step we design PD and use condition d(ii), Table 8.1 to determine K_{pd}. For PD, we decide to use the PFD/charge pump as described in Figure 7.4 (the C in the loop filter is now replaced by the filter $F_2(s)$, shown in Figure 8.5). How do we determine its required K_{pd}? There are a few alternatives. We choose to resort to phase noise consideration to obtain this value [12]. Remember that back in sub-section 8.3.1.2 we ignored phase noise from PD. Let us reintroduce this noise source in the present sub-section. From the discussion following (7.20), we have concluded that K_{pd} should be selected large enough to render its phase noise contribution from charge pump at one channel offset insignificant when compared with other sources. Hence if the synthesizer phase noise due to the PD's phase noise alone satisfies condition d(ii), Table 8.1, the K_{pd} selected must be large enough.

From (7.20), to obtain this phase noise, we need H_{ref}. What approximation should we use for this H_{ref} at $\omega_i > \omega_u$? If we use the first-order approximation as given in (8.34), then since $f_{channel} (= 1.7$ MHz) is much larger than $f_u (= 51.6$ KHz), $H_{ref} (f_{channel})$ becomes 0. This will make $S_{\theta o\_pd}$, in (7.20), equal to 0, and (7.20) will be useless in helping us to design K_{pd}. Instead we should use the second-order approximation to H_{ref} as given in (8.51). With this second order approximation, we have derived the resulting H_{ref}, which is given in (8.58). Substituting this H_{ref} from (8.58) into (7.20), and taking the logarithm on both sides, we have

$$S_{\theta o\_pd}(\omega_i)\big|_{dBc/Hz} = 10\log\left(\left|\left(\frac{\omega_u p_3 \cdots p_n M}{\omega_{channel}{}^{n-1} K_{pd}}\right)^2 \frac{\overline{i_{nd}{}^2}}{\Delta f}\right|\right) \qquad (8.127)$$

From Chapter 7, $\frac{\overline{i_{nd}{}^2}}{\Delta f}$ is the PSD of the noise current from M_1 of the charge pump in Figure 7.4, (reduced by a factor of d, the duty cycle, since the charge pump is only on for that period during steady state), and $\omega_{channel}$ is the offset frequency at which we calculate the $S_{\theta o\_pd}$. For the present design example, $F(s)$ is a second-order filter. Hence (8.127) becomes

$$S_{\theta o\_pd}(\omega_i)\big|_{dBc/Hz} = 10\log\left(\left|\left(\frac{\omega_u p_3 M}{\omega_{channel}{}^2 K_{pd}}\right)^2 \frac{\overline{i_{nd}{}^2}}{\Delta f}\right|\right) \qquad (8.128)$$

How do we design K_{pd} to satisfy this? We assume that in the present derivation, the PD is the only noise source that contributes phase noise to the synthesizer phase noise. Therefore, $S_{\theta\_syn}(\omega_i) = S_{\theta o\_pd}(\omega_i)$. First, from condition d(ii) in Table 8.1, S_{θ_syn} must be less than -102 dBc/Hz at $f_i = f_{channel} (= 1.7$ MHz), therefore, $S_{\theta o\_pd}$

must also be less than -102 dBc/Hz at $f_i = f_{channel}(=1.7$ MHz$)$. Second, we know that

$$\frac{\overline{i_{nd}^2}}{\Delta f} = \frac{\overline{i_{eq}^2}}{\Delta f} \cdot d \tag{8.129}$$

Here $\overline{i_{eq}^2}/\Delta f$ is the PSD of the noise current of M_1 in Figure 7.4. Assuming thermal noise dominates, $\overline{i_{eq}^2}/\Delta f$ is given by

$$\frac{\overline{i_{eq}^2}}{\Delta f} = \frac{4kT}{\frac{2}{3}\frac{1}{g_{m1}}} \tag{8.130}$$

By definition,

$$g_{m1} = \frac{I_{charge\_pump}}{V_{GS1} - V_t} = \frac{2\pi K_{pd}}{V_{GS1} - V_t} \tag{8.131}$$

where g_{m1}, V_{GS1} are the g_m, V_{GS} of M_1 in Figure 7.4.
Let us arbitrarily set

$$V_{GSI} - Vt = 1V \tag{8.132}$$

Since $S_{\theta o\_pd}$ has a complicated dependency on K_{pd}, rather than finding the explicit formula of $S_{\theta o\_pd}$ in terms of K_{pd}, we decide to find the K_{pd} that will result in a satisfactory $S_{\theta o\_pd}$ from trial and error. We set a trial K_{pd} as:

$$K_{pd} = 10\mu A/rad \tag{8.133}$$

We now substitute (8.132), and (8.133) in (8.131) and solve for g_{m1}. Substituting this solved g_{m1} in (8.130) we solve for $\overline{i_{eq}^2}/\Delta f$. Again substituting this solved $\overline{i_{eq}^2}/\Delta f$ and (8.125) in (8.129), we find $\frac{\overline{i_{nd}^2}}{\Delta f}$:

$$\frac{\overline{i_{nd}^2}}{\Delta f} = \frac{4 \times 4.14 \times 10^{-21}J}{\frac{2}{3} \times \frac{1}{2\pi \times 10\mu A/rad}} \times 0.001 = (0.04pA)^2/rad/s$$

$$= (0.015pA)^2/Hz \tag{8.134}$$

Substituting (8.134), (8.113), (8.121), (8.111), (8.133), and $\omega_{channel} = 2\pi \times 1.7$ Mrad/s into (8.128), we obtain

$$S_{\theta o\_pd}(1.7 \text{ MHz}) = 10 \log \left[\frac{51.6K}{1.7M} \cdot \frac{51.6K \times 3}{1.7M} \cdot \frac{1099}{10\mu A/\text{rad}} \right]^2 \cdot \frac{(0.015pA)^2}{\text{Hz}}$$

$$= 10 \log\{[2 \times 10^{-17}]\text{rad}^2/\text{Hz}\} = -166 \, \text{dBc/Hz}$$

Therefore, $S_{\theta\_syn}(1.7 \text{ MHz})$ is also -166 dBc/Hz. This is smaller than -102 dBc/Hz and hence condition d(ii) in Table 8.1 is satisfied. Accordingly, a K_{pd} of 10 μA/rad gives us an acceptable design.

8.4.5 Loop Filter: Part 2

In this sub-section

1. We substitute K_{pd} in the expression for $G(j\omega)$. Then we use the criterion $|G(j\omega_u)| = 1$ to determine the scaling factor C_3. After that we use poles and zeroes to determine values for C and R_2.
2. Finally we can check the values of R_2, C, C_3 and see if they are realizable.

8.4.5.1   Determine the Scaling Factor

We turn back to the expression of $F_2(s)$, rewritten from (8.11a) as

$$F_2(s) = \frac{s + z_1}{sC_3(s + p_3)} \tag{8.135}$$

We can now find C_3 by using (8.15). First, from the definition;

$$|G_2(j\omega_u)| = 1 \tag{8.136}$$

Then, substituting (8.15) in (8.136), we have

$$|G_2(j\omega_u)| = \left| \frac{K_{pd}K_{vco}(j\omega_u + z_1)}{M(j\omega_u)^2 C_3(j\omega_u + p_3)} \right| = 1 \tag{8.137}$$

Rearranging (8.137), we have

$$C_3 = \frac{K_{pd}K_{vco}|j\omega_u + z_1|}{M\omega_u^2|j\omega_u + p_3|} \tag{8.138}$$

Substituting (8.133), (8.95), (8.113), (8.120), (8.111), and (8.121) into (8.138) and solving, we have

$$C_3 = 2.7\text{pF} \tag{8.139}$$

Substituting (8.123), and (8.139) into (8.12b), we have

$$12 = 1 + \frac{C}{2.7\text{pF}} \tag{8.140}$$

Solving, we have

$$C = 29.7 \text{ pF} \tag{8.141}$$

Finally, substituting (8.120) and (8.141) in (8.11b), we have

$$81 \text{ Krad/s} = \frac{1}{R_2(29.7\text{pF})}$$

Solving,

$$R_2 = 410 \text{ K}\Omega \tag{8.142}$$

We have finished designing the components value.

8.4.5.2   Realizability

R_2, C, C_3 are bounded by realizability in a CMOS process. Remember that in integrated circuits, we can only realize small resistors and capacitors, because they take up chip area. Certainly, capacitors on the order of microfarads and resistors on the order of megaohms are not acceptable.

To check if our present design meets these considerations, first we assume that for our 0.6 μm CMOS technology, which has a double polysilicon capacitor available,

$$\text{cap/area} = 2 \text{ fF}/\mu m^2 \tag{8.143}$$

Hence from (8.141), (8.143)

$$\text{area of } C = \frac{29.7\text{pF}}{2\text{fF}/\mu m^2} = 1.48 \times 10^4 \mu m^2 = 121 \ \mu m \times 121 \ \mu m \tag{8.144}$$

Even though this is large, it is still within a reasonable fabrication limit. C_3 is much smaller than C and hence will be within the fabrication limit as well. For the resistor we assume that the resistivity of this process is given by

$$\text{resistance/area} = 10 \ \Omega/\mu m^2 \tag{8.145}$$

From (8.139), and (8.145), we have

$$\text{area of } R_2 = \frac{410\text{k}\Omega}{10\Omega/\,\mu\text{m}^2} = 4.1 \times 10^4 \,\mu\text{m}^2 = 202 \,\mu\text{m} \times 202 \,\mu\text{m} \qquad (8.146)$$

Again, even though this is large, it is also within a reasonable fabrication limit. Hence the components meet the realization check.

In summary, in Section 8.4 we have designed a loop filter component with $C = 29.7$ pF, $C_3 = 2.7$ pF, $R_2 = 410$ kΩ. We have also determined that the PD should have a $K_{pd} = 10 \,\mu$A/rad. Together with the VCO we have designed, which has a $K_{vco} = -15$ MHz/V, we have completed the synthesizer design.

8.5 Implementation of a Frequency Synthesizer with a Fractional Divider

As a practical example of a complete synthesizer, a 2.4-GHz frequency synthesizer was designed that uses a fractional divider to drive a dual-phase-locked-loop (PLL) structure, with both PLLs using only on-chip ring oscillators [14]. The first-stage narrow-band PLL acts as a spur filter while the second stage wide-band PLL suppresses VCO phase noise so that simultaneous suppression of phase noise and spur is achieved. A new low-power, low-noise, low-frequency ring oscillator is designed for this narrow-band PLL. The chip was designed in 0.35 μm CMOS technology and achieves a phase noise of 97 dBc/Hz at 1-MHz offset and spurs of 55 dBc. The chip's output frequency varies from 2.4 to 2.5 GHz; the chip consumes 15 mA from a 3.3-V supply and occupies 3.7 sq mm.

8.5.1 Architecture

8.5.1.1 Functional Description

The proposed architecture is shown in Figure 8.19. It starts with a fractional divider (FD) which is used to generate the reference frequencies of the channels. By the nature of the FD, these reference frequencies contain spurs. To remove/suppress these spurious tones, a variable center frequency filter implemented using a PLL is adopted. This PLL uses a narrow-band loop filter so that it has a sharp frequency response and thus can remove spurs from these reference frequencies. This PLL is denoted as the narrow-band PLL (NBPLL) and its VCO is denoted as the narrow-band VCO (NBVCO). Its output is then used to drive a high-frequency PLL. Since the required resolution of the overall synthesizer can be achieved via the FD, this PLL uses only an integer-divider.

Because the reference frequencies of this PLL are at much higher frequency than the channel bandwidth (BW), spurious tones from its phase detector are of no concern. Thus, the loop BW can be made very wide to suppress its VCO phase

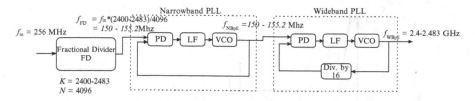

Fig. 8.19 Block diagram of 2.4GHz frequency synthesizer showing frequency planning

noise. This PLL is denoted as the wide-band PLL (WBPLL) and its VCO the wide-band VCO (WBVCO).

The fractional divider, FD, is implemented using just an accumulator to calculate phase. The FD output is:

$$f_{FD} = f_{in}\frac{K}{N} \tag{8.147}$$

The spurs from the FD is at

$$f_{spur} = f_{in}\frac{GCF(K,N)}{N} \overset{\text{worstcase}}{=} f_{in}\frac{1}{N} \tag{8.148}$$

K is the control word of the accumulator, f_{in} is the input frequency and N is the accumulator size. GCF is the greatest common factor and for worst case spur it is 1. The FD output frequency is multiplied by the divider ratio M in the wideband PLL to get the overall output frequency.

$$f_{WBPLL} = f_{in}\frac{K}{N}M = 2.4GHz \tag{8.149}$$

A small M is desirable to keep the referred noise at the input of the wideband PLL from being greatly multiplied. Its lower limit is set by f_{FD} and hence the maximum speed at which fractional divider can operate. In this design it is set to 16. From (8.148) when K is changed by 1, f_{WBPLL} should change by the minimum channel resolution (e.g 1 MHz for Bluetooth). This results in

$$1MHz = \frac{f_{in}}{N}16 \tag{8.150}$$

To reduce spur amplitude, f_{in} should be increased and is ultimately limited by the speed of the digital hardware. It is arbitrarily set to be 256MHz. Applying this to (8.150), (8.149), (8.148), gives N=4096, K=2400-2483, f_{FD}=150MHz, f_{spur}=62.5kHz. This FD output is fed into narrowband PLL and produces a filtered output at approximately 150MHz (f_{NBPLL}). The narrowband PLL has a loop bandwidth of 2kHz and cleans up the spur from FD at multiples of 62.5kHz. The narrowband PLL output of 150MHz is fed into the wideband PLL with a loop bandwidth of 10MHz to suppress the phase noise from the high frequency ring

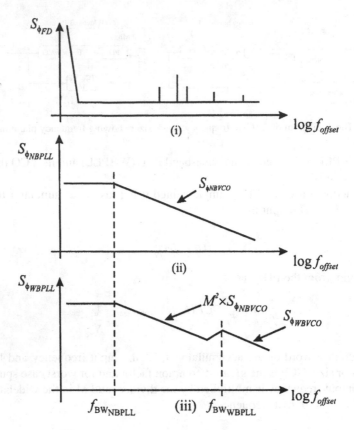

Fig. 8.20 Spectra at different nodes

oscillator. The spur frequency from the phase detector and charge pump is at a relatively high frequency of 150MHz and is not of major concern. Representative spectra, which highlight the spurs and phase noise at various nodes of the architecture, are shown in Figure 8.20.

8.5.1.2 Architectural Advantage

The major advantage of this architecture is that the phase noise of the low frequency ring oscillator in the narrowband PLL dominates the phase noise of the reference signal to the wideband PLL, and hence the overall frequency synthesizer phase noise. Since the low frequency ring oscillator has lower phase noise than the high frequency ring oscillator, the overall synthesizer phase noise is reduced. Specifically, lower frequency oscillator has lower phase noise because of the following features:

Saturation of delay cell is always achieved and no phase noise due to cycle to cycle correlation

The output of the individual delay cell in this low frequency oscillator always reaches the power supply before the end of a cycle and therefore is called a saturated ring oscillator. In comparison the output of the individual delay cell in the high frequency oscillator is never saturated and all the transistors are on all the time. Because the low frequency VCO delay cell is saturated, some of the transistors are turned off for some time during each period. Thus there is no cycle to cycle correlation. As discussed in Chapter 7, this means that phase noise of a saturated ring oscillator is lower than that of the unsaturated ring oscillator, even when all the design efforts to reduce cycle to cycle correlation in the unsaturated ring oscillator, as discussed in Chapter 7, have been applied. Therefore the low frequency VCO has lower absolute phase jitter, leading to a lower phase noise of the entire frequency synthesizer. Note that since absolute jitter, rather than relative jitter, is the proper performance metric, this advantage of the low frequency VCO over the high frequency VCO remains even after we multiply the phase noise of the low frequency VCO in the narrowband PLL by the divider ratio M (=16).

Higher Slew Rate Because of Negligible Displacement Current

At high frequency, parasitic capacitance becomes a significant factor in impacting the performance of the circuit. For the present design where the delay cell is implemented using a source coupled pair (SCP), the impact of parasitic capacitance at the common source node, C_s (C_{sb} of the diff pair and C_{gd} and C_{db} of the tail current mirror), of a conventional differential delay cell based ring oscillator is of particular importance. During every cycle of oscillation, the common source node between the differential pair would undergo significant perturbations. Since the common source voltage has to shift up every half cycle to completely turn off one side of the differential pair, one can see the source node is fluctuating at two times the frequency of the input to the differential pair. At the point of transition (i.e. differential input voltage is zero), the sensitivity of the delay cell to noise is the greatest since this is the onset of the rising/falling transition. At this time, note that the source node is at approximately its lowest potential (the parasitic capacitance C_s is discharging), this implies that the current going through the differential branches are less which in turn means less slew current is available to charge up the load capacitance. Since phase noise is inversely proportional to the amount of current available to charge up the load capacitance, one can see that the phase noise is impacted negatively by the above phenomenon. Thus in general, we want to minimize the displacement current that is spent on charging and discharging the C_s every half cycle. Note that as frequency decreases, this displacement current is less. Therefore the phase noise of the low frequency oscillator is less.

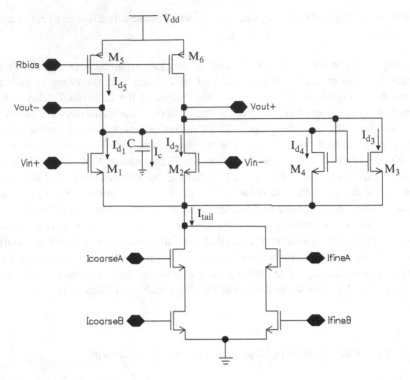

Fig. 8.21 New delay cell schematic

8.5.1.3 Circuit Implementation

The phase frequency detector (PFD), charge pump, dividers and high frequency VCO with replica bias are implemented in a differential fashion using traditional designs, like that covered in Chapter 7. The high frequency VCO, being unsaturated, should be designed according to Chapter 7 to reduce cycle to cycle correlation. On the other hand, the 150 MHz low frequency VCO needs to remain saturated. Adding more capacitance or using more stages to make it run at lower frequency while the VCO remains saturated results in poorer phase noise or more power consumption. A new delay cell is proposed to solve this problem. Figure 8.21 shows such a delay cell for the 3-stage low frequency ring oscillator. In this new delay cell, a cross coupled circuit (M_3, M_4) is added to the traditional differential delay cell. It should be noted that this cross coupled circuit is not operated in a positive feedback mode, as in traditional cross coupled circuit. Rather, this new addition allows the delay cell to operate at lower frequency while maintaining a high slew rate. It does so by stealing the current available to charge/discharge the capacitors.

We will explain how the circuit works on a low to high transition at V_{out}^-. At the beginning of this transition, M_4 is in saturation region and provides constant

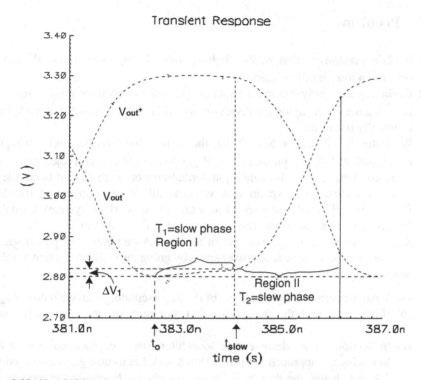

Fig. 8.22 New VCO output waveform

current, because V_{gs4} has not changed. When V_{gs1} starts to decrease, I_{d1} starts to decrease as well leading to an increase in I_c. As shown in Figure 8.22, this causes the low to high transition. However, the rate of transition is slow because only I_{d1} is reduced while I_{d4} and I_{d5} remain relatively constant. Hence we introduce a slow phase in the oscillation period which starts at t_o and ends at t_{slow}. To be precise t_{slow} occurs when V_{out}^+ starts to decrease as shown in Figure 8.22. The slow phase persists until I_{d4} starts to decrease when V_{gs4} decreases. When I_{d4} starts to decrease, the slew rate returns to normal and we enter the slew phase. Because the oscillation period is now a sum of the slow phase and slew phase, the oscillation slows down. Thus, the circuit achieves the desired effect in that it slows down the oscillation without adding extra loading elements and without slowing down the slew rate at the actual crossing. Effectively the added cross coupled circuit senses the voltage at the output, then steals or returns the current at different times during transition, therefore causing the slow phase and slew phase. The two different phases can be seen in Figure 8.22. Since this is a saturated ring oscillator, all the phase noise design considerations covered in Chapter 7 can be applied to the present case.

8.6   Problems

8.1 Explain qualitatively that the PLL behaves like a bandpass filter even though it only has a lowpass filter inside.

8.2 Explain qualitatively that in a frequency synthesizer, when we want to select a new channel via changing the divider ratio, initially the loop will not track, but eventually it will track.

8.3 We stated in sub-section 8.3.1.2 that the transfer function $H_{vco}(\omega_i)$ is a high-pass function. When expressed in ω, $H_{vco}(\omega)$ is actually a band reject function centered in ω_o. Explain how this transformation from high pass to band reject function occurs. Also explain why we can still use the concept of transfer function (which is valid only for a linear time invariant (LTI) system) when the PLL is actually a linear periodic time varying (LPTV) system.

8.4 Suppose that in the design example in Section 8.4 we tighten our power specs and we try to meet these new power specs by going through sub-section 8.4.2.1 again with a reduced I_{bias}.

   (a) What happens to the phase noise of the ring oscillator when we reduce I_{bias}?
   (b) If originally both the phase noise and spur specs are met, can we be sure now?
   (c) In Section 8.4 we designed the loop filter based on phase noise (that is, following the approach in Sub-section 8.3.1). Let us now go to Sub-section 8.4 and change the design approach: we choose to design the loop filter based on spur (that is, following the approach in sub-section 8.3.2). Using

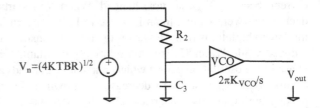

Fig. P8.1

this new approach, answer (b) again.

8.5 In Section 8.4, what potential problem will we encounter if we swap sub-section 8.4.2 and sub-section 8.4.3 (i.e., we design the loop filter before we design the VCO)?

8.6 In sub-section 8.4.4, we derived the formula for $S_{\theta o\_pd}$, for phase noise contribution from the phase detector [see (8.127)]. Derive a similar formula for $S_{\theta o\_R_2}$, for phase noise contribution from resistor R_2 in the loop filter used in section 8.4 (assume a large b).

8.7 Figure P8.1 shows a loop filter whose output is taken between R_2 and C_3. This is to replace the loop filter used in Section 8.4.

We are interested in the thermal noise contribution from resistor R_2 to the phase noise at the output of the synthesizer. Let us assume that the noise from R_2 is modeled as shown in Figure P8.1 where it is represented by a voltage source whose Laplace transfrom is $V_n(s)$ and has the same amplitude ($=\sqrt{4kTBR_2}$) as the equivalent input-referred voltage noise source. Here B is the bandwidth of interest.

(a) Derive $H_{R_2}(=\theta_o(s)/V_n(s))$, the transfer function from the resistor R_2 to the output of synthesizer. The transfer function should be expressed as a function of K_{vco}, R_2, C_3, B and other relevant parameters. Then use H_{R_2} to derive $\theta_o(s)$ as a function of $\sqrt{4kTBR_2}$.

(b) The resulting phase noise is of greatest interest in the adjacent channel region. Using the appropriate assumption, simplify the $\theta_o(s)$ expression derived in (a).

(c) Calculate the $S_{\theta o\_R_2}$ from expression in (b) in dBc/Hz at an offset frequency ω_i of 25 kHz, with $R_2 = 100$ kΩ, $C_3 = 0.0047$ μF, $K_{vco} = 5$ MHz/V.

8.8 Suppose we have a new standard that is identical to DECT except that f_o is reduced by half. Hence the specs given in Table 8.1 have their center frequency reduced by a half to 950 MHz, while everything else remains the same. To design the frequency synthesizer for this new standard, we assume that sub-

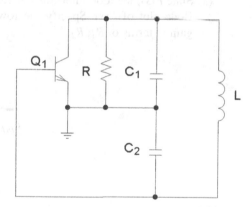

Fig. P8.2

section 8.4.2 is already carried out and a new VCO with this new f_o is designed. We further assume that this new VCO has been designed such that N, I_{bias} and $(W/L)_1$, and $V_{GS1}-V_t$ remain the same as the case in sub-section 8.4.2. Our task is to design a new loop filter. To design the new loop filter, let us assume that it has the same filter configuration as used in section 8.4.3, but with possibly different poles/zeroes locations and hence a different f_u. Find the f_u for this new loop filter and hence the new poles/zeroes locations.

8.9 We would like to use the Colpitts oscillator described in Problem 7.5 of Chapter 7 to replace the ring oscillator used in the design example in Section 8.4 and redesign the frequency synthesizer. This oscillator is redrawn in Figure P8.2. Use a reasonable assumption for the bipolar technology that is used to implement Q_1.

Since this oscillator has different phase noise behaviour, the loop filter has to be redesigned. We assume that everything else in the synthesizer remains the same.

Start with the S_{θ_vco} (f_i) derived for this oscillator in Problem 7.7, Chapter 7. Follow the design example (sub-section 8.4.3), and recalculate f_c, f_u.

8.10 Design the loop filter for a frequency synthesizer. The synthesizer should be able to produce a set of frequencies in the range from 1 to 2 MHz with a channel spacing of 10 kHz (i.e., frequencies of 1000, 1010, 1020, . . ., 2000 kHz will be generated). We will use a PFD as phase detector, whose K_{pd} is given as 1 mA/rad. Because this detector offers infinite lock range for any type of loop filter, we use the simplest of these, the passive lag filter, as shown Figure P8.3. We assume that a programmable divider is used whose ratio varies from $M = 100$ to 200. For subsequent calculation assume that the synthesizer is optimized for the divider ratio $M_{mean} = \sqrt{M_{min}M_{max}} = 141$. The VCO was given a gain $K_{vco} = 2.24 \times 10^6$ rad/s/V.

(a) State $F(s)$, the loop filter transfer function, in terms of R_1, R_2, C. Plot the Bode plot of $F(s)$. Specify the low-frequency gain and high-frequency gain in terms of R_1, R_2.

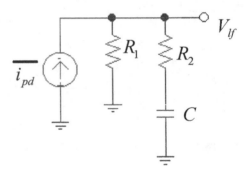

Fig. P8.3

(b) Set f_z, the zero frequency of the loop filter $F(s)$, to be 3 times the unity gain frequency f_u of the loop transfer function $G(s)$. Assume that $R_2 = 240 \ \Omega$, $C = 0.33 \ \mu\text{F}$. Calculate R_1.

(c) Calculate the phase margin of the synthesizer. Is the design stable?

(d) Assume that d (the duty cycle) of i_{spur} coming from the PFD is 0.001. Calculate the spur in dBc at the output of the synthesizer.

(e) Upon changing the divider ratio, this synthesizer has to lock (settle) to a final f_{error} within a specific time, called the lock time T_L. Assume an $f_{step}=f_{ref}= 10$ kHz and an allowable $f_{error}=1$ Hz. Calculate T_L.

References

1. T. C. Weigandt, B. Kim, and P. R. Gray, "Analysis of Timing Jitter in CMOS Ring Oscillators," in *Proc. ISCAS*, June 1994, pp. 4.31–4.35.
2. T. C. Weigandt, "Low Phase Noise, Low Timing Jitter Design Techniques for Delay Cell Based VCOs and Frequency Synthesizers" thesis, University of California, Berkeley, 1998.
3. B. Razavi, "A Study of Phase Noise in CMOS Oscillators," *IEEE JSSC*, Vol. 31, p. 334, March 1996.
4. F. Gardner, "Charge Pump Phase Lock Loops," *IEEE Transactions on Communications*, Vol. 28, No. 11, pp. 1849–1858, November 1980.
5. F. Gardner, *Phaselock Techniques*, 2nd ed., Wiley, 1979.
6. P. Gray and R. Meyer, *Analysis and Design of Analog Integrated Circuits*, 3rd ed., Wiley. 1993.
7. R. Best, *Phase Locked Loops, Theory, Design and Applications*, 2nd ed., McGraw-Hill, 1984
8. W. Egan, *Frequency Synthesis by Phase Lock*, Wiley, 1981, p. 40.
9. W. Egan, *Frequency Synthesis by Phase Lock*, Wiley, 1981, p. 95.
10. James A. Crawford, *Frequency Synthesizer Design Handbook*, Artech House, 1994, p. 122.
11. P. Gray, "An Integrated GSM/DECT Receiver, Design Specification," Design Notes, University of California, Berkeley, 1998.
12. F. Martin, "Frequency Synthesizers in RF Wireless Communications," Course Notes, 1994, Motorola, Plantation, FL.
13. T. Lee, *Design of CMOS RF Circuits*, p. 497, Cambridge University Press, 1998.
14. Z. Shu, K. Lee, B. Leung, "A 2.4-GHz Ring Oscillator Based CMOS Frequency Synthesizer with a Fractional Divider Dual PLL Architecture", IEEE JSSC, vol. 39, pp. 452–462, March 2004.

Chapter 9
Transmitter Architectures and Power Amplifier

9.1 Introduction

In Chapter 1 we focused on the blocks enclosed by dotted lines in Figure 1.1. They are the modulator, demodulator, and channel part of the communication system. We now redraw Figure 1.1 as Figure 9.1, with the transmitter back end enclosed by dotted lines.

We first review some general philosophy on deciding the back end architecture. Since the transmit architecture is important in GSM while not so important in DECT, we use it as an example in this chapter. We will refer to both GSM and DCS1800, which is a more modern and up-banded version of GSM.

We then adopt specific transmit architecture. Next the back end sub blocks are elaborated. Examples of these sub blocks include mixers, frequency synthesizers, digital to analog converters (DAC), I/Q (or RF quadrature) generator and power amplifiers (PA).

9.2 Transmitter Back End: General Discussion

9.2.1 Motivations and General Design Philosophy

As in the chapter on receiver, where the front end allows us to satisfy the boundary condition imposed by the received signal and the required SNR for the demodulator, here in the transmitter, the back end allows us to take the baseband signal and conditions it to the required power level and carrier frequency for a given standard, in this case GSM/DCS1800. Conditioning also includes proper filtering, so that the transmitted signal does not spill over and causes interference, such as adjacent channel interference.

There are a large variety of transmitter architectures, each tailor towards a particular standard. One important feature of the standard is the cell size. Since

B. Leung, *VLSI for Wireless Communication*, DOI 10.1007/978-1-4614-0986-1_9,
© Springer Science+Business Media, LLC 2011

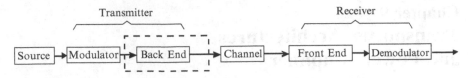

Fig. 9.1 Back end of the transmitter enclosed in dotted line

cellular standards (e.g. GSM) involve larger cell size, the required performance requirement is higher than wireless LAN standards (e.g. 802.11). On the other hand, because our underlying technology is CMOS, with low Q and low breakdown voltage, there are limitations on the type of circuits that can be employed. For example, some power amplifier (PA) design, well suited for technology with higher breakdown voltage such as GaN, is not suitable here. This limitation also impacts the resulting transmitter architecture.

Transmitters broadly fall into two classes, depending on how mathematically the complex signal, with I, Q components, is represented and transmitted. In I/Q modulator-based transmitters the I, Q signals are represented in Cartesian form i.e. as rectangular components of the complex signal. Each of the components is then converted via DAC, and the resulting analog signals are upconverted, combined, amplified by the PA and transmitted. In polar-based transmitters, the I and Q signals are represented in polar form i.e. as amplitude and phase of the complex signal. The phase can be encoded, for example, in a PLL and the amplitude encoded by varying RF signal gain of the PA, such as through the use of varying power supply as well as biasing. Although polar-based transmitters have shown promise with respect to the elimination of discrete components, such as RF filter however, there is the question of limited data rate imposed by PLL loop bandwidth and matching between the phase and amplitude path [1], as well as efficiency and output power control when applying the full modulation before the signal reaches the PA [4]. In particular it may not be straightforward to implement the required power control by varying PA gain. For the rest of the chapter we will concentrate on the I/Q based architecture.

9.2.2 Direct Conversion and Other Architectures

We start off our discussion with the direct conversion [2] (sometimes called homodyne) architecture as shown in Figure 9.2, where we denote all the circuitry enclosed by the dotted line as the back end. We can observe that the back end is attractive because of the simplicity of signal path. Note this architecture is like the reverse of that shown in problem 2.9, and with both the I, Q path explicitly shown. The modulator takes the source data and generates the I, Q signals, at baseband. They typically pass through a D/A converter, an AGC and a baseband filter. They then get upconverted to RF in a single step via two separate mixers, which has local oscillator signals LO_I and LO_Q. The I and Q signals are combined to a single sideband.

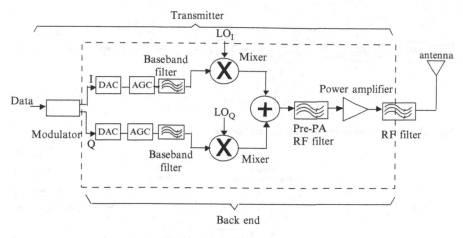

Fig. 9.2 RF transmitter using a direct conversion architecture

The output is then passed through a pre-power amplifier (PA) RF filter. The filter output is amplified by the PA, and then filtered by another RF filter, before sent off to the antenna.

Some comments on the various filters are as follows: the baseband filter (e.g. needed for reconstruction) typically is rather relaxed, with a single-stage filtering suffixes. The pre-PA RF filter is needed to reduce third-order intermodulation in the PA as well as transmitted noise. The RF filters after the PA further reduces transmit noise.

Not shown in the diagram is the frequency synthesizer, which generates the LO signal. The LO is variable to perform frequency tuning. Also not shown is the quadrature LO generator, which takes this LO signal and generates LO_I and LO_Q signals. LO_I and LO_Q are 90° out of phase.

In addition to the direct conversion architecture, there are many other types of possible I/Q based transmitter architectures. For example, like the receiver, the transmitter can also use traditional heterodyne or indirect upconversion architecture, shown in Figure 9.3. Contrary to the direct conversion architecture, the frequency conversion is done in two steps. First the baseband signal is frequency translated to a fixed IF by a first set of mixers, with a fixed LO, LO_2. This is then passed through an IF filter to remove spurious IF harmonics created by this mixing. The output is then mixed by a second set of mixers with another LO, LO_1. LO_1 is variable and performs frequency tuning. The rest is like the direct conversion architecture. LO_2 is typically fixed while LO_1 is typically varying, thus providing tuning. The architecture is more complicated. There is more filtering. There is also more mixing, which means more mixers, and extra frequency synthesizers as well to generate the additional LO signals. However, compared to the direct conversion architecture the I/Q modulation is performed at lower frequency, which means LO feedthrough is reduced.

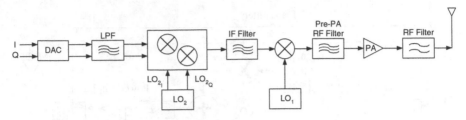

Fig. 9.3 Indirect conversion transmitter architecture

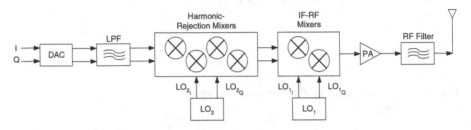

Fig. 9.4 Harmonic rejection transmitter

A modified heterodyne transmitter is reported in [3]. This dual conversion architecture, called "sliding IF" superheterodyne transmitter [17], has LO_1 harmonically related to LO_2, with both LO_1, LO_2 high in frequency (that is high IF architecture). Therefore there is no need for an additional PLL to generate LO_2. Instead it is derived from LO_2 by a divider with low division ratio.

Another modified heterodyne transmitter, called the harmonic-rejection transmitter (HRT), is also reported in [5], and is shown in Figure 9.4.

It uses a similar principle as the indirect conversion architecture, but it replaces the first mixer (with local oscillator LO_2) by a harmonic rejection mixer (HRM). HRM has been used in both transmitter and receiver [1]. To explain its operation, let us consider the I path only (and hence LO_{2_I}). From the square wave LO_{2_I}, a set of signals are derived, where each signal has the same frequency as the LO_{2_I}, but are separated uniformly in phase. To perform mixing, the input signal (I-path) is split into multiple paths. Through each path the input signal is mixed with one of these LO_{2_I} derived signal via a basic mixer, which acts simply as a multiplier. This is to be repeated by the input signal in the Q path. Note the HRM in Figure 9.4 splits signals into 2 paths. With both I, Q paths that means there is a total of 4 such basic mixers. This is to be contrasted to the indirect conversion transmitter, shown in Figure 9.3, where the input signal in each of I, Q path is mixed via only one such basic mixer, with LO_{2_I} and LO_{2_Q}, respectively, meaning only a total of 2 such basic mixers. Returning to HRM in Figure 9.4, the outputs from all these paths are then combined to form the output of the HRM. Mathematically this results in cancellation of selected harmonic of LO_{2_I} and LO_{2_Q}. For example (and again considering only the I path and hence LO_{2_I}), let us assume this time 3 derived waveforms are generated. First one is LO_{2_I} itself. The second one is like the first one, but whose phase is advanced by 45° and amplitude 0.707 times the first one, denoted as $LO_{2_{I+45}}$. The third one is like the first one, but whose phase is delayed by 45° while amplitude is still 0.707 times the first one, denoted as $LO_{2_{I-45}}$. Mixing (multiplication) in the HRM then means a delay of 135° at third harmonic of input and a delay of 225° at fifth harmonic of input are generated. The output, which is the sum, will then see cancellation of the third and fifth harmonic, and harmonic rejection is achieved. The same situation applies to the Q path. When it is used in the transmitter, this means the baseband signal will not be upconverted to around these harmonics. Rather, it will be upconverted to just around the fundamental. Thus the PA, generally nonlinear, will not generate intermodulation products between the upconverted baseband signals at the fundamental and harmonics. Accordingly, the IF filter and pre-RF filter, shown in the indirect conversion transmitter in Figure 9.3, is no longer needed, as evident by their absence in Figure 9.4. This, of course, is because the original reason for these filters is to filter out baseband signal upconverted to around harmonics of LO. A more detail explanation of this aspect will be given later in 9.4.1.3, after the introduction of PA nonlinearity, and also in 9.4.5, which shows the trade-off between the PA nonlinearity and these filter suppression, thus highlighting one of the advantages of the harmonic rejection transmitter.

Another difference with the indirect conversion architecture is the tuning aspect. As mentioned above, for the indirect conversion architecture, LO_2 is typically fixed while LO_1 is typically varying, thus providing tuning. On the other hand in Figure 9.4, the opposite is true, with LO_2 varying and provides tuning while LO_1 is fixed. In general the phase noise of a frequency synthesizers is usually higher when the LO it generates is higher in frequency, as well as being variable. Thus LO_1 in Figure 9.3 has a high phase noise requirement, while both LO_1 and LO_2 in Figure 9.4 have moderate phase noise requirement and on chip implementation becomes possible. Because of the high phase noise in LO_1 in Figure 9.3, the transmitted noise also increases, and both the pre-PA RF filter and the RF filter are adopted to suppress it. With a moderate phase noise in LO_1 in Figure 9.4 and hence lower transmit noise, only the RF filter is used to suppress the transmit noise.

Finally there is also the frequency translation architecture [11], shown in Figure 9.5. Like the indirect architecture, the baseband signal is frequency translated to a fixed IF, at f_1-f_2 in this architecture. This is then filtered by a BPF to remove unwanted IF harmonics. The IF signal is then upconverted to RF, at f_2. Unlike the indirect architecture, where the IF signal is upconverted to RF via a mixer, here the upconversion is done by a PLL, which internally has a mixer in the feedback path. This feedback mixer takes an input from a fixed oscillator, at frequency f_1. It then mixes the PLL output with this f_1 signal, feeds the resulting mixed output back, and ignoring for the time being the feedback LPF, compares this mixed output with the f_1-f_2 input signal of the PLL, via the phase detector. When the PLL is in lock, the mixed output is forced to have frequency also at f_1-f_2. Working backwards along the feedback mixer, since its output has frequency f_1-f_2, and we already know one of its inputs has frequency f_1, its "other" input's frequency would be f_2. Therefore the PLL output, or the feedback mixer's "other" input, has frequency f_2. Returning

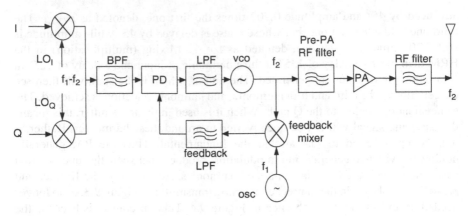

Fig. 9.5 Frequency translation architecture

now to the function of the feedback LPF, which we have so far ignored, we will discuss its function by starting with the output of PLL, which we have just shown has frequency f_2. Now mixing, via the feedback mixer, with f_1, its output has frequencies f_1-f_2 and $f_1 + f_2$. The feedback LPF is designed to filter the $f_1 + f_2$ frequency component. Thus the feedback signal to the PD is indeed f_1-f_2, consistent with the discussion above. In essence the PLL acts as a "mixer", and mixes the two input signals: one at $f_1 - f_2$ and one at f_1, and output the desired signal at f_2. The rest follows the indirect architecture.

The above discussion brought us to the end of surveying transmitter architectures. Next we will move on and discuss the implementation of the transmitter. For the rest of the chapter we will concentrate on direct conversion architecture to help illustrate such circuit implementation.

Referring back to Figure 9.2, in direct conversion architecture the filter design is similar to that in the receiver chapter, although the requirement is different. This will be discussed more later, for example, in section 9.4.5. One way to design the D/A converter is to generalize the design of 1-bit internal D/A converter design (used inside a sigma-delta A/D converter in chapter 6) to a multi-bit general purpose D/A converter. Again the choice of D/A converter can be rather general. Whereas in chapter 6, the D/A converter is a switch capacitor type D/A converter, DAC used in transmitter can also be resistor string based. As for the mixer design, the techniques presented in chapter 4, 5 can be used. For LO generation, the frequency synthesizer technique presented in chapter 7, 8 can be used to generate the primary LO signals. Quadrature LO, LO_I and LO_Q, are then generated from it. The new blocks in Figure 9.2 are then

a) Quadrature LO generator
b) Power amplifier

The rest of this chapter concerns their design.

9.3   Quadrature LO Generator

For direct conversion architecture, single sideband modulation is used and quadrature LO is needed. Not only do we want LO_I and LO_Q to differ by 90°, it is also desirable that both signals (typically square waves) have 50% duty cycle. Some design issues for these circuits include oscillation frequency, amplitude imbalance and phase imbalance.

9.3.1   Single Ended RC

A single-ended RC quadrature generator is shown in Figure 9.6a. It consists of feeding the LO to two sections: an RC section and a CR section.

When LO is applied to R_1, C_1, the circuit's response is:

$$\phi(f) = \frac{j2\pi C_1 R_1}{1 - j2\pi f C_1 R_1} \tag{9.1}$$

R_1, C_1 were chosen to give a 45° phase shift between LO and LO_I at the frequency of operation (the carrier frequency), as shown in Figure 9.6b.

LO is also applied to C_2, R_2. The circuit's response is:

$$\theta(f) = \frac{1}{1 - j2\pi f C_2 R_2} \tag{9.2}$$

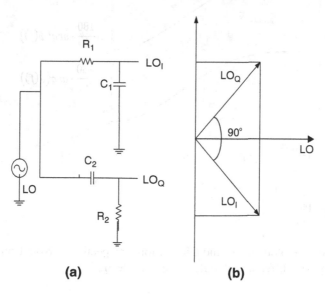

(a)   **(b)**

Fig. 9.6   Single ended RC quadrature LO generator (a) circuit (b) relative phase of 90°

R_2, C_2, were chosen so that there is a $-45°$ phase shift between LO and LO_Q at the frequency of operation, as shown in Figure 9.6b (if $R_1=R_2$, $C_1=C_2$, the $-45°$ phase shift at the carrier frequency just follows from symmetry. Together there is a phase shift of $90°$ between LO_I and LO_Q. In practice, process variation on R and C causes phase and amplitude errors. Accordingly this circuit is not suitable when there is tight quadrature requirement on phase and matching requirement on amplitude. In addition, the input impedance of this generator is determined by RC. For high input impedance (to avoid loading LO circuit), large R and small C are desirable. However at RF frequency, parasitic capacitance associated with R starts to become significant in loading the LO circuit. Furthermore as C becomes smaller, mismatch becomes more pronounced. As usual, drivers (typically common source amplifier or source followers) can be used. Finally instead of connecting LO_I, LO_Q directly to the mixers, buffers can be used for interfacing. This is especially important if the mixers are passive mixers, which require rail-to-rail LO_I, LO_Q swing. With a passive RC/CR network, its inherent loss will be made up for by the buffers to provide such swing. However these buffers stage usually consume fair bit of current and the extra phase noise introduced can be significant.

Numerical Example 9.1
Design a single-ended RC quadrature LO generator such that it gives a $90°$ at 1GHz.

Let us select the following value: $C=C_1=C_2=47pF$, $R=3ohm$, Substituting in (9.1), (9.2) there is a phase shift of $90°$ between LO_I and LO_Q at 1GHz. To show this let us look at the resulting phase plot in Fig. 9.7:

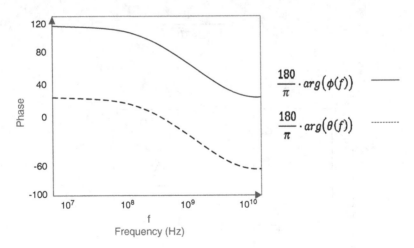

Fig. 9.7 Phase Plots

Here θ, ϕ are the plots of RC and CR sections, respectively. Note from the phase plots, their phase difference at 1GHz is $90°$, as desired.

The resulting amplitude plot in Fig. 9.8 is

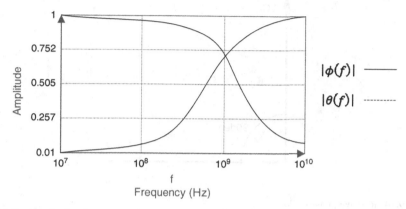

Fig. 9.8 Amplitude plots

From the amplitude plot, at 1GHz, the RC and CR section have identical amplitude, as desired. This amplitude, when referenced to dc, presents a loss of 3dB, as expected.

9.3.2 Single Ended LC

A single-ended LC circuit is shown in Figure 9.9.

In this circuit, LO_I is taken directly from LO. The circuit response is:

$$\frac{v_{LO_Q}(\omega)}{v_{LO_1}(\omega)} = \frac{(\omega)^2 C_1 L R_2}{(\omega)^2 L R_2(2C_2) - R_2 - j\omega L} \tag{9.3}$$

Next we note that at resonance, the phase shift between LO_Q and LO_I is shown in Figure 9.10, which is at 90°. Thus we pick the value so the circuit resonates at the carrier frequency.

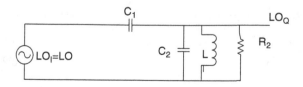

Fig. 9.9 Single ended LC quadrature LO generator

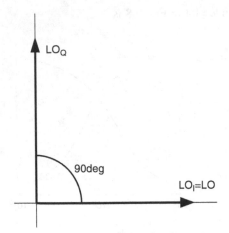

Fig. 9.10 Relative phase of 90°

Once again process variation causes phase and amplitude errors. The resonant frequency (and thus the frequency at which we want a 90° phase shift) can be tuned by varying, for example, C_2. This can be achieved by implementing C_2 as an on chip varactor. The issue of loading, discussed in RC quadrature generator, remains.

Numerical Example 9.2

Design a single-ended LC quadrature LO generator such that it gives a 90° at 1GHz.

Let us select a value the following value: $C_1 = 0.5pF$, $C_2 = 2.7pF$, $R_2 = 10ohm$, $L = 4.7nH$, then from (9.3), there is a phase shift of 90° between LO_I and LO_Q at 1GHz. To show this let us look at the resulting phase plot:

Note from the phase plot in Fig. 9.11, its phase at 1GHz is 90°.

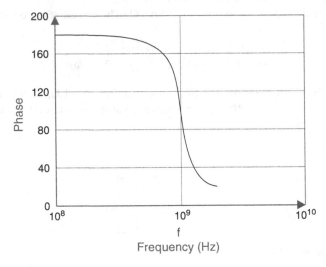

Fig. 9.11 Phase plot

The resulting amplitude plot in Fig. 9.12 is:

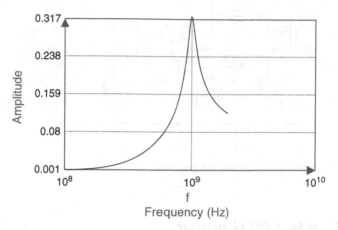

Fig. 9.12 Amplitude plot

From the amplitude plot in Fig. 9.12, at 1GHz, there is resonance, as expected. At resonance the phase shift across C_1 is 90°. The Q_L chosen was 3.3, which is typical on chip IC spiral inductor Q's.

9.3.3 R-C with Differential Stages

A differential implementation of the single-ended RC quadrature LO generator shown above can also be realized. This is shown in Figure 9.13.

$$\text{Set LO as } LO = A_{LO}\sin(\omega_{LO}t), \text{ then}$$

$$LO_1 = (2R_2/R_1) \times (A_{LO}\sin(\omega_{LO}t)) \tag{9.4}$$

$$LO_Q = (2R_2C_1) \times (d/dt(A_{LO}\sin(\omega_{LO}t))$$

$$= \omega_{LO}(2R_1C) \times (A_{LO}\cos(\omega_{LO}t))) \tag{9.5}$$

Since LO_I, LO_Q depends on $\sin(\omega_{LO}t)$ and $\cos(\omega_{LO}t)$, respectively, thus they are 90° out of phase.

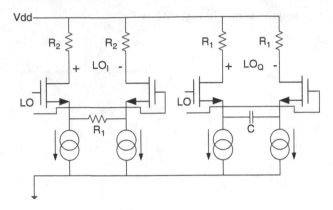

Fig. 9.13 Differential RC quadrature LO generator

9.3.4   Polyphase I/Q Generator

Another extension to the single-ended RC quadrature LO generator shown above can also be realized. This is achieved by having each section realizing not 90°, but 90°/N phase shift, where N is the number of RC sections (N=1 in the single-ended RC quadrature LO generator). This is called polyphase quadrature generator [8, 11][1], since internally multiple phases corresponding to 90°/N, are generated. The advantage of polyphase over single-ended RC quadrature LO generator is that phase and amplitude mismatch due to R, C process variation is reduced significantly.

The case for N=2, and with LO designed for 1GHz, is shown in Figure 9.14. The circuit in Figure 9.14 has 4 inputs. Two of them come from LO (plus and minus if LO has differential output, plus and ground if LO has single-ended output). The other two inputs are connected to ground. This is shown in Figure 9.15. Its phase response is shown in Figure 9.16.

There are 2 traces here that correspond to the phase of LO_I and LO_Q. Note that at 1 GHz, the phase difference is 90°. The RC values of the first stage are designed for 900MHz center frequency, and the RC values of the second stage for 1100MHz center frequency. By skewing the values, it is possible to achieve a broader band response around 1GHz. Because the generator contains multiple RC/CR sections, successive sections' impedance can increase progressively. This can alleviate the loading problem.

In summary, the quadrature generators in this section provide phase splitting without any frequency division, and they are passive circuits. These imply high linearity, low noise, and low power consumption, which make them ideal for many designs.

---

[1] Polyphase concept is also used in receiver [7], for image rejection

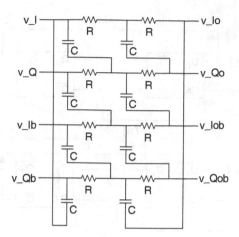

Fig. 9.14 Polyphase quadrature LO generator

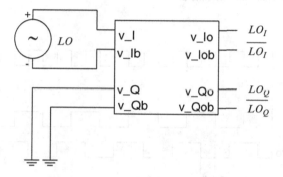

Fig. 9.15 Input to polyphase quadrature LO generator

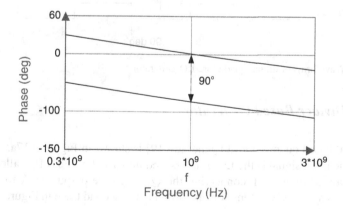

Fig. 9.16 Phase plots for polyphase I/Q generator

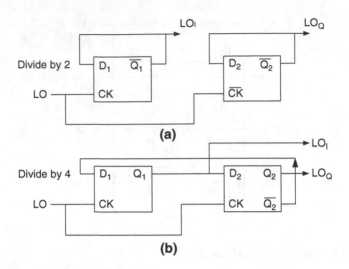

Fig. 9.17 Digital quadrature LO generator

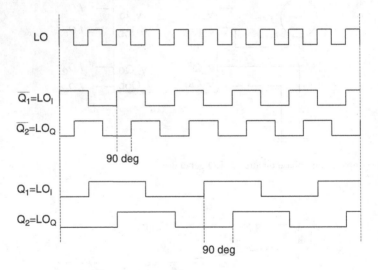

Fig. 9.18 Waveforms of digital quadrature LO generator

9.3.5   *Divider Based Generator*

The digital based quadrature LO generator [9] is shown in Figure 9.17a.

Here the input signal is the LO. It is first fed into a T flip flop (FF), called T-FF1. This is realized by a D-FF connecting the $\overline{Q_1}$ (negative output) back to its input. This realizes a divide by 2 function, as shown in the second trace in Figure 9.18, and in phase.

Meanwhile LO is fed into the $\overline{ck}$ (rather than CK) of the second T-FF, T-FF2, and edge triggers it (in essence this is equivalent to inverting LO (which gives it a 180° phase shift) and edge triggers T-FF2). Again the output is divided by 2 and is $\overline{Q_2}$. The phase inversion, meanwhile, is also divided by 2, so that compared to $\overline{Q_1}$, it has a phase shift of 90°, as shown in the third trace in Figure 9.18. Thus quadrature LO is generated.

Figure 9.17b further generalizes this by feeding the output of $\overline{Q_2}$ (rather than $\overline{Q_1}$) back to D1. In addition this time LO is fed into the CK (rather than $\overline{ck}$) of the second T-FF. This circuit realizes also quadrature LO, but it is divided by 4, rather than 2. The waveforms are shown in the last 2 traces in Figure 9.18.

The D-FF used in these circuits is typically current mode based. The bandwidth of operation depends, among other things, on the load impedance. Typically broadband operation is desired, since the delay at all harmonics of the LO signal is maintained, and phase mismatch is minimized. For broadband operation, the load can be resistive, (while minimizing parasitic capacitance), or inductive.

One advantage of the digital based generator is that the 50% duty cycle feature is inherent. In addition, amplitude and phase imbalance are typically small. As for phase noise, the divide by 2 and divide by 4 configurations have a 6-dB and 12dB improvement, respectively. On the divide by 2 circuit, since the 2 FF are triggered by two different clock edges, clock skew can be a problem and lead to mismatch. This is not an issue in the divide by 4 circuit as the triggering is done by the same clock edge. Also since the LO_I, LO_Q are at a different (lower) frequency than LO, the impact of feedthrough and unwanted uncoupling is not as acute, although the LO has to be generated at a higher frequency.

9.4 Power Amplifier Design

9.4.1 Specs

9.4.1.1 General Discussion

Like LNA in chapter 3, power amplifier, PA, has specs for its gain G, NF and IIP_3.

To demonstrate how this is obtained, we use GSM as an example. Figure 9.19 is the power output specification [11] for GSM (880MHz-915 MHz) as well as transmit spurious in the transmitter [14]. We further assume the PA is used in a homodyne architecture, as shown in Figure 9.20, where the filters and switch are assumed to have the loss as indicated.

From Figure 9.19, it can be seen that the power output is 2W (33dBm) (Tx band). Phase accuracy of modulated carrier is 5° rms and 20° peak.

In addition spurious from transmitter has to be less than -162dBc/Hz at the Rx band(935MHz-960MHz) so as not to interfere with its own received signal. Also it has to be below -104dBc/Hz at the Rx band of DCS1800(1805MHz-1880MHz) so as not to interfere with the DCS 1800 Rx band.

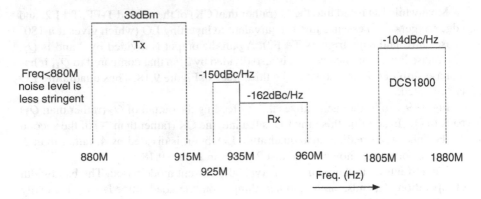

Fig. 9.19 PA output power level for GSM

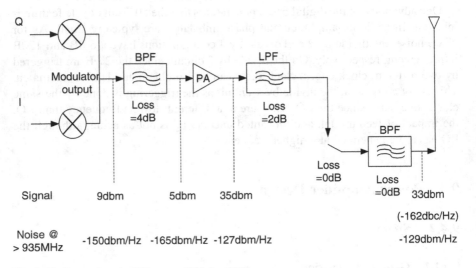

Fig. 9.20 Example gain, NF specs of PA for GSM

9.4.1.2 G and NF Specs

Next, referring to Figure 9.20, we know that the power output level, (from Figure 9.19), has to be 33dBm. Let us further assume at the output of the I, Q modulator, signal is 9dbm.

To achieve this, G has to be:

$$G = 30\text{dB}. \tag{9.6}$$

To see this, going along the transmitter chain:

At the output of the BPF, signal = 9dBm−4dB (from loss) = 5dBm. Thus the amplitude of input to the PA, denoted as A_{in}, is

$$A_{in} = 5dBm. \tag{9.7}$$

At the output of the PA, signal level = A_{in}(=5dBm, from (9.7))+G(=30dB, from (9.6)) = 35dBm

At the antenna, signal level = 35dBm−2dB(from loss)= 33dBm

This satisfies the required output level of 33dBm.

Meanwhile transmit noise in the receive band needs to be kept below −129dBm/ Hz from 935MHz to 960MHz. This is because from Figure 9.19 spurious from transmitter has to be less than −162dBc/Hz at the Rx band (935MHz-960MHz). With signal level i.e. carrier at 33dBm, −162dBc/Hz (with respect to carrier) means −162dBc/Hz + 33dbm or −129dBm/Hz.

To carry on the noise analysis, let us further assume at the output of the I, Q modulator, noise at frequency larger than 935MHz (Rx band) is less than −150dbm/Hz.

To achieve this, NF needs to be:

$$NF = 8dB \tag{9.8}$$

To see this, going along the transmitter chain:

At the output of the BPF, noise = −150dBm/Hz−15dB (from frequency response which equals −15dB@935MHz) = −165dBm/Hz

At the output of the PA, noise = -165dBm/Hz + G(=30dB, from (9.6)) + NF (=8dB, from (9.8)) = −127dBm/Hz

At the antenna, noise = −127dBm/Hz−2dB(from loss)= −129dBm/Hz, satisfying the specs.

In summary, in this example the specs for G is 30dB and NF is 8dB.

Next let us discuss IIP_3.

9.4.1.3 IIP3 Specs

We take note that nonlinearity of the PA, as quantified by its IIP_3, is going to affect the GSM specs on total transmit error. Below we look at some of those mechanisms.

IIP3 Specs Due to AM-PM Conversion

One possible mechanism is that it causes AM-PM conversion [6] and that the tolerated amount of the resulting PM is limited by the phase error specs of GSM.

Referring to the example architecture in the I/Q modulator as shown in Figure 9.20, the I, Q baseband signals are modulated on the carrier and applied to

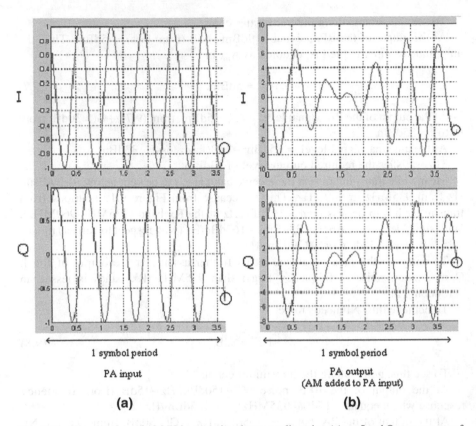

Fig. 9.21 Visualization of AM-PM conversion due to nonlinearity. (**a**) are I and Q components of the input. (**b**) are I, Q components of the output, with AM impressed on the input, which passes through a PA with nonlinearity.

the PA. Shown in Figure 9.21a are visualizations of I and Q components of the PA input. One symbol period of the modulated carrier waveform is shown. The envelope is practically constant (around the arbitrarily set value of 1) and the modulation is on the phase. The circles highlight the values of the I, Q components at the end of the symbol period. Going through the calculation, this gives a phase of 220 degree. Next we add AM to the PA input. We then pass it to the PA. For visualization, we arbitrarily set the PA gain to around 8 and also add an exaggerated level of nonlinearity. In addition the amount of AM is highly exaggerated, so that the envelope on PA output, in Figure 9.21b, varies a lot. Again using the values highlighted by the circles at the end of the symbol period, the phase is 180 degree. Thus an error of 40 degree is made. Also as we look at consecutive symbols this phase error changes. To capture this, we have both the peak phase error and an RMS (root mean square) phase error.

Now let us see how this AM to PM conversion mechanism arise in the example architecture shown in Figure 9.20. We first simplify Figure 9.20 so that BPF, LPF,

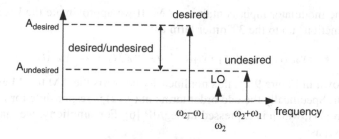

Fig. 9.22 Ratio of wanted to unwanted sideband

switch are ideal with loss/gain of 0db. From 9.2.2 if we have only tone modulation (i.e. I, Q components are a single tone at ω_1), we have:

$$I \text{ baseband signal} = A_s\cos(\omega_1 t), \quad Q \text{ baseband signal} = A_s\sin(\omega_1 t) \qquad (9.9)$$

Let us represent the LO signals for the I, Q paths (at ω_2) as:

$$LO_I = A_c\cos(\omega_2 t), \quad LO_Q = A_c\sin(\omega_2 t) \qquad (9.10)$$

The resulting modulator output is:

$$A_sA_c(\cos(\omega_1.t).\cos(\omega_2.t) + .\sin(\omega_1.t)\sin(\omega_2.t)) \qquad (9.11)$$

This can be written as $\frac{A_sA_c}{2}\cos((\omega_2 - \omega_1)\cdot t)$. This is single side band (SSB) modulation, which produces a single tone at ω_2-ω_1.

In practical implementation, mismatch between I and Q path can happen in the transmitter, just like in the receiver [15]. Going back to Figure 9.2, which shows the practical realization of Figure 9.20, there is a DAC for the I path as well as one for Q path, and there is imbalance between the two, due to the mismatch in circuit components. Similarly, there is a mixer for the I path as well as one for Q path, and there is also imbalance. In addition there is also imbalance due to the quadrature LO generator, described in 9.3. This is because the generator does not generate LO_I and LO_Q that are exactly 90° out of phase. The consequence of the I, Q mismatch is that the resulting frequency spectrum no longer has a single tone at ω_2-ω_1, but has an image. This is shown in Figure 9.22, where, in addition to the desired SSB output at ω_2-ω_1, there is an undesired and smaller component, at ω_2+ω_1[2].

This undesired component means the modulator input contains AM. In general, even though the single tone is actually the message, I, Q mismatch is still going to

---

[2] Also shown is a frequency component at ω_2, which may come, for example, from LO feedthrough

result in the modulator input containing AM. If we approximate the PA nonlinear transfer function[3] up to the 3rd order term as:

$$v_{out}(t) = \alpha_0 + \alpha_1 \cdot v_{in}(t) + \alpha_2 \cdot v_{in}^2(t) + \alpha_3 \cdot v_{in}^3(t) \qquad (9.12)$$

then as shown in Figure 9.21, this nonlinearity converts the AM to PM and causes phase error. Specifically the odd order terms in (9.12) is responsible for AM to PM conversion [11], which is expressed in deg/dB [6]. For simplicity, we may want to represent v_{in} as:

$$v_{in}(t) = A(t) \cdot \sin[(\omega_2 - \omega_1) \cdot t + \phi(t)] \qquad (9.13)$$

Here $\phi(t)$ represents the message. $A(t)$ contains the $A_{desired}$ term, where $A_{desired}$ is the amplitude of the desired signal at ω_2-ω_1 (shown in Figure 9.22), or the RF amplitude. $A(t)$ also contains the modulating signal term.

To learn a bit more about what happens when v_{in} goes through the nonlinearity, as represented by (9.12), and generates v_{out}, let us look at the visualization in Fig. 9.23. Instead of writing the nonlinear transfer function as a polynomial, as in (9.12), a compact way is to represent it is the use of a tanh (hyperpolic tangent) function, so that $v_{out}(t) = \tanh(\alpha \cdot v_{in}(t))$. Here α controls the amount of nonlinearity, with α increasing as the PA becomes more nonlinear. We also use low frequency signals to simplify analysis, so that the message again becomes a tone. Thus the AM modulated v_{in} is represented as $A(t) \cdot \sin[(\omega_2 - \omega_1) \cdot t]$. For illustration, we exaggerate the level of modulation, so v_{in} is written as $(\sin(2\omega_1 \cdot t) \cdot \sin((\omega_2 - \omega_1)t))$, with $2\omega_1$ as the modulating frequency. v_{out} is now given as:

$$v_{out}(t) = \tanh(\alpha \cdot \sin(2\omega_1 \cdot t) \cdot \sin((\omega_2 - \omega_1)t))$$

In the visualization, we set $\alpha=2$, ω_2-ω_1 about 10 times $2\omega_1$.

Plotting we have Fig. 9.23, where AM modulation (exaggerated) is evident.

Next we perform an fft on v_{out}. The frequency spectrum is shown below in Figure 9.24, where the fundamental and 3rd order intermodulation product is shown to be around $-17db$ below the fundamental i.e. $IM_3 = -17db$(from this IIP_3 can be calculated). In general as α (set equal 2 here) increases, this increases.

Having used the above visualizations to clarify how nonlinearity in a PA causes AM-PM conversion in the example architecture shown in Figure 9.20, let us next try to estimate the amount of conversion and hence the required IIP_3 that satisfies the specs.

Given:

For GSM (which uses GMSK), the phase accuracy of modulated

carrier is (5°rms (20°peak) in Tx phase error).   (9.14)

---

[3] In [18], more advanced model using Volterra series, like in chapter 4, 5 are used to model the PA. This includes capacitive effect, which further moves the zero crossing and aggravates the situation.

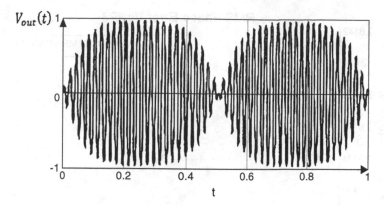

Fig. 9.23 Visualization of v_{out} when an AM modulated v_{in} is applied to a PA whose nonlinearity is represented by a tanh function: waveform plot (1 symbol period)

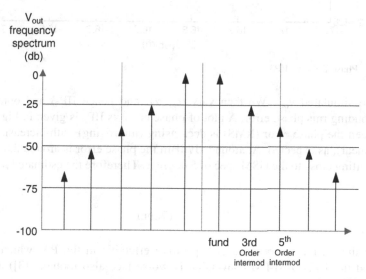

Fig. 9.24 Visualization of v_{out} when an AM modulated v_{in} is applied to a PA whose nonlinearity is represented by a tanh function: frequency spectrum (fundamental and first few intermodulation components)

From Figure 9.20 signal level of input to PA=5 dBm i.e. $A_{desired}$= 5 dBm. The amount of mismatch between I, Q path is selected, with the resulting undesired signal (shown in Figure 9.22) giving an AM modulating signal whose amplitude is 1/10 of $A_{desired}$. If we further set the example architecture in Figure 9.20 so that the resulting $\omega_1 = 45$MHz, $\omega_2 = 900$MHz (i.e. $\omega_2 = 20\omega_1$), this will allow us to set up the v_{in} given in (9.13). Representing the PA by (9.12) we then simulate v_{out}. The simulation is first run for a given α_1, α_3, and the rms phase error is calculated

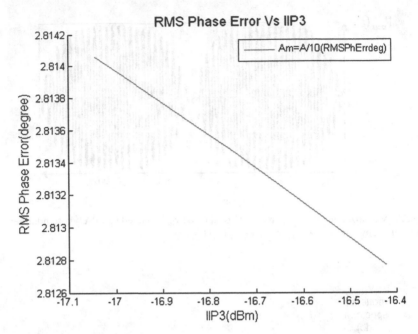

Fig. 9.25 Phase Error vs IIP3

from the simulated v_{out}. We then vary α_1, α_3 (and hence IIP$_3$) and obtain the corresponding rms phase error. A plot of phase error vs IIP$_3$ is given in Fig. 9.25:

As seen the phase error (RMS) is decreasing (improving) with increasing IIP3 (more linear), as expected. At around -17dBm the phase error is around 2.8 degree and is getting close to the GSM spec of 5 degrees. Therefore the estimated required IIP$_3$ is:

$$IIP_3 = -17dbm \qquad (9.15)$$

Note that in reality, because of capacitive effective in the PA, which is not modelled in (9.12), AM-PM conversion is worse (see also footnote [3]) and the required IIP$_3$ is higher.

IIP3 Specs Due to Distorted LO

The limitation on IIP$_3$ described in 9.4.1.3.1 is present for the architecture shown in Figure 9.20, regardless of the exact circuit implementation. Often the local oscillator does not generate a perfect sine wave and thus it has higher order harmonics. Coupled with nonlinearity of the PA, this is going to cause further problem [5]. Here the nonlinearity intermodulates unwanted signals and reduces the SNR.

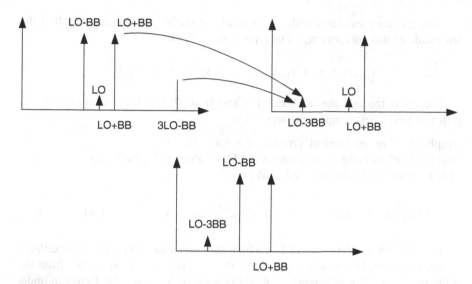

Fig. 9.26 IIP3 due to LO 3rd harmonic mixing

We start with the desired signal, whose frequency is at LO+BB, as shown in Figure 9.26. Here LO is the local oscillator fundamental frequency and BB is the frequency of the baseband signal. Now typically the local oscillator does not generate a perfect sine wave and thus it has higher order harmonics. Let us turn our attention to its 3rd harmonics, 3LO. During the upconversion of the baseband signal via mixing, this 3LO component mixes with the BB component and creates a frequency component at 3LO-BB. Both the LO+BB and 3LO-BB components are present at the input to PA. As a result of the PA's nonlinearity, they intermodulate, and a component at LO-3BB is generated (LO+BB component mixes with 3LO-BB component, forms LO-3BB component by the formula: (3LO-BB)-[2× (LO+BB)]=LO-3BB). This, if not properly filtered, the LO-3BB component looks like noise to the desired signal at LO+BB. The above is shown in Figure 9.26.

From (9.7) A_{in}=5dBm. We now calculate the allowed IIP$_3$.

We treat the PA as a nonlinearity, whose input signal has two frequency components: ω_1 (=LO+BB) and ω_2 (=3LO-BB) and denote their amplitudes as $a_1@\omega_1$ and $a_2@\omega_2$

$$\text{Thus}: \ a_1@\omega_1 = A_{in} = 5\text{dBm} \qquad (9.16)$$

We now assume LO is a square wave, so that the amplitude of the 3LO component is approximately 1/3 that of the amplitude at LO. We further assume that the upconverted baseband signals also follow the same ratios, so that:

$$a_1/\ a_2 = 1/3. \qquad (9.17)$$

Now ω_1 intermodulates with ω_2 to form the ω_1-$2\omega_2$ component. To find the amplitude of this component, let us represent the PA as:

$$v_{out}(t) = \alpha_0 + \alpha_1 \cdot v_{in}(t) + \alpha_2 \cdot v_{in}^2(t) + \alpha_3 \cdot v_{in}^3(t) \qquad (9.18)$$

They pass through the nonlinear PA and from (9.18), the amplitudes of v_{out} different frequency components are:

Amplitude of ω_1 component's output $= \alpha_1 \times a_1 \times A_{in}$
Amplitude of ω_2 component's output $= (3/4) \times \alpha_3 \times a_2 \times A_{in} \, (a_1 \times A_{in})^2$
This ratio should be the required SNR i.e.

$$20\log[(3/4) \times \alpha_3 \times a_2 \times A_{in}(a_1 \times A_{in})^2] / \, [\alpha_1 \times a_1 \times A_{in}] = SNR \qquad (9.19)$$

To find SNR we start from (9.14), which states the phase accuracy of modulated carrier is $5°$ rms. In general, noise causes the received vectors to deviate from the reference vector. The differences, called the error vectors, have random amplitude. Numerous ways to characterize the amplitude of the error vectors (which depends on SNR) have been reported and a series of definitions of the error vector magnitude (EVM) given in [10]. For ease of explanation, we go through the following approximations and relate the amplitude of the error vectors to SNR, and hence transmit phase error, below. With a transmit phase error of 5 degree rms this means, in the worst case, the transmitted vectors(which is random), is tolerated to deviate from the reference vector by 5 degrees in an rms sense. For simplicity of calculating the SNR, let us assume, in an rms sense, the transmitted vectors and error vectors are represented as a single transmitted vector as well as a single error vector, henceforth denoted as the transmitted and error vector. The transmitted vector is constructed by rotating the reference vector by 5 degrees (clockwise or counter-clockwise). We further assume the error vector is given by the vectorial difference between the transmitted vector and the reference vector. Typically the magnitude of the error vector is small compared to the reference vector and thus the magnitude of the transmitted and reference vector are about the same. Hence the transmitted vector can be viewed as obtained from the reference vector simply by rotation, with the reference vector serving as the radius of the circle. Furthermore since the rotation angle is small (5 degree is small compared to 360 degrees), the magnitude of the error vector is approximated by the length of the arc on that part of the circle between the tips of the transmitted and reference vectors. Let us now denote the error vector's amplitude as amp_error and the transmit vector's amplitude as amp_xmit. Since the radius of the circle is given by amp_xmit, and the length of the arc given by amp_error, by definition we have:

$$amp\_error = [(5 \text{ degree}/360 \text{ degree}) \times (2\pi)] \times amp\_xmit \qquad (9.20)$$

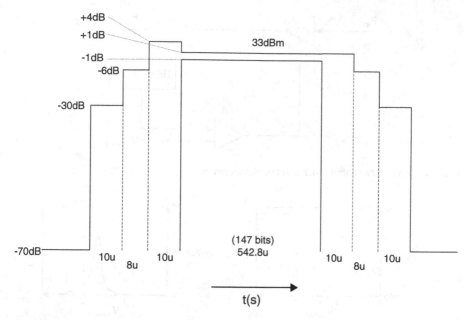

Fig. 9.27 Power level ramping for GSM

We now further assume SNR, in db, is roughly given as the ratio of amp_xmit to amp_error, in a log sense. Thus

$$SNR = 20 \times \log[(\text{amp\_xmit}/\ \text{amp\_error})]$$
$$= 20 \times \log[(360\text{degree}/5\text{degree}) \times (2\pi)] = 21\text{db} \qquad (9.21)$$

Substituting (9.19) in (9.21) we have:

$$20\log[(3/4) \times \alpha_3 \times a_2 \times A_{in}(a_1 \times A_{in})^2]/\ [\alpha_1 \times a_1 \times A_{in}] = 21\text{db} \qquad (9.22)$$

Substituting (9.16)-(9.17) in (9.22), we can find the ratio of α_1/α_3, which is simply IM_3, as:

$$IM_3 = 0.267, \text{ or about } -11\text{db} \qquad (9.23)$$

Finally we have:

$$IIP_3|_{dbm} = A_{in}|_{dbm} - \left(IM_3|_{db} \right)/2 = 10.5\text{dbm} \qquad (9.24)$$

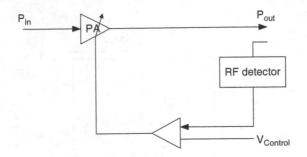

Fig. 9.28 PA power control block level implementation

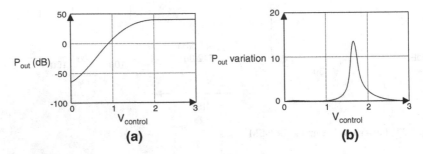

Fig. 9.29 PA power control characteristic

9.4.2   *Power Output Control*

To achieve a power output level, specs typically includes one for power level ramping. Shown in Figure 9.27 are the specs for GSM.

Notice that there is also power control specs. For GSM it is from 33dBm to 5dBm output power, or 28dB range, with accuracy to ±3dB at lower power levels. Shown in Figure 9.28 is a block diagram illustration of how power control, in general, is achieved.

In essence, power control is achieved by sensing the output power via the RF detector(basically extracting the envelope of the carrier), comparing it with a control signal ($V_{control}$) and feeding the result back to vary the gain of the PA. A typical P_{out} versus this $V_{control}$ characteristic or transfer function (TF) is shown in Figure 9.29, where it is seen that as $V_{control}$ increases, P_{out} initially increases, but eventually levels off. This TF can be quite nonlinear, and a typical variation with respect to $V_{control}$ is also shown in Figure 9.29. This variation can be a problem in a closed loop circuit (the present case) if the variation is large. Typically the gain variation can be as much as 100:1, and so loop stability needs to be analyzed carefully.

9.4.3 PA Design Issues

Next we note that unlike LNA, which amplifies signal, a power amplifier (PA), transfers power from dc to RF, rather than amplifies the power.

Since V_{dd} is decreasing for scaled technology, thus R decreases. Therefore we need a matching network at output of PA. This matching network is similar to one discussed in LNA. An example will be shown below.

Most of the CMOS PA [22] have the following issues. One issue is oxide breakdown. This limits the available voltage swing at the drain of the transistor. As a result the load impedance seen by the transistor needs to be very low in order to deliver the required amount of power, without damaging the device. Furthermore, the current level must rise in proportion with the reduced voltage swing. Since CMOS has a poor transconductance, the sizes of transistors can be quite large in order to generate the required output power. In general, we can use matching network and increase V_{dd}. Circuit technique such as cascoding can also be used. Several PA can also be combined, each with small P_{out}.

Another issue is power efficiency. Power added efficiency (PAE) is defined as:

$$PAE\% = \frac{100 \times P_o(RF)}{P_{in}(RF) + P_{dc}} \qquad (9.25)$$

In other words, it is the ratio of output power to the sum of DC input power and RF input power. Maximum PAE depends on the class of PA. Class A amplifier has the low efficiency (50%), with efficiency increasing as one moves to class B (78%) and then class C. On the other hand the linearity decreases. Because it is linear, class A PA design is more straightforward, while nonlinear class B, class C's analysis is more involved.

Finally integrated PA has more problems with thermal runaway. In particular, if power level is high, the heat generated may not get dissipated quick enough. This will significantly change the operating point of the device, and may also cause stability problem.

9.4.4 Class A Amplifiers

First, a class A amplifier [12] is examined. This is shown in Figure 9.30.

L_{RFC} is a choke inductance for biasing. L_s is the parasitic inductance. The matching network can be realized as in [12]. For example, Match 2 in Figure 9.30, transforms R_L, the 50Ω load from the antenna, to R, as shown in Figure 9.31, via the Q of LC network. This network consists of a series C_s, a parallel L_p.

Another configuration consists of a series L_s, a parallel C_p network.

Next, the load line of the PA circuit is given by Figure 9.32.

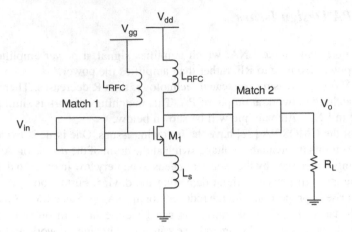

Fig. 9.30 Class A power amplifier

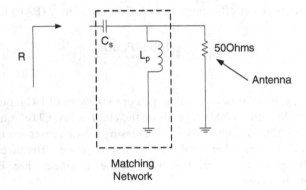

Fig. 9.31 Matching network for PA

The average dc power consumption is given by

$$P_{DC} = I_{DO} \times V_{DD} \tag{9.26}$$

The optimum drain resistance is

$$R_{L\_opt} = \frac{V_{DD}}{I_{DO}} \tag{9.27}$$

The maximum RMS power delivered to the load is given by

$$P_{out} = \frac{V_{DD}}{\sqrt{2}} \times \frac{I_{DO}}{\sqrt{2}} = \frac{V_{DD} \times I_{DO}}{2} \tag{9.28}$$

Fig 9.32 Load line of class A
power amplifier

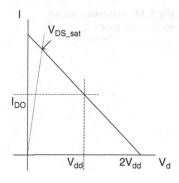

From Figure 9.32, the maximum peak signal current and voltage swing will not be equal to I_{DO} and V_{DD}, respectively, due to the V_{DS_sat} of M_1. The effect of the R_s of M_1 (due to e.g. contact resistance, interconnect), and L_s(package + bond wire) will also cause voltage drop. Therefore

$$PA \text{ efficiency, } \eta = \frac{P_{out}}{P_{dc} + P_{in}} < 50\% \qquad (9.29)$$

Ground bounce is due to M_1's source current into R_s and L_s. Some amount of resistive degeneration in the source is desirable to provide negative feedback and increase stability. Voltage drop across inductance L_s is undesirable and should be minimized.

Ground bounce:

$$V_{ground\_bounce} = R_s I_D(\omega) + j\omega L_s I_D(\omega) \qquad (9.30)$$

Note that the voltage across L_s is in quadrature to the signal current.

If the input signal to M_1 is sinusoidal, with amplitude V_{ip}, then the drain current equal the source current I_s as shown in Figure 9.33, and is given by:

$$I_D(\omega) = \frac{V_{ip} \sin(\omega t)}{R_s + j\omega L_s} \qquad (9.31)$$

and the output voltage by

$$V_o(\omega) = \frac{R_L V_{ip} \sin(\omega t)}{R_s + j\omega L_s} \qquad (9.32)$$

Maximum peak to peak output voltage at the drain of M_1 is given by

$$V_{o\_max}(\omega) \approx V_{dd} - V_{DS\_sat} + I_s(R_s + j\omega L_s) \qquad (9.33)$$

Fig 9.33 Maximum output swing with parasitic resistance, inductance

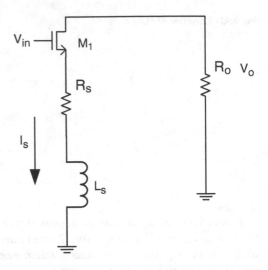

With the load identified, the stability of the PA has to be checked. S-parameters should be used. Unconditional stability can be ensured if the following conditions [16] are satisfied:

$$K_S = \frac{1 - |S_{11}|^2 - |S_{22}|^2 + |\Delta|^2}{2|S_{21}S_{12}|} > 1; \quad |\Delta| < 1 \qquad (9.34)$$

where $\Delta = \det([S]) = S_{11}S_{22} - S_{12}S_{21}$.

A circuit is defined as unconditional stable if there is no passive source and passive load combination that can cause the circuit to oscillate [18].

Alternatively unconditional stability can be ensured if the following condition [18] is satisfied:

$$\mu_1 = \frac{1 - |S_{11}|^2}{|S_{22} - S_{11}^*\Delta| + |S_{21}S_{12}|} > 1 \qquad (9.35)$$

where again $\Delta = \det([S]) = S_{11}S_{22} - S_{12}S_{21}$.

[18] also provides geometrical explanation in terms of mapping (e.g. mapping between input/output reflection coefficient and load /source reflection coefficient) that can help to guide the design in achieving stability.

Numerical Example 9.3

Consider designing a class A RF single stage PA. Assume the transistor can deliver 2W output (challenging for CMOS) with the following parameters:

$$V_{dd} = 4.8V, \ I_{DO} = 1A. \ L_s = 0.1nH, \ R_s = 0.2\Omega. \ V_{ds\_sat} = 0.6V$$

From (9.27), we have $R_{L\_opt} = 4.8\Omega.$ \qquad (9.36)

Fig. 9.34 Reduction in
supply voltage and impact on
PA design

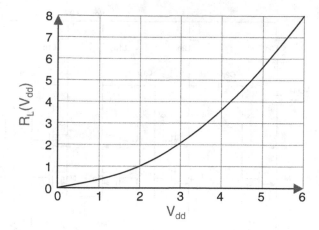

If we ignore $V_{ds\_sat}=0.6V$ and source voltage drop, the maximum output power
would be (from (9.28)):

$$P_{out} = \tfrac{1}{2}\, I_{DO} V_{dd} = 2.4W \qquad (9.37)$$

From (9.29), $\eta=50\%$.

At 1GHz, the maximum output power would be less than this, and is given by:

$$P_{o\_max}(\omega) \approx \frac{1}{2} I_{DO}\left(V_{dd} - \sqrt{V_{DS\_sat}{}^2 + I_{DO}{}^2\left(R_s{}^2 + \omega^2 L_s{}^2\right)}\right) = 1.95W \quad (9.38)$$

In order to obtain higher power in class A, the DC operating (bias) current I_{DO},
would have to be increased to compensate for the extra voltage drop i.e. the slope of
the ac load line would be increased.

The small signal voltage gain is given by (if L_s is ignored)

$$G = \frac{R_o}{R_s} = \frac{4.8}{0.2} = 24 = 27.6dB \qquad (9.39)$$

Thus the G and power level of this design is slightly below the specs of the
example PA for GSM (shown in Figure 9.20). The actual voltage gain at 1GHz is
less due to device capacitance, which has not been included. Nevertheless the large
amount of gain at high frequency can be a problem, and the amplifier design needs
to be checked for potential instability over frequency.

Plot in Figure 9.34 is R_L versus V_{dd}. Using (9.27), for a given power output P_o,
we can derive

$$R_L(V_{dd}) = \frac{\frac{V_{dd}^2}{\sqrt{2}}}{P_o}$$

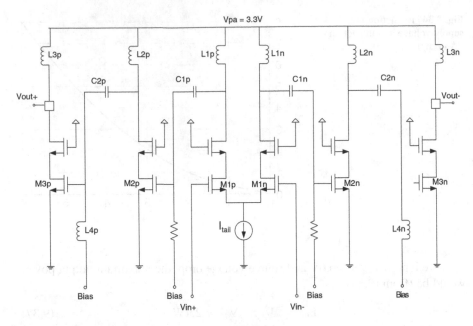

Fig. 9.35 Example 3 stage Class A power amplifier

We then plot R_L versus V_{dd}. The result shows that as V_{dd} decreases, R_L decreases. This R_L may not be matched to the 50Ω output. Thus we need a matching network, as discussed above.

An example core (without the matching network) class A PA [3] design is given in Figure 9.35. This consists of three stages to give it the necessary gain. The first stage is a differential pair, and operates in class A. Cascode devices are used to mitigate breakdown problem since the voltage supply is 3.3V. The gates of these devices are connected to the power supply. The output is fed into L1n, L1p, which serves as inductive load. They are on chip inductors. This is then ac-coupled via C1n, C1p to the second stage (M2n, M2p) and then into the third stage (M3n, M3p). The second and third stages are common source amplifiers, and are biased through the resistors to keep them in class A operation. Level shifting between each stage is achieved through capacitors C1n-p to C3n-p. L4n, L4p are on chip inductors, and are added to resonate with gate capacitances of M3n, M3p. This allows the level shifting capacitors C2n-p to be kept below 2pF. There is also a power control loop (not shown), performing power control function, as described in 9.4.2. This loop, like Figure 9.28, includes a RF peak signal detector, a comparator and 24dB of adjustable transmitter gain. The power is adjusted in steps, as shown in Figure 9.29, in steps of 0.5dB. $V_{out}{}^{+}$ and $V_{out}{}^{-}$ are fed to an off chip balun. This PA delivers a 22dBm peak output power and 17.8dBm average power.

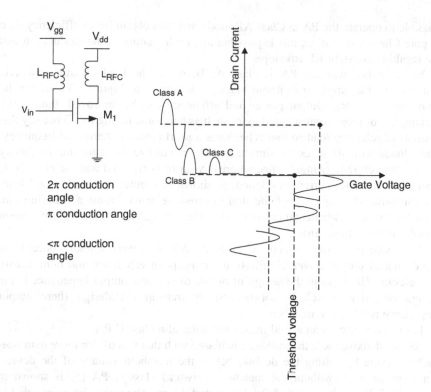

Fig. 9.36 Class AB/B/C power amplifier

9.4.5 Class AB/B/C Amplifiers

To improve efficiency, PA does not need to conduct all the time. This leads us from class A amplifier to class AB amplifier. Thus, example, the amplifier for class A (Figure 9.30) is redrawn here (without matching network) in Figure 9.36.

As shown the difference between different classes depends on the bias condition, for the same circuit. On the right hand side of the diagram is I-V plot of the transistor. When gate voltage (set up via V_{gg}) is below the threshold voltage, drain current is zero. The voltage at which the gate should be biased to operate the PA in class A is the rightmost dotted line. As the biased voltage moves towards the left, operation becomes class AB, class B, then class C. Conduction angle represents the % of a cycle when the PA conducts (i.e. drain current larger than zero). For class A it is 2π or 360° since the PA conducts all the time. For class B it is π or 180° since the PA conducts half the time. Class AB is somewhere in between. Class C is below π or 180°. Since the conduction angle goes down, the amplifier is off for some part of the cycle. This means the dc power dissipation, P_{dc}, is reduced and hence efficiency increases, according to (9.29). This, however, also means PA is more nonlinear. In spite of being more nonlinear, in the GSM system, though, it is possible to operate the PA in Class AB mode and thus obtain better efficiency than in pure Class A operation; this is possible due to the nature of the modulation and the resulting constant RF envelope.

Note, so far, whether PA is class A, B, or C, the load remains resistive. This means, the amplifier (without biasing), is similar to Figure 9.33, where the load consists of R_o. Output power and efficiency can be improved, though, by making use of more sophisticated load, such as harmonic tuning [6]. Basically this consists of selecting load so that it becomes a short circuit at harmonic frequency. Such load, will allow us to simultaneously maximize fundamental frequency voltage and current swings. Thus, whereas with resistive load and for class B, C operation, the drain current is clipped, as shown in Figure 9.36, with tuned load, the fundamental frequency of the drain current remains almost a full sinusoid. This maximizes fundamental current swing (high frequency component of current remains distorted though).

As a side note, the concept of class A, AB, C shares similar concepts in the design of output driver. Similarly the concept of efficiency and nonlinearity are relevant. However in the design of output driver, low output impedance is the design objective, which is not necessarily true in PA design (here output impedance needs to be matched).

Let us now discuss an actual implementation of a class C PA.

As stated above, here the transistor is biased such that it is off for more than one-half the cycle by setting the dc bias below the threshold voltage of the device. An example core (without the matching network) class C PA [5] is shown in Figure 9.37. It is a three-stage fully differential design. The stages are ac-coupled in order to allow the input transistor of each stage to be biased independent of the output of the previous stage. Since the PA was to be used in an integrated environment, the previous stages would likely not be able to drive a heavy load. Therefore three stages were used in order that the first stage would not load down the output of the mixer, and also gives it the necessary gain. The first stage is a differential pair and operates in a class-A fashion, with a variable tail current source that controls the gain and thus the output power level. The final two stages are also differential pairs, but with tail node connected directly to ground, rather than to a current source. Thus by setting the proper bias at the gates of the input differential pair, they can be biased to be on for less than half of the cycle i.e. in class C operation. The second and third stage also use a cascode structure with the top transistor biased at the supply in order to reduce the stress on the oxide due to voltage excursions at the outputs of those stages. At the input to the final stage, a different tuning technique was used to peak the gain at the desired frequency (in much the same way as the narrowband LNA of chapter 3). Due to the high current drive required and the poor transconductance of CMOS, the output transistor size was very large (18300u/0.35u). Using a single spiral inductor (as discussed in chapter 3) to tune out all the capacitance between the second and third stages requires a very small value of inductor (on the order of 0.1nH). This inductor was quite difficult to implement without parasitic trace inductance affecting the tuning at that node. Furthermore, inductors were used across the differential cascode node in the second and third stages, in order to reduce the current required to charge and discharge the

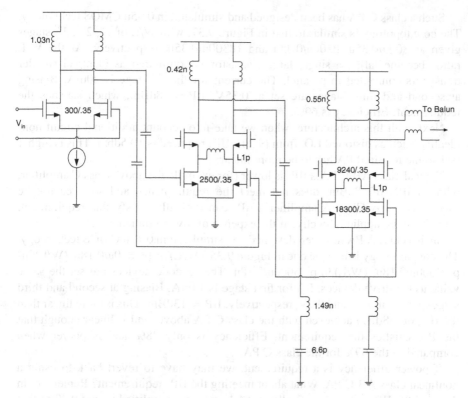

Fig. 9.37 Class C power amplifier

capacitors at those nodes and boost the efficiency. The inductors used at the output of the first and second stages as well as those at the cascode nodes were implemented as spiral inductors on chip. The other inductors used in this circuit were implemented as bond wires, in order to take advantage of their higher Q. The inputs of the second and third stages were biased at 0.35V.

9.4.6 Choice of Class A vs AB/C Amplifiers

There are many issues ranging from efficiency, power output level, required linearity etc. that determine this choice. One of the important considerations is whether the required linearity allows one to use class AB/C or one has to use class A PA. We again use the required IIP_3 derived above for the architecture in Figure 9.20 as an example.

For this architecture, from (9.15), IIP_3 of around -17dBm is required. A Class AB/C amplifier, even though nonlinear, can typically satisfy this requirement. With its better efficiency, it is an attractive choice.

Such a class C PA has been designed and simulated in 0.35u CMOS technology. The core topology is similar to that in Figure 9.37, with W/L of 1st, 2nd, 3rd stages given as 800u/0.35u, 2500u/0.35u and 18300u/0.35u, respectively. As the W/L ratios become large, a single large transistor is configured as identical smaller transistors connected in parallel. The current for the 1st stage is 20mA. Biasing at second and third stages are set at 0.35V. $IIP_3=$ -8dBm, which satisfies the requirement. Efficiency is 69%.

Next, with this architecture, when we take into account additional circuit non-ideality, such as distorted LO, from (9.24), IIP_3 required is 10.5dbm. This is higher and so the required PA has to be more linear.

Typical class AB/C PA's IIP_3 is lower (for example the above class C amplifier, with an IIP_3 of -8dBm, does not meet the requirement), and so it cannot be used directly. A Class A amplifier's IIP_3 can typically satisfy this requirement. Thus it can be applied directly, at the expense of lower efficiency.

Such a class A PA has been designed and simulated in 0.18u CMOS technology. The core topology is like the one in Figure 9.35. (W/L)1n-p=800u/0.18u, (W/L)2n-p=2500u/0.18u, (W/L)3n-p=5000u/0.18u. The cascode devices are set the same value as the drive devices. I_{tail} for first stage is 14mA. Biasing at second and third stages are set at 0.5V and 0.9V, respectively. $IIP_3=$ 13dBm. This is more linear than the IIP_3 of -8dBm achieved with the class C PA above, and is linear enough that the PA satisfies the requirement. Efficiency is only 28% and is poorer when compared to the 69% for the class C PA.

If power efficiency is a requirement, we may have to revert back to using a nonlinear class AB/C PA. What about meeting the IIP_3 requirement? Remember in deriving the IIP_3 specs, specifically in 9.4.1, we have simplified Figure 9.20 so that BPF is ideal with a loss of 0db, and thus is assumed to have a flat frequency response. In practice this filter, which is before the PA, serves as a Pre-PA RF filter [5] and attenuates the frequency component at 3LO-BB. If we implement this with a two pole filter, we can provide a 19db suppression at 3LO-BB. Then the required IIP_3 of 10.5dbm is relaxed by 19db to -9.5dbm. Again a Class AB/C amplifier can typically satisfy this requirement (for example, the class C PA designed and simulated above, with an IIP_3 of -8dBm, satisfies this requirement). Thus there is a trade-off between the design of PA (e.g. class A vs class C) and design of filter (e.g. 1-pole versus 2-pole). This filter, however, likely is a discrete filter because component using CMOS integration has low Q, and is thus costly. An alternative modification is with the use of the harmonic rejection transmitter architecture [5], which does not require this filter. This architecture rejects the 3LO and thus less intermodulation product is present, thus reducing the required IIP_3.

9.4.7   Class E Amplifiers

To improve efficiency even further, one can use switching mode PA. Even though there are a variety of switching mode PA (e.g. Class D, class S etc.), vast majority of

Fig. 9.38 Class E power
amplifier

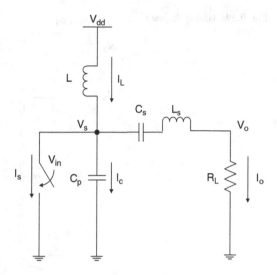

attention was devoted to class E scheme, because it is simple to implement and also
has simple design relationships[6, 21]. Shown in Figure 9.38 is a conceptual picture
of a class-E PA [21].

The transistor is represented as a switch. C_p includes its output parasitic
reactance, the stray capacitances and an explicit capacitor. Its value should be set
so as not to short circuit all the voltage harmonics. $L_s C_s$ form a resonant circuit that
acts as a bandpass filter. It is designed so that the center frequency is the RF
frequency. The load is represented by R_L (following [21], we assume resistive
load only, although in practice there can be an inductance in series with it [6]).

To explain the operation, let us start with V_{in}, a square wave at the RF frequency,
as shown in Figure 9.39. Therefore it opens and closes the switch at the RF
frequency. During the ON phase, V_s is shorted to ground and stays there.
Since V_s remains constant, the voltage across the capacitor C_p, V_c ($=V_s$) remains
constant and from the relation $I_c=Cp(dV_c/dt)$, we have $I_c=0$. Now jumping to the
output, since L_s, C_s forms a bandpass filter with center frequency at the RF
frequency, then we assume the output voltage V_o contains just the fundamental
frequency at the RF frequency i.e. all harmonics are filtered and V_o is just a
sinusoid. Let us further assume the phase of the sinusoid is set so that it starts
from negative, as shown in Figure 9.39. Since V_o and V_{in} are designed to have the
same frequency, thus during ON phase, V_o stays negative. Since V_o is developed
across a resistive load, R_L only, the current I_o has the same phase, and stays negative.
Next we observe that V_L, the voltage across the inductor, which equals $V_{dd}-V_s$, is
simply V_{dd} since V_s was already shown to be zero during this ON phase. Thus from
the relation $I_L=\int (V_L /L)dt= I_L=\int (V_{dd} /L)dt$ we have I_L increasing linearly.

Fig. 9.39 Timing diagram

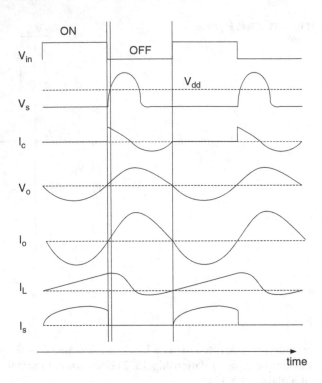

Then the current through the switch, I_s, given as $I_s = I_L - I_c - I_o = I_L - I_o$, rises above zero and stay above zero, as shown.

During the OFF phase, switch is open, $I_s = 0$. Then current is directed so that I_c jumps to a positive value and charges up the capacitor. Thus V_s ($= V_c$), from the relation $V_c = \int (I_c / Cp)dt$, starts to rise from 0. When V_s increases above the power supply V_{dd}, though, the voltage drop across L turns negative and from relation $I_L = \int (V_L / L)dt = \int ((V_{dd} - V_s) / L)dt$, I_L starts to decreases, as shown. Meanwhile V_o rises above zero and since V_o and V_{in} are designed to have the same frequency, thus during the OFF phase, V_o stays positive. Then the current I_o also rises above zero and stays positive.

In summary, looking at power dissipation across the PA i.e. the switch, during the ON phase, its voltage $V_s = 0$ and during the OFF phase, its current, $I_s = 0$. Therefore at no point in time are both values nonzero. That means the power, while equals the product of the two, is zero all the time. This is the reason why class E PA has high efficiency.

In practice, we can phase/frequency modulate V_{in}. Thus the switch is not just supplying power from supply to the load, but also amplifying, via modulation, the input signal to the output.

9.5 Problems

9.1 A modified direct conversion architecture is shown in Figure P9.1, where the oscillator frequency is offset from the carrier frequency (set equal to f_{LO_2} $\pm f_{LO_1}$). (a) What is the advantage over direct conversion architecture (b) What is the difference with an indirect conversion architecture?

9.2 (a) For the direct conversion architecture, explains how the pre-PA filter reduces potential third-order intermodulation in the PA.

    (b) For the indirect conversion transmitter architecture, explain its performance advantages over the direct conversion transmitter, apart from what is given in the chapter.

9.3 For the harmonic rejection transmitter, explain how it offers advantage over the indirect upconversion transmitter in terms of reducing the potential third-order intermodulation in the PA.

9.4 What are some of the potential drawbacks of using passive methods such as RC filter, polyphase filter as quadrature LO generator?

9.5 For the differential RC quadrature LO generator, derive the relation between R_1, ω_{LO}, C and R_2 if LO_I and LO_Q have equal amplitudes.

9.6 What are some of the loss mechanisms of a polyphase filter based quadrature LO generator? A quadrature LO generator is shown in Figure P9.2 that tries to combat this. How does it minimize some of the loss?

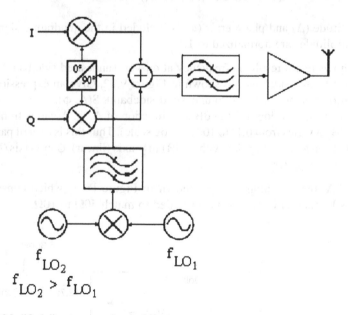

Fig. P9.1 Modified direct conversion architecture

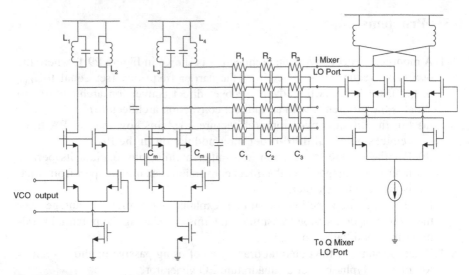

Fig. P9.2 Quadrature LO generator

9.7 In a quadrature LO generator, because of mismatch in the circuits, there are amplitude and phase mismatch between LO_I, LO_Q. Assume amplitude and phase mismatch is represented as follows:

$$\text{I: } \cos(\omega_1 t), \; LO_I\text{: } \cos(\omega_2 t)$$

$$\text{Q: } (1+\Delta)\sin(\omega_1 t + \phi), \; LO_Q\text{: } \sin(\omega_2 t)$$

Amplitude (Δ) and phase error (ϕ) is included in the quadrature signal path, all amplitudes are normalized to 1.

a) Shows that it results in frequency at $\omega_1 + \omega_2$ (undesired side band) and $\omega_1 - \omega_2$ (desired sideband) as shown in Figure 9.22. Give an expression of the ratio of error of desired to undesired sideband, $S(\Delta, \phi)$.

b) Plot S (take 10log of it) in dB as a function of Δ, ϕ. Δ varies from 0dB to 3dB. ϕ varies from 0.1° to 100° (in log scale). Thus this is a set of parametric plots on a linear-log plot, with S (dB) in y-axis (linear), ϕ in x-axis (log), and Δ as parameters.

9.8 In a PA, the matching network shown in Figure P9.3 (which consists of a series L, parallel C network) is provided to match 50Ω to 10Ω.

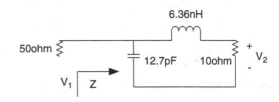

Fig. P9.3 Matching network

Plot the frequency response of the above circuit i.e. $20\log V_2/V_1$ in dB vs frequency (linear-log plot). The frequency span should be from 10MHz–1.5GHz.

From the frequency response, at what frequency does the matching occur? The frequency response of the above circuit can be approximated by the frequency response of the following circuit shown in Figure P9.4. Plot the frequency response of this approximated circuit i.e. $20\log V_2/V_1$

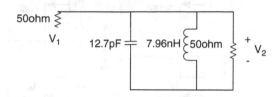

Fig. P9.4 Circuit approximating matching network in frequency response

Compare the two frequency plots in terms of a) center frequency b) bandwidth c) Q d) general shape

9.9 Calculate the p-p sinusoidal input voltage for $HD_3=1\%$ in the circuit shown in Figure P9.5.

FET data:

NMOS: LD=0.1V, VTO=0.7, GAMMA=0.5, PHI=0.6, KP=70E-6, LAMBDA=0.03, CGD0=0.3E-9, CGS0=0.3E-9, CJ=0.4E-3, MJ=0.5, TOX=25E-9, CJSW=0.4E-9, MJSW=0.5, PB=0.6

PMOS: LD=0.1V, VTO=0.7, GAMMA=0.5, PHI=0.6, KP=70E-6, LAMBDA=0.03, CGD0=0.3E-9, CGS0=0.3E-9, CJ=0.4E-3, MJ=0.5, TOX=25E-9, CJSW=0.4E-9, MJSW=0.5, PB=0.6

PMOS WELL: CJSO (FF) $= 0.1 \times$ (area bottom μm^2) $+ 1.1 \times$ (perimeter μm)
MOS dimensions:

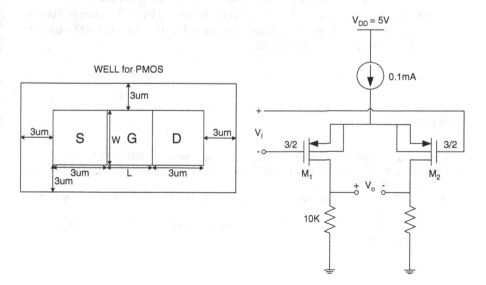

Fig. P9.5 Circuit used to investigate distortion in one stage of PA

Recalculate if W/L ratio of M_1 and M_2 is changed to 30000/2. Show that it is impractical to achieve the same HD_3.

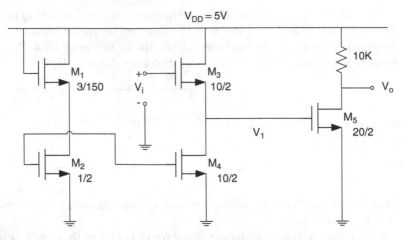

Fig. P9.6 Circuit used to investigate distortion in cascading stages in a PA

9.10 For the circuit shown in Figure P9.6 the dc voltage V_i is adjusted for $V_0 = 2Vdc$.

a) Calculate bias I_D and V_{GS} for each device.
b) Calculate HD_2 and HD_3 (in dB) in V_1 and V_0 for a sinusoidal-signal voltage of 0.2V rms at V_0. FET data as in 9.9 but add $R_S = 25k\Omega/W(\mu m)$ in series with each source. Neglect charge storage and Early effect.
   Hint: Start by deriving a power series for V_i as a function of V_1. Then derive V_1 as a function of V_i. Neglect body effect in M_5 but not in M_3.
c) Repeat the problem but M_5 W/L ratio is increased by 4000 times to 80000/2, and also decrease the drain resistance R_d to 2.5ohms (4000 times). Also make I_{M_5} increase by 4000 times.

References

1. Laskar et.al, Advanced Integrated Communication Microsystems, Wiley/IEEE, 2009
2. Zhang P. et.al "A Direct Conversion CMOS Transceiver for IEEE 802.11a WLANs" Digest of ISSCC, Feb 2003 pg. 354–355
3. Su D. et.al "A 5GHz CMOS Transceiver for IEEE 802.11a Wireless LAN" Digest of ISSCC, Feb 2002 pg. 92–93
4. Sowlati T. et.al, "Quad-band GSM/GPRS/EDGE polar loop transmitter" Digest of ISSCC, Feb 2004 pg. 186–187

5. Weldon, J., R Narayansaswami, Gray, P.R et.al, "A 1.75-GHz Highly Integrated Narrow-Band CMOS Transmitter with Harmonic-Rejection Mixers" IEEE Journal of Solid State Circuits, Vol. 36, No. 12, pg. 2003–2015, December 2001

6. Paolo Colantonio, Franco Giannini, Ernesto Limiti, High Efficiency RF and Microwave Solid State Power Amplifiers, Wiley, 2009

7. Behbahani F. et.al, "CMOS Mixers and Polyphase Filters for large Image Rejection", IEEE Journal of Solid State Circuits, Vol. 36, No. 6, pg. 873–887, June 2003.

8. Crols J. Steyaert M., "A Fully Integrated 900MHz CMOS Double Quadrature Downconverter", IEEE International Solid State Circuit Conference, Feb 1995, pg. 136–137

9. Magoon R., Molnar A., "RF Local oscillator Path for GSM Direct Conversion Transceiver with True 50% Duty Cycle Divide by Three and Active Third Harmonic Cancellation", IEEE Radio Frequency Integrated Circuits Symposium, June 2002, pg. 23–26

10. R. Hassun, M. Flaherty, R. Matreci, M. Taylor, "Effective evaluation of link quality using error vector magnitude techniques," Proceedings of Wireless Communications Conference, Aug., 1997, pp. 89–94

11. Frank Carr-ISCAS 1996, special session on "VLSI for wireless applications"

12. I. D. Robertson(edited), MMIC Design, IEE, 1995

13. G. Gonzalez, Microwave Transistor Amplifiers, 2nd ed., Prentice Hall, 1997.

14. ETSI European digital cellular telecommunications system, radio transmission and reception, GSM 05.05

15. Lam, N., Leung B., "Dynamic Quadrant Swapping Scheme Implemented in a Post conversion Block for I,Q Mismatch Reduction in a DQPSK receiver", IEEE Journal of Solid State Circuits, Vol. 45, No. 2, pg. 322–337, February 2010.

16. Donald Woods, "Reappraisal of the Unconditional Stability Criteria for Active 2-Port Networks in Terms of S Parameters", IEEE Transactions on circuits and systems, vol. 23, no. 2, February 1976, pg.73–81

17. A. Behzad, "Wireless LAN Radio Design", IEEE ISSCC Tutorial, February 15, 2004, pg.74

18. Marion Lee Edwards, and Jeffrey H. Sinsky, "A New Criterion for Linear 2-Port Stability Using a Single Geometrically Derived Parameter", IEEE Transactions on microwave theory and techniques, vol. 40, no. 12, December, 1992, pg. 2303–2311

19. DVB-T Power Amplifiers Modelling and Compensation Carole Raynall, Denis Masse', Pierre KasserI, Jean-Pierre Cances2 and Vahid Meghdadi2 ITDF, Metz, France; 2University of Limoges, Limoges, France

20. H. Krauss, C. Bostian, F. Raab, Solid State Radio Engineering, New York, John Wiley & Sons, In., 1980

21. Tsai, K., Gray, P.R, "A 1.9-GHz 1-W CMOS Class-E Power Amplifier for Wireless Communications" IEEE Journal of Solid State Circuits, Vol. 34, No. 7, July, pg. 962–970, July 1999

22. Chen J., Niknejad, A. M., "A Compact 1V 18.6dBm 60GHz Power Amplifier in 65nm CMOS" Digest of ISSCC, Feb 2011 pg. 432–433

Index